GROUND ANCHORS AND ANCHORED STRUCTURES

PETROS P. XANTHAKOS

Consulting Engineer
Washington, D.C.

A Wiley-Interscience Publication
JOHN WILEY & SONS, INC.
New York / Chichester / Brisbane / Toronto / Singapore

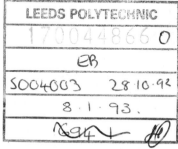
In recognition of the importance of preserving what has
been written, it is a policy of John Wiley & Sons, Inc. to
have books of enduring value published in the United
States printed on acid-free paper, and we exert our best
efforts to that end.

Library of Congress Cataloging in Publication Data:

Xanthakos, Petros P.
  Ground anchors and anchored structures / Petros P. Xanthakos.
    p.   cm.

  "A Wiley Interscience publication."
  Includes bibliographical references and index.
  ISBN 0-471-52520-0

  1. Anchorage (Structural engineering) I. Title.
TA772.X36 1991
624.1'5—dc20   90-44291
                                       CIP

Printed in the United States of America

10 9 8 7 6 5 4 3 2 1

To my wife

PREFACE

Since the strengthening of the Cheurfas dam in Algeria in 1934, where 1000-ton anchors were used to transfer tension to underlying sandstone, anchorage technology and practice have evolved to become a viable interdisciplinary technique utilized in almost every part of the world. Anchors, in both rock and soil, enhance civil engineering works where it is necessary to support, stabilize or transfer loads acting on a wide spectrum of structures, foundations, and slopes.

Ground anchors and anchored systems have become, therefore, interdependent, and their interaction cannot be optimized unless this interdependence is recognized and considered in the analysis and design. Interestingly, the abundance of existing rules, code clauses, standards and guidelines has provided the incentive in our effort to simplify the complexities, relax the differences and develop a unified approach not only by necessity but also as a matter of expediency. Thus, this book treats anchorages as a direct application of the laws of statics and the theories governing the transfer of load, while emphasis is placed on designs that are safe and have reasonable cost.

As has been our experience with slurry walls, anchorages are likely to remain more of an art than a science, like most underground structural systems, because of the many uncertainties associated with this type of work. The "open end" concept takes, therefore, precedence and suggests that, as a field of technology, the technique will continue to evolve although a great deal is disseminated now and will enable us to articulate its advantages and limitations.

The book is organized essentially in two parts. Part 1 consists of the first seven chapters, which are developed in the same logical order in which anchorages should be designed and constructed. Chapter 1 is an introduction to the fundamentals of the technique. It also articulates its relevance to ground engineering problems while cautioning when anchorages should not be used, and provides an overview of statistical and regional data.

Chapter 2 deals with anchor systems, components, installation and construction details. Anchors are characterized and grouped according to serviceability, permanence, prestressing level, method of load transfer, and type of ground. This chapter continues with a discussion on anchor parts and assemblies, types of tendon, their mechanical and physical characteristics, and tendon behavior under stress. Anchor installation and construction are reviewed in detail, including the associated operations such as drilling, flushing, water testing and waterproofing, tendon preparations and homing, grouts and grouting, and the basics of anchor stressing and jacking. Construction limitations are analyzed in a practical context, and contingent factors are related to details and solutions envisaged at planning and tender stages. Chapter 3 supplements the material in Chapter 2 by presenting special anchor systems such as extractable anchors, compression-type anchors, multibell anchors, and regroutable units.

The transfer of load and its relation to the modes of failure and anchor load capacity is analyzed in Chapter 4. This topic is first treated theoretically, and then subjected to practical guidelines based on the observed performance of anchorages. Deterents to rational design are identified since they tend to inhibit the validity of idealized assumptions and thus influence the accuracy of predictions with regard to field performance. The analysis covers the failure mechanism of steel tendons, failure of grout/tendon bond, and failure of ground/grout bond for anchorages in rock, sand and clay. Creep, long term loading and repetitive loading are also discussed in the context of time-dependent effects and cyclic shear stresses.

Chapter 5 deals with design considerations of anchors. Among these are ground and site investigations necessary to ensure the feasibility of a proposed anchorage installation, legal aspects, and stability of a mass of ground around anchorage. Methodologies are presented for selecting fixed anchor zone, anchor spacing and inclination, anchor type, length and diameter, and estimating the lock-off load. This chapter also presents a review of loads acting on anchors, and recommends factors of safety. A design procedure is also developed based on a step-by-step approach.

The topic of corrosion and its protection is quite important particularly with permanent anchors, and imposes general requirements to be considered in the design stage. The mechanism and types of corrosion are discussed in Chapter 6 in conjunction with the aggressivity of environments, aggressive circumstances and the associated risk. The objectives of corrosion protection are defined and explained since they dictate the protection level and the requirements of the protective system. The general conclusion is that in permanent installations all anchor components should have double protection against corrosion. With this criterion, several sections of the chapter review various protective systems and combinations therefrom, and articulate on what constitutes double protection. A brief survey of corrosion incidents and case histories is included to document anchorage performance.

Part 1 is completed with Chapter 7, which deals in detail with anchor stressing, testing programs and evaluation standards that normally constitute acceptance criteria. Relevant topics are precontract tests, acceptance tests of production anchors, and basic on site suitability tests, with typical examples for each category. Since anchors are also installed in soils susceptible to creep, basic and acceptance tests should be mandatory where creep behavior may govern anchor working load. Service behavior and acceptance criteria are analyzed with reference to interpretation methods and to those aspects of anchor behavior that should be monitored, i.e., relaxation or creep.

Chapter 8 introduces the second part of the book, and deals with uses and applications. I have chosen to divide anchor-structure groups according to the structural system. Thus, distinct groups are vertical walls, intermittent structures, anchor/shotcrete supports, massive structures, and free-standing anchorages. Among the uses and applications are the improvement of slope stability, dam strengthening and restoration, soil preconsolidation and soil heave control, anchorages for concentrated forces, anchorages to secure caverns, anchors for tunnels, anchorages for underpinning, anchoring of excavation supports, anchoring of foundation structures, and waterfront installations.

Chapter 9 deals with the design aspects of anchored structures. Anchors are considered in the analysis of dam stabilization by prestressing, soil reinforcement in soft ground tunneling, soil preconsolidation by prestressing, and control of swelling in rock tunnels. The analytical treatment is extended to rock caverns and tunnels supported by anchors, and for a wide spectrum of configurations and uses. Rock mechanics and deformations are quite relevant in this case, since underground work typically is done in media which are stressed so that any opening will cause changes in the initial state of stress. Methods chosen for presentation and review are the semiempirical approach based on observed cavern performance presented by Cording et al. in 1971; the Q system (rock mass quality) introduced by Barton et al. in 1977; rock tunnel reinforcement by equivalent support method presented by Bischoff and Smart in 1977; spiling reinforcement techniques advanced by Korbin and Brekke in 1978; rock tunnel supports based on the convergence-confinement theory, including the NATM, developed mainly in the 1980's; elastic theories for openings in competent rock based on the Kirsch solution; rock anchoring using the exponential formulation presented by Lang and Bischoff in 1981; and the development of block theory introduced by Goodman in 1989. Chapter 9 continues with the analysis of rock slopes supported by anchorages; structures resisting concentrated forces; foundation mats and rafts; and consideration of dynamic loads. The fundamentals of anchored walls are reviewed in detail, and include procedures for estimating lateral earth stresses and deformations; guidelines for control of movement; general stability of the ground/anchored wall system; estimation of anchor loads; and analysis of anchored walls by finite element methods. The effect of wall stiffness, anchor stiffness and prestressing is analyzed and quantified in examples of parametric studies. Among the miscellaneous topics reviewed at the end of the chapter are numerical procedures, underpinning considerations, and the applicability of limit state in the design of anchored walls.

Design examples of practical value and reasonable simplicity are presented in Chapter 10, with special emphasis on engineering judgement since this will continue to be an indispensable tool in assessing solutions. These examples include the selection of design criteria for an anchorage project; rock slope stabilization by prestressing; the development of a design and testing program; examples of anchor supports for tunnels; the basic design of an anchored wall using limit equilibrium analysis; the design of an anchored slurry wall with analysis of ground/support stability; a methodology demonstration of anchored wall design by elastic-plastic methods; and an example of limit state design.

Long term behavior and past experience always provide valuable tools for new designs although they should not be used as general basis for interpretation. Thus, Chapter 11 balances the theoretical theme of the book with examples and case histories from observed performance of anchored systems. These include a typical monitoring program; observations on the effect of a single anchor failure in a group of anchors; results from long term performance of an anchorage project; surveillance programs and field studies of rock slope stability and spiling reinforcement in rock tunnels; studies on freezing and thawing effects; behavior of anchored walls; and field studies of special anchorages.

Anchorage technology and practice have resulted from the dedicated and persistent work of researchers, practitioners and contractors throughout the world. This contribution is acknowledged through the numerous references included in each chapter.

PETROS P. XANTHAKOS

Great Falls, Virginia
October 1990

CONTENTS

6   CORROSION AND CORROSION PROTECTION   240

7   STRESSING, TESTING, AND ACCEPTANCE CRITERIA   297

11   OBSERVED PERFORMANCE OF ANCHORED STRUCTURES

GROUND ANCHORS AND ANCHORED STRUCTURES

CHAPTER 1

STATE OF THE ART

1-1 FUNCTION OF GROUND ANCHORS

Using anchors in civil engineering is a comparatively recent development, initially conceived in conjunction with the suitability of rock as foundation material and later expanded to accommodate almost any type of soil. The resulting technique produced a wide variety and range of applications.

A ground anchor functions as load carrying element, consisting essentially of a steel tendon inserted into suitable ground formations in almost any direction. Its load-carrying capacity is generated as resisting reaction mobilized by stressing the ground along a specially formed anchorage zone. This arrangement is shown schematically in Fig. 1-1 together with the basic components of the system. These components include the head, the free length, and the bond length. The latter is intended to interact with the enveloping ground materials in order to transfer the load; whereas the free length remains unbonded and thus free to move within the soil environment.

As structural devices, anchors usully are attached to ground supports at their head. The anchor tendon is installed in special boreholes in a wide variety of soils or rock. This involves complex and highly specialized procedures, which require careful tendon manufacture and assembly, anchor hole drilling, anchor homing, and a variety of associated operations such as grouting, stressing, quality controls, and monitoring. Furthermore, in permanent works and for any installation in aggressive soil environment all anchor components must be protected against corrosion attack. The associated longevity and sustained performance of permanent anchorages have resulted in sophisticated protection and monitoring techniques, field-tested

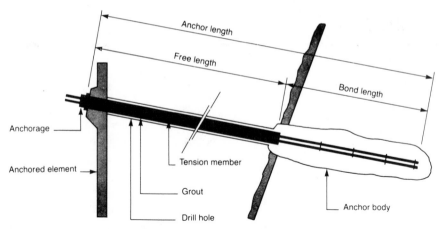

Fig. 1-1 Schematic presentation of a ground anchor showing the three main components: anchorage (anchor head), free length, and fixed length (bond length).

and evaluated. Nonetheless, this subject remains under review and continuous reassessment.

Progress in materials technology has been compatible with the improvement of construction techniques. Thus, cement grouts available today can attain high strength within a few hours after injection. Instrumentation techniques have also been improved to enable site staff to keep anchor performance under close observation and assessment.

As part of the structure, a ground anchor contributes to the overall stability and interaction of the soil–structure system. The anchor function, however, is manifested in a load–tendon deformation pattern that is complex and hardly susceptible to an exact analysis. In this context the solution of ground engineering problems where anchorages are involved is based on semiempirical approaches.

1-2 ORIGIN AND FIRST APPLICATIONS

Historically, the origin of anchorages can be traced to the end of last century. Frazer (1874) has described tests on wrought-iron anchorages for the support of a canal bank along the London–Birmingham railway. Anderson (1900) has documented the use of screw piles to restrain floor slabs against flotation.

One of the earliest and most impressive applications was the strengthening of the Cheurfas dam in Algeria, pioneered by Coyne in 1934. This gravity structure, shown in Fig. 1-2, was built of conventional masonry materials in 1880 but was partially destroyed in 1885 following a serious flood. The dam was rebuilt in 1892, but in the early 1930s it showed signs of foundation instability. Structural integrity was restored by the use of vertical 1000-ton

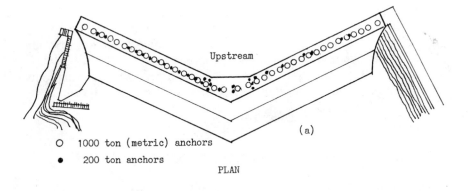

Upstream

(a)

O 1000 ton (metric) anchors

• 200 ton anchors

PLAN

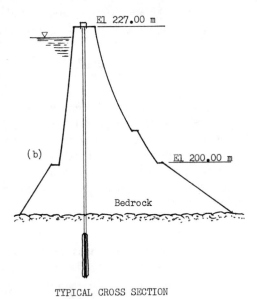

El 227.00 m

(b)

El 200.00 m

Bedrock

TYPICAL CROSS SECTION

Fig. 1-2 Cheurfas Dam in Algeria: (a) general plan; (b) section through main structure showing the anchorage.

capacity anchors placed at 3.5-m intervals, and then stressed by hydraulic
jacks between the crest of the dam and the lower part of the cable head.
The anchors transfer the tension to sandstone, approximately 15 m below
the base of the dam. During drilling of the holes and the homing of the cables
difficulties were encountered because of poor quality of the masonry structure, the presence of swelling marls in the ground, and the presence of water
in the body of the dam. These conditions contributed to the initiation of
corrosion attack, and in 1965 several of the anchors had to be replaced.

Following the demonstration by Coyne, the manufacture of dependable
high-tensile steel wire and strand together with improvements in grouting

and drilling methods led to the postwar development of ground anchors mainly in France, Germany, Sweden and Switzerland, and later England. During the 1950s anchors were first used to support deep excavations. Today, anchorage practice is common in most parts of the world, including the United States, for both rock and soils, and current methods can produce high-capacity anchors in stiff clays as well as in fine sands and silts.

1-3 CURRENT DEMAND FOR ANCHORAGES

Anchoring techniques have evolved mainly in the last 30 years, and it is interesting to mention that between 1953 and 1972 at least 35 patents were taken out for work in soil and rock, some with many similarities and common characteristics.

The ground anchor market at present appears quite sizable, yet it should not be considered fully developed in terms of types and number of applications. With the completion of major projects in construction work here and abroad, anchor market temporarily reaches a limit level of activity. It appears, therefore, that this type of work will have to be expanded since current construction volumes are not sufficient to compensate for the decreased activity following programs already completed. Furthermore, the unique characteristics of underground projects and their associated dependence on the general economic conditions make anchorage work highly susceptible to market fluctuations.

In general, the use of anchorages is favored by the following factors:

1. Development of conventional prestressing and miniaturization of active loads, allowing the preloading of structure.
2. Production of high-speed drills and injection methods under pressure in alluvial terrains.
3. Extension of anchor longevity and improved monitoring methods, both enhancing permanent construction.
4. Ability of anchors to resist tensile forces.
5. Production of high-capacity anchors that can be used in stiff clays and in fine sands and silts. In fact, only the very soft and compressible soils do not readily lend themselves to anchorage systems, since in these conditions the cost per unit of resisting force increases to levels that make anchors rarely competitive with other methods.
6. Demand for more and deeper urban excavations often below the groundwater table, including underground parking, gradually prompting the abandoning of traditional bracing techniques in favor of anchorages.

7. A remarkable technological awareness and judgment, which has enabled engineers and contractors to successfully respond to special or exceptional project conditions, even when a relevant precedent was lacking.

Anchor Groups and Uses

In terms of anchor response to loading, the method may be regarded as a special application of prestressing in foundation and ground work. On the other hand, anchorages can be grouped into three main categories in terms of ground terminology and according to the geology and topography of the site: (a) soil anchors, covering 70–80 percent of the market; (b) rock anchors, representing 10–20 percent of the market; and (c) marine anchors, also installed in fluvial environment or aggressive water, accounting for about 10 percent of the market.

The clear predominance of soil anchors indicates the frequency of soil substratum in the majority of urban and industrial excavations, consisting mostly of alluvium, silty sand and clay. The most common and frequent uses, whether temporary or permanent, are as follows:

Soil Anchors

1. Support retention systems in deep excavations.
2. Anchor and stabilize foundation slabs subjected to uplift caused by groundwater or heave.
3. Preconsolidate unstable soils to increase their bearing capacity.
4. Provide reactions for pile load tests.
5. Compensate and balance the effect of overturning forces in power transmission towers, special roofs, ski jumps, and mobile homes.
6. Tiedown underground storage tanks.
7. Provide lateral support of tunnel walls in cut-and-cover excavations.
8. Stabilize deep slabs of nuclear structures.
9. Carry out remedial, salvage and repair work.

Rock Anchors

1. Protect and stabilize rock formations and slopes.
2. Support underground rock cavities and galleries, where anchors replace timber and steel supports.
3. Raise and strengthen large dams, often quite expensive and inconceivable without the use of high-capacity anchors.

4. Anchor abutments of cableways, television masts, and bridge abutments where large tension forces must be transmitted to the ground.
5. Consolidate mine shafts and other special structures.

Marine Anchors

1. Protect coastal structures and defenses.
2. Stabilize reclaimed areas.
3. Protect river embankments and navigation canals.
4. Strengthen sea and fluvial facilities.
5. Protect oil jetties.

1-4 TIEBACKS AND TIED-BACK WALLS

Tiebacks are essentially similar to the anchor systems described in the foregoing sections, and the ground, rock, or soil is again the medium where the loads are transferred. Tied-back walls can be flexible or stiff, in which case they approach a vertical mat acted on by the concentrated loads induced through prestressing of the tiebacks and by an appropriate soil response. Earth pressures are thus manifested by the actual prestressing level and are less dependent on the particulr state of a soil. These forces, both active and passive, persist unchanged with time unless the wall undergoes displacement relative to the ties. On the other hand, by tensioning the tiebacks to a predetermined level the ground behind the wall can be maintained in the same condition before the excavation commenced.

Unlike tiedown structures, tied-back walls normally are constructed in vertical formation. Their stability is derived from the action of tiebacks that distribute the prestressing load at the wall–soil interface as lateral earth pressures. Structurally, their thickness can be less than of a conventional wall owing to the close spacing of point supports, and footings are not needed. However, since the tiebacks are installed at an angle to the horizontal in order to reach suitable earth layers for the anchor zone, the result is a vertical reaction imposed on the wall which must be resisted by base bearing or side shear along the back face.

Tiebacks eliminate interior obstructions allowing free excavation and fast mechanical earth moving. In these conditions, the construction of interior supports and intermediate floors is better and faster completed especially on irregular or congestal sites.

Tied-back walls can be constructed to their full depth in a single phase, such as the diaphragm wall shown in Fig. 1-3, or in a process similar to underpinning as shown in Figs. 1-4 and 1-5.

The latter walls are processed in a progressive construction of horizontal strips, vertical ribs, and concrete filling. Tiebacks are inserted and stressed as a new layer is added.

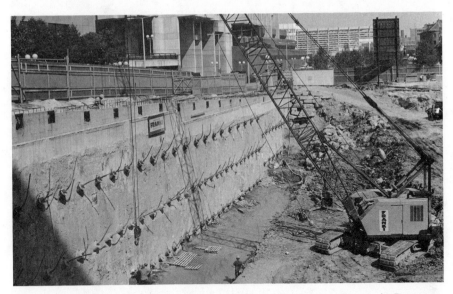

Fig. 1-3 A tiedback diaphragm wall for the Sixty State Street Tower in Boston.

1-5 COMPATIBILITY WITH GROUND ENGINEERING PROBLEMS

In a broad context an anchorage is used to mobilize the shear strength, and often the passive resistance, of a soil. An anchorage can thus receive earth loads acting on a structure, for example, a retaining wall, and transfer these loads back to the ground. Likewise, tensile forces can exist at the foundation of a modern structure as a result of the redundancy and unique arrangement of loading combinations, and since most soils are very weak in tension, these

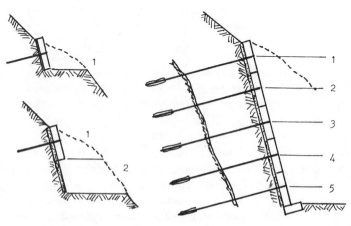

Fig. 1-4 Anchored wall built in horizontal strips in a downward construction process; this process is feasible in relatively stable ground.

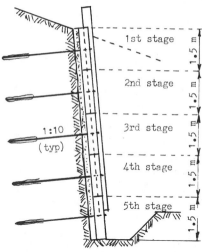

Fig. 1-5 Construction of a retaining wall by progressive underpinning, Wingreis, Switzerland; (a) view of the wall showing the vertical tie beams and the filler concrete in between; (b) section through the wall showing cover with concrete cladding. (VSL–Losinger.)

forces must be transferred to the ground indirectly. Furthermore, through posttensioning it is possible to induce a predetermined reaction in order to maintain an active zone of soil in compression, or provide the required level of preload on a retention system to cause inward ground movement that will compensate outward movement during excavation. Anchors can also be used where massive weight is needed to balance exterior upward pressure merely by replacing this weight by tensile forces transferred to the ground.

A special application of prestressing is for structures carrying heavy permanent loads. For example, a large foundation mat can be more economically designed using posttensioned tendons inserted eccentrically in a profile that will counterbalance the stresses induced in the same member when the structure is loaded and these loads are transferred to the soil underneath. This arrangement is exemplified in radially cantilevered foundations of towers and tall chimneys.

The foregoing principles are demonstrated in the following categories of problems.

Soil Preconsolidation. Prestressed ties can be used to apply a compressive force directly to soil causing it to consolidate and improve its stability under new loads.

Figure 1-6 shows this aplication for relatively soft layer of clay. A group of ties are anchored into bedrock and subsequently prestressed, producing the same effect as consolidation by preloading. The method has the advantage of scheduling the prestressing sequence and level according to the results to be achieved and with minimum disturbance to nearby facilities. The application is, however, costly but the cost is often justified by time savings. Typical examples are found in the construction of airport pavements on sites of recent landfills. Another example of soil preconsolidation is shown in Fig. 1-7. In this case, the eccentricity of the resultant load causes the tilted structure to return to the upright position.

The Nature of Uplift. Vertical anchors can be used where an underground structure is subjected to base heave or to uplift caused by external hydrostatic pressure and where it is not feasible to increase the dead load to balance these forces.

Figure 1-8 shows a structure founded below the groundwater table and thus subjected to uplift. The base is held down using tension anchors. Where a relatively impervious layer exists close to the base, an alternate solution

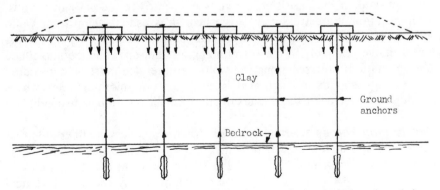

Fig. 1-6 Preconsolidation of a soft soil by the use of prestressed anchored ties.

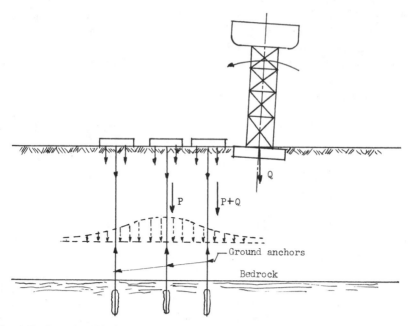

Fig. 1-7 Preconsolidation of a soil to return a tilted structure to upright position.

is to separate the base slab from the perimeter walls and extend the latter into the impervious layer to form a cutoff combined with sumping just below the base level.

If the uplift is moderately small and the ground is suitable for anchoring, the use of sealed ties without prestressing may be sufficient. With large uplift pressures, however, appreciable prestressing will be needed, especially if the ties are relatively long. The prestressing is likely to consolidate the soil beneath the slab, especially where this downward movement is not resisted, and cause settlement, which can result in a loss of prestress until equilibrium is established. If the base slab is rigidly connected to the perimeter walls, the application of prestress may cause shear stresses to develop at the wall–soil interface opposing further prestressing or it may cause the entire structure to move relatively to the surrounding ground, and thus the magnitude and direction of stresses along the wall–soil interface must be considered in the analysis. This interaction is likely to become more complex where swelling or heave occurs and tends to approach an elasto–plastic condition.

Overturning Forces. Certain types of structures have a configuration that causes them to be loaded eccentrically, and this often results in large overturning moments. An example is the ski jump shown in Fig. 1-9. The overturning effect of the free cantilever is counteracted by the use of vertical anchors installed at the base. Because of the unusually large overturning

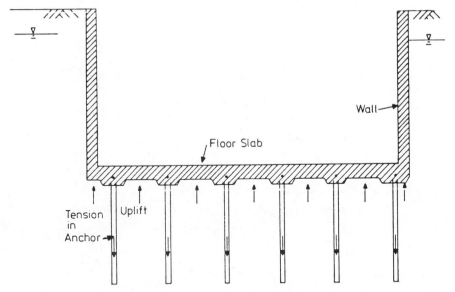

Fig. 1-8   Use of anchors to hold down an underground structure subjected to uplift.

moments and the resulting high-capacity anchors, this solution is more feasible in competent rock.

Other important classes of tiedown structures are power transmission towers anchored vertically to balance wind and nonsymmetrical cable loads, and concrete or masonry dams anchored to balance overturning caused by uplift or excessive hydrostatic pressure after foundation deterioration or as a result of raising the crest of the dam.

Rock Cavities. Conventional rock bolts produce independent action of individual rock blocks, thereby increasing the shear capacity and stiffness of the rock mass. In small-section galleries, mines, and tunnels they serve to prevent rock falls from the roof and the sides. In some instances this may be a temporary safety measure until the permanent lining is installed, or it may be the only rock support in the completed structure. Bolting is often cheaper and more convenient than lining or interior supports.

However, the excavation of large underground chambers in rock for tunnels and caverns causes considerable changes in the state of stress with associated strains and deformations, and in the worst case this can lead to collapse. In these conditions prestressed rock anchors create a self-supported underground opening through the application of a stressing force in the surrounding zone, and provided dislocation of surface material is prevented (Rabcewicz, 1957). Stabilization with prestressed rock anchors is distinguished from other methods in that prestressing creates an active arch in the rock mass, and the natural function of this arch makes the opening

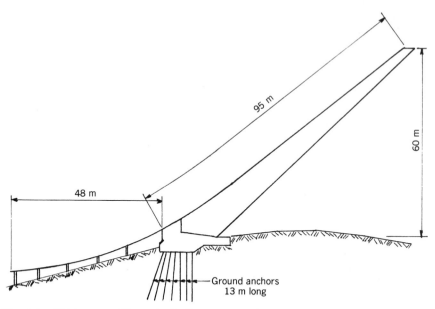

Fig. 1-9   Ski jump at Oberstdorf, Germany, anchored into rock to offset the cantilever overturning effect.

self-supported. The chamber continues to adapt itself to rock deformation until the entire mass becomes stable.

The shear failure theory governing this behavior is based primarily on a reciprocal relationship of required lining resistance and rock deformation, and secondarily on the fact that the time behavior of the rock mass is fundamental for predicting the response of the opening. In a fashion that is known as the convergence–confinement principle, after a cavity has been made the forces in the readjustment process of the surrounding rock are carefully controlled according to a competent stress distribution and rock yielding theory (Rabcewicz, 1973). In any excavation there exists a reciprocal relationship between pressure relief and radial deformation around the opening, and as the ground deforms stresses near the opening decrease. Theoretical equilibrium is reached between support resistance and ground reaction at an optimum point but before loosening of the ground occurs.

The foregoing concept, advocated for a long time, is exemplified in the support of underground openings where shotcrete and prestressed anchors are combined with rock to form a composite structure. Adjustments in the construction procedures are possible and practical through continuous control of rock and support system behavior by means of measurements of geomechanical interpretation. Elegant applications are in conjunction with the NATM (New Austrian tunneling method) widely used where preengineering of a tunnel is often impracticable because of great depth or other factors inhibiting a complete design.

An example of a large underground cavern is shown in Fig. 1-10 for the powerhouse at Waldeck II pumped-storage scheme, West Germany. This cavern is 106 m long, 54 m deep, and 33.4 m wide. The only feasible way of keeping this opening stable was by means of a self-supporting vault using prestressed rock anchors. The stress conditions around the cavity were determined by photoelastic analyses modeled on a computer program that provided the basis for determining anchor load and length.

Slope Protection. The prestressing of an unstable slope can increase its shear strength by improving friction, a process similar to preconsolidation. An example is shown in Fig. 1-11, where a deteriorated slope is stabilized by prestressed anchors, and the ground becomes stable to support the bridge loads. The prestressing is applied gradually to avoid high interstitial pressures on the soil in the initial stage and any probable adverse effects (Cestelli-Guidi, 1974). The process is slow and often expensive, but it can ensure the stability of structures founded on slopes that deteriorated without altering other soil characteristics. Likewise, the prestressing of rock slopes can improve their quality, particularly along the joints and fissures.

Deep Building Excavations. The assessment of deep basement construction usually involves an analysis of the estimated cost in relation to the expected use and demand for underground space. A deep basement implies a foundation at a lower level and a corresponding reduction in its cost, especially where the soil-bearing capacity increases with depth, but as the excavation becomes deeper the cost of ground support and its bracing is likely to increase.

Tiebacks are suited to almost any deep basement configuration. They may be temporary if a permanent interior bracing is contemplated or where the

Fig. 1-10 Cavern for powerhouse at Waldeck II pumped-storage scheme. (VSL–Losinger.)

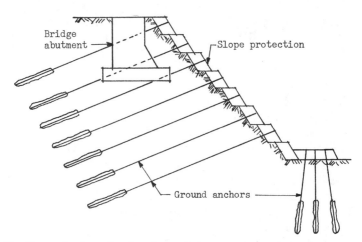

Fig. 1-11   Strengthening of a deteriorated slope supporting a bridge foundation by the use of anchors.

institutional arrangements cannot guarantee the permanence of the anchorage zone, or permanent if the intended use of the space precludes other forms of bracing. Applications cover the range from the anchored bulkhead to the multianchored wall, where the use of tieback becomes a construction convenience.

Long Excavations. These are common in the construction of underground transportation facilities or utility tunnels in open cuts. For relatively narrow openings it may be practical to use interior cross bracing or struts and cross walls, especially if the excavation involves temporary and permanent ground supports. Where the width of the cut increases considerably, interior bracing becomes expensive and tiebacks may offer an economically more attractive solution.

Underpinning of Structures. In a general context underpinning is the insertion of a new foundation or support below an existing to transfer the load to a lower level. Underpinning is typically required if an excavation adjacent to an existing foundation is carried down to a level lower than the existing foundation. The same result can be achieved in a different manner by combining ground strengthening with lateral bracing. This case is illustrated in Fig. 1-12, and involves a deep excavation adjacent to an existing subway tunnel of the Paris Metro. In lieu of underpinning, the subway structure was protected, as the excavation was carried down and below the base of the tunnel, by two diaphragm walls built as shown and tied together with four rows of tiebacks. The ground underneath the tunnel was injected with a suitable grout to increase its strength. The diaphragm walls were constructed from an intermediate excavation level.

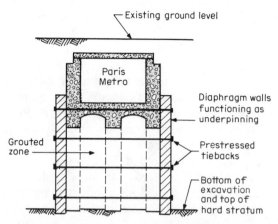

Fig. 1-12 Use of diaphragm walls in conjunction with grouting to underpin an existing subway structure.

Reinforced Earth. Unlike reinforced concrete, reinforcement in earth is used to increase the strength of weak soil material. The primary requirement is the development of friction between soil and reinforcement to give the composite material an apparent cohesion. Thus, the objective is to unify a mass of earth through the use of reinforcement of an appropriate length and disposition. Quite often, the use of inclined reinforcing strips provides better stability than is possible with horizontal strips. Notable applications are mentioned in conjunction with tunnel arches, small earth dams, and reinforced soil slabs to solidify weak zones or voids in the ground.

Special Cases. The useful functions of anchors in inducing prestressing, mentioned at the beginning of this section for large foundations, is also compatible with the control of deflections. This feasibility can be exploited in the design of continuous foundations.

A special case of prestressing is, for example, a long portal frame on poor soil under heavy loads and seismic potential, which will probably require a foundation beam to connect its two columns. This beam must be sufficiently rigid for a uniform pressure distribution. However, a similar condition of undeformability is obtained by a parabolic profile of prestressed tendons along the main axis of the beam producing an effect opposite to that caused by the soil ressure. The two effects compensate each other, and the resulting uniform soil pressure distribution ensures a uniform settlement.

1-6 CONSTRUCTIONAL FEASIBILITY AND REQUIREMENTS

Why Choose an Anchorage?

Ground anchors constitute a versatile construction system that, if properly used, can offer advantages in ground engineering. This versatility is derived num the following considerations: (a) ground anchors can accommodate variations in the soil environment and can be positioned to suit local site conditions; (b) they complement the use of soil and rock as foundation materials to support structures and slopes, whereas their tensioning provides additional useful information as to the condition of those materials; (c) the space required for their installation is minimum within the excavation site, and once the anchors are placed and stressed, there is no further obstruction to the next excavation level; and (d) in special or unusual conditions anchors may be the only method of support.

Ground anchors may be considered in the following context: (a) as integral part of the design concept of a project, (b) as a solution to problems developed following unexpected conditions during the course of construction, and (c) as remedial measure to improve or rehabilitate deteriorating structures.

However, in choosing an anchorage consideration must be given to the variability and severity of conditions affecting underground work, and the anticipated advantages must be judged in conjunction with the associated ground engineering problems. Difficulties, for example, will be encountered if the ground is not entirely suitable for the transfer of load from the anchor tendon, where agressive materials exist but remained undetected, or where statutory requirements canot be met.

An important factor of efficient anchor design and construction is adequate knowledge of soil conditions at the site. This requires more than a conventional soil investigation, discussed in detail in other sections, and may cost considerably more. Lack of this knowledge will impede the designer in assessing the exact degree of ground restraint available in the grout injection zone, and thus the full potential of the anchor system will not be realized.

In a geologic context, an anchorage can encounter the entire range of conceivable conditions, from tectonic layers to valleys and groves, from glacial formations to recent sediment fills, and from rock formations to soft layers. The engineer, therefore, is often confronted with soil investigation problems sometimes considered unsurmountable. Furthermore, in many sites anchors must be drilled and sunk into problematic overburden layers including flowing and erratic blocks. It must be emphasized, however, that if these difficult conditions did not exist anchorages would not have been required at all, hence choosing ground anchors only when conditions are ideal or favorable is very unlikely and often academic.

Mandatory Requirements. When a detailed ground investigation is completed, a comprehensive design will follow, including static and dynamic loads, anchor spacing and location, anchor capacity and load transfer length, overall stability, and service life requirements (Littlejohn, 1979), all discussed in detail in subsequent sections.

On the other hand, anchorage planning, design, and construction should be mandated to specialists because of the many unique and innovative solutions that are feasible with a given project. Furthermore, it is fair to say that, in spite of impressive applications in both temporary and permanent works, ground anchor technology remains in an active state of development and still lacks a unified practice. This is evident in the variety of design, construction and testing concepts that exist and can be identified, and it places additional emphasis on the value of specialty advice.

Anchor Choice. The selection of a suitable anchor for a given project requires complete knowledge and understanding of the variety of anchor systems. New types of anchors are continuously developed and introduced in the market, in response to a complex construction endeavor and in order to improve performance and reduce costs. Some are found suitable for certain applications, and some are considered practical within a restricted range of ground conditions and loading combinations. On the other hand, different practices and local economic constraints often preclude a unique solution based on value performance, and an improper choice can thus lead to later problems with technical and economic implications. Given the construction site and the use of an anchorage, the problem of anchor selection is not simple and is further complicated by the many factors affecting anchor performance.

As anchor construction continues to attract more interest it will remain a specialist operation, and despite the diffusion and wide dissemination of anchorage technology, a great deal remains to be learned about the subject. In this context, failure to give proper attention to the choice of a suitable anchor system can lead to deficiencies with potential structural implications and damage.

Construction Workmanship and Testing

This factor has a decisive influence on anchor service and performance. The construction of an anchorage is a sequence of several separate operations: hole drilling, tendon manufacture, anchor installation, grouting, testing, and stressing. The workmanship factor inherent with each of these phases tends to diminish ability to predict anchor performance solely on the basis of empirical rules and guidelines. Nonuniform ground conditions, on the other hand, can go undetected in routine site investigations resulting in considerable variations in anchor performance particularly with close anchor spacing. Hence, in order to ensure and confirm the workmanship of an anchorage installation, it is common to introduce quality controls as mandatory requirements during construction, specify performance tests, outline acceptance criteria for long-term behavior, and carry out a monitoring program during the service life of the anchorage.

Most anchors must be installed through a relatively thick overburden layer in order to reach stable formations where they are fixed. In these conditions, the objective of the anchoring technique focuses on longer and higher-capacity anchors; hence it greatly depends on the requirements of the drilling process. It appears, therefore, that two most difficult phases in the entire work are drilling the borehole and injecting the grout. Some of the many questions confronting the engineer before the first steps are taken must be answered quickly in order to be of any real use. For example, an answer must be provided about the necessity of an exact soil investigation as opposed to drilling a test hole; the stability of the soil materials and their probable collapse during drilling; the necessity of rotation core drilling in order to obtain a representative profile of the soil layers, direction, and thickness; or the necessity of a casing.

Whereas this is a typical example of sample questions raised before construction, it must be understood that whether an anchor will carry its load after it is installed and prestressed depends first on the geologic conditions and second on the injection technique. On the other hand, ample experience with this work is necessary for a successful primary injection and a possible postinjection to ensure that the anchoring force will be solely transferred to the adhering section of the anchor. Since there is seldom a direct procedure for checking the operation on a step-by-step basis, stressing and loading the anchor is mandatory in order to verify its capacity following its installation. Thus, values compiled from operational sequence of load applications constitute the most reliable record, and suggest the importance of a testing program prior to the activation of an anchorage installation.

Special Problems

In built-up areas, an occasional disadvantage of an anchorage is the ground movement associated with deep excavations. Earth unloading causes stress relief in the ground inducing a three-dimensional movement. In soft or in heavily overconsolidated clays as well as in clay shales the largest movement usually is in the horizontal direction. In some instances this movement has been found to extend laterally to a zone from 5–10 times the excavation depth. When economic considerations and other site conditions preclude the extension of the anchorage beyond this zone, the anchor length must be formed within a volume of ground prone to deformation. In these conditions a tied-back wall is likely to move, sometimes continuously, as result of movement of the ground in which the support is anchored, and unless this movement can be tolerated it may have unacceptable effects on surroundings. Problems of this nature, discussed in detail in subsequent sections, have been noted in overconsolidated stiff London clays.

A second example of special problems is the change in anchor prestressing caused by internal redistribution of stresses and associated strains in the soil following progressive excavation and tieback installation. These changes are likely to be greater with increasing number of rows of anchors. It is, however, possible to measure the amount of force remaining in a prestressed anchor generally by repositioning the same equipment. This monitoring permits the contractor to adjust the anchor load and restress the anchor if necessary. Likewise, it is possible to monitor the movement of the anchor head. If the anchor slips excessively, indicating appreciable loss of prestress, it may become inoperative requiring additional units in the same vicinity.

In some instances it may not always be possible to determine immediately whether an anchor has failed, since creep results in a loss of load over a period of days. Furthermore, replacing a high-level anchor that has failed can be difficult and expensive.

1-7 STATISTICAL AND REGIONAL DATA: BRIEF REVIEW

Anchorage Practice in the United States

By the late 1960s the advantages of anchorages were demonstrated in deep excavations and underground operations, but the high cost of labor and persisting organizational problems confined the use to relatively unsophisticated installations of moderate to low capacity. Thus, until the early 1970s tieback installations here remained almost exclusively the concern and endeavor of contractors. This trend changed rapidly when anchorage work was expanded following an expansion in the underground construction market, and as practical ways were found to reduce labor costs and improve the quality of the work. An account of the early American practice is given by White (1970, 1974a).

Among the early large examples is the anchorage for the Atlantic–Richfield office building in Los Angeles, completed in 1969. Approximately 300 tiebacks were used to support the excavation, which reached a height of 112 ft (34 m). The tiebacks carried a load of 43 tons (86,000 lb), and some were removable. In one stage the construction was combined with the excavation of the Security Pacific National Bank Building, shown in Fig. 1-13. Both excavation and underpinning exemplify the principles discussed in foregoing sections.

The anchorage project which marked the beginning of wider uses and applications is probably the World Trade Center in New York, shown in Fig. 1-14. The perimeter diaphragm wall has a total length 950 m (3100 ft), and is keyed into underlying rock. The 70-ft excavation required six rows of tiebacks that remained in service until the floor slabs were in place. Thereafter, the tiebacks were distressed. Their considerable inclination to the horizontal (45°) in some instances produced vertical reactions exceeding the available rock strength resulting in minor instability problems (White, 1974b). The tiebacks had a maximum load capacity 300 tons (600,000 lb). The entire excavation, shown in Fig. 1-14, is crossed by the PATH (Port

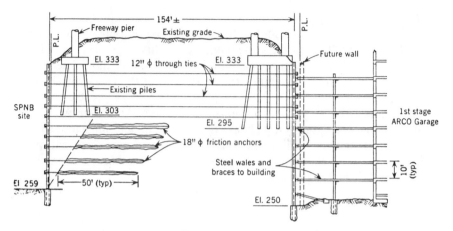

Fig. 1-13 Cross section through Security Pacific National Bank protective system, also showing tiebacks of the Atlantic–Richfield excavation, Los Angeles. (From White, 1974a.)

Authority Trans-Hudson Corp.) tubes that were completely exposed and underpinned.

Maximum Attained Load. Table 1-1 shows data for several anchorage installations, representative of the 1970s. Evidently, the highest loaded soil anchors are for the spoil disposal confinement structure at Kenosha harbor by the U.S. Army Corps of Engineers, with a test load of 240 tons (480,000 lb) (Nicholson Anchorage Co., 1979). Likewise, the highest loaded rock anchors are for the Newburg–Cannelton dams in Indiana, sustained test loads of 750 tons, which is in most cases the upper limit for rock. Anchor capacity in rock depends mainly on the available capacity of prestressing equipment and hydraulic jacks, whereas in soil this capacity relates primarily to the method of load transfer to the soil materials.

Statistical and Cost Data. Permanent rock anchors probably were not used until 1965, whereas permanent soil anchors followed later. Potential problem areas in permanent installations relate to corrosion protection, long-term creep, large extension of the anchor-head load, and development of pullout resistance in a potentially active zone. The number of anchors that normally would be expected to fail or perform unsatisfactorily during proof load tests in a typical job has been reduced from as high as 20 percent to as low as one percent, and it is now common to complete a job without anchor test failure.

Incidents requiring corrective measures are related to some form of failure or impaired service of the anchored or tied structure, such as excessive lateral or downward movement caused by premature excavation or over-

Fig. 1-14 General view of the World Trade Center, New York, showing the excavation, anchored diaphragm walls, and steel framing.

TABLE 1-1   Tieback and Anchor Loads Attained in U.S. Installations

Capacity (tons)	Soil Type	Location	Comments
300	Rock	World Trade Center, New York	Design load
240	Clayey glacial till	Kenosha harbor, Wisconsin	Test load
750	Rock	Newburg & Cannelton Dam, Indiana	Test load
150	Fine sand	Cobian Plaza, San Juan, Puerto Rico	Test load
200	Clay shale	Medical Center, Pittsburgh, Pa.	Test load
186	Sand and gravel	Power Station, Shippingport, Pa.	Test load
80	Sand and rock	Underground Gymnasium, Chicago	Permanent

loading. Radical structural failures of anchorages or anchored structures have been sporadic, if not rare.

The usual U.S. practice stipulates a preproduction test program for permanent anchors of high capacity or in difficult ground. For exmple, this program is necessary for anchor loads in excess of 200 kips (kilopounds) in sand, 240 kips in sand and gravel, 400 kips in soft rock, and 800 kips in hard rock. Permanent anchors in clay will require a preproduction test to determine suitability against long-term creep.

Anchor costs vary widely, and the main factors determining this variation are anchor capacity, type of soil, and construction conditions at the site. Reference to 1985 cost data for rule-of-thumb estimates shows that in good to average conditions anchors can be installed for $15–$23 per ton of working load and for relatively low capacity. For example, in 1985 a 75-ton anchor was likely to have a cost $1500 installed. Anchor cost is also quoted as price per foot of length. In good conditions, the 75-ton anchor will probably cost $30 per foot for an average length of 50 ft, but this cost may be much higher for higher-capacity anchors or in difficult ground. Thus, a 150-ton anchor can probably be installed for about $45 per foot (1985 price index) or as high as $60 per foot in unfavorable conditions.

In some instances, it has been found structurally expedient to use anchors but suitable anchoring layers did not exist close to the ground surface. In this case, contractors drilled through the overburden in order to reach rock, sometimes 100–150 ft below surface. Anchors have been installed in drilled holes more than 200 ft deep (60 m), which is presently the limit for an economical installation.

A relatively recent permanent anchorage installation is the George Westinghouse undergroud gymnasium in Chicago, (Xanthakos, 1977) completed in 1982, shown in Fig. 1-15. The posttensioned diaphragm walls are braced at the top by the roof framing and at the bottom by rock anchors. Competent rock just below excavation level allowed the economical use of anchors in lieu of a rock socket.

Anchorage Practice in Europe

In Europe, considerable work on anchorages during the postwar period was disseminated and eventually led to broad uses and applications, especially with the introduction of standards, recommendations, and codes of practice. These, in spite of variations and lack of technical uniformity, referred specifically to rock and soil anchors and covered a variety of topics in design, construction and testing.

Germany and Austria. A main postwar development is the Bauer system, introduced in West Germany in 1958. According to this method, discussed in detail in subsequent sections, a steel rod is anchored directly into a 8-cm-diameter borehole while a suitable cement suspension is injected into

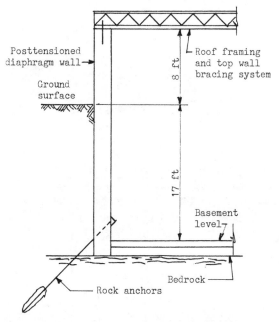

Fig. 1-15 Diaphragm walls and rock anchorages for George Westinghouse Underground Gymnasium, Chicago.

the hole. Bauer (1966) reports that by the end of 1965 about 30,000 units had been installed. A later account is provided by Ostermayer (1974) shown in Fig. 1-16 for both temporary and permanent anchors.

The Bauer anchor is based on the principle that grouted anchorages are more efficient if they can be installed in boreholes of relatively small diameter, 8–14 cm (3–6 in), and carry working loads of above average intensity, 30–50 tons in cohesive soil and 40–80 tons in cohesionless material. In this respect, the solution is to use tendons of high tensile strength so that for the same borehole size the anchor can carry a greater load.

For deep basement enclosures bracing with anchors often replaces internal bracing, whereas for subway construction in open cut in Munich anchored walls were found more economical than internally braced walls for excavations wider than 12 m (40 ft) (Ostermayer, 1974). Examples from anchorage practice in Germany are shown in Fig. 1-17. Part (a) shows a foundation slab subjected to uplift. The anchorage shown in (b) exemplifies the use of anchor tension ropes of suspension bridges in tent-type roofs, whereby dead-weight foundations are substituted with ground anchors incorporating soil weight. For the Olympic roof in Munich this arrangement allowed a 20 percent reduction in the cost of foundation.

In Austria, prestressed rock and alluvium anchors have been used since the mid-1960s, particularly in the mountainous Alpine region and the western

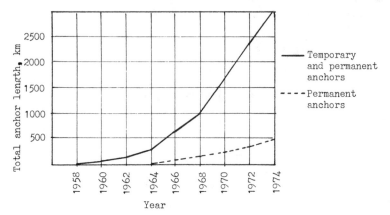

Fig. 1-16 Total length (as of 1974) of grouted anchors installed by the Bauer system since 1958.

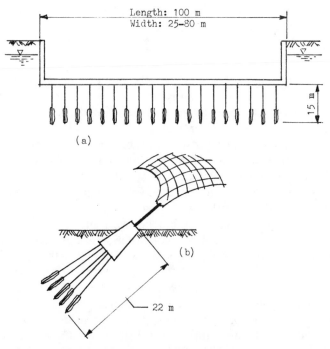

Fig. 1-17 Examples from the use of permanent ground anchors in Germany: (a) anchored foundation slab for the Tivoli building in Munich; (b) anchorage for the Olympic tent-type roof in Munich.

sections of the country. The construction method shown in Figs. 1-4 and 1-5, also known as the "top-to-bottom technique," has been found technically attractive in unstable rock formations, especially those including clay lenses, which can provide a potential sliding plane during excavation. This solution is applied with the intent to replace conventional bracing using a contact pressure through prestressing, acting as a retaining wall and roughly equal to the previously existing one so as to minimize any disturbance of the natural balance. Thus, relatively small forces can stop huge earth and rock masses from sliding.

Switzerland. A locally developed system is the VSL anchor (Grivelli, 1969a) discussed in detail in subsequent sections. In principle, the VSL practice tends to erase the distinction between rock and soil anchors, but recognizes the apparent distinction between temporary and permanent anchors and the specific requirements for corrosion protection.

The method shown in Figs. 1-4 and 1-5 is particularly adaptable to road construction in the Alpine region and where considerable cuts into a hillside are required (Grivelli, 1969b). Where the rock surface displays discontinuities such as seams and fractures, the depth of each progressive strip is limited to 1.5 m (5 ft), and the wall is anchored using one anchor per section. The support is thus structurally disconnected, and retains the advantage of free movement as each anchor is stressed without inducing secondary stresses or deformations to adjoining sections.

The principle of a self-supported underground structure using prestressed rock anchors, mentioned briefly in Section 1-5, was initially applied to the Hongrin underground pumped storage station completed in 1970 on the shores of Lake Geneva at Veytaux. In lieu of the traditional lining, the cavern was supported with prestressed rock anchors and shotcrete (Buro, 1970). This application might be regarded as the progenitor of the convergence–confinement theory and the NATM introduced in the late 1970s.

France. The evolvement of ground–support systems is evident in France, and has paralleled the demand for underground space in built-up areas. Ground anchors and tiebacks have contributed to this activity since 1953 (Forth, 1966). Deep excavations may be carried out in progressive underpinning, with soldier piles and lagging, or in conjunction with cast-in-place and prefabricated slurry walls (Fenoux, 1971). Institutional and legal problems are usually approached with incentives aimed at extending the benefits of a proposed anchorage installation to all parties involved.

A representative anchorage installation for urban development is the Halles Forum project in Paris shown in Fig. 1-18. This involved the construction of a subway station for the regional transit system with a four-level forum above, built in an open-air plaza. The construction requirements along each side of the project were different, and are reflected in the different solutions and support systems. The wall shown in part (a) consists of steel

Fig. 1-18 (a–b)

Fig. 1-18   The Halles Forum Project, Paris: (a) soldier piles and concrete lagging; (b) buttress-type slurry wall on piles; (c) slurry wall on piles. Anchorages were used for lateral bracing. (Bachy)

soldier piles with concrete lagging, and is temporary since the excavation was eventually extended onto the other side. The wall shown in (b) carries a new roadway built as overhang, and the large cantilever moments required buttresses. The system shown in (c) is a slurry wall built in alternate panels. Tiebacks were used for bracing, and in spite of their temporary nature they were protected against corrosion because of the long (4–5 years) planned construction period (Bachy, 1978).

Great Britain.   It is conceivable that anchors were in use before 1966 for pile testing reactions and for stabilizing coal mine roadway floors (Gillot and Mielville, 1964). In 1966 high-capacity anchors were used for the excavation of a key wall in Bristol. Systematic anchorage work was probably introduced with the Bauer system, which in the late 1960s was under license in construction work of the Second Mersey Tunnel (Jackson, 1970). Subsequently, several other systems were introduced in Great Britain. Among them is the UAC anchor used in the defense scheme at Oxfordness, which at the time of its construction probably involved the largest anchorage installation in the world (McKay, 1970). At the same time ground anchors were introduced

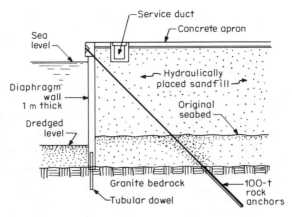

Fig. 1-19   Quay wall for the Peterhead (England) harbor development.

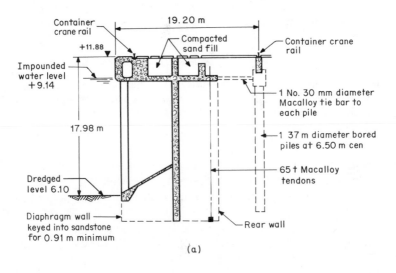

(a)

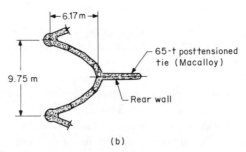

(b)

Fig. 1-20   Details of Seaforth dock wall. Liverpool: (a) typical cross section; (b) plan. (From Agar and Irwin-Childs, 1973.)

for permanent work in London clay (Concrete, 1970), and in conjunction with diaphragm walls.

As the technique became standard practice in a wide range of structures and applications, it attracted considerable interest among engineers (Hanna and Leonard, 1969). Attention was focused on certain theoretical aspects, such as transfer of load, creep, and corrosion protection, and semiempirical design rules were developed (Littlejohn, 1970). Anchorages became a prime topic in the September 1974 conference of the Institution of Civil Engineers.

A wide variety of unique applications are found in waterfront structures and installations such as piers, quays, and wharves. Examples are shown in Figs. 1-19, 1-20, and 1-21. For the Seaforth dock, the long back-fly wall improves resistance to overturning, and the degree of stability is further increased by the tiedown efffect of the permanent anchors.

Anchorage Practice Elsewhere

In Australia anchorages have been used in deep building excavations, cliff faces, and foundation slabs.

In North and South America anchorage practice is concentrated mainly in large metropolitan areas. In Caracas, for example, most of the underground retaining structures and ground supports built since the early 1970s have permanent anchorages (Petrini and Roca, 1974). In Brazil, anchorages

Fig. 1-21   View of Seaforth dock arch wall, Liverpool.

were introduced probably in 1957, and among the significant works is the anchorage test program carried out in the clayey soil of Sao Paolo in connection with the construction of the subway (Da Costa Nunes et al., 1969). In Canada notable examples of anchor uses are for foundations in fissured rock (Schousboe, 1974).

In the Far East, anchorages are used mainly in deep building excavations and construction in open cuts; hence this activity is concentrated in large metropolitan areas such as Tokyo, Osaka, and Singapore. In most cases, however, the tiebacks serve temporarily because of the unavailability of permanent anchorage zones. Removable anchors are thus popular by necessity, and various techniques are used to produce free gripping extractable strands (Yamada, 1978).

REFERENCES

Anderson, C. W., 1900: "Hydraulic Pile Screwing," *Proc. Inst. Civ. Eng.,* **139,** 302–307, London.

Agar, M., and F. Irwin-Childs, 1973: Seaforth Dock, Liverpool; Planning and Design, Proc. Inst. Civ. Eng., Lond., 1, p. 54.

Bachy, 1978: *The Halles Forum Development*, Special Communications.

Bauer, K., 1966: "The Injection Bauer Anchor System," *Trans. Civ. Eng. J.* (47), 1265–1273, Zurich.

Buro, M., 1970: "Prestressed Rock Anchors and Shotcrete for Large Underground Powerhouse." *Civil Eng.*, ASCE, May.

Cestelli-Guidi, C., 1974: "Ground Anchors." *Prestr. Concr. Found. Ground Anchors*, 7th FIP Congress, pp. 4–6. (New York).

Concrete, 1970: Wandsworth Bridge Southern Approach. Concrete, April 1970, London.

Costa Nunes, A. J., N. T. Chiossi, and T. C. Maggi, 1969: "Anchorage Tests in Clays for the Construction of the Sao Paolo Subway," *Proc. 7th Intern. Conf. Soil Mech. Found. Eng.*, Specialty Session 15. pp. 120–125, Mexico City.

Fenoux, Y., 1971: "Deep Excavations in Built-up Areas." *Travaux*, Aug.–Sept., Paris.

Forth, H., 1966: "The Anchoring of Structures," *Travaux*, April–May 1966, Paris.

Frazer, T., 1874: "Experiments on the Holding Power of Earth and the Strength of Materials," *Corps of Royal Engineers Papers*, New Series, Vol. 22.

Gillot, C. A., and A. L. Mielville, 1964: "The Use of Wire Cables in Anchoring Structures to Solid Strata." *The Mining Engineer* (47), Aug. pp. 645–655.

Grivelli, G., 1969a: *Prestressed BSI Rock and Alluvium Anchors*, Losinger and Co., Bern, Switzerland.

Grivelli, G., 1969b: "Application of Underpinning Method of Construction of Anchored Retaining Walls," *Proc. 7th Int. Conf. Soil Mech. Found. Eng.*, Speciality Session 15, pp. 191–194, Mexico City.

Hanna, T. H., and M. W. Leonard, 1969: "Some Design and Const. Considerations on the Use of Anchorages and Tiebacks,'' Piling Committee, Inst. Civ. Eng., Jan. 1969, London.

Jackson, F. S., 1970: "Contractor's Experience with Ground Anchors," *Ground Anchors, The Consulting Engineer (Cons. Eng.)*, London.

Littlejohn, G. S., 1970: "Development of Soil Anchorage Systems," *Ground Anchors*, Con. Eng., May 1970, pp. 9–12.

Littlejohn, G. S., 1979: "Ground Anchors, State-of-the-Art," Symposium on Prest. Ground Anchors, Concrete Society of South Africa, Johannesburg, Oct.

McKay, A., 1970: "Ground Anchors at Oxfordness," *Civ. Eng.*, London, February.

Nicholson Anchorage Co., 1979: "Anchorages in the U.S.A.," In-House Report, Bridgeville, Pa.

Ostermayer, H., 1974: "Construction, Carrying Behavior and Creep Characteristics of Ground Anchors." *Proc. Diaphragm Walls and Anchorages*. Inst. Civ. Eng., London.

Petrini, S., and A. M. de la Roca, 1974: "Permanent Anchored Walls in Caracas. Venezuela," *Proc. Diaphragm Walls and Anchorages*. Inst. Civ. Eng., London.

Rabcewicz, L. V., 1957: "Modeliversuche mit Ankerung in Kohasionslosem Material," *Die Bautechnik*, 34.

Rabcewicz, L. V., 1973: "Principles of Dimensioning the Supporting System for the NATM," *Water Power* (March)-88.

Schousboe, I., 1974: "Prestressing in Foundation Construction," *Proc. 7th FIP Congress*, May–June, pp. 75–78, New York.

VSL–Losinger, 1972: "Rock Anchors for Waldeck II," W. Germany, In-House Report, Jan. 1972.

White, R. E., 1970: "Anchorage Practice in the United States," Ground Anchors," The Cons. Eng., May, pp. 32–37, London.

White, R. E., 1974a: Prestressed Tendons in Foundation Construction," *Prest. Conc. Found. and Ground Anchors*, 7th FIP Congress, May–June, pp. 25–32.

White, R. E., 1974b: "Anchored Walls Adjacent to Vertical Rock Cuts. *Proc. Diaphragm Walls Anchorages*, Inst. Civ. Eng., London.

Xanthakos, P. P., 1977: "Anchored Diaphragm Walls in Chicago," In-House Report.

Yamada, K., 1978: "Ground Anchor Market in Japan," VSL Intra-Co. Symposium.

CHAPTER 2

THE ANCHOR SYSTEM: COMPONENTS AND INSTALLATION

2-1 BASIC CONSIDERATIONS

In general, anchor capacity and performance are influenced by three main factors: (a) ground characteristics, especially shear strength; (b) installation techniques, particularly the method of fixing the bonding zone; and (c) the workmanship attained in the field. For permanent installations, problems are almost certain to arise if the development of bond between steel tendon and injected grout is not as predicted. It appears, therefore, that these considerations often tend to inhibit the complete preengineering of an anchorage installation solely on the basis of standard rules and procedures. An essential supplement to the technical background is ample practical experience with the various anchor systems, especially the potential construction problems associated with this type of work.

For a preliminary analysis and feasibility study of a proposed anchor installation, the first factor to be considered is the nature and strength of foundation materials. Some types of soils may not be suitable at all for anchorage work, whereas loose or soft materials may preclude an economical utilization of the method. For example, in soft soil the maximum design load is often limited to 30 tons irrespective of the tendon type, but in rock the attainable load is several times higher. Furthermore, ground conditions at the same site often encompass a wide variation of materials from soft soil to rock, and thus selection of anchor type and load without actual in situ verification becomes impractical and academic.

In general, soil anchors offer a good solution to ground engineering problems if they are installed in stiff clay, or dense silts, sands, and gravels. In most instances, the installation is adequate without altering the soil characteristics. Anchor capacity may range from low to medium, and depends mainly on the soil strength. The fixed anchor zone, or bonded length, is formed in a competent zone and is created by a straight shaft, multiple bells, or end bearing plates. Rock anchors, on the other hand, are usually bonded by means of a straight shaft and can resist much higher loads since their capacity is derived from the much higher shear strength of rock materials. Furthermore, rock anchors can be installed in almost any inclination, and the hole is drilled without the need of a casing. Apart from these differences, emphasized mainly in the past, the distinction between soil and rock anchors has no longer any technical merits and tends to be superseded. Commonly, the same anchor system with the same essential features and details can now be used for either application and in a variety of ground conditions.

The main details for each type are determined by the design requirements of the project, and are developed in conjunction with the size of tendon, the drilling and grouting methods, and the shape of the grout body. Furthermore, the wide variety of rock formations—often ranging from clay shales to limestones, chalks, and highly fractured rocks—dictates a corresponding variety of techniques used in bonding the tendon.

Marine anchors and their applications are considered beyond the scope of this book. This group includes various types but predominantly deadweight, grouted tendon, free-fall, driven, direct embedment, and embedded suction anchors. A complete summary is given by Taylor et al. (1975).

Another category includes anchor systems utilizing tiedown techniques to secure ground facilities. Examples are mobile homes, powerline poles, and military installations such as large tents, inflatable structures, and special membranes installed to cover poor soil and provide landing pads. In these cases, an anchorage zone is created by a triangular-shaped member, by helix and multihelix combinations, or by expandable screwdown plates. Useful reviews are given by Kovacs (1977) and by A. B. Chance Co. (1975)

2-2 ANCHOR GROUPING AND CLASSIFICATION

Temporary and Permanent Anchors

The division between temporary or short-term, and permanent or long-term, anchors is arbitrary at best and often academic. Temporary anchors are, by virtue of their limited durability, devices of a temporary nature that will become useless and inoperative beyond a certain stage in the work program, irrespective of the time lapse between their installation and the stage when they become unnecessary. Permanent anchors, unlike temporary ties, are devices which, by virtue of their long durability, will maintain the stability of a project on a permanent basis.

The two groups have different requirements. The planning of temporary installations usually involves a mandatory structural analysis and design, dimensioning and testing of bonded anchors, and finally their in situ tensioning to confirm the load carrying capacity. For permanent anchors, further requirements must be satisfied mainly in the arrangement and protection of system components, all discussed in detail in subsequent sections. Several codes specify the duration of temporary service as 2 years, but this guideline should be accepted with caution and full understanding of its limitations, and where soil conditions are fully known and controllable. More conservative specifications recommended by this author make the following distinction:

- Temporary installations—where the anchors will remain in service for less than 6 months. During this period it is highly unlikely that a corrosion process of detrimental magnitude will be initiated; hence neither corrosion protection nor monitoring are required.
- Semipermanent supports—where the anchors will be used for 6–18 months. In this case, corrosion protection may not be specified, but some monitoring of anchor performance is advisable and should be included in the construction program.
- Permanent support—where anchors must function longer than 18 months. In this case, corrosion protection and monitoring are mandatory, and should be provided according to the longevity characteristics of the installation.

It appears from the foregoing remarks that the requirements of a permanent anchorage installation cannot be determined strictly on the basis of time during which performance is needed. Equally essential are the ground conditions and corrosive tendencies of the soil environment in which the anchors are placed. Additionally, in determining allowable working loads long-term stability and creep characteristics must be considered together with possible reduction of bonding ability along contact surfaces. These considerations are particularly important in soils composed mainly of silt or clay that may become subject to remolding or loss of pore water pressure, resulting in a gradual loss of stress in the tendon. On the other hand, an agressive or hostile environment can initiate a corrosion process leading to considerable loss of load-carrying capacity.

Active and Passive Anchors

Active anchors, also called "prestressed," apply initial force to the structure thus supported, irrespective of the final natural soil–structure interaction, and according to the actual prestress level (Cestelli-Guidi, 1974). This force is introduced with the use of jacking devices, and will persist with time unless the structure undergoes displacement relative to the anchor itself. Passive anchors, also called "dead," are not prestressed but respond to loading only when the structure thus supported begins to move following excavation on one side. As the excavation proceeds, a natural soil–structure interaction is initiated and maintained. More common is the use of intermediary anchors where a fraction of the full potential prestress load is applied, usually between $\frac{1}{2}$ and $\frac{2}{3}$ of the design load. The concept of active and passive anchors is shown schematically in Fig. 2-1 as a function between the externally applied tensile force, and the relative displacement as excavation is completed on one side of the support. Since anchors are rather long and flexible members, it follows that relatively large movement of the anchor head must occur in order to stretch the tendon and thus develop the full load. The use of initial prestress serves to reduce this movement while taking into account the excavation and anchoring stages.

Active anchors are useful where it is necessary to consolidate slopes and foundation beds by prestressing, as shown in Figs. 1-6 and 1-7. The passive type is suitable where interstitial pressures can develop, in which case the anchor assumes the function of a nail. Intermediary anchors are used frequently in conjunction with the lateral bracing of earth retaining structures where large movement must be prevented, but a small displacement necessary to reach and maintain an active state of stress is also compatible with the construction conditions at the site.

A tentative classification of anchors according to the above functions and degree of tensioning is summarized in Table 2-1. The details and procedures shown for each type are general, and variations are conceivable and often necessary, particularly in the grouting methods.

Anchor Grouping According to the Method of Load Transfer

Most anchorage installations are completed with the so-called cement grout injection anchor, discussed in detail in the following sections. In this case, anchor pullout capacity largely depends on anchor geometry for a given set of ground conditions, but is also influenced by the configuration and size of

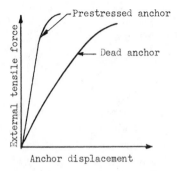

Fig. 2-1 Active (prestressed) and passive (dead) anchors. Relative displacement as function of applied force.

TABLE 2-1 Classification of Ground Anchors According to Function and Degree of Tensioning

Type of Ground Anchor		Steel of the Tie	Type of Anchorage on the Structure	Anchoring Zone on the Soil	Phases of Grouting
Active		Pretensioned	Anchoring device for prestressed steel	Limited to a terminal length	1. Injection of the terminal zone (anchorage of the wires in the rock) 2. Injection for bonding ducts to soil 3. Grouting of the ducts
Passive	Simple tie or strut	Not pretensioned	Ordinary end anchorages for ties or struts	Extended over the length	1. Injection for bonding ducts to soil 2. Grouting of the ducts
	Tie of prestressed concrete	Pretensioned			Bonding of tie to ground
Intermediary		Partially pretensioned	Anchoring device for prestressed steel	Limited to a terminal length	1. Injection of the terminal zone (anchorage of the wires in the rock) 2. Injection for bonding ducts to soil 3. Protection of the wires (eventually without grouting)

From Cestelli-Guidi (1974).

the anchor zone. Accordingly, there are four main groups, characterized by the mechanism of stress transfer from the fixed anchor zone to the surrounding ground, for which construction methods and design rules are available. The four types are shown in Fig. 2-2, and even though each type is more suitable under specific ground conditions the choice is often dictated on a regional basis.

Type A. This is characterized by a tremie-grouted straight shaft cylindrical hole of a uniform diameter, which may be lined or unlined according to the requirements of hole stability. This type is suitable in rock as well as in very stiff to hard cohesive layers where it is most commonly used. The load transfer is by shear resistance mobilized along the ground–grout interface.

Type B. With this type, the anchor zone is created as an enlarged cylinder formed in a grouted borehole under low injection pressure (usually <1 N/mm^2 or 145 psi) using a lining tube or in situ packer. In this process, the actual effective diameter of the fixed zone is increased with some minimum disturbance to surrounding earth materials as the grout permeates through the pores or natural fractures under injection pressure normally less than the total overburden pressure. The enlarged cylinder

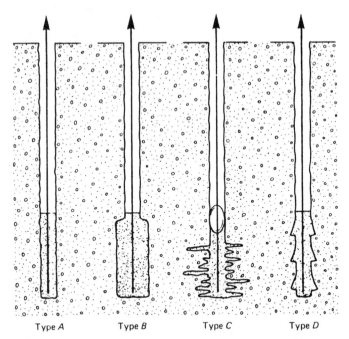

Fig. 2-2 Anchor grouping according to the method of load transfer. Four types of fixed anchor zone for grout injection anchors.

is suitable in soft fissured rock and coarse alluvium, but many contractors use it also in fine-grained soils. In the latter case, the cement particles will not always invade the small soil pores, but under pressure the grout will compact the soil locally to increase the effective diameter and thus shift the contact surface of maximum resistance along this enlarged diameter. For type *B* resistance to withdrawal begins with side shear, but end bearing at the upper end may be eventually mobilized and contribute to the ultimate capacity.

Type C. In this case the grout is injected under high pressure (>2 N/mm^2 or 290 psi), forcing the cement particles to penetrate the soil irregularly and thereby enlarge the anchor zone through hydrofracturing of the ground mass. A grout root or fissure system is thus produced beyond the core diameter of the borehole as shown. This anchor is suitable primarily in cohesionless soil, although it has been successfully used in stiff cohesive deposits (Littlejohn, 1980b). The design is based on an assumed uniform shear along an appropriate diameter at the fixed zone, or it may have a semiempirical origin based on tests. It is not always clear under what injection pressure ranges the grout will produce type *B* or *C* anchorage, given the same soil conditions and ground characteristics, but in many instances a composite system may be produced, incorporating the features of both types.

Type D. As in type *A*, the borehole is tremie-grouted, but it includes a series of enlargements (bells or underreams) formed mechanically in the fixed anchor zone. This type is common in stiff to hard cohesive deposits, where underreams as large as four times the sizes of the borehole have been cut successfully. Pullout capacity is derived primarily from side shear, but plug and end bearing also increase resistance to withdrawal.

2-3 ANCHOR ASSEMBLY AND PARTS

The schematic presentation of the anchor assembly shown in Fig. 1-1 distinguishes the following three main parts and components.

Fixed Anchor Length. This is also referred to as "bonded length" or simply "anchor body." It represents the design length of an anchor along which the tensile force is transmitted by bond to the surrounding ground, according to one of the mechanisms shown in Fig. 2-2.

In general, the fixed length is produced by cement grout injection, and this is valid for both rock and soil anchors. However, in addition to providing a load-transmission zone, the grout also performs the all important function of protecting the prestressing steel against corrosion. For permanent anchors, further protection is necessary and afforded by means of a corrugated sheath that ensures the isolation of the tendon steel from the soil environment. A simple section of an anchor body is shown in Fig. 2-3(a) and (b) for a temporary and a permanent anchor respectively. The transfer of load in (a) is from the steel tendon to the grout and then to the ground. In (b), however, the load is transferred from the tendon steel to the grout, then to the sheath, then back to the grout in the zone outside the sheath, and finally to the soil at the grout–soil interface.

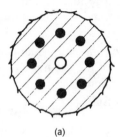
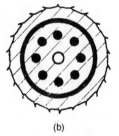

(a)   (b)

Fig. 2-3 Typical cross section through the anchor body for temporary (a) and permanent (b) anchors.

Free Anchor Length. This is referred also as "free tendon length." It represents the portion of the anchor between the start of the fixed length and the anchor head. No load transfer is assumed or assigned to this length, since it is intentionally isolated from the surrounding ground and thus remains free to move during the anchor–soil interaction. Therefore, the main requirement imposed on the free anchor length is freedom to elongate under prestressing without hindrance. Additionally, the method of forming this section of the anchor must also accommodate the general requirements imposed on the installation.

Anchor Head. Sometimes referred to as "end anchorage" or "stressing anchorage," this is the end part and component that transmits the tensile load from a loaded anchor to the ground surface or structure, or vice versa. Through a simple mechanical interlock, the head fastens the anchor to its structural support, but also serves to introduce the prestressing force to the anchor. In this context, the anchor head constitutes a main feature of an anchor system, together with the stressing mechanism.

Most manufacturers of anchor supplies have developed standard anchor head details. These, however, should be checked to ensure structural compatibility with the particular configuration of the fixed anchor length, load-carrying capacity, actual prestressing level and method of introducing the prestressing force, and other relevant considerations.

General Requirements of Anchor Performance. Anchors and their components must satisfy various requirements, arising from the intended use and application, service life, type of ground, magnitude and type of loads, and other special considerations. Thus, for the usual applications, an anchor must satisfy any or all of the following (VSL–Losinger, 1980):

1. The functional life of the anchor can be extended, requiring a high degree of protection against corrosion and mechanical damage.
2. The prestressing can be applied in stages.
3. The anchor may have to be destressed and restressed again.
4. The anchor and its components can be load-tested, the prestressed load locked off after this operation, and the remaining force in the tendon measured as needed.
5. The anchor may have to be installed in such a manner that any occurrence of transverse displacement of the ground will not induce excessive secondary stresses in the tendon.
6. The force in the tendon remains central.

It must be emphasized that these requirements are frequently interdependent. For example, corrosion protection, which employs corrugated and smooth sheathing, facilitates restressing and testing of the anchor, and thus ensures the permanent effectiveness of the stressing length. This is important, for example, in case of unexpected strain occurring because of a fissure in the rock in the stressing length, caused possibly by blasting operations after the installation of the anchor. If this strain is distributed over the entire stressing length, it will have only a limited effect on the stresses in the tendon. If, however, the anchor after stressing is bonded to the rock, such local fissures can cause overstrain in the steel at this location and lead to unavoidable failure (Kern and Herbst, 1974).

2-4 ANCHOR TENDON

Comparative Description of Tendon Types

A tendon usually consists of a bar, wire or strand, used singly or in groups. The quoted tensile strength covers a broad range from 1200 to 2000 N/mm^2. Data on the technical characteristics and exact mechanical response under load are available from manufacturers and suppliers. Variations therefrom are possible and should be expected, particularly in the size and cross-sectional area, ultimate strength, elastic limit and relaxation loss, and development of mechanical bond. Additionally, tendons must comply with applicable prestressing steel standards. In the United Staes these include the Tentative Recommendations for Prestressed Rock and Soil Anchors of the PCI, whereas most European countries involved in the design and construction of anchorages have adapted Euro-Standard 138-79. However, the applicability and interpretation of data and guidelines as well the choice of a suitable tendon with an appropriate factor of safety demand individual assessment and judgment. This decision is further influenced by cost factors, fabrication and transport considerations, corrosion protection requirements, design load, and permissible stress levels.

Bars, plain or threaded, provide the simplest type of tendon. They can be more readily protected against corrosion, and for shallow or low-capacity installations they usually cost less. They possess appreciable stiffness, which facilitates placing by allowing the tendon to be handled in manageable lengths and with minimum risk of mechanical damage, and in certain conditions the bar itself can be used as the drill rod. Threaded bars are easily connected to the embedded anchorage and the anchored structure by means of deformations providing an interlock with the grout body without other bonding requirements. Furthermore, the threaded portion allows partial stressing, restressing, and quick checking or releasing of the load. The relatively large residual elongation of bars after stressing allows appreciable ground movement to occur before anchor failure is induced. Finally, the stiffness of the bar helps keep the grout tube undistorted and facilitates the unimpeded flow of grout in the anchor zone.

However, wire and strand offer distinct advantages with respect to tensile strength, easiness of storage, fabrication, and transportation to the construction site. Their higher flexibility allows easy handling at the site and favors their use in limited space. They possess higher elasticity inherent with the higher tensile strength, which results in lower creep loss or load increase following ground movement. With the advent of significant improvements in prestressing techniques and equipment in the early 1970s, the use of strand gained popularity among both designers and contractors. Several surveys (FIP, 1974a) have shown that strand tends to be accepted more often than bars and wire even in countries where the basic material cost is higher.

Most contractors agree that a smaller tendon diameter may result in a lower material cost per unit of prestress force, but they caution that direct cost comparisons for the supply of tendon material in a given country and region can be misleading since the total cost of tendons also reflects fabrication, installation and stressing costs.

Tendon types are discussed in the following sections.

Bars

Bar anchors are used more in North America and Germany. They are manufactured in sizes from $\frac{1}{4}$ in (6.4 mm) to $1\frac{3}{8}$ in dia. (35.8 mm) with usual increments of $\frac{1}{8}$ in (3.2 mm). Common sizes are 1 in, $1\frac{1}{4}$ in, and $1\frac{3}{8}$ in. dia. (26.5, 32, and 36 mm). Steel strength quoted by various suppliers normally is 835/1030 and 1080/1230 N/mm², or 121/149 and 157/176 ksi (kilopounds per square inch), yield strength/ultimate tensile strength ratios (Stump–Vibroflotation, 1982). These characteristics produce tendon failure loads up to 280 kips (1250 kN).

In the United States, PCI codes specify high-alloy steel bars, either smooth or deformed. However, in the United Kingdom B.S. 4486 specifies low-alloy steel from 20 to 40 mm dia. Stainless-steel bars from 10 to 32 mm dia. have been available in most markets, but they are used with caution owing to limited data on relaxation characteristics.

Single bars are preferred for relatively short and low-to-medium capacity anchorages; hence they are used in large quantities in soils where the anchor force that can be developed by the soil has the same order of magnitude as the monobar anchor. Multiple units are used in sophisticated combinations and in conjunction with compression tubes and elaborate end bearing plates for heavier loads. These plates are usually tapped to take the bar thread. They can be designed either to supplement the bonded bar by giving a positive end in the same way as the nut, or to take the full bar force in direct bearing on the grout. Groups of four and five bars are used successfully, but clusters require relatively larger anchor holes and are more difficult to handle.

Threaded bars can be either partially or fully provided with deformations. In the simplest form, the bar has one end threaded. This requires a minimum hole size, but the ultimate carrying capacity of the bar is likely to be only slightly higher than that of the plain bar because of the fine pitch of the thread. With continuous thread, the transfer of load is achieved by means of threaded connections, or mechanical bond as in a deformed bar. Multiple threaded bars are anchored in one plate as shown in Fig. 2-4(a) and are stressed simultaneously by one stressing jack, which allows better control of elongation. However, with multiple bar anchors the forces cannot be transferred to the ground in one simple grouting operation, and special procedures are necessary to grout under pressure several times depending on the soil conditions.

Wires

Prestressing wire generally is produced from cold-drawn plain carbon steel. The material can be left "as drawn" or "prestraightened" through a process involving a stress-relieving heat treatment that improves elasticity and imparts to the steel a normal relaxation behavior, or a hot stretching treatment that also produces the same beneficial results. In the United States, wires

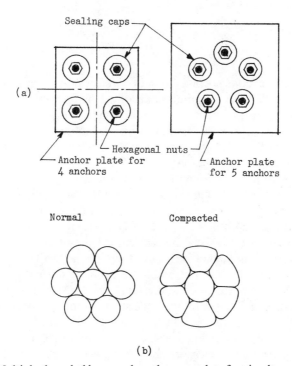

Fig. 2-4   (a) Multiple threaded bars anchored on one plate for simultaneous stressing; (b) typical 7-wire strand anchor cross section showing the normal and compacted form.

must conform to ASTM A421 "Uncoated Stress-Relieved Wire for Prestressed Concrete." Elsewhere, the material conforms to local standards, for example, British Standard 2691 (1969) in England. Elaborate combinations of quenched and tempered wire ribbed during hot rolling have produced varieties predominating in some countries, for example, Germany.

Tendon wire varies in diameter, the most common size ranging from 5 to 8 mm, and has a usual ultimate tensile strength 1670 N/mm^2, or 242 ksi (Littlejohn and Bruce, 1977; Stump–Vibroflotation, 1982). In general, it is used in groups of 10–100.

It is interesting to mention that the Soviet Industry (Shchetinin, 1974) manufactures wires with ultimate tensile strength capacity from 1375 to 1865 N/mm^2 to meet applicable Soviet Codes. A popular size for anchors is 5-mm wire, with 1670 N/mm^2 ultimate tensile strength, 184,000 N/mm^2 elastic modulus, and 6.8 percent relaxation at 1000 h. Shchetinin recommends wire tendons because they eliminate potential torsional and bending problems often arising with strand anchors.

Within the usual range of applications, there appears to be no limit to the number of wires that can be grouped to comprise an anchor. Soletanche reports that 660 No. 5 mm wires were used per anchor at the Cheurfas Dam repair work in 1934.

Strand

This type of tendon consists of a group of wires usually from 4 to 20, arranged in a helical form around a common axis of a straight wire, and in diameters of 12.7 and 15.2 mm, respectively. Exceptions are possible, and it is interesting to mention that for a retaining wall on Interstate Highway I-96 near Detroit 54 wires 12.7 mm dia. were used to produce anchors with 7010 kN (1570 kips) ultimate tensile capacity.

The 7-wire strand is common here and abroad, usually available in sizes of 13 mm (0.5 in), 15 mm (0.6 in), and 18 mm (0.7 in); 19-wire strand is also common and available in sizes of 22.2, 25.4, 28.6, and 31.8 mm. The quoted ultimate tensile strength is from 1570 to 1765 N/mm^2 (228–256 ksi), but occasionally strengths up to 2000 N/mm^2 can be produced.

Likewise, strand is made from cold drawn plain carbon steel. The 7-wire strand is stress relieved after stranding to produce normal relaxation characteristics. Other types include low-relaxation strand produced by a potential stabilization process whereby tensile stress is applied to the strand during the stress relieving operation. On the other hand, strand can be produced in a compacted form, or dry process, whereby about 20 percent more of the nominal cross-sectional area is occupied by steel (Littlejohn and Bruce, 1977) with respect to the ordinary strand so that higher loads can be sustained. This type of strand has also low relaxation characteristics. A typical 7-wire strand is shown in Fig. 2-4(b) for both the normal and the compacted form.

2-5 TENDON CHARACTERISTICS

Basic Definitions

The choice of working loads and permissible stress levels usually is made with regard to (a) mechanical strength, (b) elastic properties, (c) creep response, and (d) relaxation behavior. These characteristics are defined and discussed in this section in terms of their effect on the safe load-carrying capacity of the anchor, and are reviewed in subsequent sections in the context of load transfer, stressing and testing, and long-term monitoring.

Characteristic Strength. In general, permissible stress levels and working loads for a given type of anchor are quoted in terms of the specified characteristic strength, for which the notation f_{pu} is commonly used. This is the guaranteed limit below which not more than 5 percent of the test results fall, and none of these is less than 95 percent characteristic strength. Instead of the term "guaranteed limit," other equivalent notations used as frequently are "ultimate tensile strength," already referred to in this text, and "guaranteed minimum ultimate tensile strength."

Proof Stress. It is common here and abroad to specify tendon stresses in terms of some elastic limit, such as 0.1 and 0.2 percent proof stress. Inasmuch as the stress–strain curve of high-tensile steel does not have a definite yield point, the proof stress is defined as the stress at which the applied load causes a specified permanent elongation, such as 0.1 and 0.2 percent. It is interesting to note that the term T_G used by the French Code to designate the tension corresponding to the elastic limit actually is measured as the 0.1 percent proof stress, or the point at which the permanent elongation reaches this value. This should not be confused with the 0.2 percent proof stress used by the British and other codes to designate the elastic limit. Data provided by material suppliers show that the 0.1 percent proof stress may be 3–5 percent lower than the 0.2 percent proof stress, which is roughly equal to 87 percent characteristic strength f_{pu}. From this it follows that 0.1 proof stress used by the French Code approximately corresponds to 0.96×0.87 or 83.5 percent f_{pu} (Littlejohn and Bruce, 1977). This correlation is very useful in comparing working stresses and factors of safety. (See also Table 4-1.)

Elastic Modulus. The usual range of the modulus of elasticity for the three types of tendon is shown in Table 2-2 for various grades and with approriate remarks. (See also Section 4-2.)

These values are quoted by tendon steel suppliers and manufacturers, and it is conceivable that there is an error of ±5 percent in their derivation reflecting unavoidable variations in testing and recording procedures. A further difference in E values is also evident between laboratory test length

TABLE 2-2 Modulus of Elasticity^a for Prestressing Tendons

Type	Remarks	Average E (N/mm^2)	Average E (ksi)
Bar	—	165,000–175,000	23,910–25,360
Wire	Mill coil	192,000	27,825
Wire	Prestraightened	201,000	29,130
Strand	Regular normal relaxation	198,000	28,700
Strand	Regular low relaxation	200,000	30,000
Strand	Normal relaxation	197,000	28,550
Strand	Superlow relaxation	198,000	28,700
Strand	Dyform	195,000–198,500	28,260–28,770

^{a} The modulus of elasticity shown here corresponds to the elastic limit shown in Fig. 4-1; hence, it is the proportionality constant defining the linear portion on the stress–strain diagram.

and the relatively long length at the site, and comparison under controlled conditions shows that this may account for a variation of ± 2.5 percent in the values thus obtained. Finally, it must be emphasized that among the test features having the greatest influence on the E values are the rate of testing and the nature of repetitive loading, as well as the characteristics of deformability behavior between a single wire, a strand, and a cluster of strands. Thus, it is very unlikely that the E of strand will be the same as the E of tendon consisting of the same strand. In this respect, Littlejohn and Bruce (1977) have quoted the following values for the prestressing steel used at the Wylfa nuclear generating station:

$$E_{strand} = 183,000–195,000 \text{ N/mm}^2$$

$$E_{tendon} = 171,000–179,000 \text{ N/mm}^2$$

It is evident that E tendon is lower than E strand, but no general relationship can be established correlating the two values. The obvious reason for this difference is that during loading strand wires tend to unwind, with the corresponding effect depending on the restraining capacity of the stressing system.

Creep Response. In simple terms, "creep" is a change in strain of the tendon with time under constant stress. When a tendon is in tension under constant load, slow plastic deformations can occur even at a stress level below the elastic limit, and this is true for anchors installed in both cohesive and uniformly grained noncohesive soils. The relationship between creep displacement and time can be described as an exponential function of a straight line on a semilogarithmic plot (Ostermayer, 1974).

Differences in creep under constant and repeated load generally exist, but are not completely understood. Caution is thus necessary where the number of cycles of repeated loading during the lifetime of a structure is expected to be very large, in which case the possibility of cumulative effects should be considered. Creep is reviewed and discussed in more details in subsequent sections.

Relaxation Behavior. Stress relaxation is a decrease of stress, and a corresponding loss of load in the tendon, with time while the tendon is held under constant strain. This behavior is manifested by the gradual replacement of elastic strain by plastic strain causing the subsequent relaxation of elastic stresses. Like creep, relaxation is a function of the logarithm of time, and its extent depends mainly on the treatment of steel during its manufacture, the temperature conditions, and time. It can cause a loss of load of 5–10 percent, most of which occurs during the first few hours of loading (Antill, 1967; Bannister, 1959). Hence, a deliberate initial temporary overloading of the tendon is often useful and serves to reduce overall relaxation loss by compensating for the rapid initial loss. Littlejohn and Bruce (1977) quote a load loss after one hour of the order of 50–60 percent of that at 100 h, which again is about 80 percent of that at 1000 h. The loss at 1000 h is about one-half that at 5–8 years.

Stress relaxation increases rapidly with temperatures above 20°C, and thus in warm geologic environments both the ground and groundwater temperatures should be monitored. Furthermore, the tendency of wires to unwind during stressing can be limited by the restraining effect of jacking systems, and in this respect stressing devices preventing rotation during jacking are preferred.

Figure 2-5 shows stress relaxation data for tendons of bar, wire, and strand, all stressed to 70 percent of the ultimate tensile strength. For a typical alloy steel bar the stress loss at 1000 h is about 4 percent but double that at 100,000 h (not shown on the graph). The lowest stress loss evidently is for

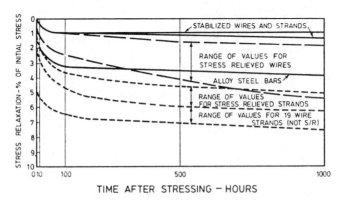

Fig. 2-5   Stress relaxation of tendons at 20°C from initial stress of 0.7 ultimate tensile strength., (From Antill, 1967.)

stabilized wires and strand, whereas the highest is for wire strand not stress-relieved. The graphs show that beyond a certain time point, generally 100 h after stressing, the relationship between percentage relaxation and time becomes a linear function, but becomes even flatter after 1000 h.

The elastic limit, corresponding to the 0.2 percent proof stress, increases from as-drawn wire to normal relaxation and then to low relaxation wire, as shown in Fig. 2-6, which plots stress–strain curves for 7-mm-dia. plain wire (Allen, 1978). Ponts B_1, B_2, nd B_3 represent 0.2 percent proof stress for low relaxation, normal relaxation, and as-drawn wire, respectively, and are 90, 85, and 75 percent the specified characteristic strength.

Relation between Ultimate and Working Stresses

Recommendations and design procedures for steel tendons, including the mechanism of failure, are reviewed and discussed in detail in Section 4-2, in conjunction with the general problem of anchor design and failure. In order to provide an introductory and expedient demonstration of how characteristic strength and working stress have been correlated by various codes, engineers, and contractors, this section presents a general survey of data from domestic and foreign practice.

Allowable stresses for steel tendons show wide disparity but also definite trend to increase the factor of safety against ultimate failure (Littlejohn and Bruce, 1977). This is evident, particularly in installations that involve high-capacity anchors and repeated loading, or in unfavorable conditions such as poor quality rock and corrosive environment.

Tables 2-3, 2-4, and 2-5 show design stresses (working) and factors of safety for various types and sizes of tendon; namely, bar, wire, and strand, respectively. This information is for permanent anchors only and is fairly

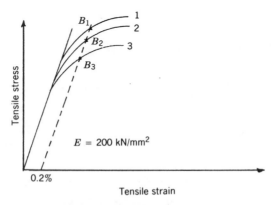

Fig. 2-6 Stress–strain curves for plain wire strand, 7 mm diameter. Curves 1, 2, and 3 are for low relaxation, normal relaxation, and as-drawn steels, respectively. (From Allen, 1978.)

TABLE 2-3 Design Stresses and Factors of Safety for Bar Tendons

Bar	Working Stress (% Ultimate)	Test Stress (% Ultimate)	Measured Safety Factor	Ultimate Safety Factor	Source
28-mm Lee Macalloy	70	—	—	1.43	Britain—Banks (1955)
32-mm Macalloy	56	84	1.5	1.79	Britain—Jackson (1970)
32-mm hollow	54	64	1.2	1.85	Sweden—Nordin (1968)
35-mm	50	75	1.5	2	USA—Drossel (1970)
22-mm HS	47	52	1.1	2.1	USA—Koziakin (1970)
HS bars	—	—	1.5	—	USA—Wosser and Darragh (1970)
35-mm Bauer	44	54	1.2	2.27	USA—Larson et al. (1972)
27-mm Dywidag	55	58	1.06	1.82	Japan—Construction Ministry (1969)

From Littlejohn and Bruce (1977).

TABLE 2-4 Design Stresses and Factors of Safety for Wire Tendons

Wire (mm)	Working Stress (% Ultimate)	Test Stress (% Ultimate)	Measured Safety Factor	Ultimate Safety Factor	Source
5	64	74	1.36	1.57	Britain—Morris and Garrett (1956)
7	63	69	1.1	1.59	Britain—Gosschalk and Taylor (1970)
7	66	79	1.2	1.52	Switzerland—VSL (1966)
8	68	82	1.2	1.47	Switzerland—VSL (1966)
8	50	65	1.3	2.0	Switzerland—Moschler and Matt (1972)
6.4	60	—	1.08	1.67	Canada—Golder Brawner Assocs. (1973)
6.4	60	70	1.17	1.67	USA—Eberhardt and Veltrop (1965)
7	60	62	1.03	1.67	Australia—Rawlings (1968)

From Littlejohn and Bruce (1977).

TABLE 2-5 Design Stresses and Factors of Safety for Strand Tendons

Strand (mm)	Working Stress (% Ultimate)	Test Stress (% Ultimate)	Measured Safety Factor	Ultimate Safety Factor	Source
15.2	55	61	1.1	1.82	Britain—Ground Anchors Ltd. (1973)
15.2	58	80	1.37	1.71	France—Soletanche (1968)
12.7	48	57	1.2	2.1	Switzerland—VSL (1966)
12.7	30	73	2.43	3.3	Switzerland—Sommer and Thurnherr (1974)
12.7, 15.2	60	—	—	1.67	Canada—Golder Brawner (1973)
12.7	65	80	1.23	1.54	Canada—Golder Brawner (1973)
15.2	50	80	1.6	2.0	Canada—Golder Brawner (1973)
12.7	52	78	1.5	1.93	USA—White (1973)
12.7	60	80	1.33	1.67	USA—Buro (1972)
15.2	59	79	1.34	1.69	USA—Schousboe (1974)
12.7	60	85	1.42	1.67	Australia—Langworth (1971)

From Littlejohn and Bruce (1977).

representative of anchor practice here and abroad. Beside the apparent disparity, it appears that there is definite trend toward higher working stresses for wire, followed by strand and bar tendons; hence the factor of safety against rupture is thus in the inverse relation. Testing the anchor to 1.5 times the working stress does not appear to be the general rule at present, and very often contract anchors are overprestressed by an amount roughly corresponding to long-term load loss, usually 10 percent. There is no correlation between ultimate safety factor, defined in terms of the ultimate tensile stress, and the measured safety factor, which is direct function of the field test load selected (see also Chapter 4).

Manufacturers and suppliers of steel tendon usually provide data on all aspects and strength characteristics, which are useful for anchor design purposes. Such a summary is shown in Fig. 2-7, and includes strand of nominal diameter 13 mm (0.5 in) and 15 mm (0.6 in). This strand is also used for prestressed concrete applications. The constituent wires of the strand are cold drawn. After laying-up, the strand is stress-relieved or stabilized. The main physical and mechanical features, such as cross-sectional area, ultimate tensile strength, elastic limit, and relaxation loss are likely to vary with the manufacturer, and this variation may be as high as 5 percent.

2-6 ANCHOR HEAD

The anchor head, described briefly in Section 2-3, is one of the three main component parts of the anchor system. The assembly generally includes a stressing head, wedges, and a distribution bearing plate used to transfer the

Diameter		13 mm (0.5'')		15 mm (0.6'')	
Type of strand		normal	super	normal	super
Nominal diameter	mm	12.5	12.8	15.2	15.5
Nominal steel area	mm²	93	99	139	140
Nominal weight per m	kg	0.74	0.78	1.10	1.10
Minimum breaking load	kN	165	184	244	261
Minimum load at 0.2% offset	kN	146	165	218	222
Modulus of elasticity	N/mm²	1.95 x 10⁵		1.95 x 10⁵	
Relaxation after 1000 h at 20° C from 70% initial load	normal relax.	7%		7%	
	low relax.	2.5%		2.5%	

(a)

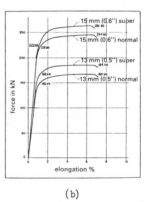

(b)

Fig. 2-7 Strand data summary: (a) values of strength characteristics; (b) stress–strain diagrams for the steels shown in (a). (From Losinger–VSL.)

load to the structure. A protective cap may also be fitted over the anchorage to accommodate accessibility and surveillance. The parts and accessories of the anchor head usually are designed and developed by the prestressed concrete industry or by specialty anchorage contractors, and their details are standardized to suite the type of tendon selected. With heavier loads secondary distribution systems are added, and these include concrete blocks or steel walings. These are useful in preventing excessive concentrated shear and direct bearing stresses on the anchored structure.

The head must be set concentrically with the tendon, and normally is fitted with minimum tolerance, which should not exceed ±5 mm. An angular deviation between tendon and anchor head in the axial position of more than 3° will affect load-transfer efficiency. The assembly should allow access for grout injection tube preferably in a central axis position. A common problem caused by excess deviation is wedge pullin, and this can be avoided if the wedges are homed within a 5-mm-depth band (Littlejohn, 1980a). Recommended tolerances are shown in Fig. 2-8.

The anchor head assembly must be workable within the general requirements mentioned in Section 2-3, which are specified in advance so that the assembly can be designed and detailed accordingly. Furthermore, the anchor head may have to be detailed to accommodate the following anchor functions:

1. The tendon can be stressed and the load locked off with a magnitude of up to 80 percent, and sometimes 90 percent, the specified characteristic strength.
2. The anchor can undergo load adjustments up and down according to the tensioning specifications.

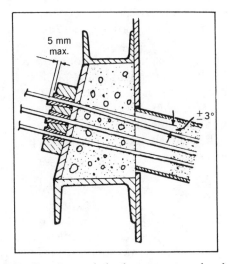

Fig. 2-8 Recommended tolerances at anchor head.

3. All strands in the tendon can be stressed simultaneously, but locked off individually by wedges in the conical bores of the anchor head.
4. The force in the tendon can be verified by check-lifting and for small extension loss it can be recovered by shimming or thread-running.

Further design requirements are facilitated by anchors than can perform the following: (a) the anchor can be load-tested, but needs not be posttensioned; (b) it can be load-tested and posttensioned, but tension cannot be released or relaxed; and (c) the anchor needs to be neither posttensioned nor load-tested. Whether any or all of the foregoing requirements must be satisfied by an anchor system will depend on the functions and service time of the anchored structure.

Figure 2-9(a) shows typical anchor head details for STUP strand or wire anchor. In this case, the head is secured by gripping wedges or truncated cones, pressed against the strands or wires and forced into tapered holes in the steel bearing plate. After tendon stressing, the individual strands are locked into position. Figure 2-9(b) shows a simple head device for a steel bar, consisting of a bearing plate and nut against a concrete pad or block.

2-7 ANCHOR HOLE DRILLING

Anchorage construction should always be carried out by skilled and experienced contractors specializing in this types of work, and be supplemented with competent on-site inspection. The various phases in their actual sequence are anchor hole drilling and flushing, water testing, tendon preparation and installation, grouting, stressing and testing, and finally corrosion protection. These phases are discussed in the same sequence beginning with anchor hole drilling, except corrosion protection, which is reviewed in a separate chapter because of its important effects on anchor performance and longevity.

The remarkable progress in anchorage construction, which has enabled anchoring in variable ground conditions and produced techniques suitable for complex applications, has nonetheless demonstrated that anchors are sensitive to poor workmanship. It is interesting to note, however, that the majority of problems are related to the grouting stage and methods, but some anchor failures have been attributed to poor tendon preparation or improper drilling and flushing.

Factors Affecting Drilling

Productivity and Efficiency. Drilling rates and efficiency determine productivity; hence they influence total cost. Efficiency is maximized by selecting the most suitable drilling method, and this choice usually is made after a consideration of several factors, including the type of ground, site

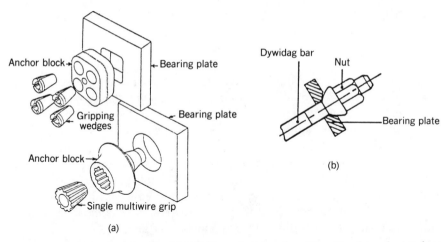

Fig. 2-9 Anchor head details: (a) STUP-type anchor head for strand or wire tendon; (b) anchor head for Dywidag bars.

accessibility and topography, hole geometry and size, scale of drilling operations, type and capacity of anchor, availability and suitability of flushing medium, local labor costs, and construction restrictions at the site.

Drilling rates are predicted after a study of machine characteristics, bit and flushing medium properties, and relevant ground parameteres. A prior knowledge of drilling rates in similar conditions is essential (Littlejohn and Bruce, 1977), and unavailability of such data will deprive the engineer of the opportunity to chose alternative operational procedures if necessary.

Constructional Requirements. In the majority of cases anchor holes are drilled in a near horizontal through to a downward vertical direction. Occasionally, holes must be drilled in an upward direction. In addition to these geometric requirements, the drilling method should be chosen to satisfy the following conditions:

1. There should be minimum disturbance to the surrounding ground. In this respect, all flushing should be avoided in weak, finely grained soil, and where buildings are present control of flushing pressures is necessary to avoid hydrofracture effects.
2. Hole stability should be maintained, and loss of ground should not exceed significantly the volume of the specified drill hole. To satisfy this condition, it is customary to check the volume of material removed during drilling.
3. Loosening of the borehole walls in cohesionless soil should be avoided, whereas in cohesive soil and sensitive rock appreciable changes in water content and smoothing of the borehole surface can be detri-

mental to the transfer of load. Matt (1981) reports that certain clays and marly rocks will swell or soften in the presence of drilling fluid, and recommends treating water drilling fluids with antiswelling additives.

4. If drilling is carried out in ground under artesian water pressure, it will be necessary to counteract this conditions and prevent ground outflow through the borehole during drilling because of the possible damaging effects on surroundings.

5. In general, drilling should be completed in a manner that allows direct detection of the ground, and registers any major changes in ground characteristics, especially if these deviate substantially from the set on which the design of the project was based. It is well known, for example, that vibratory drilling of casing can have compaction effects on loose granular soils with unavoidable settlement.

Drilling Methods

The most important single factor influencing the choice of drilling method is the type of ground, namely, rock or soil. For each type various drilling systems are available, each designed to handle normal variations in ground parameters. Thus, a hole can be drilled with the use of rotary, percussive, or rotary–percussive equipment. Occasionally, vibratory driving techniques are suitable. Diamond core drilling is seldom used because of the high cost and the risk of reducing the bond on account of the smooth hole surface thus produced. Any drilling machine and procedure can be used provided it satisfies the foregoing requirements, and also produces a hole that has the specified dimensions and tolerance and is free of obstructions and protrusions. In general, anchor boreholes are 75–150 mm in diameter, but in many instances much larger shafts can be drilled (Schousboe, 1974).

Certain simple rules can help engineers understand the most promising features as well as the inherent limitations of a given drilling system, hence its suitability for a specific type of ground. For example, anchoring in soil where the borehole will be stable against collapse during drilling can be carried out with a rotary drill rig equipped with a continuous flight auger. Where water bearing cohesionless soil overlies clay layers, a rotary–percussive rig may be used to advance the hole in this zone and is sealed in the cohesive layer. The hole is, then, completed with the use of a continuous flight auger. In alluvium deposits the hole is always cased over its full length, and the casing is slowly withdrawn as the anchor hole is grouted.

Rock of high compressive strength responds favorably to a percussive bit and the accompanying chipping–crushing action. However, weakly bonded hard rock is likely to respond to percussive action more like a ductile material than a brittle one. In this case, the rock is more efficiently drilled with a wear-resistant rotary drag bit (Littlejohn and Bruce, 1977).

Percussive Drills. These accomplish penetration by the action of an impulsive blow, usually exerted from a chisel or wedge-shaped bit. Repeated applications of high-intensity, short-duration force disintegrates hard material provided the blow is sufficiently large. Mawdsley (1970) summarizes three main types:

Type A, with a compressed air-powered drifter driving standard coupled drill rods.

Type B, with a compressed air-powered drifter driving special coupled drill rods which also act as the anchor.

Type C, with an independent rotation compressed air-powered drifter simultaneously driving coupled drill tubes and drill rods, also known as "Atlas Copco overburden drilling method."

Hammer drills, in which the hammer remains at the surface, are used to drill holes up to 125 mm in diameter. Special down-the-hole (DTH) tools, in which the hammer is always immediately above the bit, are reported to drill holes up to 750 mm in diameter.

Figure 2-10(a) shows a type *C* percussive drill in operation. Using a combination of percussion, rotation, and high-pressure water flushing, a drill rod and an outer tube penetrate the overburden together. All these operations are independently controlled by the driller, so that various types of overburden can be penetrated by changing the drilling action as soon as a new formation is reached. For anchoring in bedrock, the drill rod continues alone when this stratum is reached, as shown in (b), and until a suitable depth is reached. Then the drill rods are replaced by a plastic hose through which grout is injected. After anchor insertion the drill tubes are withdrawn and the anchor is tightened. For anchoring in the overburden, the drill rods are replaced by the anchor when the proper depth is reached. Grout is injected while the steel tube is withdrawn, as shown in (c). When the anchorage zone is fully injected, the remaining drill tubes are withdrawn and the grout is allowed to harden.

Rotary Drills. Rotary drills impart their action through a combined axial thrust (static action) and a rotational torque (dynamic action). Mawdsley (1970) distinguishes two main types:

Type D, auger driving with coupled flight augers.

Type E, normal rotary drilling with flush coupled drill rods and usually a drag bit as the cutting component.

Augers are often used in self-supporting materials that include stiff-to-hard clays, marls, and soft rock. This catergory includes standard continuous flight augers for normal open-hole drilling; continuous flight augers with

Fig. 2-10 Type *C* percussive drill, Atlas Copco drilling method: (a) drill in operation; (b) anchoring in bedrock; (c) anchoring in the overburden.

hollow couplings to permit water, bentonite, or cement grout to be pumped into the hole; and hollow-stem augers with a removable center bit allowing sampling duirng the drilling stage.

Open-hole rotary and auger drilling, sometimes with belling at the bottom, are usually used in cohesive soils, for a hole size ranging from 6 to 24 in (150–600 mm). Continuous flight, hollow-stem augers are usually 8–15 in dia. with a $2\frac{1}{2}$–4 in hollow center. In most instances, the tension member is inserted into the hollow-stem auger prior to drilling. The auger is rotated advancing a detachable bit to the required depth, and then the bit is removed while the grout is injected. Two drilling machines of this type are shown in Fig. 2-11.

Rotary–Percussive Drills. These are combinations of the two types described. Their action is primarily derived from (a) an axial thrust of lower magnitude than that of rotary drills; (b) a torque lower than a rotary drill but much higher than a percussive drill; and (c) impact, of a magnitude usually lower than that of a percussive tool. The rotation mechanism may be powered by the impact mechanism or by a separate motor.

For rule-of-thumb correlation of drilling methods and type of drill with the set of pertinent conditions, and for the applicability of drilling methods to different rock types reference is made to Table 2-6(a) and (b).

Small Drills. In the United States these are usually used in the midwestern and eastern part of the country, especially in the Washington, D.C.–Baltimore area. They are of two basic types: (a) crawler-mounted percussive drills suitable for rock and with appropriate modifications for earth anchors and (b) rotary drills such as those used for soil exploration and sampling. The primary requirement are compactness, stability and ability to drill almost horizontally (White, 1970).

With percussive drills the procedure involves drilling a hole at an angle of about 15°, insert up to 12–5 mm 7-wire strands, grout the hole, and then test to 120–150 percent the working load. If the ground exhibits tendency to cave, a casing of 100 mm internal diameter is driven with the percussive tool, often to the full depth of the hole, and as the grout is injected, the casing is withdrawn to the slip plane.

Figure 2-12 shows a compact rotary drill employing the lost points. The drill rod is flush-jointed inside as well as outside. Usual outside diameter is 60–70 mm. The drill is water-jetted through a swivel, and the water is circulated through the flush-joint drill pipe and out into the ground through holes in the lost point. Strands are inserted after the hole is completed, and grouting is carried out as the casing is withdrawn.

Choice of Drilling Method. For a preliminary selection of a suitable drilling system, the following guidelines are useful:

Fig. 2-11   Augered and open-hole drilling operations (VSL-Losinger)

1. Ground strata to be drilled, anchor type, and capacity will initially determine the length and diameter of the hole, and thus establish a range of suitable drilling methods and equipment.
2. Several contractors recommend that for holes up to 100 mm dia. and 60 m (200 ft) long, percussive tools are preferable for most rock strata. Rotary methods are suggested for deeper holes or poor rock conditions.
3. In hard rock, type *A* drill will invariably be the first choice, whereas in soft strata this method is clearly excluded.
4. For drilling in rock that has alternating hard and soft material (collapsible zones) the use of a rotating eccentric bit has proved successful, since it underreams the rock and permits the use of one-size casing.

TABLE 2-6 (a) Choice of Drilling Method and Machinery to Fit Pertinent Conditions

	Percussive (A)	Percussive (B)	Percussive (C)	Rotary (D)	Rotary (E)
Drill string	Standard coupled rods, separate anchor	Coupled rods also act as anchor	Coupled drill tubes and rods used simultaneously from same drive adapter Atlas Copco OD method	Coupled flight augers	Standard rotary drilling tubes
Drilling machine	Wagon drill with drifter or crawler drill with independent rotation drifter, compressed-air-powered	Wagon drill with drifter or crawler drill with independent rotation drifter, compressed-air-powered	Special independent rotation drifter mounted on heavy-wheeled chassis or crawler, compressed-air-powered	Standard auger drill capacity of torque and pull down dependent on hole size and depth; Diesel–hydraulic power; chassis powered wheel or crawler designed for drilling of shallow angle holes; wheeled or skid mount possible	Rotary rod drill or diamond drill; performance about 2000 ft-lb torque, 5 ton pulldown 0–500 rpm; diesel–hydraulic power; chassis-powered wheel or crawler designed for drilling of shallow angle holes; wheeled or skid mount possible
Suitable strata	Self supporting rock only; few feet of overburden possible with aid of stand pipe	All materials	All materials provided drill tubes are uncoupled when rock is encountered and drilling continued alone with rods	All self-supporting soft material such as clay and chalk; not rock; not non-self-supporting material such as sand and gravel unless casing is used	All soft materials such as clay, sand, and gravel; also soft and medium rods; not hard rock
Anchor	Multistrand rope or single rod	Special coupled rods	Mutlistrand rope and single rod	Multistrand rope most common; single rod also possible	Single rod most common as in Bauer system; multistrand rope possible where ground is self-supporting
Flushing medium	Normally air, but water could be used	Invariably water but air occasionally useful	Water; air used very rarely	None	Water; air used very rarely

From Mawdsely (1970).

TABLE 2-6   (b) Application of Drilling Systems to Rock (*Continued*)

Method	Resistance to Penetration of Rock			
	Soft	Medium	Hard	Very Hard
Rotary—drag bit	X	X		
Rotary—roller bit	X	X	X	
Rotary—diamond bit	X	X	X	X
Percussive	X	X	X	X
Rotary—percussive	X	X	X	

From Paone et al. (1968).

5. The DTH hammer method sometimes has one serious disadvantage: whereas the method is less prone to jamming than the ordinary percussive drills, where jamming occurs the financial consequences are far greater because of the far more expensive hammers.

6. In built-up areas there has been a tendency to abandon the use of percussive tools in favor of rotary drills, and this is mainly due to restrictions on acceptable noise levels and occasional vibratory effects.

Drilled Hole Diameter.   For the usual applications the range of the hole diameter will be from 75 to 150 mm (3–6 in). In collapsible soil, anchor work will invariably require a casing, drilled or driven into the ground to the specified depth. The weight of the casing as well as the drilling and handling problems associated with larger casings will limit the optimum drill size to 6 in (150 mm). Another usual size is $3\frac{1}{2}$-in outer diameter (OD) (90 mm) used with percussive drills, and 5-in OD (125 mm) used with rotary methods. Several techniques have been developed for installing soil anchors without the use of casing, and the most popular is a hollow stem auger. Uncased holes should not be used in the following cases: (a) in sites with difficult or limited access; (b) in difficult soil conditions, or highly variable ground characteristics; or (c) in urban or built-up areas where undermining of structures and services is conceivable.

Drilling Rates.   The rate of drilling becomes a pertinent factor for anchorage work in rock. From the practical standpoint, which is also the contractor's view, there should be simple procedures for predicting penetration rates, particularly in percussive and rotary–percussive drilling. One test consists basically of fracturing rock samples by impacting them with a falling weight. The corresponding damage is rated by screening the broken sample. Since the test is relatively simple and does not require elaborate equipment, it can be carried out as routine procedure in the field and thus produce useful results in one day. Investigators report that satisfactory results have been obtained in correlating field penetration rates with the data obtained from this test for rotary–percussive drills (Unger and Fumanti, 1972) and for per-

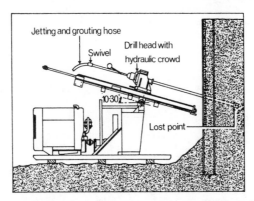

Fig. 2-12 A compact rotary drilling rig able to drill almost horizontally.

cussive tools (Schmidt, 1972). Complete details on the test are given by Paone et al. (1968).

More recent work (Van Ormer, 1974) relates penetration rates to rock mass and material properties. Pertinent factors are texture (porous to dense fine), hardness (1–10 on the Moh scale), breaking characteristics (brittle to malleable), and geologic structure (solid to laminated). In all four groups, the first named is drilled faster. Table 2-7(a) provides comparative data for rock hardness, indicating also the usual range of percussive and rotary drilling. The drilling rate index relative to 1.0 (for solid, homogeneous Barre Granite) is shown for various rocks in part (b). In this case, Barre Granite is used as reference index (assigned value of 1.0) bbecause of its even texture, hardness, and the resulting consistent drilling. Table 2-7(b), however, does not consider the secondary structure of the rock mass. On the other hand, recorded differences between actual (measured) and predicted (based on physical properties) drilling rates can be accounted for considering the unavoidable variations in these properties over the hole length. Among these, rock material and mass anisotropy are known to affect drillability, but little work has been done on this subject. The general effect of rock mass structure on drilling rates is shown comparatively in Table 2-7(c) (Van Ormer, 1974).

Whereas solid formations generally constitute good drilling media, seamy and broken formations retard drilling rates as tedious, extended supervision is necessary to avoid loss of flushing capacity, loss of drill string, and bit sticking problems.

Flushing

All particles and byproduct materials from the bit should be removed quickly and completely. The most common flushing media are water and air or a bentonite slurry. Air is most efficient and is best used in dry ground. In confined space, however, the use of air requires caution, and should pref-

TABLE 2-7 (a) Relative Hardness Index of Common Rocks

Mineral or Rock	Hardness	Scratch Test
Diamond	10.0	
Carborundum	9.5	
Sapphire	9.0	
Chrysoberyl	8.5	
Topaz	8.0	
Zircon	7.5	
Quartzite	7.0	
Chert	6.5	Quartz
Trap rock	6.0	Quartz
Magnetite	5.5	Glass
Schist	5.0	Knife
Apatite	4.5	Knife
Granite	4.0	Knife
Dolomite	3.5	Knife
Limestone	3.0	Copper coin
Galena	2.5	Copper coin
Potash	2.0	Fingernail
Gypsum	1.5	Fingernail
Talc	1.0	Fingernail

Percussive drilling spans Topaz (8.0) to Galena (2.5). *Rotary drilling* spans Schist (5.0) to Talc (1.0).

(b) Drilling Rate Index for Various Rocks

Characteristics	Comparative Drilling Speed[a]	Rock Material
Hardness—1–2 Texture—loose Breakage—shatters	≥1.5	Shales Schist Ohio sandstone Indiana limestone
Hardness—3–4 Texture—loose grained to granitoid Breakage—brittle to shaving	1.0–1.5	Limestone Dolomites Marbles Porphyries
Hardness—4–5 Texture—granitoid to fine grained Breakage—strong	0.6–1.0	Granite Trap rock Most fine-grained igneous Most quartzite Gneiss
Hardness—6–8 Texture—fine grain to dense Breakage—malleable	≤0.5	Hematite (fine-grained, gray) Kimberly chert Taconite

[a] Barre granite is used as the standard for determining a comparative drilling speed of 1.0 because of even texture, hardness, and consistent drilling.

(c) Effect of Rock Mass Structures on Drilling Rates

Rock Mass	Nature of Fractures	Drill Rate
Massive	—	Fast
Stratified	Perpendicular to drill rod; >1.2 m apart, clean	Fast–medium
Laminated	Perpendicular to drill rod; <1.2 m apart, clean	Medium
Steeply dipped	Small angle to drill rod, 1.2 m apart, clean	Slow–medium
Seamy	Various inclinations to drill rod; close, open fractures	Slow

From Van Ormer (1974).

erably be introduced with reverse circulation because of the health hazard of dust particles.

Water flushing improves generally ground conditions, and is best used in sticky clayey soil. Its sweeping action cleans the sides of the hole for a stronger bond at the grout–ground interface. In soft rock such as marls, chalk, and fissured shales, water should be used with caution and under competent advice because of possible softening effects. Water is also the common flushing medium below the natural water table and for diamond drilling. Bentonite slurry flushing is not very common, but is used successfully in certain countries including France for open hole drilling through silts and sand overlying rock. Its suspending power keeps individual earth particles in its volume and facilitates their removal, while its sealing action keeps the hole from collapsing.

Regardless of the expected efficiency of the flushing process, it is usual to provide a sump length of 1–2 ft (0.3–0.7 m) at the bottom of the borehole for collecting debris, beyond the design length. After the hole is drilled and flushed out, it should be sounded to detect the presence of any foreign materials. If the probe is satisfactory, the top of the hole is plugged in order to be protected and remain free of falling debris. The flushing medium may be introduced through the drill rods and the drill bit, and then return to the surface between the rods and the walls of the hole, a process called "normal circulation." Alternatively, the flushing medium may follow the opposite way in a process called reverse circulation.

Local variations in the ground conditions, sometimes occurring within meters, can have considerable effect on anchor performance, and in this respect it is recommended to keep records regarding the groundwater and flushing medium. Additional qualitative data on ground conditions can be obtained by recording drilling rates and extent of bit blocking, and by observing changes in the amount and composition of flush return.

Hole Deviation and Tolerance

With anchor hole spacing relatively wide (>2 m or >7 ft), deviations in hole alignment are not very serious and may be tolerated. With closer spacing, or in difficult ground where considerable deviation can occur (e.g., steeply inclined bedding planes, extensively fractured rock, or presence of boulders), hole alignment should be monitored.

Misalignment usually originates from two sources: (a) initial incorrect setting of the drill and (b) deviations of the hole from the correct initial line and angle during drilling. The condition in (a) is avoided and checked by the use of a spirit level and profile, or by the use of special mats. Deviation during drilling usually does not begin from a single cause; hence it is more difficult to control. It may be caused by the use of rods that are too thin, from excessive thrust, presence of fissures, and rock discontinuities. These problems are less serious with short-to-medium-length anchors, and they are noticed more frequently in vertical downward holes. With inclined horizontal holes, the rods are apt to lie on the lower side under their own gravity, causing the bit to upturn slightly. Hence, inclined angle holes sometimes tend to follow the configuration of a shallow curve with its chord the true inclined direction.

Quoted tolerances among various authors and codes reveal considerable differences of opinion, and provide little guidance on maximum permitted variations. These tolerances (measured as deviation of anchor hole from the specified center divided by the length, or by the angle of deviation) range from 0° 28' to a maximum of 2° 30' permitted by the South African Code.

In order to rectify this matter, the following angular tolerance is recommended:

2° for widely spaced anchors (spacing >2 m)

1° (or 1–50) for closely spaced anchors (spacing <2 m)

Furthermore, where ground conditions dictate these tolerances may be modified, but this should be mutually agreed with the designer.

2-8 WATER TESTING AND WATERPROOFING BY PREGROUTING

In rock formations, on completion of drilling the borehole is tested for watertightness by measuring the rate of water loss or gain. The test is intended to disclose the possibility of probable grout loss during injection into formations where fractures have been encountered or suspected. This loss of grout material from around the tendon in the fixed anchor zone can be detrimental to the load transfer and aggrevate corrosion attack. Prior to testing the finished hole is thoroughly flushed with clean water until the water outflow emerges clear. The water test is then carried out while adjacent holes are observed to detect possible interhole interaction.

The test must provide reasonable data for water loss or gain to justify the need for waterproofing by pregrouting. A criterion generally accepted is that for normal cement particles to be lost to a fissure, the latter must be larger than 250 μm, although experimental studies suggest a value closer to 160 μm (Littlejohn and Bruce, 1977). A single 160-μm fissure under an excess head of one atmosphere produces a flow rate 3.2 liters/min/atm (Littlejohn, 1975), whereas a lower fissure width of 100 μm gives rise to a flow rate 0.6 liters/min/atm. The same author considers these values a reasonable threshold for water loss for ordinary Portland cements and for fine-grained cements, respectively.

Water tests can be carried out over sections, for example, the fixed anchor zone, with the help of packer injection techniques, and this method is preferred to falling head tests since it provides more detailed information. In many instances, however, falling head tests are cheaper and quicker. If the water loss exceeds the limiting values, pregrouting is necessary to waterproof the hole. Where a measured water gain is associated with artesian conditions, it must be counteracted by back pressure prior to grouting. If the flow cannnot be established in this manner, pregrouting should be carried out irrespective of the magnitude of the water gain.

In the context of design requirements water testing is not standard practice or routine procedure. Furthermore, there is an acknowledged variation in the limiting flow rates used as criteria for pregrouting requirements. Thus, judgment and caution should be exercised, based on the recommendations by Littlejohn and Bruce (1977), summarized as follows:

1. For a better understanding of the relationship between flow rate and single fissure, reference is made to Fig. 2-13 expressing in graphical form the theory of flow in fissured formations (Baker, 1955).

2. If the water loss disclosed by the test in the borehole exceeds 3.0 liters/min/atm for a test duration not less than 10 min, pregrouting should be considered necessary.

3. The flow rates shown are minimum values since they apply to single fissures. Thus, larger limiting values are acceptable if several fissures exist (thickness <160 μm), provided this is confirmed at the site.

4. Since permissible flow rates are related to excess head, the location of the water table in relation to the section under investigation must be known in order to make appropriate allowance for the driving or excess head inducing flow at the section.

5. In fine fissures high applied pressures are likely to induce turbulent flow, create high pressure gradients, and open up natural fissures, all causing deterioration of the ground. These changes in the local ground

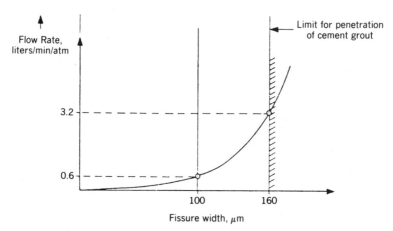

Fig. 2-13 Variation in flow rate through a single fissure as function of fissure width. (After Baker, 1955.)

environment must stay within the limits of an inconsequential range by keeping the applied pressure-inducing flow as small as possible.

Waterproofing by pregrouting is done with cement grout tremied into the hole from the base upward. Redrilling usually takes place within 24 h, and the water test is repeated. If pregrouting is not successful in waterproofing the hole, pressure grouting may be necessary in order to force the grout into the rock fissures. The composition of the cement grout should be essentially similar to that for the subsequent anchor grouting, since this process has a consolidating effect on the rock mass. Chemical grouting should not be used without the consent of the engineer because of possible deleterious effects on the anchorage and the corrosion protection.

Waterproofing by pregrouting, irrespective of the magnitude of the water loss or gain, may be avoided if an appropriate grouting method is used for the fixed anchor length. For example, multistage high-pressure grouted anchors achieve the waterproofing of the surrounding rock and the grouting of the fixed anchor length in the same operation, provided the corrosion protection requirements are satisfied (Matt, 1981).

2-9 TENDON PREPARATION AND INSTALLATION

Anchors are fabricated in a workshop or in the field by trained personnel and under competent supervision. During manufacture, handling, and installation, anchors and their components should remain clean and free of any mechanical or structural defect, and also be continuously protected against corrosion.

Storage and Handling

Local codes and standards give guidelines and recommendations on the subject of storage and handling. In general, steel for anchor tendons is stored indoors in clean and dry conditions. If tendons must be left outdoors, they should be stacked off the ground and completely covered by a waterproof tarpaulin, fastened so as to permit air circulation through the stack. If the relative humidity exceeds 85 percent, as in marine tropical areas, the steel is protected by wrappings impregnated with a vapor-phase inhibitor powder preventing air flow. Severe corrosion is known to occur under humidity exceeding the 85 percent level. On the other hand, a uniform normal rusting can actually improve bond at the tendon–grout interface, but all loose rust should be completely removed. Severely pitted tendons, an indication of serious localized corrosion, should be rejected, particularly if they consist of small-diameter multiwire strands or threaded sections of bars.

Bars usually are stored in straight lengths, and wires and strands in coils of diameter at least 200 times the diameter of the prestressing steel. Tendons should not be dragged across abrasive surfaces or through deleterious materials; nor should they be accessible to weld splash. Kinked or twisted wire should be rejected.

After a tendon is cut to proper length, its ends are treated to remove or smoothen sharp edges that can damage the protective sheathing. Minor damage to the threaded section of a bar often is repaired by the use of a file, but extremely damaged threads should be rejected.

Anchor Fabrication and Assembly

After cleaning, bar tendons are lightly oiled. Subsequently, they are checked to ensure that the bars are properly screwed into couplers and that the full thread engagement is obtained in nuts and tapped plates. Multistrand or multiwire tendons normally require longer fabrication time, and may have to be unravelled to facilitate cleaning and degreasing the bond length, and then return the wires to the correct way.

Centralizers and Spacers. A typical detail of a centralizer for a single bar tendon is shown in Fig. 2-14(a). In this case, the centralizer keeps the bar centrally located in the borehole and thus ensures a uniform grout cover in the fixed anchor zone. With flexible bars, properly spaced centralizers help minimize the sagging effect of the steel between support points. The arrangement of centralizers should also take into consideration the shape of the hole (e.g., the location of underreamed bells) and the extent to which the ground may be prone to disturbance during insertion of the anchor. Besides these foregoing factors, centralizer spacing will depend mainly on the bar stiffness. Centralizers should be fixed firmly to the anchor tendon in order to avoid displacement or distortion during homing operations.

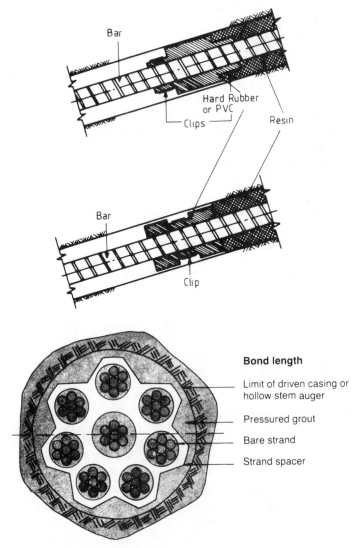

Fig. 2-14 (a) Bar centralizer details for anchor fixed to rock using resin as bonding medium; (b) spacer detail for an anchor in soil.

Spacers, usually made of steel or plastic, are used in both the free and fixed sections of multicomponent tendons, that is, systems consisting of multiple bar combinations, wires and strands. A typical spacer detail for a strand anchor in soil is shown in Fig. 2-14(b). Like centralizers, spacers help to maintain anchor components parallel and in their correct alignment, and thus prevent contact friction from generating between them. This is particularly important in the free length of long anchors where tangling or rubbing of individual bars, wires, or strands resulting from a distorted design geometry and initial alignment can cause the loads to dissipate during stressing. Furthermore, extremely high stress concentrations may be generated, especially under the top anchor head, and rupture of individual elements can thus occur (Littlejohn and Bruce, 1977). These problems are avoided by the use of spacers, usually placed 4–8 m apart.

In the fixed anchor zone spacers serve three primary purposes: (a) to centralize the tendon system in the borehole for an adequate and uniform grout cover, which enhances corrosion protection and provides good grout bond at the borehole interface; (b) to provide a positive grip for the tendon and the grout without restricting the flow of the latter in the hole in order to completely penetrate the space between tendon units for full cover, a condition ensuring efficient transmission of bond stress; and (c) to help prevent contamination of the tendon parts such as clay smear. Spacers in this zone can also be used in conjunction with intermediate fastenings to form nodes and waves, intended to provide a more positive mechanical interlock between tendon and surrounding grout.

There remains some uncertainty with respect to the exact effect of spacers and centralizers on the load-transfer and efficiency characteristics of the fixed anchor zone, and much will be learned on this subject from ongoing experimental work. Thus far, experience suggests that these devices facilitate the load transfer process mainly in two ways, initially by ensuring a continuous compact grout cover over the fixed length, and they by creating a chain of compression rings along the tendon axis, which cause it to act as one unit and thus mobilize shear resistance along the entire fixed zone. On the other hand, while the method of forming nodes and waves ensures the geometry necessary for adequate grout cover, the practice of unraveling strands followed by bushing of the wires creates a random geometry that may not always guarantee efficient load transfer.

In many instances, it is feasible and practical to combine the characteristics of spacers and centralizers into one unit, and this trend reflects the relatively large number of design and construction details for the fixed anchor length available in the industry. These details, however, are not standardized, and they may depend on the method of grout placement and on whether the strands are arranged parallel or waisted at suitable intervals.

The considerable variance in anchor practice is evident in Table 2-8, which provides data on the pitch of spacers along the fixed anchor length. The apparent variation in the actual pitch dimension (0.5–2.0 m) suggests that only limited work has been carried out to study the effect of pitch or spacer design on load transfer in the fixed zone.

Anchor Installation (Homing)

Homing should be done as soon as possible, since it is advantageous to complete drilling, anchor installation, and grouting on the same day. A delay

TABLE 2-8   Pitch of Tendon Spacers in the Grouted Fixed Anchor Length

Pitch of Tendon Spacers (m)	Remarks	Source
0.5	Cheurías dam	USA—Zienkiewicz and Gerstner (1961)
0.5	3-m fixed anchor	Czechoslovakia—Hobst (1965)
0.8	Multiwire tendons	France—Cambefort (1966)
0.6	VSL anchors	Switzerland—Losinger SA (1966)
0.8–1.6	Multistrand tendons	Britain—Littlejohn (1972)
2.0	Multistrand tendons	Italy—Mascardi (1972)
0.5–2.0	Dependent on "stiffness" of tendon system	Germany—Stocker (1973)
1.5–2.0	Conenco (Freyssinet) anchors	Canada—Golder Brawner Assocs. (1973)
1.8	7.3-m fixed anchor	USA—Chen and MacMullan (1974)
2.0	8-m fixed anchor (12 No. 15.2-mm strands)	Britain—Littlejohn and Truman-Davies (1974)
0.5	Multiwire tendons	USSR—Shchetinin (1974)

From Littlejohn and Bruce (1977).

from drilling to grouting with the hole open can be cause for ground deterioration, particularly in overconsolidated fissured clays and soft rock.

Immediately prior to its installation the anchor is inspected and checked for possible damage to its components and protective system. The homing will largely depend on anchor length and weight. In practice, any method can be used provided the tendon is lowered at a steady controlled rate. With very flexible tendons, a drum is used from which the tendon is unreeled into the hole.

For tendons weighing in excess of 200 kg, mechanical handling equipment is recommended as manual handling tends to be difficult and hazardous. For cased holes, it is recommended to use a funnel or a circular entry pipe at the top of the hole to guide the tendon as it passes the sharp edge at the top of the casing, and avoid possible damage. At the beginning of construction one tendon should be withdrawn after homing to check the efficiency and integrity of spacers and centralizers, and also detect possible damage. Littlejohn and Bruce (1977) also note that if pregrouting is done below the water table, grout dilution can occur if the tendon is lowered too quickly.

Inasmuch as it is seldom practical to introduce time restrictions, it is always essential to coordinate drilling and tendon installation with the subsequent grouting operation in order to minimize construction effects. As already mentioned, some ground types are prone to time dependent changes of their properties, and this can be avoided by minimizing the time between drilling, tendon homing, and grouting. If a delay is unavoidable, the thole should be plugged to prevent entry of foreign materials, and where the ground is prone to swelling tendon homing and grouting should follow drilling as quickly as possible.

2-10 GROUTING

General Requirements.

The choice and design of a suitable grout system first depends on the ground conditions in which it is to be placed, and then on the setting time, strength, and intended functions of the grout. Low-cost materials chosen for this application include a wide range of conventional hydraulic cements, often of a special variety. The general requirements are defined in approriate specifications or in local codes and standards. Predesigned and ready-mixed grout materials delivered bagged to the site are becoming increasingly popular since they help avoid delays and produce a consistent mix.

Grouts in general will perform any or all of the following functions: (a) holding the anchor tendon to the ground by forming a load-transfer zone, which is the fixed anchor length—in this case the grout may be injected before or after tendon homing, but prior to stressing (primary grout); (b) bonding the tendon to a capsule, which can be done simultaneously with bonding to the ground, or subsequent to the bonding of a capsule to the ground; (c) filling the void space within and around the tendon to augment protection against corrosion, which can be done simultaneously with (a) or as a second stage after stressing (secondary grout)—however, with the development of restressable anchors that have the free length decoupled (greased or sheathed), the complete injection may be carried out in a single-stage grouting operation; and (d) filling voids or fissures in the ground prior to tendon installation where pregrouting is necessary.

Durability. The choice of grout should further consider the aggresivity of the ground toward the grout, and the aggresivity of the cement toward the tendon steel. Similar emphasis is placed on the selection of a grout that can function as a metal proofing agent.

Under normal conditions most cement grouts are durable. However, in the long and short term severe and quick deterioration can occur in adverse environmental conditions, such as chemical attack in the presence of dissolved sulfates or acids contained in groundwater, and under extreme temperature fluctuations (Littlejohn, 1982). These effects are magnified if there are deficiencies in grout quality, for example, low density and high permeability. Whereas no positive guidelines exist on grout design for durability, it is evident that a minimum cement content is essential to ensure reasonable durability under the expected conditions of exposure. Grout defense against chemical attack is further improved by the use of rapid hardening, sulfate-resisting cements, and especially low-heat varieties.

From these brief remarks it follows that grout performance should be assessed first in the context of load transfer and then in terms of resistance to corrosion attack.

Composition and Materials

Cement. It is essential to specify only fresh cement and insist on ideal storage conditions. Partial dehydration or carbonation can lead to particle agglomeration and reduction in postmix hydration, observed with old age cement or poor storage. In order to avoid stress corrosion on the steel tendon, the cement should have a chlorine content from chlorides not exceeding 0.02 percent by weight, or sulfur from sulfides not exceeding 0.10 percent by weight.

Ordinary Portland cement (type I) may suffice in certain cases, but it has low resistance to chemical attack. If this is anticipated, a sulfate-resisting (type II), or a rapid-hardening variety (type III) will be required. The use of high alumina cement appears now more restricted worldwide, and is confined mainly to test anchors and temporary anchors with service life less than six months because of the high heat of hydration and problems of reversion.

For most common cements the maximum practical size (99 percent passing) ranges from 44 to 100 μm. These particle sizes limit penetration of cement grouts to soils with permeability less than 5×10^{-6} cm/s, or fissures in rock of width less than 160 μm unless fracturing pressures are used.

Water. Any water suitable for drinking, except for the presence of bacteria, is generally acceptable for cement grout formulation. Water containing sulfate (>0.1 percent), chloride (>0.5 percent), sugars, or suspended matter is dangerous and techically unsuitable, especially for applications involving high-strength prestressing steel, or where the steel tendon is in contact with the grout. Where doubt exists as to the quality of the water, appropriate tests should be carried out to assess the water suitability.

The water/cement ratio, unlike any other factor, is the most important single item influencing grout properties and characteristics. Initially, the water/cement ratio must be sufficiently high to give workability and fluidity (flowability) as the grout is pumped into the borehole, yet low enough to prevent bleeding and shrinkage. Low ratio is also essential for high strength, structural continuity, and where the grout must perform as waterproofing and anticorrosion medium. The extent to which some of these effects are related to the water/cement (W/C) ratio is shown graphically in Fig. 2-15.

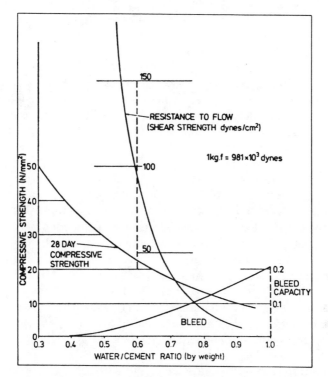

Fig. 2-15   Effect of water content on grout properties.

Based on an international survey, quoted W/C ratios cover the range from 0.35 to 0.55, and evidently the higher values are used in sandy alluvium deposits. The consensus of opinion is that a suitable W/C ratio is between 0.40 and 0.45. This range will ensure sufficient pumpability for grout placement in small-diameter boreholes, and it will also impart to the grout continuity and strength.

Admixtures. These are occasionally recommended to produce a grout mix with low bleeding characteristics (<0.5 percent), to ensure fluidity, and to control shrinkage and setting time.

In Europe, inert fillers such as ground quartz, limestone dust, fine sand, and sawdust have been added to mixes used primarily to waterproof or consolidate boreholes (pregrouting) prior to redrilling. The main reason is economy. These fillers are hardly suitable in grouts used for tendon bonding.

With respect to anchor grouts, chemical admixtures may offer certain advantages, but their compatibility with the cement type must be checked prior to use preferably by trial mixes. Different types of admixtures should not be included in the same grout. For example, admixture of calcium chloride should not be used in sulfate resisting or high alumina cement. Table

2-9 shows admixtures commonly added to cement grouts for anchor work. Besides these chemicals and their action on grout properties, work by Geddes and Soroka (1964) shows that aluminum-based expanding agents improve grout workability while increasing the confined strength. This effect increases the bonding capacity of the grout, confirmed by a reduction in bond transmission length. Whereas other investigators have also favored the use of aluminum powder, caution is clearly indicated because of the great sensitivity of grout properties to the percentage of admixture added.

Some contractors, for example, Nicholson et al. (1982), caution that expansive agents can reduce grout strength through unrestrained expansion in open boreholes, and also question the effect of released hydrogen upon the brittleness of the tendon steel.

It is evident that broad agreement exists on the potential effects of chemical admixtures. For example, chloride bearing compounds are prohibited in anchor grout work in most European countries and in the United States. Furthermore, most codes stipulate admixture use only when it can be demonstrated that it will enhance the quality of grout (B.S. CP 110; Mascardi, 1973; Hilf, 1973; White, 1973; ACI, 1971). From these remarks, it can be concluded that the use of admixtures in anchor grouts still remains an art. If a new mix is introduced containing admixtures, a complete set of technical data should be established and become available to designers.

Grout Crushing Strength. Sufficient grout strength must be attained for bond at the grout–tendon and grout–ground interfaces. A usual measure is the unconfined compressive strength F_u at 7 days and at 28 days. Variables affecting grout strength are, in the sequence of their importance, the W/C ratio, the pore ratio of set grout, the type of cement, and the presence of admixtures.

Considering only the parameter w (W/C ratio) and disregarding all other factors, the unconfined compressive strength may be approximated from Abram's equation

$$F_u = \frac{A}{B^{1.5w}} \tag{2-1}$$

where F_u = unconfined compressive strength of grout
A = strength constant = 14,000 lb/in^2
B = dimensionless constant depending on cement type at age of test
w = W/C ratio

For type I cement at 28 days, $B = 5$. Full strength is manifested under complete hydration; hence Eq. (2-1) is valid for $w > 0.3$ and for grout subject to minimum bleeding, specifically, $w < 0.7$.

Under normal curing conditions and excluding chemical attack, a set grout continues to gain strength, and does not reach the ultimate until approxi-

TABLE 2-9 Common Cement Admixtures for Anchor Grouts

Admixture	Chemical	Optimum Dosage (% of Cement by Weight)	Remarks
Accelerator	Calcium chloride	1–2	Accelerates set and hardening
Retarder	Calcium lignosulfonate	0.2–0.5	Also increases fluidity
	Tartaric acid	0.1–0.5	May affect set
	Sugar	0.1–0.5	strengths
Fluidifier	Calcium lignosulfonate	0.2–0.3	
	Detergent	0.5	Entrains air
Expander	Aluminum powder	0.005–0.02	≤15% expansion
Antibleed	Cellulose ether	0.2–0.3	Equivalent to 0.5% of mixing water
	Aluminum sulfate	≤20	Entrains air

From Littlejohn and Bruce (1977).

mately one year from placement, as shown in Fig. 2-16. Cements with low hardening rate have a tendency to reach a higher ultimate strength because of slower formation of denser gel during initial setting (Littlejohn, 1982). Type I grout attains a 28-day strength of approximately 60–70 percent the ultimate, but for type III grout the same proportion of ultimate strength is reached in only 7 days. Strength development curves as function of the W/ C ratio are shown in Fig. 2-17 for types I and III grout. It is interesting to note that for W/C ratios exceeding 0.6, the two curves almost coincide, meaning that types I and III both attain equivalent strength development.

Considerble disparity exists with regard to specified grout strengths. Minimum code requirements stipulate a compressive strength in excess of 17 N/mm^2 (about 2500 psi), but with prestressing steel quoted 7-day strength values are as high as 30 N/mm^2 (or 4350 psi). The PCI (1974) recommends a 7-day strength 24 N/mm^2 (about 3500 psi). Where rapid controlled acceleration in strength development is an important requirement, it can be achieved by mixing finely ground cement and calcium chloride. This grout, however, has low tensile strength and exhibits brittle characteristics.

Mixing. This operation influences the quality of the set grout, particularly its strength. For good mixing, the following guidelines should be followed:

1. The cement and admixtures should be measured accurately by weight.
2. Water and admixtures should be added to the mixer before the cement.

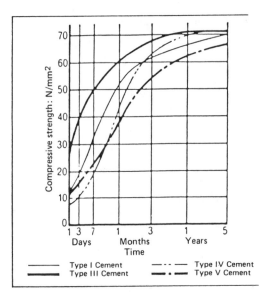

Fig. 2-16 Strength increase with time for set grouts. (From Littlejohn, 1982.)

3. Mixing time for each batch should be long enough to produce a mix of uniform composition. Mixing time varies according to the type of mixer, but it usually is 2–3 min.
4. Mixing by hand should not be attempted.
5. All mixing equipment and pumps should be clean and well maintained.

Grouting Methods

Modes of Grouting Application. Grouting can be accomplished by two distinct modes: two-stage and single-stage injection.

Two-Stage Grouting. This process, mentioned briefly in previous sections, involves first the injection of a primary grout to create the bond zone in the fixed anchor length, and after tendon stressing a secondary grout is introduced in the free length zone mainly for corrosion protection of the tendon. For anchors in rock, the primary grout may be preplaced or postplaced with respect to tendon homing. Postplacing is advantageous with large tendons and poor rock, and probably the only choice for very shallow holes or anchors inclined upwards.

The primary grout extends usually 2 m (7 ft) beyond the designated fixed anchor length in order to inhibit crack formation in the proximal end of the anchorage during stressing. If the primary grout is preplaced, the tendon should be homed no later than 30 min after injection. There is a difference of opinion as to whether the tendon should be left static after homing, in

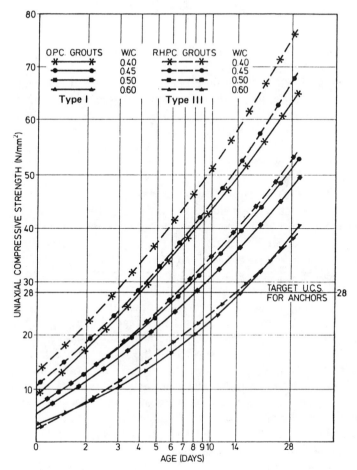

Fig. 2-17 Strength development curves as function of the water/cement ratio for types I and III grouts. (From Littlejohn, 1982.)

view of some problems with grout–tendon bond development experienced even in cases where the tendon has been correctly inserted.

Secondary grouting is better accomplished with a mix of the primary composition. However, some investigators (Mitchell, 1974) recommend the use of sand, gravel, or weak grout in backfilling the free length in order to ensure complete freedom of tendon movement, a practice favored in North America.

Two-stage grouting offers construction convenience, but has the following disadvantages:

1. An additional interface is created at the top of the fixed zone where the two grouted sections meet as construction joint, and becomes a prime target for corrosion attack.

2. Because of the potential of grout escaping to the ground, it is difficult to estimate and check the grout quantity required in the fixed zone.

3. The process is time-consuming and laborious.

Single-Stage Grouting. In this process the borehole is filled in a single continuous operation; hence the functions of the grout are achieved simultaneously. However, unless the free anchor length is carefully greased before sheathing, the final load applied to the head as prestressing may not be entirely transmitted to the intended fixed zone because of possible friction in the free anchor length.

Injection Methods.   Grouting always begins at the lower end of the section to be grouted. If the anchor slopes upward, provisions should be made for venting the hole during the operation. For proper filling, air and water should be allowed to escape. The grout should never reach and maintain contact with the structure being anchored, since the anchor force will not be completely transferred to the ground but instead a prestressed column will be produced.

Before starting grout placement, all pipes and their joints should be cleaned and checked for airtightness. Each stage of injection should be performed in one continuous operation, and at no stage should the end of the grout pipe be lifted above the surface of the grout, or a cold joint will be produced. If grouting is interrupted or delayed beyond the setting period, the tendon should be withdrawn, the grout removed by flushing or redrilling, and the grouting stage repeated. Grout should be tremied at a steady rate, and the pipe withdrawn slowly and at frequent intervals during the operation. Where casing is used, further coordination is necessary between grout injection and withdrawal of the casing.

Grouting Pressures.   In general, grouting pressures are recommended by the specialist contractors, and may be followed with the stipulation that, where necessary, test trials will be carried out before a construction procedure is accepted. The concensus of opinion is that high grout pressures are not necessary for anchors in intact rock, but very useful in badly fissured rock or in soil. The quoted range is 0.30–0.70 N/mm^2 (45–100 psi). Practical and economic considerations often set the maximum grouting pressure at 3 N/mm^2 (about 435 psi), and there is no evidence at present that higher pressures will produce any real benefits.

Figure 2-18 shows the relationship of ultimate anchor capacity to grout pressure suggested for the Soletanche–Tamanchu high-tensile steel tendon, sealed by grouting at high pressures from a central "tube-a-manchette," also known as the "IRP" (acronym derived from the French phrase "injection repeteés en pression") system. From these graphs, it appears that pullout resistance is closely dependent on the grouting pressure, and this dependence becomes greater for loose or soft soil (see also Section 3-4).

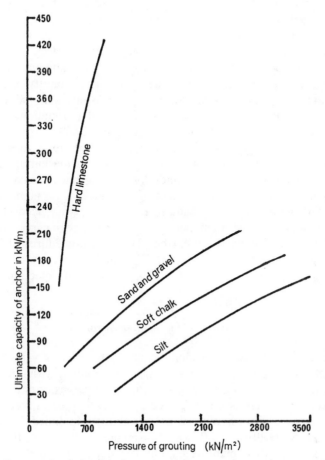

Fig. 2-18 Proposed relationship between ultimate anchor resistance and grouting pressure. (From Soletanche, 1970.)

Grouting pressures, however, considered appropriate for a given project are largely a matter of conjecture. Excessive pressure must be avoided to prevent distress in the ground or disturbance causing damage to adjacent facilities. Grouting pressures depend also, besides site geology, on the method of load transfer in the fixed anchor zone. Thus, rules of thumb do not have much significance, but it is customary to consider a pressure in the region of 0.02 N/mm² per meter of overburden.

Quality Control. In general, when a new mix is introduced, its adequacy for the intended purpose is established through various tests carried out to determine and record its properties. The following information is obtained: (a) W/C ratio and type of cement; (b) admixture type and concentration; (c) flow reading or viscosity (through flowmeter, flow cone, or viscometer); (d) crushing strength (obtained by testing two tubes at 3, 7, 14, and 28 days); and (e) data on expansion, shrinkage, bleed, and final setting time.

At any time the quality of grout depends on several factors, and variations in the specified properties can occur because of (a) inadequate mixing and improper grouting, (b) variations in the quality and quantity of grout constituent materials, and (c) variations from inconsistencies in the testing procedures and the recording of measurements. Useful information on the effect of these factors is given by Littlejohn and Bruce (1977).

Quality controls include fluidity and specific gravity tests during the fluid stage, setting time and bleed measurements during the curing stage, and cube crushing strengths at 7, 14, and 28 days. Chemical contamination of the grout is detected by measurement of pH values. The number and frequency of these tests is not standardized and may vary according to site conditions and job requirements, but it is good practice to carry out the tests on a daily basis. Emphasis should be placed on procedures allowing the grout to be assessed prior to injection. The quality control program is supplemented by recording the quantity of grout injected, injection time, and grouting pressure.

Resins

Synthetic resins develop considerable strength, often several times the strength of cement grouts, and exhibit structural continuity that makes them suitable for use as bonding materials. Varieties of polyester or epoxy-based resins have been tried as bonding media in rock bolting and conventional anchoring, particularly where quick holding ability is intended. Experimental work (Pearson, 1970) shows that a resin-bonded bolt system forms an anchorage zone, as with cement-base grout, transferring the load to the ground along the contact area. The cured resin composition develops an ultimate strength of the order of 110 MN/m^2 in compression, 60 MN/m^2 in tension, and about 500 MN/m^2 in shear, combined with a quick curing time. Likewise, steel tendon anchors may be surrounded by resins for the transfer of load. Whereas a hardening period is required with cement grouts before a load can be applied, with resin the hardening process is a matter of minutes.

An important characteristic of resin-bonded bolts is their plastic behavior and yield under stressing beyond the failure point. Pullout in this case does not lead to sudden rupture of the system, but the member begins to yield slowly. On tension release a permanent elongation remains, but the bolt can be reloaded to almost the original failure value, as shown in Fig. 2-19. This cycle can be repeated several times with more extension but without any significant loss of initial strength.

Where it is necessary to avoid direct transfer of stress in the upper zone close to the surface and create a free anchor length in this area because of weak ground or friable material, common practice is to coat this length of the fully resin-grouted bolt to break adhesion of the steel to the resin. As

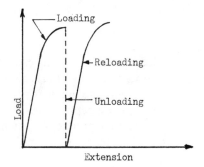

Fig. 2-19 Load extension curves and reloading of resin-bonded bolt.

anticorrosion materials, resins are favored in the sense that they bond and yet expand with the anchor, a behavior that inhibits formation of cracks.

Resins are manufactured in cartridges, consisting of a pack containing a reinforced polyester resin component and a catalyst, isolated from each other by a skin interface to prevent reaction between the two components until required. The cartridge can be inserted in a wide range of borehole sizes, where it may be readily pushed to the extremity at any angle above or below the horizontal. Reactions does not take place until the anchor bolt is rotated through the cartridge, breaking the skin and mixing the two components to start the curing process. The mixed resin fills the area around the bolt and bonds the member to the ground within 10–20 min. Stressing can be introduced within one day from installation.

In anchor works resins have been used mainly for short applications. In the absence of relevant standards and owing to the present limited knowledge with respect to long term performance, their use should be considered with caution and under competent supervision. Tests should be mandatory to study and assess the following characteristics: (a) resin strength for the permanent transfer of stress, (b) suitability of the bonding medium and its inertia to ground conditions, (c) the degree of ductibility necessary to ensure compatible response with the load–extension characteristics of the tendon, (d) non-shrinking characteristics of the material on setting, (e) the extent of fluidity prior to setting, necessary for the material to fill the void completely, and (f) creep response under service load conditions.

2-11 ANCHOR STRESSING AND JACKING

Stressing an anchor after installation by jacking is carried out to confirm its competence and test its capacity to carry the prescribed load (see also Chapter 7). This is mandatory to ensure satisfactory service performance, and it is equally true for both prestressed (active) and dead (passive) anchors. The conditions during stressing should be fairly representative of the actual field conditions; for example, complete debonding should exist along the free anchor length. During the test, a load–extension curve is obtained and used to interpret the expected behavior of the anchor.

A standard approach to load testing procedures and scope is presented in the following sections, together with available records and published data. It includes, but is not limited to, tests and quality controls for (a) precontract component testing (test anchors), (b) acceptance tests of production anchors, (c) long-term monitoring of selected production anchors, (d) special test anchors, and (e) testing and monitoring of the overall anchor–ground–structure system.

Proof loading immediately following the installation provides a satisfactory check of the design load, and establishes a measured factor of safety. Any significant errors made either in the design or introduced in the construction stage are disclosed, and potentially dangerous and costly problems are avoided. It is common practice to proof load each anchor to 1.25 and 1.50 the working load for temporary and permanent anchors, respectively.

The most common, suitable, and frequently used stressing method is by direct pull. Since strand is more common in anchor tendon, multistrand and monostrand direct-pull jacks are available on the market on a commercial basis and in a wide variety of operational characteristics and load capacities. Monojacking performs single strand stressing whereby individual tendon units are tensioned separately. Multistrand units stress all strands simultaneously, but the strands can be locked off individually by wedges in the conical holes of the anchor head. This can be done in one operation, which facilitates cyclic loading tests. Where the maximum force in the bar unit does not exceed about one third the characteristic strength, stressing by torque is usually allowed.

Specialty anchor contractors normally maintain standard stressing anchorage units, that can be easily modified to meet special design requirements. Thus, stressing devices are available for surveillance anchors, anchors that may have to be restressed at a later time, and for anchors that must be detensioned and restressed (see also Chapter 7).

2-12 CONSTRUCTION LIMITATIONS

General Overview

The construction procedures and stages reviewed in Sections 2-7 through 2-11 suggest that it is possible, at least in a theoretical context, to install ground anchors in almost any ground type, and that the anchors will perform satisfactorily. However, load capacity and high installation costs often will preclude construction in soft and organic clays or other materials of similar characteristics, and in severely decomposed or fissured rocks. In some instances, it may be difficult to form anchors unless a particular anchor system has been previously used successfully under the same conditions. Anchor hole construction will encounter difficulties in very fine sands and silts, clay shales, marls, and chalks susceptible to rapid softening during borehole drilling. In these conditions, the borehole forming method should produce maximum roughness along the walls of the hole, and the flushing technique should not cause the surrounding materials to become loose or soft. Occasionally, it will be necessary to resort to special chemical grouts to permeate finely grained soil in the fixed anchor length, or combine chemical and cement grouts in stage grouting, but limited knowledge or field work is presently available on this process. An alternate approach is the use of multistage controlled grouting in the fixed anchor zone using a tube-a-manchette for soft rock and aluvial deposits. These techniques combined with controlled grout pressures may result in a greater range of anchor capacities than can normally be achieved.

Exceptionally long boreholes can be very expensive even when the driling machine has extra torque capacity. Furthermore, tendon homing becomes more difficult, and extra care is necessary to provide the correct alignment of the borehole.

Site access is not necessarily a major obstacle to anchoring, but where space and access are limited they add considerably to the cost. Site access is important when anchoring in very steep slopes or in very congested areas, and in this case use of small drills is mandatory.

Contingent Factors

Engineers should be cautioned that it is not always possible to proceed with and implement construction methods and details envisaged at design and tender stages. In this context, the following comments are a useful supplement to the principles and procedures reviewed in the foregoing sections.

Tendon Preparation and Assembly. Many contractors agree on having the completed tendon factory-prepared and delivered to the site ready for homing. They caution, however, that this is not practical where flexibility to cope with changing ground conditions is desired. For example, if some anchors must be extended, it will be difficult to ensure complete corrosion protection along the free length where the coupling is located, and this problem will be more serious for a multistrand tendon.

The alternative of using factory prepared components assembled at the site is satisfactory provided adequate covered facilities are available for final assembly. For site assembly, the polyvinyl chloride (PVC) sheating and grease must be completely removed from the bonded anchor length, and the method used can have marked effect on the bond at the tendon–grout interface. To overcome this problem, some contractors have expedited the use of noding effects, thereby creating mechanical interlocks. During tendon handling and lifting for the homing operation, the possibility of undetected damage to protective sheathing or to the tendon itself is strong, especially when dragging the tendon and when attaching the assembly to cranes. Any encapsulations at the site should be protected from frost during curing period.

All materials used in the anchor should ideally have compatible elasticity characteristics. Complete similarity in elastic modulus under service loads is seldom possible, but it is important to consider and investigate favorable combinations of steel and grout systems. Contractors have reported high incident rate of failures (debonding) with the use of epoxy resins.

Grouts and Grouting. Cost-saving incentives prompt manufacturers and contractors to consider new combinations of materials. For example, cement-base grouts and corrugated plastic sheath are continuously marketed for use in the fixed anchor zone and as double corrosion protection. These combinations usually are tested and fully documented, but for immediate applications it is seldom feasible to assess their long-term behavior.

Many contractors consider grout specifications too rigid and single-scope-oriented, since they do not cover the contingency of consistency changes to cope with changing ground conditions. More criticism is directed toward the absence of a generalized and conclusive procedure to assess the effect of grout pressure on ultimate crushing strength. The suggestion is that a difference should exist in a particular grout mix of the same composition between a normal cube and a sample of grout injected into a borehole under pressure. It has been pointed out that pressurized anchors have successfully been tested when cube results were very low.

More demand for higher-capacity anchors places more requirements on drilling procedures, since the increase in tendon size means a larger borehole if the recommended grout cover is to be maintained. Whereas these aspects are discussed in subsequent sections in terms of load transfer and corrosion protection, it is interesting to mention in the context of this critique that minimum grout cover requirements are from 5 to 20 mm. In alluvium deposits the thickness of grout cover is mostly unpredictable, whereas in rock it is a function of the competence and strength of all rock materials and grout. Thus, where doubt exists, the practical solution is to drill larger holes for a greater grout cover, hence more strength in the grout column, but with larger holes problems may arise where weak overburden must be drilled and cased before competent zones are reached. In view of possible local limitations in drilling capability, the selection of a greater number of anchors with lower load capacity may in some instances be economically more attractive.

Drilling. Some contractors suggest that deviation and tolerance in borehole drilling should be considered on the basis of actual soil conditions rather than fixed rules. The contention is that checking for correct alignment is expensive, and correcting misalignment is even more expensive.

If the design specifies close anchor spacing, hole deviation due to ground conditions (e.g., obstacles and bedding planes) should be investigated with test anchor installation. If the specified tolerance is found to be unattainable, one solution is to modify the design by varying anchor inclination. It appears, at least at present, that there is no commercial instrument available that allows borehole checking without inducing considerable cost and delay to the construction process.

Some contractors contend that ramming methods are not suitable where ground conditions vary excessively, and they recommend drilling techniques that return spoil and cuttings to the surface. Whereas this should not replace normal site investigations, it can nonetheless detect and disclose major variations in ground conditions.

Flushing. Most contractors favor water and air with open return to the surface to prevent hydrofracturing effects. The use of bentonite and other supporting fluids appears to be loosing popularity, the contention being that the borehole cannot be adequately cleaned (as is the case with, e.g., large bored piles annd slurry walls); hence there may be loss of frictional resistance along the bonding interface. Little is known, however, on this subject either from laboratory or from full-scale tests.

In installations where the anchor entrance level is below the water table, a situation occurring frequently in practice, drilling at this level may cause sufficient ingress of water and fines to create cavities behind the structure, sometimes of concern. When this condition is anticipated, preventive methods should be available at the very beginning of construction.

2-13 EXAMPLES OF ANCHOR SYSTEMS USED IN NORTH AMERICA

Figure 2-20 shows a single corrosion protection anchor in rock. This is suitable for temporary installations in fairly aggressive environment where some protection against corrosion is desired. The maximum ultimate capacity, or guaranteed limit, of the steel tendon is 2147 kips for the 52-strand anchor. In sound rock, drilling usually is carried out by percussive methods with air powered down-the-hole (DTH) hammer. In less competent rock, rotary drills are more suitable. Core drilling is reportedly used through concrete where large quantities of reinforcing steel are present. The drill hole diameter shown may vary according to the exact in situ rock characteristics and drilling equipment available. The prestressing steel has an ultimate tensile strength 270,000 psi, and is produced and tested according to ASTM A416. A single, $\frac{1}{2}$-in dia., 7-wire strand unit has a guaranteed ultimate strength 41.3

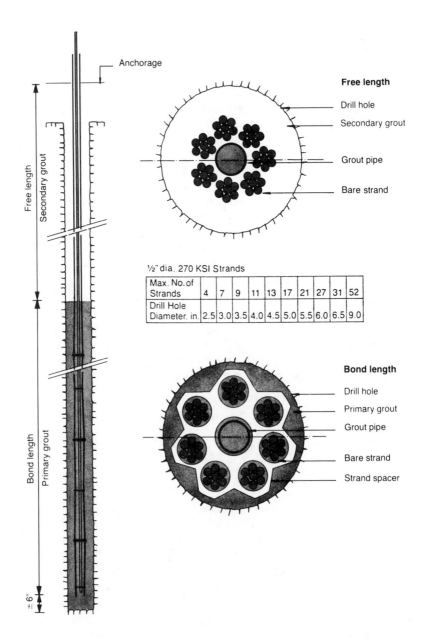

Anchorage

Free length

Drill hole

Secondary grout

Grout pipe

Bare strand

½" dia. 270 KSI Strands

Max. No. of Strands	4	7	9	11	13	17	21	27	31	52
Drill Hole Diameter. in.	2.5	3.0	3.5	4.0	4.5	5.0	5.5	6.0	6.5	9.0

Bond length

Drill hole

Primary grout

Grout pipe

Bare strand

Strand spacer

Free length

Secondary grout

Bond length

Primary grout

±6"

Fig. 2-20   Typical rock anchor with single corrosion protection. (VSL.)

kips, a proof load 35.1 kips at 0.1 percent extension, and an approximate modulus of elasticity 28,000,000 psi.

A double-corrosion-protection anchor in rock is shown in Fig. 2-21, which is suitable for use in aggressive environment for permanent installations. Drilling methods are essentially similar to those for the anchor of Fig. 2-20 except that the drill hole diameter is larger to accommodate the double-protection process. In both cases, the primary and secondary grout is assumed to be carried out in a watertight hole. If pregrouting is necessary, the hole must be redrilled.

An augered soil anchor is shown in Fig. 2-22. It has a straight shaft; hence it is a type *A* anchor in Fig. 2-2. It is suitable in very stiff to hard cohesive layers where it is possible for the hole to remain open after drilling. The single protection corrosion suggests a temporary use in aggressive soil environment, similar to Fig. 2-20, but for permanent use the corrosion protection shown in Fig. 2-21 is applicable. The lower ground strength usually limits the normal ultimate capacity of this anchor to 413 kips or 10 strands. Drilling for this anchor usually involves the use of a continuous-flight auger, or a kelly bar auger. Diameters may vary from 6 to 24 in for a straight hole where bond is developed by direct shear at the grout–ground interface.

In unstable ground, a drilled or driven casing is mandatory to the required depth, usually of a diameter of 3–6 in. When the hole is complete, the casing is cleaned, the tendon is inserted, and the hole is pressure grouted over the anchor zone as the casing is withdrawn.

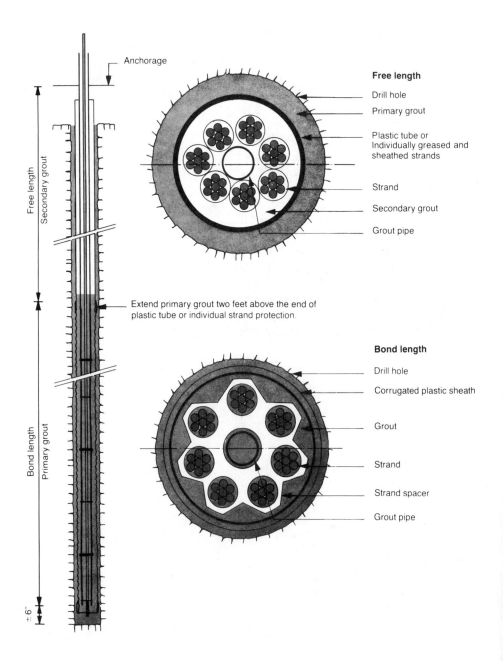

Anchorage

Free length

Drill hole

Primary grout

Plastic tube or
Individually greased and
sheathed strands

Strand

Secondary grout

Grout pipe

Extend primary grout two feet above the end of
plastic tube or individual strand protection.

Bond length

Drill hole

Corrugated plastic sheath

Grout

Strand

Strand spacer

Grout pipe

Free length

Secondary grout

Bond length

Primary grout

± 6"

Applications:

This anchor is used where the environment is aggressive and protection against corrosion is necessary. The maximum ultimate capacity of the standard anchor is 2,147 kips.

Drilling methods:

Drilling methods corresponding to those for a single corrosion protected rock anchor are employed. In the case of either the single or the double corrosion protected anchor, it is very important that the anchor is installed in a hole that is watertight. If water tests as described in the guide specification fail, the hole must be grouted and redrilled.

The following points are suggested for consideration when selecting a prestressing unit:

- Working force
- Loss of prestress
- Allowable stresses in prestressing steel
- Drill hole diameter
- Bond length

The $\frac{1}{2}$″, 7-wire strand for prestressing application has an ultimate strength (f_s') of 270.000 psi and is produced and tested in accordance with the requirements of ASTM A 416.
Physical properties of $\frac{1}{2}$″ strand are as follows:

Guaranteed ultimate strength	41.300 lb.
Yield strength (at 0.1°₀ extension)	35.100 lb.
Approx. modulus of elasticity	28.000.000 psi
Min. elongation at rupture	3.5°₀ in 24 inches

$\frac{1}{2}$″ dia. 270 KSI Strands

Max. No. of Strands	4	7	9	11	13	17	21	27	31	52
Drill Hole Diameter, in.	4.5	5.0	5.5	6.0	6.5	7.0	7.5	8.0	8.5	11.0

The diameter of the drill holes is given for information only. It may vary according to the type of anchor, characteristics of rock (loading capacity) and drilling equipment available.

When conditions demand that the drill hole be lined, then these dimensions correspond to the internal diameter of the casing.

Intermediate and larger units are available.

Fig. 2-21 Typical rock anchor with double corrosion protection. (VSL.)

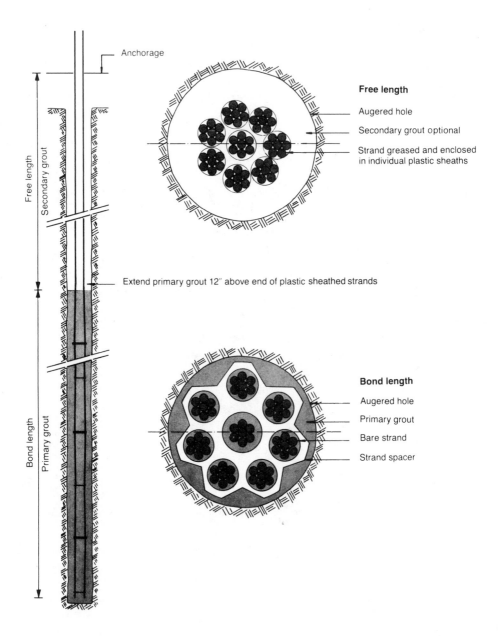

Anchorage

Free length

Secondary grout

Bond length

Primary grout

Free length

Augered hole

Secondary grout optional

Strand greased and enclosed in individual plastic sheaths

Extend primary grout 12″ above end of plastic sheathed strands

Bond length

Augered hole

Primary grout

Bare strand

Strand spacer

Applications:

This anchor is used where the environment is aggressive and protection against corrosion is necessary. The maximum ultimate capacity of the standard anchor is 2,147 kips.

Drilling methods:

Drilling methods corresponding to those for a single corrosion protected rock anchor are employed. In the case of either the single or the double corrosion protected anchor, it is very important that the anchor is installed in a hole that is watertight. If water tests as described in the guide specification fail, the hole must be grouted and redrilled.

The following points are suggested for consideration when selecting a prestressing unit:

■ Working force
■ Loss of prestress
■ Allowable stresses in prestressing steel
■ Drill hole diameter
■ Bond length

The $\frac{1}{2}$", 7-wire strand for prestressing application has an ultimate strength ($f_s{}'$) of 270,000 psi and is produced and tested in accordance with the requirements of ASTM A 416.
Physical properties of $\frac{1}{2}$" strand are as follows:

Guaranteed ultimate strength   41,300 lb.
Yield strength (at 0.1°₀ extension) 35,100 lb.
Approx. modulus of elasticity   28,000,000 psi
Min. elongation at rupture   3.5°₀ in 24 inches

$\frac{1}{2}$" dia. 270 KSI Strands

Max. No. of Strands	4	7	9	11	13	17	21	27	31	52
Drill Hole Diameter, in.	4.5	5.0	5.5	6.0	6.5	7.0	7.5	8.0	8.5	11.0

The diameter of the drill holes is given for information only. It may vary according to the type of anchor, characteristics of rock (loading capacity) and drilling equipment available.

When conditions demand that the drill hole be lined, then these dimensions correspond to the internal diameter of the casing.

Intermediate and larger units are available.

Fig. 2-22 Typical augered soil anchor with single corrosion protection. (VSL.)

REFERENCES

A. B. Chance Co.: *Encyclopedia of Anchoring*, In-House Report, Centrella, Missouri. (1975)

ACI Committee 212, 1971: "Guide for Use of Admixtures in Concrete," *J. Am. Conc. Inst.*, Sept., pp. 646–676.

Allen, A. H., 1978: *An Introduction to Prestressed Concrete*, Cement and Concrete Assoc., Wexham Springs, Slough, Great Britain.

American Society of Testing Materials: *Uncoated Stress-Relieved Wire for Prestressed Concrete*, 1974. ASTM U.S.A.

Antill, J. M., 1967: "Relaxation Characteristics of Prestressing Tendons," *Civ. Eng. Trans., Inst. Eng. (Australia)*, **7** (2), 151–159.

Baker, W. J., 1955: "Flow in Fissured Formations," *Proc. World Petroleum Cong.*, Rome, Section II, pp. 379–393.

Banks, J. A., 1955: "The Employment of Prestressing Technology on Allt-na-Lairige Dam," *Proc. 5th Int. Cong. Large Dams*, Paris, pp. 341–357.

Bannister, J. H. L., 1959: "Characteristics of Strand Prestressing Tendons," *Struct. Eng.*, **37** (3), 79–96.

British Standard 2691, 1969: *Steel Wire for Prestressed Concrete*, British Standards Inst., London.

British Standard, CP 110, 1972: *The Structural Use of Concrete*, British Standards Inst., Part I, London.

Brunner, H., 1972: "Praktische Erfahrungen Beiden Hangsicherungen in Nekenhoff," *Strabe und Veruchr*, **9**, 500–505.

Buro, M., 1972: "Rock Anchoring at Libby Dam," *West. Const.* (March), 42, 48, 66.

Cambefort, H., 1966: "The Ground Anchoring of Structures," *Travaux*, **46**, April–May, 15 pp.

Cementation Co., Ltd., 1962: "Anchor Stressing," In-House Report No. 10, Croydon, Surrey, Great Britain.

Cestelli-Guidi, C., 1974: "Ground Anchors," *Proc. Techn. Session on Prest. Concrete Found. and Ground Anchors*. 7th FIP Cong., pp. 4–6, New York.

Chen, S. C., and J. G. McMullan, 1974: "Similkameen Pipeline Suspension Bridge," *Proc. ASCE, Transp. Eng. J.*, **100** (TEI), 207–219.

Compte, C., 1971: *Technologie des Tirants*, Inst. Research Found., Kolibrunner/Rodio, Zurich, p. 119.

Drossel, M. R., 1970: "Corrosion Proofed Rock Ties Save Building from Earth Pressures," *Constr. Meth.* (May), 78–81.

Eberhardt A., and J. A. Veltrop, 1965: "Prestressed Anchorage for Large Trainter Gate," *Proc. ASCE, J. Struct. Div.* **90** (ST6), 123–148.

FIP, 1974a: *Recommendations for Approval, Supply and Acceptance of Steels for Prestressing Tendons*, Cement and Concrete Assoc., Lond.

Geddes, J. D., and I. Soroka, 1964: "The Effect of Grout Properties on Transmission Length in Grout-Bonded Post-tensioned Beams," *Mag. Conc. Reg.*, **16** (47), 93–98, Great Britain.

Golder Brawner Assocs., 1973: "Government Pit Slopes Project III: Use of Artificial Support for Rock Slope Stabilization," Parts 3.1–3.6, Unpublished, Vancouver, Canada.

Gosschalk, E. M., and R. W. Taylor, 1970: "Strengthening of Muda Dam Found. Using Cable Anchors," *Proc. 2nd Cong. Int. Soc. Rock Mech.* Belgrade, pp. 205–210.

Ground Anchors, Ltd., 1973: *The Ground Anchor System*, 28 pp., Reigate, Surrey.

Hilf, J. W., 1973: "Reply to Aberdeen Questionnaire," Unpublished Report, Aberdeen, Scotland.

Hobst, L., 1965: "Vizepitmenyek Kihorgonyzasa," *Vizugi Koziemenyek*, **4**, 475–515.

Jackson, F. S., 1970: "Ground Anchors-The Main Contractor's Experience," *Suppl. Ground Anchors, Cons. Eng.*, May.

Jorge, G. R., 1969: "Devices to Destroy in Place Temporary Anchorages, Rods, Wires, or Cables," Specialty Session No. 15, *Anchorages Especially in Soft Ground, Proc., 7th Intern. Conf. Soil Mech. Found. Eng.*, Mexico City.

Kern, G., and T. Herbst, 1974: "Rock and Soil Anchors," *Proc. Techn. Session on Prest. Concrete Found. and Ground Anchors*, 7th FIP Cong., New York, pp. 19–24.

Kovacs, W. D., 1977: *Soil and Rock Anchors for Mobile Homes, A State-of-the-Art Report*, Center for Building Techn., National Bureau of Standards, Washington, D.C.

Koziakin, N., 1970: "Foundation for U.S. Steel Corporation Building in Pittsburg, Pa.," *Civ. Eng. Pub. Works Rev.* (Sept.), 1029–1031.

Langworth, C., 1971: "The Use of Prestressed Anchors in Open Excavations and Surface Structures," Australian Inst. Mining and Metallurgy (Illwarra Branch). Symposium on Rock Bolting, Feb. 17–19, Paper No. 8, 17 pp.

Larson, M. L., W. R. Willette, H. C. Hall, and J. P. Gnaedinger, 1972: "A Case Study of a Soil Anchor Tie-back System," *Proc. Spec. Conf. on Performance of Earth and Earth Supported Structures, Purdue Univ., Indiana* **1** (2), 1341–1366.

Littlejohn, G. S., 1972: "Ground Anchors Today—A Foreword," *Ground Eng.,* **6** (6), 20–23.

Littlejohn, G. S., 1975: "Acceptable Water Flows for Rock Anchor Grouting," *Ground Eng.,* **8** (2): 46–48.

Littlejohn, G. S. 1980a: "Ground Anchors, State-of-the-Art," *Civ. Eng. Surveyor* (Aug.), London.

Littlejohn, G. S., 1980b: "Design Estimation of the Ultimate Load-Holding Capacity of Ground Anchors," *Ground Eng.*, (Nov.), Great Britain.

Littlejohn, G. S., 1982: "Design of Cement-Based Grouts," *Proc. Conf. on Grouting in Geotechnical Engineering*, ASCE, New Orleans.

Littlejohn, G. S., and D. A. Bruce, 1977: "Rock Anchors—State of the Art," *Ground Eng. Found. Pub.*, Essex, England.

Littlejohn, G. S., and C. Truman-Davies, 1974: "Ground Anchors at Devonport Nuclear Complex," *Ground Eng.* **7** (6), 19–24.

Longbottom, K. W., and G. P. Mallet, 1973: "Prestressed Steels," *The Structural Engineer (Struct. Eng.)*, London, pp. 455–471.

Losinger SA, 1966: "Prestressed VSL Rock and Alluvium Anchors, *Techn. Brochure*, Bern, 15 pp.

Mascardi, C., 1973: "Reply to Aberdeen questionnaire," Unpublished Report.

Matt, P., 1981: "Recommendations on Prestressed Ground Anchors," VSL-Losinger Report, Bern.

Mawdsley, J., 1970: "Choice of Drilling Methods for Anchorage Installations," *Ground Anchors, Cons. Eng.*, London, pp. 5–6.

McGregor, K., 1967: "The Drilling of Rock," C. R. Books Ltd., London.

Mitchell, J. M., 1974: "Some Experiences with Ground Anchors in London," *Proc. Diaphragm Walls and Anchorages Conf.*, Inst. of Civ. Eng., London, pp. 128–133.

Morris, S. S., and W. S. Garrett, 1956: "The Raising and Strengthening of the Steenbras Dam (and Discussion)," *Proc. ICE, Part 1*, **5**, (1), 23–55.

Moschler, E., and P. Matt, 1972: "Felsanker und Kraftnessanlage in der Kaverne Waldeck II," *Schweizerische Bauzeitung*, **90** (31), 737–740.

Nicholson, P. J., Uranowski, and P. T. Wycliffe-Jones, 1982: "Permanent Ground Anchors, Nicholson Design Criteria," U.S. Dept. Trans., Fed. Highway Admin., Washington, D.C.

Nordin, P.O., 1968: "In situ Anchoring," *Rock Mech. Eng. Geol.*, **4**, 25–36.

Ostermayer, H., 1974: "Construction, Carrying Behavior and Creep Characteristics of Ground Anchors," *Proc. Diaphragm Walls and Anchorages Conf.*, Inst. of Civ. Eng., London, pp. 141–151.

Paone, J., H. F. Unger, and S. Tandenand, 1968: "Rock Drillability for Military Applications," Final Contract Report AD 671671 for Army Research Office, U.S. Dept. Interior, Bureau of Mines, Minneapolis.

PCI Post-Tensioning Committee, 1974: "Tentative Recommendations for Prest. Rock and Soil Anchors," PCI, Chicago.

Pearson, L. L., 1970: "Resin Based Fixings for Anchorages," *Cons. Eng.* (May), London, pp. 16–17.

Rawlings, G., 1968: "Stabilization of Potential Rockslides in Folded Quartzite in Northwestern Tasmania," *Eng. Geology*, **2** (5), 283–292.

Schmidt, R. L., 1972: "Drillability Studies-Percussive Drills in the Field," Bureau of Mines, Report of Investigations 7684, U.S. Dept. Interior, Washington, D.C.

Schousboe, J., 1974: "Prestressing in Foundation Construction," *Proc. Techn. Session on Prest. Concrete Found. and Ground Anchors*, 7th FIP Cong., New York, pp. 75–81.

Shchetinin, V. K., 1974: "Investigation of Different Types of Flexible Anchor Tendons for Stabilization of Rock Masses," *Gidrotekhnicheskoe Stroitel, Stuo.* **4** (April), 19–23.

Soletanche, 1968: "La Surrelevation du Barrage des Zardezas sur l'oued Saf Saf," Unpublished Report, 4 pp.

Soletanche, 1970: "Other Types of Anchor," *Ground Anchors, Cons. Eng.* (May), London, pp. 12–15.

Sommer, P., and F. Thurnherr, 1974: "Unusual Application of VSL Rock Anchors at Tarbela Dam, Pakistan," *Proc. Tech. Session on Prest. Concrete Found. and Ground Anchors*, 7th FIP Cong., New York, pp. 65–66.

Stocker, M. F., 1973: "Reply to Aberdeen Questionnaire (1972)" (Unpublished).

Stump–Vibroflotation, 1982: "Permanent Ground Anchors, Design Criteria," U.S. Dept. Transportation, Federal Highways Administration, Washington, D.C.

Taylor, R. J., D. Jones, and R. M. Beard, 1975: *Handbook for Uplift Resisting Anchors*, Civ. Eng. Laboratory, Naval Const. Battalion Center, Port Hueneme, Calif.

Unger, H. F., and R. R. Fumanti, 1972: "Percussive Drilling with Independent Rotation," Bureau of Mines, Report of Investigations 7692, U.S. Dept. of Interiors, Washington.

Van Ormer, H. P., 1974: "Determining Rock Drillability," *Rock Products* (Feb), pp. 50–51.

VSL, 1966: "Ground Anchors," In-House Report, Bern.

VSL–Losinger, 1980: "Soil and Rock Anchors," Company Report, Bern.

White, C. G., 1965: "A Rock Drillability Index," Thesis, Colorado School of Mines, Colorado.

White, R. E., 1970: "Anchorage Practice in the United States," *Ground Anchors, Cons. Eng.* (May), London, pp. 32–37.

White, R. E., 1973: "Reply to Aberdeen Questionnaire," Unpublished Report.

Wosser, T., and M. Darragh, 1970: "Tie-Backs for Bank of America Building Excavation Wall," *Civ. Eng.* (N.Y.), **40** (3), 65–67.

Yamada, K., 1978: "Ground Anchor Market in Japan," Paper Prepared for VSL Intra-Co. Symposium.

Zienkiewicz, O. C., and R. W. Gerstner, 1961: "Stress Analysis of Prestressed Dams," *Proc. ASCE J. Power Division,* **87** (PO1), Part 1, 7–43.

CHAPTER 3

SPECIAL ANCHOR SYSTEMS

3-1 REMOVABLE AND EXTRACTABLE ANCHORS

A frequent problem with temporary installations in built-up areas is the un-availability of unrestricted land for placing the anchorage zone. This problem is compounded by local codes often prohibiting any new construction in the upper 20 ft of public property, presumably to allow installation of future underground utilities. In this case, a mandatory requirement before granting permission to install any anchors below private land is that the anchors can easily be pulled out after use and without disturbance to surroundings. An alterante option to anchor removal is the use of detensionable anchors. These meet all the requirements of the restressable head discussed in other sections, and in addition the tendon can be detensioned in a controllable manner while the anchorage is in service.

Removal and Extraction Techniques

In general, these are based on the use of mechanical, chemical, physical or electrical action on the anchor system (Jorge, 1969). Mechanical methods are used for bar tendons, and include washover with a drill, unscrewing with special couplings, and sectioning by a predetermined failure where the free length meets the fixed anchor body, which remains in the ground. Chemical methods are based on some acidifying process or the involvement of electrolytic corrosion initiated in critical parts of the tendon. In many instances they have been successful, but the process is still in the experimental stage.

Rey (1978) distinguishes two basic systems of extraction: (a) anchors where the bond length is removed and (b) anchors where the bond length remains in the ground.

Anchors Where the Bond Length Is Removed

Use of Explosives. In this case, the bond between the steel tendon and the enveloping cement grout is broken by detonating a charge along the fixed anchor length.

The explosion tube, usually unplasticized polyethylene, is placed along the center of the anchor and sealed at both ends. The strands around it must be parallel and firmly fixed in place. Strand overlaps or sharp bends should be avoided since they can squash the explosion tube and inhibit the subsequent introduction of the charge. At the stressing end, the tube should preferably pass through the anchor head. After the tendon is stressed, the tube is cut off at the anchor head, and the assembly is closed. The entire arrangement is shown in Fig. 3-1.

A detonating chord containing an appropriate amount of explosive per unit length is used as the charge. The length of the chord is about 1.5 ft greater than the bond zone. Rey (1978) reports that in cohesive soils a greater charge is necessary than in alluvium. The charge also depends on anchor capacity and tendon size. Success of the operation is better ensured if a low-sensitivity detonator with sufficiently long wires is used, and couplings are avoided in the region of the anchor.

Before the detonating chord is inserted, the explosion tube should be checked for obstructions along its length, and this is done with a thin wire having a ball attached to its end. The detonating chord usually is lubricated with soft soap. Water may be present in the explosion tube, but in this case the charge should be detonated within one hour.

During blasting, the anchor can remain prestressed if conditions permit. The momentum from the energy release throws the anchor partly or completely out of the borehole. If a detensioned anchor is blasted, its extraction is carried out with pulling equipment.

The main advantage of this method is the high probability of success provided the preparations for blasting are carefully made. On the other hand, the requirement to employ a blasting specialist tends to raise the cost. Furthermore, clients and engineers are often biased against this method, because of possible, and often speculative effects on surroundings. Thus, before blasting is considered, vibratory effects should be assessed in conjunction with soil type, geometry, and structural condition of nearby buildings, and resonance frequency.

Splitting of Anchor Bulb. By fitting a cone to an additional strand, the latter is pulled to break and disintegrate the grout bulb. The cone with a compression fitting is within the bond length near the bottom of the borehole,

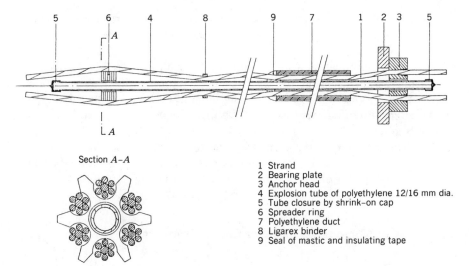

1 Strand
2 Bearing plate
3 Anchor head
4 Explosion tube of polyethylene 12/16 mm dia.
5 Tube closure by shrink–on cap
6 Spreader ring
7 Polyethylene duct
8 Ligarex binder
9 Seal of mastic and insulating tape

Fig. 3-1   Extractable anchor, explosive method. (From Rey, 1978.)

and the strand holding it is placed at the center of the anchor. Alternatively, small rings are slid onto the center wire of the additional strand. When the outer wires are in position, a bulge will appear at each ring. The additional strand is encased in a polyethelene sleeve along the bond length and protected by insulating tape against bonding to the cement grout. For anchor extraction, this strand is tightened with a jack until the process bursts the anchor bulb. The main strands are then pulled out, first with the jack and then by hand.

Some contractors maintain reservations about this method, and consider it labor-intensive and risky for the operatives.

Temperature-Dependent Anchor. This is available under a Japanese patent, whereby the bond produced by a temperature-dependent synthetic resin is destroyed. The heat necessary along the bond length for this purpose is provided by closing an electric circuit in the prestressing steel.

Anchors Where the Bond Length Is Not Removed

Induction. With this method, an induction coil fitted at the transition between fixed length and stressing (free) zone induces heat in the steel when it is supplied with a high-frequency alternating current. This heat alters the strength characteristics of the steel tendon and reduces its tensile limit to a value below the actual applied stress, thus causing the strands to fail and break.

The coil, usually copper wire, has a diameter of 1.0–1.3 mm and a number of turns from 200 to 230. For standard electrical insulation, the entire coil is impregnated with epoxy resin, which also keeps the moisture out and acts as mechanical protection. At each end of the coil, an eternite tube about 50 mm long is mounted and the coil is wound with asbestos tape. The entire assembly is held together with a shrink-on sleeve [see Fig. 3-2(a) and (b)].

During tendon stressing, the electrical conductor is protected from breaking by fitting loose within the stressing length. The tendon steel is heated up by alternating field generated by the high-frequency alternating current (usual frequency 500 Hz, voltage 280 V, and current 10 A).

Contractors report that the heating time necessary to cut the strands is 3–7 min. However, in field applications success of the method may be inhibited if water can enter the induction zone, thereby retarding the heat process.

Examples of Removable and Extractable Anchors

In North America, bar tendons of low to medium capacity are often removed by the so-called gripnut-and-plate method. The bar tendon at the low end is provided with threads while a nut is screwed to it. The nut is welded to

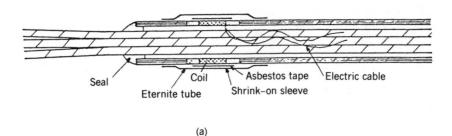

(a)

No. of strands	d	D	L	Wire ϕ
3	30	52	50	1.05
4	35	57	50	1.20
5	42	64	50	1.20
6	42	64	50	1.30
7	42	64	50	1.30

Dimensions are in mm

(b)

Fig. 3-2 (a) Anchor assembly with induction coil for extractable system; (b) induction coil details and data. (From Rey, 1978.)

an end bearing plate used to transfer the pull from the rod to the grout column by direct compression. The steel rod is greased and wrapped with paper so that it can be rotated and unscrewed at the end. When the grout has set, the rod is turned sufficient times to break any bond with the grout. When the anchor is no longer needed, usually 1–2 years after installation, the steel rod is completely unscrewed and withdrawn, leaving the nut, bearing plate and grout in place (White, 1970). Since no effective contact is assumed at the tendon–grout interface for bond development, the load-carrying capacity depends on the direct bearing between the end plate and the grout column.

In France, contractors use a melting process initiated with oxygen-gas cutting. Self-igniting oxygen hoses are lowered through a special tube attached to the tendon until they reach the desired level of sectioning. The heat transforms the steel into a liquid slag and also melts and disintegrates the grout locally. The entire anchorage is thus separated and the cut part is extracted with the use of a block and tackle exerting sufficient pull.

In Japan, most temporary anchors are extractable. Anchor removal is carried out by mechanical methods inducing failure to certain parts of the tendon. According to one method, individual metal gripping pieces are securely attached to the low end of each strand, as shown in Fig. 3-3(a), whereas a bearing plate transfers the pull to the grout by compression. Likewise, the strand is lubricated along its full length to inhibit bond resistance. The gripping sections are detailed to be stripped from the low end of each strand merely by failing under pullout exceeding the strength of the connection, but without exceeding the ultimate anchor load. Each strand is thus overstressed individually, eventually completing the extraction of the anchor (Yamada, 1978).

Another extractable anchor used in Japan is the so-called sliding-wedge anchor. This has a wedge-shaped member tapered in the direction of pullout. The tendon strands are surrounded with sheath, again to prevent bond with the grout. After homing of the tendon, the hole is grouted using a mix containing, in addition to cement, filler materials of low rigidity in order to affect formation of an environment similar to voids and air bubbles. These materials may be foamable polysterene or cork of suitable porosity. The anchor can be removed by pulling the tension steel tendon with a jack, and as the attached sliding wedge is withdrawn it exerts a disintegrating action on the grout materials around the tendon, thereby detaching the anchor steel from the grout body. This arrangement is shown in Fig. 3-3(b). Evidently, this method works on the condition of a grout system that will fail before the steel tendon; hence the full capacity of the latter is not utilized.

Assessment of Extraction Techniques

Among the methods currently available and described in the foregoing sections, many contractors consider blasting a useful and practical solution. However, the usual response to its use in anchorage sites in built-up areas,

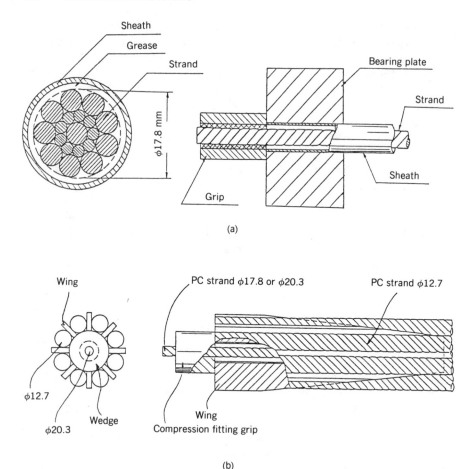

Fig. 3-3 (a) Arrangement of gripping piece, bearing plate, and strand for extractable anchor; (b) siding wedge anchor. (From Yamada, 1978.)

particularly from the public and owners of adjoining properties, involves hesitation, reluctance, and often an element of prejudice, which in some instances can be dispelled by reasoned arguments. In the general context of anchor removal, blasting is essentially the application of shock waves electrically released to produce sufficient cracks and fracture within the injection cavity and thus destroy the adhesive section of the anchor. Competent blasting experts are satisfied with the harmlessness of the detonating fuse claimed in this case, and agree that it does not denote the dangerous action normally implied by bursting charges and firing and blasting operations.

A case of successful removal of anchors, each consisting of 6 or 7 strands, at a building site in Vienna is reported by Straus (1978). The area immediately surrounding the anchorage site was closed to traffic during the operation, but vibration and noise levels were nominal. Following detonation, the anchors were removed by manual pulling, hoisting, and jacking. During prestressing of the strands, both ends of the shock wave pipe were carefully sealed to avoid clamping between strands. It is reported that anchor parts and chipped wedges poped out as much as 9 m (about 30 ft) from the insertion point, whereas prestressed anchors were easier to remove than passive tendons. The contractor removed the anchors one at a time in order to keep vibrations and ground disturbance to a minimum.

3-2 COMPRESSION, COMPRESSED BOND AND COMPRESSED TUBE ANCHORS

These anchor systems were developed in the late 1970s as an attractive alternative to the more conventional types. Whereas the initial cost may in some instances be higher, they can offer improved economy since they afford better corrosion protection and thus can last longer.

Compression Anchors

For the anchor types presented in Chapter 2, the applied load transferred to the fixed length by the stressing mechanism at the anchor head induces a pulling force starting at the top and progressively moving downards toward the low end. In this manner shear, and occasionally normal stresses are developed at the tendon–grout interface and migrate along this length until they dissipate at the low end after the entire pulling force in the tendon is absorbed.

In this interaction the grout column is subjected to tension, more near the top and less near the low end of the fixed zone, with corresponding tension cracks. Where the long-term effectiveness of the corrosion protection system is in question, and this in spite of double-protection systems, the alternative is to consider a compression type anchor.

Figure 3-4 shows the details of a commercially available, standard compression anchor. A pressure pipe contains the tendon in the fixed zone, and is fitted with an endplate also attached to the tendon. As the latter is

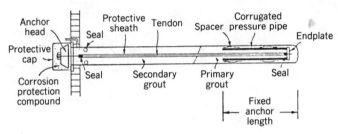

Fig. 3-4 Schematic presentation of compression-type anchor.

stressed, the force is transmitted from the endplate to the grout in an upward direction, hence the grout column is subjected to compression with a corresponding elimination of tension cracks.

Compressed Bond Anchors

This works on the same principle as the compression anchor shown in Fig. 3-4. The steel tendon is mechanically connected to the lower part of a steel tube, which is sealed to the ground by pressure grouting, as shown in Fig. 3-5. With this arrangement both the exterior steel tube and the grout column are subjected to compression, whereas the steel tube protects the tendon against corrosion.

The outer protective tube comprises two zones: (a) the sealed length (fixed zone) made of steel and equipped with rubber sleeves for grouting and (b) the free length made of plastic tubing and connected to the sealed length by a waterproof coupling. Since the lower 3 or 4 m of the fixed zone of the tube resist almost the entire pulling force in compression, this part is made of higher-strength steel because of buckling potential.

The installation is completed in the following sequence: (a) drilling the borehole and installing the outer tube; (b) sealing the fixed anchor zone outside the tube using pressure grouting through the sleeves; (c) installing

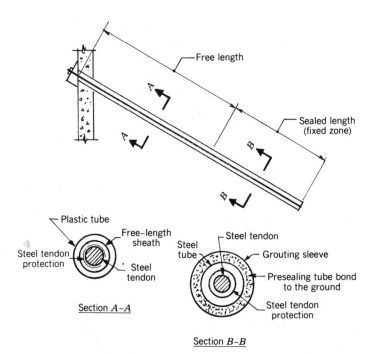

Fig. 3-5 Schematic presentation of a compressed bond anchor.

the steel bar and mechanically attaching it to the lower end of the tube; and (d) tensioning, testing, and lockoff. The manufacturer of this anchor claims that complete protection is afforded to the steel bar (Pfister et al., 1982), but as an additional precaution the annular space between the tube and the bar is filled after lockoff with a cement-base grout. If necessary, the steel tendon may be removed (by unlocking it from the tube) after testing for additional pressure grouting to increase bond at the ground–grout interface.

This anchor is commercially available in bar diameters of 26 and 32 mm.

Compressed Tube Anchors in Rock

The mobilization of bond along the rock interface of a tube in compression to transfer forces from the anchor tendon to the rock medium is based on the principles discussed in the foregoing sections. The concept of the compressed tube anchorage is shown in Fig. 3-6.

Improved bond characteristics of a grouted tube subjected to confined compression within a rock medium, as compared with a bar or tube in tension, are related to Poisson's effect manifested by lateral expansion or contraction. This effect becomes significant when the axial stresses in the steel tube approach the yield limit and Poisson's ratio increases from 0.3 to 0.5 (Ivering, 1981).

From an elastic analysis of an embedded tube it is known that the transverse deformations of the wall both outside and inside are positive when the tube is subjected to compression. This is shown in laboratory tests confirming that failure is approached as the walls begin to bulge, as shown in Fig. 3-7. When the axial compression stress enters the range beyond the yield limit, a considerable amount of resistance is still offered by the tube. On the other hand, owing to the lateral expansion of the tube in compression, the slip between the tube and the enveloping grout is significantly reduced. The analytical approach to this subject is discussed in subsequent chapters treating the load transfer of anchors.

A commercial version of compressed tube anchor is shown in Fig. 3-8. The load transfer to the anchor medium is intended primarily through deep corrugations on the walls of the tube. The design and presence of grout

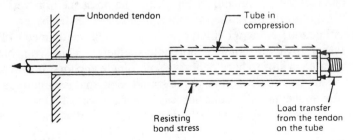

Fig. 3-6 A compressed tube anchor in rock; schematic presentation.

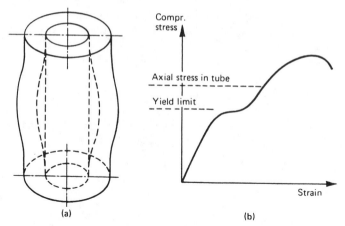

Fig. 3-7   Plastic failure of a thick-walled tube in compression; (a) local bulging; (b) failure mechanism and stress–strain diagram showing the range of axial stress. (From Ivering, 1981.)

injection vents allows the annulus inside the tube to be also grouted simultaneously with the grout injection of the outside space, suggesting that part of the load transfer may take place along the anchor length.

For added protection the tendon can be covered with a paste and inserted into a plastic sheath. After stressing, the tendon can be removed for inspection.

When sound rock is available, the compressed tube anchors offer two advantages: (a) they require a smaller borehole for the same size and tendon capacity, whereas the fixed bond (transmission) length can be shorter; and (b) the appearance of tension cracks in the grout is less likely because of the inherent compression, which improves protection against corrosion.

3-3   MULTIBELL (UNDERREAMED) ANCHORS

These deviate from conventional anchors in the method of forming the fixed anchor zone (see type D anchor in Fig. 2-2). The presence of reverse cones or bells of solid grout alters the mechanism of load transfer by producing a shear resisting zone along a cylinder projected along the diameter of cones or bells rather than the nominal borehole diameter. This arrangement is better attainable in stiff to hard cohesive soil. Single anchors, however, with underreams have been cut and tested in sound rock, and techniques are also available for applications in cohesionless materials. Satisfactory resistance to pullout has been obtained with the use of underreamed anchors in sandstone overlain by difficult and varying layers of silt, clay, gravel, and waterbearing ground (Soletanche, 1970). After drilling through the overbur-

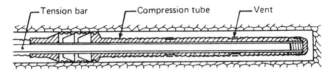

Fig. 3-8 Compressed tube anchor by Stump Duplex system (Weber, 1966); available for applications in rock.

den, a 5-m penetration into rock with underreams produced a sustained load capacity 2000 kN (450 kips) in the presence of wet marl.

Borehole drilling and underreaming requires special devices and tools. For a given soil type and anchor capacity, the underreams should be formed with complete control of the correct bell spacing, number and dimensions. The construction shown in Fig. 3-9 begins with borehole drilling using a rotary drill with continuous-flight auger in clay strata. In the fixed zone, the underreaming device consists of an expandable cutter tool with several (as many as eight) hinged blades mounted on a chassis attached to the bottom end of a drill string. The string is inserted into the borehole, already drilled to the correct anchor length, and while the tool is slowly rotated, the blades are forced to open gradually expanding the underreams to the required size.

It is better to have as many underreams formed simultaneously as practicable in order to avoid delays in the grouting stage and prevent ground softening or cavitation. The clay cuttings produced in the process are brought up by circulating flushing water. Prior to grouting, physical tests can confirm the complete removal of spoil from the underream space. The grout should fill the annular space continously and completely, since any ungrouted pockets in the hole will reduce load transfer and capacity.

Underream dimensions, spacing and number are chosen with respect to load capacity as well as practical considerations. The latter include anchor inclination and depth, soil properties, and availability of equipment at the site. The spacing, number and size of blades should be consistent with the optimum combination of underreams. A suitable underreaming device can be opened and closed mechanically at a controllable rate, sufficiently slow so that spoil and cuttings do not lodge in excavated space. Figure 3-10 shows the shape and dimensions of a multibell underreaming device chosen for an anchorage installation in London clay, with attained working loads up to 800 kN (180 kips).

3-4 REGROUTABLE ANCHORS

In relatively poor soils a single grout injection carried out within the pressure limits discussed in Section 2-10 will produce anchors of moderate capacity. This operation can be combined with pregrouting to consolidate the soil and also avoid loss of sealing material during normal grouting injection. How-

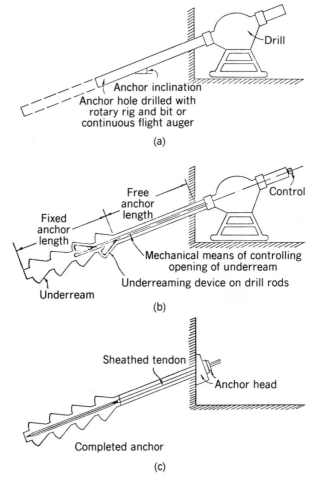

Fig. 3-9   Multibell underreamed anchors: (a) drilling; (b) underreaming; (c) homing (completed anchor).

ever, in adverse soil conditions involving plastic clays or marls, loose silts, badly fissured rocks, and soils of poor mechanical properties, problems are conceivable in the transfer of load as soon as the first prestressing is applied. These problems will persist if there are no remedial measures and provisions for additional grouting.

Regroutable anchors are intended to provide this remedy, and thus reconcile poor soil conditions with a higher-capacity system. This application requires a special grouting technique that allows the anchor to be grouted in stages, with increasing pressures and controlled grouting lengths at any time before or after the initial prestressing of the tendon. After the first grout has set, a second grouting stage follows under much higher pressures, which

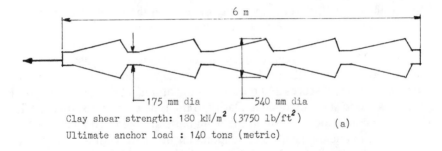

Clay shear strength: 180 kN/m^2 (3750 lb/ft^2)   (a)
Ultimate anchor load : 140 tons (metric)

Fig. 3-10 Multibell underreamed anchor; (a) dimensions, details, and ultimate carrying capacity; (b) extracted section of the anchor showing the underreams. (From Bastable, 1974.)

can be further increased for a third application. Thus, an anchor failing its first load test can be regrouted to restore its load capacity or increase it.

The concept of a load-to-failure test after initial grouting, to be followed by regrouting, has the advantage of allowing the factor of safety to become equivalent to the ultimate factor of safety since the load used to measure this factor is actually applied to failure. In this context, any small movement of the anchor can be induced by applying a measurable pull in stages, after which the anchor regains its capacity by regrouting.

Flexibility in the grout injection process with respect to location of grouted areas and volume of grout ensures the optimum impregnation of the anchorage zone with the correct grout constituents, whereas stage grouting allows soil consolidation at higher pressures. Reported capacities of regrouted anchors are from 50 to 250 tons (100–500 kips), and are attained in horizontal or angular inclinations.

Apart from the physical and mechanical improvement afforded to the anchor system, regrouting offers also the following advantages: (a) opportunity to inject small fissures that may develop in the grout column during tensioning, (b) a twofold protection of the steel against corrosion, and (c) feasibility of sectioning the core for the purpose of extracting the free length.

Figure 3-11 shows a regroutable anchor developed and used by Soletanche, known as the IRP anchor (an abbreviation for "injection repeteés en pression"). The fissures and root formation attained by the injected grout in the fixed zone show distinct similarity to the type C anchor shown in Fig. 2-2. As in a conventional system, the main parts are the fixed zone, the free length and the anchor head. The tendon can be made with a bar, wire or strand, but the details are different for each type as shown in Fig. 3-12.

For the IRP anchor, a centrally located sleeve grout pipe or tube fitted with perforations and grout valves runs along the entire length. The anchor head can be positioned outside the supported structure or incorporated in the concrete wall. Along the free length, wires or strands are held by spacers around the central grout pipe, and the entire assembly is enclosed in a plastic casing that allows free steel elongation during stressing. The annular space inside the plastic casing in the fixed length can be filled with neat grout after stressing, and if the anchor is permanent cement grout is injected inside and outside the casing for corrosion protection. If the tension steel must be withdrawn for inspection, the inside of the casing is filled with anticorrosive deformable resin.

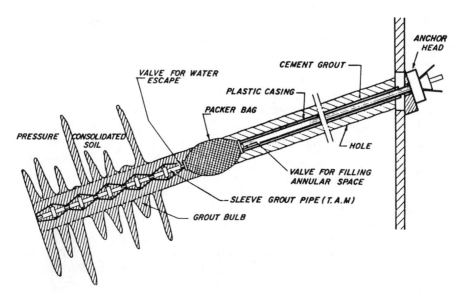

Fig. 3-11   Schematic presentation of regroutable anchor; the IRP anchor used by Soletanche.

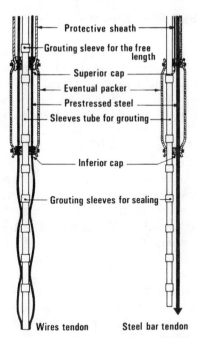

Fig. 3-12 Regroutable IRP anchors; details for wire and steel bar tendon. (Soletanche.)

Inflatable Packer. A special feature of the anchor shown in Fig. 3-11 is the inflatable packer bag, which separates the free length from the fixed bond length. The bag is 1 m long. When in place it is inflated, and thereafter grouting of the fixed zone is carried out through valves in the inner plastic grout pipe, which open to expose the perforations through which the liquid grout flows (Jorge, 1970). The grout pipe is provided with a double-packer system that allows the grout in the fixed zone to be injected in 1-ft lengths.

After the first grouting, the inner plastic casing is flushed and cleaned so that the grout pipe can be reinserted at a later time to perform stage grouting at a higher pressure. In general, two grout injections are sufficient, but it is feasible to repeat stage grouting up to pressures of 500 psi (3.5 N/mm^2) until the specified anchor capacity is attained. The grout mix consists of cement and admixtures for early strength development.

Since the main advantage in this case is ability to regrout after installation and first tensioning and progressively increase anchor capacity, the regroutable anchor is adaptable to a wide variety of conditions. However, installation costs will vary and increase according to the number of grout injections, grouting pressure, anchor capacity, and soil conditions. Thus, regroutable anchors should be expected to cost more, but this can be justified where other solutions are not feasible.

3-5 PRESSURE BULB SOIL ANCHORS

These are essentially similar to the type *B* anchor shown in Fig. 2-2. They are suitable where loose sand or gravel endanger the stability of an open hole after drilling. The usually attained ultimate load capacity in these soil conditions is 300 kips. Whereas a bulb anchor is more often considered temporary, it can be used on a permanent basis by providing double corrosion protection.

Pressure bulb anchors can be installed in two ways. First, a casing is driven to the required depth by an air track. The tension tendon is inserted in the casing, and as the latter is retracted the hole is pressure grouted using pressures generally less than the total overburden pressure. In the second method, the tension member is attached to the stem of a hollow-stem auger, and is advanced simultaneously with the hole. The auger is then withdrawn and the hole likewise grouted.

A commerically available pressure bulb anchor is shown in Fig. 3-13. It consists of $\frac{1}{2}$-in-dia., 7-wire strand with an ultimate tensile strength 270,000 psi, produced and tested according to ASTM A-416. Each strand has thus guaranteed ultimate strength 41,300 lb and a yield strength at 0.1 percent extension 35,100 lb. It is available with 4 or 7 strands, with casing or auger center hole diameter $2\frac{1}{2}$ and 3 in, respectively.

Tubfix Anchors. These combine the features of pressure bulb anchors with regroutable and compressed bond systems. The construction is carried out according to the tube-a-manchette technique, and the tube–ground bond is improved by repeated grout injection (Compte, 1971).

The installation involves the following stages, shown in Fig. 3-14(a):

1. Drill the anchor hole.
2. Insert a steel tub provided with slots and rubber sleeves over the fixed (bond) length.
3. Backfill the hole with a cement–bentonite mix under gravity to ensure hole stability for the subsequent stages.
4. Apply multistage cement pressure grouting using a double packer, to form the fixed zone. The grout will replace the cement–bentonite slurry in this area by gravity.
5. Allow the cement grout to harden, usually for 4–14 days.
6. Stress the steel tube by pulling, test and lockoff at specified load. Backfill the steel tube in the free length with a cement mix.

The tubfix method can also be applied with the same steps shown in Fig. 3-14(a), except that after stage 5 a steel bar is inserted in the tube and sealed at the lower end. The system in this manner is similar to compressed bond anchor. This arrangement is shown in Fig. 3-14(b).

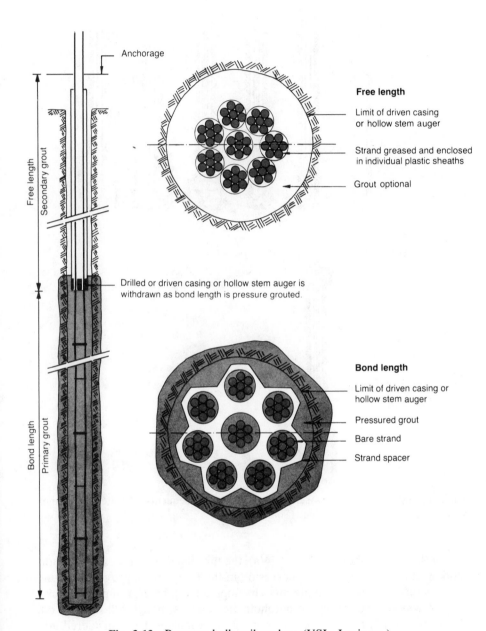

Fig. 3-13 Pressure bulb soil anchor. (VSL–Losinger.)

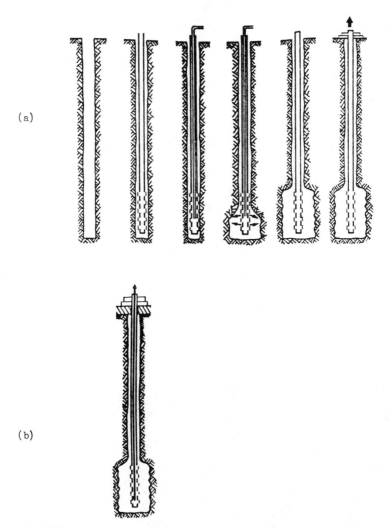

Fig. 3-14 (a) Tubfix anchor—schematic construction sequence; (b) alternate design with both a tube and a steel bar. (Soletanche.)

For the anchor shown in Fig. 3-14(a), the tube itself constitutes the anchor tendon. Its protection against corrosion in the free zone is provided by the surrounding cement–bentonite mix enveloping the tube as a ring, the thickness of which depends on the borehole diameter. Although full reliance on this protection is not justified, in normal environment it is considered adequate for long-term service because of the extra thickness of the tube wall and its low grade of steel (about 40 kg/mm^2 elastic limit), which reduces the severity of corrosion attack. With the modified version shown in Fig. 3-

14(b), the steel bar is fully protected in the tube and covered by a cement grout filling the steel tube in stage 6.

3-6 ANCHORS FOR SPECIAL CONDITIONS

Remedial Anchors. When an anchor fails to carry the test load, the result is downgrading of the working load and overstressing the remaining anchors in the group. This problem can be avoided if one or more additional anchors are installed to rebalance te load distribution. In these conditions, it becomes necessary to drill new holes between anchors already in place.

This is feasible if (a) the spacing of anchors already in place allows re-positioning of equipment and the associated installation activities, (b) a hole can be cut in the wall to accommodate the drill and anchor head assembly, and (c) the time for construction and grouting of the new anchor can be kept to a minimum. A further factor to be considered is the care that can be reasonably exercised during the operation to avoid disturbance that can damage or distress existing anchors.

Anchoring in Weak Soil. Anchoring techniques in weak or poor soil must enable the following: (a) drill a hole with minimum ground disturbance and (b) produce an anchorage zone that can resist the same load as the tendon. Among the drilling methods available, rotary tools are more versatile since they can facilitate fully cased holes to almost any desired depth and with limited access or headroom. These machines also can drill below the water table, and preserve cuttings for inspection and confirmation of soil strata.

In soils ranging from old river alluvium to loose sand, marls and soft clays, careful consideration of the anchor types discussed in the foregoing sections can ensure anchorage zones suitable to resist loads of sufficient intensity to make the installation cost effective. Anchor assemblies consist of an external metal tube for the length of the fixed zone, whereas parallel wires and strand constitute the tendon. The external tube is bonded to the soil by grouting over its full length in two phases. First, low-pressure grouting is used to seal the lower part of the anchorage, after which the tube is bonded to the soil by a second phase grouting. The anchor tendon is introduced either by injection or mechanically.

Buttonhead Anchorages. These are examples of anchor systems suitable within a specific design context. Buttonhead anchorages are commonly made up of high-tensile steel wire in standard sizes of 5, 6, and 7 mm dia. (in the United States $\frac{1}{4}$ in), witih a minimum ultimate tensile strength of about 1570 N/mm^2, and arranged in parallel lengths. There can be several (as many as 30) wires in a tendon. Each wire terminates in a cold-formed buttonhead, after it passes separately through a machine-finished anchorage fixture.

When used as a rock anchor, the assembly is provided with a fixed head anchor at one end, and a stressing head at the other. A leather diaphragm acts as grout seal, whereas a grout injection pipe is placed between the diaphragm and the fixed end anchor. The assembly is inserted in the rock with the fixed end first, and grout is injected to fill the space up to the diaphragm. After the grout has set the temorary injection pipe is removed, and the tendon is stressed through the simultaneous stressing of all the wires. The use of buttonhead in this case allows the simultaneous development of ultimate force in each wire.

Figure 3-15(a) and (b) shows anchor view and schematic presentation of the BBRV system. All individual wires have buttonheads bearing against a special head. The assembly has a supporting ring screwed over it and bearing against the load distribution plate as shown. During stressing, an extension part is coupled to the threaded head. After stressing the support ring is

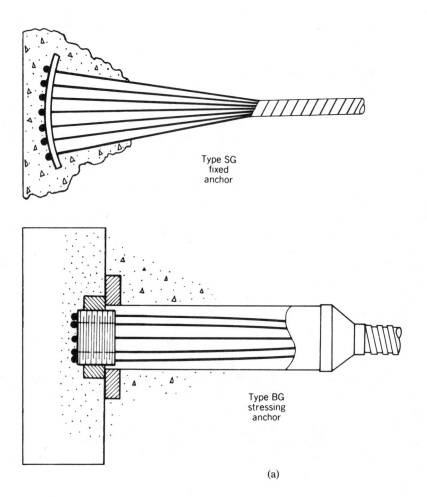

Type SG
fixed
anchor

Type BG
stressing
anchor

(a)

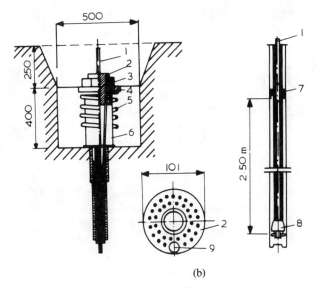

1 - grout pipe, 2 - anchor head, 3 - supporting ring, 4 - load distributing plate,
5 - spiral reinforcement, 6 - sheath of the hollow for the insertion of head
prior to the stressing of the cable, 7 - sealing collar, 8 - anchor base,
9 - deaerating pipe.

Fig. 3-15 Example of buttonhead anchorage—the BBRV system; (a) anchor view;
(b) schematic presentation.

screwed up to the bearing plate with minimum slip and sealing loss, which
allows to mobilize maximum potential force in every tendon.

The button head system claims a higher initial cost per unit weight of
tendon, but very often it yields a lower cost per unit of force delivered
because of wire length uniformity and uniform stress distribution in the tendon. Where large forces must be attained, this system should be given ample
consideration.

3-7 CAISSON-TYPE ANCHORS

These are large-diameter tiebacks installed in cohesive soil. The method is
quite similar to he caisson procedure, also known as "bored piles," commonly used in conventional foundation work. Even though large-diameter
anchors do not offer any technical advantages, their choice is often dictated
by regional trends and contractor availability. The hole is drilled with a
conventional digger equipped with a kelly bar with hydraulic crowd, which
is useful in rapid advancing of the hole. Anchor inclination can be as high
as 30° to the horizontal. For deeper holes (≤100 ft) telescoping kelly bars
are necessary (White, 1970).

Anchor hole diameter ranges from 300 to 400 mm. The usual drilling tools are flight plain augers, hollow-stem augers, and belling buckets for underream work. The latter can enlarge the shaft to two to three times the original diameter. This configuration is commonly used in cohesive soils, hence it is very popular in the Los Angeles area, where this soil type prevails.

These anchors are also installed in caving formations. In this case, a large drill is combined with a continuous flight hollow-stem auger, which has the advantage of casing the hole. The rod tendons are inserted into the hollow stem before the auger is withdrawn, and as the latter is pulled out without rotation the hole is grouted filling the cavity being formed.

A plate with threaded nut attached to it remains at the bottom of the hole and bears against the hardened grout as shown in Fig. 3-16. In order to keep the bearing plate from being pullout with the auger, a short length of the continuous flight auger is also welded to the bearing plate, and in this manner it acts as screw anchor.

3-8   THE INJECTION BAUER ANCHOR SYSTEM

This anchor is one of the early cement grout injection types to be used in Europe and subsequently in North America. A special feature is the use of a single high-strength steel bar that is inserted into a hollow rod. Drilling is possible by percussive, rotary, or rotary–percussive tools (Bauer, 1966).

Using lost drill bits, hollow drill poles are driven to the required length while flushing is used to clean the hole and enlarge it for better bond. The tie bar is then inserted and connected to the drill bit, which forms an integral part of the anchor. With the entire assembly in place, the poles are pulled back slightly by the drill bit allowing the cement grout injection process to begin and fill the annular space between the rod and drill pole. The drill poles should fit tight against the surrounding soil so that even at high injection pressures (often 500 psi) injected material does not flow back along the outer wall of the poles.

As the injection process continues the poles are retracted by a certain length so that the outlet opening is positioned for the injection of the next section. Higher pressures are used in the fixed anchor zone where some cement grout will penetrate the soil. As the drill poles are recovered, the remaining of the borehole in the free length is filled with lean cement grout under a low injection pressure as protection against corrosion.

Since the tie bar is not inserted until the drill pole has reached the bottom, it is possible to assemble the latter in single-tube sections. On the other hand, the advancing tube is only slightly larger in diameter than the steel bar itself, and this arrangement results in borehole sizes of relatively small diameter, usually 8–14 cm (3–6 in). The structural integration of the tie rod and the drill bit allows the transfer of load both by bond and by direct bearing

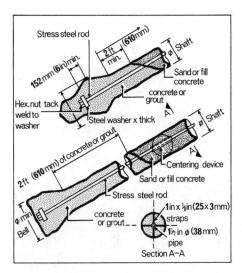

Fig. 3-16 Caisson-type anchor used in the United States for cohesionless soils.

on the grout column, hence in this respect the Bauer system is similar to the compression anchor described in the foregoing sections.

Initially, production and application range of this anchor were confined to loose sand and gravel. For 1-in-dia. bars, the average working load is 30 tons (60 kips) and requires a load-transfer zone length from 10 to 15 ft. The same anchor has been successfully used in firm cohesive soil, where failure loads up to 80 tons (160 kips) have been sustained using larger holes (≤ 15 cm) and higher injection pressures. A modified version of the Bauer anchor used in the construction of the Second Mersey Tunnel in England is schematically shown in Fig. 3-17.

3-9 VERTICAL ANCHORS

Frequently, ground engineering problems rely on solutions that are based on the use of vertical anchors installed in soil or in rock. Examples are soil preconsolidation, structures and foundations subjected to uplift or built in swelling soils, and special classes of structures that must resist overturning moments.

Most of the anchor systems described in the foregoing sections are suitable for this application in conjunction with vertical drilling techniques. Equally practical and economical, provided suitable ground conditions exist, is the use of conventional drilling tools to advance a hole to the desired depth and install a shaft provided with an expanded base. The drilled hole is then backfilled with cement grout or lean concrete. Anchors installed in this man-

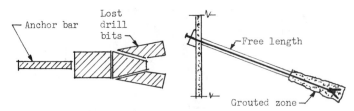

Fig. 3-17 Schematic presentation of modified Bauer anchor used in the Second Mersey Tunnel in England.

ner transfer the load to the ground by shear resistance along the walls of the borehole and direct bearing on the expanded base.

Burried Expansion Plate and Shaft Anchor in Soil. Figure 3-18 shows a commercial vertical-type anchor suitable for installations in soil. This anchor has a head and a shaft as in conventional anchorages. At the base, however, it has two plates directly above one another, with a diameter 6 in or larger. The installation involves drilling a hole of the same diameter or

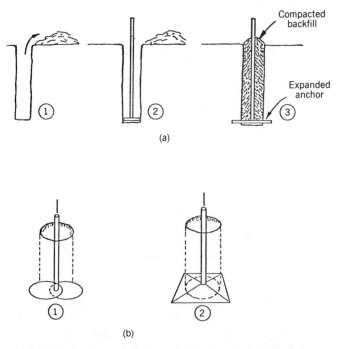

Fig. 3-18 Buried expansion plate and shaft anchor: (a) installation procedure (1—drill hole, 2—insert anchor, 3—expand anchor and backfill); (b) available types (1—eccentric plates, 2—expanding screwdown plates).

slightly larger. After homing, the head and shaft are rotated by 180°, and this transmits the same rotational movement to the bottom plate relative to the plate above. This is possible by keeping the bottom plate fixed to the shaft, and the top plate detached from it. On completion of this rotation, the initial set of two concentric plates has expanded to a set of two overlapping circular plates covering a length approximately 1.5 times their diameter, as shown in Fig. 3-18.

A second type, shown in Fig. 3-18(b) as an expanding screwdown plate, is likewise installed with similar drilling methods. Once in place, the anchor rod is twisted, causing the bottom part to expand into undisturbed soil. The installation is completed by placing and compacting backfill materials or cement grout.

Expandable Rock Anchor. An example of this anchor type is shown in Fig. 3-19. In this case, the hole is drilled slightly larger than the assembly in the closed position, using auger drills or other suitable bits to penetrate existing rock material. After the hole is cleaned, the anchor assembly is-

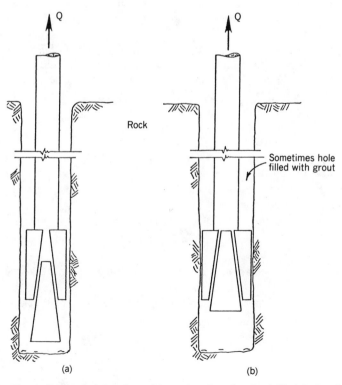

Fig. 3-19 Expandable rock anchor; (a) anchor homing in drilled hole; (b) anchor with components in expanded position.

inserted and positioned in the hole. Better results are achieved if the anchor is centrally located within the hole. The anchor head is then rotated forcing the two anchor components to slide on each other thereby increasing their overall diameter. The rotation of the anchor continues until the specified installation torque is reached and the expandable sections of the anchor bear and fit tightly against the sides of the hole. The installation is completed by grouting the borehole.

REFERENCES

Bastable, A. D., 1974: "Multibell Ground Anchors in London Clay," *Prest. Concrete Found. and Ground Anchors*, 7th FIP Cong., May, New York, pp. 33–37.

Bauer, K., 1966: "The Injection Bauer Anchor System," *Trans. Civ. Eng. J.* (47), 1265–1273, Zurich.

Compte, C., 1971: "Technology of Tiebacks," Inst. for Engineering Research Found., Kolibrunner-Rodio.

Ivering, J. W., 1981: "Development in the Concept of Compression Tube Anchors," *Ground Eng.*, March, Found. Publ. Ltd., Essex, pp. 31–34.

Jorge, G. R., 1970: "The Regroutable IRP Earth Anchor Tiebacks," Specialty Session No. 15. *Anchorages Especially in Soft Ground, Proc., 7th Int. Conf. Soil Mech. Found. Eng.*, Mexico City.

Jorge, L. L., 1969: "Devices to Destroy in Place Temorary Anchorages, Rods, Wires, or Cables," Specialty Session No. 15, *Anchorages Especially in Soft Ground. Proc., 7th Int. Conf. Soil Mech. Found. Eng.*, Mexico City.

Pfister, P., G. Everson, M. Guillaud, and R. Davidson, 1982: "Permanent Ground Anchors, Soletanche Design Criteria," Fed. Highway Admin., U.S. Dept. Transp., Washington, D.C.

Rey, H. P., 1978: "Extracting of Soil Anchors," VSL Intra-Co. Symposium. Soletanche, 1970: "Various Types of Anchors," *Ground Anchors, Cons. Eng.* (May), London, pp. 12–15.

Soletanche, 1970: "Other Types of Anchors," *Ground Anchors, Cons. Eng.* (May), London, pp. 12–15.

Straus, R., 1978: "Totally Removable Anchors," VSL Intra-Co. Symposium.

Weber, E., 1966: "Injection Anchor, the Stump Bohr A.G. System," *Schweizerische Bauzeitung,* **84** (6, Feb.), 11.

White, R. E., 1970: "Anchorage Practice in the United States," *Ground Anchors, Cons. Eng.* (May), London, pp. 32–27.

Yamada, K., 1978: "Ground Anchor Market in Japan," Paper Prepared for VSL Intra-Co. Symposium.

CHAPTER 4

THE TRANSFER OF LOAD AND MODES OF FAILURE

4-1 GENERAL CONSIDERATIONS

From the theoretical and practical standpoint, understanding the mechanism of load transfer is essential for the design of an anchorage that has adequate factor of safety and satisfies implicit economic criteria. This topic must be analyzed in the context of theory, and then confirmed by empirical record. Caution, however, is necessary to distinguish and identify the limitations of our present knowledge.

Anchor load transfer theories often are based on idealized assumptions, and where conditions are different results can be misleading and questionable, and this is more serious when theory is freely applied to nonhomogeneous ground conditions. A different approach is to infer the transfer of load from the wide variety of design rules available, all claiming origin from full-scale tests and general field experience. Since no empirical procedure can claim general applicability, routine reference to these rules can often lead to crude approximation of safe and economical design.

Several topics are still partly investigated and incompletely understood, hence they present a deterent to rational design. Such topics inlcude load transfer mechanisms in nonidealized media, grout pressure limits beyond which the associated effects are minimal, fixed anchor load–displacement relationships with parameters including all variable causes, and serviceability safety factors meaningful and consistent with test and actual loads. The effects of construction techniques and workmanship on anchor pullout capacity are quite obvious and inhibit our efforts to make accurate predictions for field performance. Thus, anchor design is supplemented by a mandatory testing program to confirm the load carrying capacity of an anchorage.

The Concept of Failure. In general, theories and design methods assume that a mass of soil will fail along slip lines or shear planes, postulating a failure mechanism, and then engage the relevant forces in a stability analysis. For the fixed anchor zone configurations presented in the foregoing sections, two basic load-transfer mechanisms cause ground resistance to be mobilized as the anchor undergoes displacement under load application. The first is side shear (adhesion or friction), commonly called "bond," to be followed by end bearing where suitable configurations exist and when sufficient movement occurs. Accordingly, anchors can fail in localized shear as long as the continuity of the surrounding ground is not disrupted. General failure occurs when the shear planes are fully mobilized or under significant deformations progressively reaching the ground surface. The latter, however, is very unlikely for slenderness ratios exceeding 15, which is commonly the case with the small anchor diameters required for this work (Littlejohn, 1980).

In general, the analysis of the load resistance of an anchor must consider the following:

1. Mechanism of failure as load is transferred from one medium to another in the anchor–soil system.
2. Ground characteristics at failure.
3. Area roughness and configuration of potential failure interfaces.
4. Stress conditions, namely, type of stress, magnitude, and direction, occurring along the failure interface when failure is initiated.

An anchor can fail or become inoperable in one of the following modes:

1. By structural failure (rupture or sectioning) of the steel tendon and its component parts.
2. By bond failure (slippage) at the tendon–grout interface.
3. By shear failure along the contact surface of grout and ground.
4. By failure within soil or rock supporting the anchorage.
5. By crushing or bursting of the grout column around the tendon.
6. By displacement or excessive slippage of the anchor head.
7. By gradual long-term deterioration rendering the system inoperative.

Under overloading or during pullout tests, any of these failure mechanisms may prevail or take precedence, hence it is clearly not feasible to design, proportion, and construct anchors in which all parts, when intensionally overloaded, will collapse or fail simultaneously. Usual anchor practice dictates selection of anchor components and analysis of potential failure modes under an appropriate factor of safety, consistent with the actual known strength or the associated degree of risk.

The Concept of Anchor. Anchor types to which this section applies are the standard forms of tendon discussed in Chapter 2: bar, wire, and strand. Unless otherwise noted, the present analysis is valid for the prestressed cement grout injection anchor with straight borehole, including multistage pressure grouted, regroutable, and restressable anchors in soil and rock.

The Concept of Loading. Short-term static loading is considered in the context of this analysis. Excessive static loading will cause an anchor to fail. Excessive loads can be induced by pull applied to the anchor during testing or at lockoff stage. Overloading can also result from wrong excavation sequence, additional surcharge from construction materials and equipment, or adjacent excavation and operations depriving the ground anchorage zone of strength.

Short-term loading does not imply only loads of short duration. It does imply, however, that such loads cause failure that is not related to time-dependent effects. The anchor will fail as soon as excessive load is applied irrespective of its duration; hence the latter is not a factor in the analysis of the cause of failure. Time-dependent effects are discussed with creep.

Unlike short-term loading, repetitive or cyclic loading has time-dependent effects. These can cause decrease in strength or even failure, and are analyzed under different procedures (see Section 4-11).

The Concept of Anchor Design. For preliminary purposes, the design of anchorages may be confined to simple determination of an upper limit of fixed anchor length, and then assume boundary conditions at failure to confirm its adequacy. This may be sufficient for determining the suitability of a proposed anchorage. Final design is much broader and may include the following objectives:

1. Select anchor inclination.
2. Identify suitable tendon types, size and configuration.
3. Determine horizontal spacing of anchor heads, and vertical distance of anchor rows.
4. Estimate fixed and free anchor length.
5. Estimate anchor resistance to static or cyclic loading.
6. Specify a suitable anchor testing program.
7. Select and detail corrosion protection system.
8. Check the overall anchor–structure stability.

4-2 STEEL TENDON: FAILURE MECHANISM AND ANALYSIS

Failure of Steel in Tension

When a load is applied, the steel tendon is stressed in tension. If the load exceeds the strength of the material, failure will occur by excessive yielding followed by sectioning. A typical stress-strain diagram for normal high-tension steel is shown in Fig. 4-1. The curve has no definite yield point; however, there are three characteristic points. One is the ultimate tensile strength f_{pu}, also called "characteristic strength" in Section 2-5. The elastic limit T_G corresponds to a specific permanent elongation, in this case 0.1 percent used by the French Code. The third point is the elastic limit E_s, which defines the proportionality constant in the linear portion on the stress–strain diagram.

Allowable working load in the steel tendon is determined from characteristic strength f_{pu} or the elastic limit T_G with an appropriate factor of safety. Where the soil is susceptible to creep, tension in the steel may be further limited to prevent large creep deformation at anchorage level, and this can result in allowable load less than the normal working load. Further reduction in working load is not warranted (for example to control cracking and ensure crack-free sections). The relatively low modulus of rupture of ordinary grout $(0.7 \sqrt{f_c'})$ implies that this strength is reached and exceeded quickly and at fraction of the working load; hence cracking in the grout column is likely to occur at an early stage of load application.

Certain codes specify allowance for possible reduction in load-carrying capacity of the steel tendon at the head, end block, and connection points (couplings) generally associated with some yielding at these locations. Smaller working loads are also justified if conditions unusually severe are anticipated and can cause reduction in the effective cross-sectional area of the steel with time. These topics are discussed in detail in subsequent sections.

Analysis Considerations

In spite of the emphasis generally placed on bond failure, yielding and fracture of steel tendons is not uncommon. This will happen especially where the grouting process attains excellent workmanship, which, combined with ample soil strength, prevents failure in these media, and where large unexpected loads can stress the tendon beyond capacity.

Based on a rational prestressing doctrine, it is good practice to limit the maximum temporary (test) load to a fraction of the guaranteed ultimate tensile strength. The ACI Code, for example, specifies maximum temporary load at 80 percent f_{pu}, whereas some European codes are more conservative and limit the test load to 75 percent f_{pu}. In the United States, prestressing practice specifies lockoff or transfer load at 70 percent f_{pu}, which, with

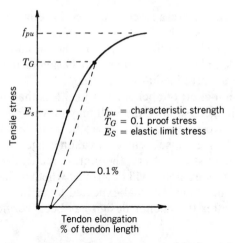

Fig. 4-1 Stress–strain response for normal high-tesnion steel.

nominal allowance of 15 percent for long-term relaxation, results in a final effective prestress force of 60 percent f_{pu}. This practice apparently should not be applied to anchors. Indeed, if the design stipulates anchor testing to 150 percent the working load (taken as 60 percent f_{pu}), it follows that the procedure is clearly inconsistent unless the test tendons are oversized.

Failure incidents of strand tendons have been reported at about 85 percent f_{pu} (Nicholson et al., 1982). By definition, as many as 5 percent in a group of anchors may have actual ultimate tensile strength 95 percent f_{pu}, and this is in accordance with most codes. If the deficient anchors in the group are tested to 80 percent f_{pu} and noting that no procedure is available to identify the deficient anchors, it is obvious that the tests will induce a high risk of failure. It is safer, therefore, to specify test loads that do not exceed 75 percent f_{pu}. A similar inconsistency arises where test programs of production anchors call for their stressing to 150 percent the design load but without preselection of test anchors. In this case, the working load should be not more than 50 percent f_{pu}.

It is always necessary to estimate beforehand the actual maximum resultant load that may act on the anchorage during service. However, it is poor practice, and often unsafe, to group these loads according to a particular category and increase the allowable stress according to the probability of occurrence of each group. Loads that must be included in the resultant force are pressures from the wall and soil response, potential surcharge, seismic forces, pressure from water-level fluctuations, and loads of various origins.

In certain anchored structures, for example, retaining walls, anchor loads increase considerably as excavation reaches final depth, particularly in the upper row of multitiered supports. In this case, the stipulated working load may be much greater than the resultant earth load in the initial stages. If the initially applied prestress force is quoted in terms of the working load (equal to or greater than this load), it will cause the wall to move toward the ground excessively, particularly where the ground is soft and the retaining system flexible. This problem can be avoided by a lockoff force consistent with the earth load at this stage.

However, final lockoff of anchors at a load less than the design working load is not advantageous, since the wall may move excessively toward the excavation. If this problem arises, it can be remedied by restressing the anchors involved at a subsequent excavation stage. Equipment that can apply prestressing in stages offers, therefore, practical advantages.

Similar design requirements will determine whether the lockoff force will be greater or smaller than the final expected load. In practice, the lockoff force is decided after the anchor design has been verified by testing (see also Chapter 5).

Working Stresses

In the past, the majority of codes recommended working stresses of the order of 62.5 percent f_{pu}, with a corresponding factor of safety 1.6. Recently, however, engineers recognized the necessity for more conservative design. Thus, according to current recommendations, $0.625 f_{pu}$ is the working stress for steel tendons in temporary installations, whereas for permanent anchors a working stress $0.50 \, f_{pu}$ is now accepted for design purposes. Thus, the factor of safety against ultimate tensile failure of the steel tendon is 1.6 for temporary and 2.0 for permanent anchors.

These working stresses are accepted under normal conditions, and are related to the potential failure mechanism of the steel tendon and its constituent materials. They should not be used to estimate working loads in terms of failure values determined from in situ tests or from soil creep characteristics. On the other hand, caution is appropriate under special conditions, and if necessary the associated factors of safety should be increased.

Procedures for Selecting Tendons

If F_t is the maximum estimated tensile force in the direction of anchor, suitable anchor units are selected by correlating this force with the guaranteed limit f_{pu} or the elastic limit T_G, under the appropriate factor of safety.

This correlation is provided by the following expressions.

For permanent anchors

$$f_{pu} = \frac{F_t}{0.5} \qquad (4\text{-}1)$$

or

$$T_G = \frac{F_t}{0.6} \qquad (4\text{-}2)$$

Equation (4-1) is based on U.S. practice and most foreign codes. Equation (4-2) relates the design load with the 0.1 percent elastic limit T_G, and is used by the French Code.

Likewise for temporary anchors

$$f_{pu} = \frac{F_t}{0.625} \qquad (4\text{-}1a)$$

or

$$T_G = \frac{F_t}{0.75} \qquad (4\text{-}2a)$$

All symbols and notations designate forces rather than stresses. With values of f_{pu} or T_G calculated from the foregoing equations, anchor units are selected by reference to design tables and technical aids available from manufacturers and suppliers for bar, wire, and strand tendons in standard units and combinations. These aids also include data on tendon characteristics, cross-sectional area, commercial lengths, couplings, anchor-head details, and stressing equipment.

The simple diagram shown in Fig. 4-2 is an example of technical aid that allows preliminary anchor selection by direct reference. The use of the diagram is self-explanatory. Characteristic data, tendon properties, and anchor capacity should be confirmed and guaranteed by the tendon manufacturer or supplier. Figure 4-2 supplements the data shown in Fig. 2-7 (VSL–Lossinger, 1978).

Likewise, anchor selection can be made by reference to a table such as Table 4-1, if the design load is known. This table conforms to anchor design requirements contained in the French Code (Pfister et al., 1982). For example, for a design load F_t 90 tons (metric), or 200 kips, Table 4-1 offers two selections, strand unit 10T13 (10 strands, 13 mm dia.) and 7T15 (7 strands, 15 mm dia.)

Table 4-1 also shows that the elastic limit T_G at 0.1 percent elongation is 81, 88, and 89 percent f_{pu} for bar, strand, and wire, respectively. The elastic limit (point E_s on the stress–strain diagram of Fig. 4-1) is 71 percent f_{pu} for bar and strand, and 67 percent f_{pu} for wire, for steel conforming to French standards. Considering the three types of steel tendon, the average T_G value is about 85 percent f_{pu}. Thus, the average working load according to French practice is $0.6 \times 0.85 \ f_{pu} = 0.51 \ f_{pu}$, practically the same as for U.S. and other European codes (see also Section 2-5).

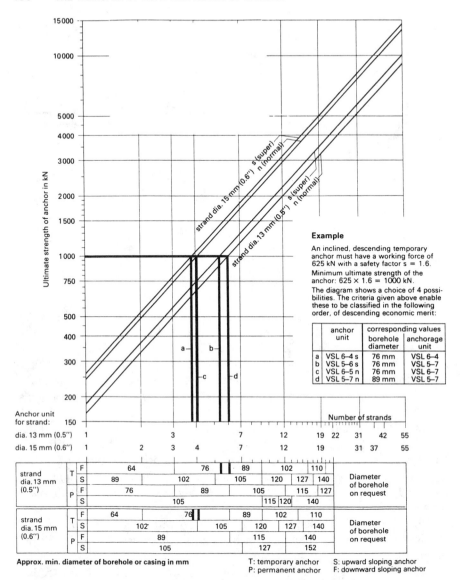

Fig. 4-2 Use of diagram for preliminary selection of suitable steel tendon unit. Strand diameter 13 and 15 mm (0.5 and 0.6 in). (From VSL–Losinger, 1978.)

4-3 FAILURE OF GROUT–TENDON BOND AND SAFE BOND LENGTH

General Bond Considerations

In conventional reinforced concrete, for plain bars bond resistance is caused by maximum bond stress over a short length where adhesion is about to fail and a lower friction drag over the length where adhesion has failed. Initially

TABLE 4-1 Mechanical Characteristics and Cross-Sectional Area of Bars, Strands, and Wires[a]

	Type	A (mm^2)	%/ton	E_s (ton)	T_G (ton)	F_{pu} (ton)	Class of Steel (French Standards) (kg/mm^2)	F_T (tons) for Permanent Anchors
Bars	Ø26DY[b]	551	0.086	41	47	58	85–105	28
	Ø32DY	804	0.059	60	68	84		41
	Ø36DY	1.018	0.047	76	87	107		52
Strands	1T13[c]	93	0.525	12	15	17		9
	2T13	186	0.262	24	30	34		18
	4T13	372	0.131	48	60	68		36
	6T13	558	0.087	72	90	102		54
	7T13	651	0.075	84	105	119	163–185	63
	8T13	744	0.065	96	120	136		72
	9T13	837	0.058	108	135	153		81
	10T13	930	0.052	120	150	170		90
	11T13	1.023	0.048	132	165	187		99
	12T13	1.116	0.044	144	180	204		108
	1T15	139	0.351	18	22	24		13
	6T15	834	0.058	108	132	148		79
	7T15	973	0.050	126	154	173		92
	8T15	1.112	0.044	144	176	198		105
	9T15	1.251	0.039	162	198	222		118
	10T15	1.390	0.035	180	220	247	153–175	132
	11T15	1.529	0.032	198	242	272		145
	12T15	1.668	0.029	216	264	296		158
	13T15	1.807	0.027	234	286	321		171
	14T15	1.946	0.025	252	308	346		184
	15T15	2.085	0.023	270	330	371		198
	16T15	2.224	0.022	288	352	395		211
	17T15	2.363	0.021	306	374	420		224
	18T15	2.502	0.019	324	396	445		237
	9T18	2.007	0.019	266	297	349		178
	12T18	2.676	0.014	354	396	465	148–194	237
Wires	6W8[d]	301	0.165	30	39.6	44.4		23
	8W8	401	0.124	40	52.8	59.2		31
	10W8	502	0.099	50	66	74	131–148	40
	12W8	604	0.083	60	79.2	88.8		47

ton = 1000 kg = 2205 lb

[a] Direct selection of suitable tendon unit fromt he design load F_T. Data conform to French Code. *Key to notation:* F_{pu} = ultimate tensile load; T_G = 0.1 percent elastic limit; E_s = elastic limit (proportionality constant); F_T = design load.

[b] Dywidag bars.

[c] Strands: 8T13 = 8 strands Ø13 mm.

[d] Wires.

From Pfister et al. (1982).

the bond strength depends on adhesion, but even after adhesion is broken friction between the steel and concrete continues to provide bond resistance.

With deformed bars this mechanism is changed. Adhesion and friction still assist the interaction of steel bars and concrete, but most of bond resistance is provided by the interlocks. The bond strength is provided primarily by bearing of the lugs on concrete and by shear strength of concrete along a column between the lugs. An idealized representation of bond in terms of its components is shown in Fig. 4-3.

For reinforced concrete the bond stress u is computed, assuming uniform distribution along the bond length, from the following expression

$$u = \frac{f_s A_s}{\Sigma_o L} = \frac{P}{\Sigma_o L} \tag{4-3}$$

where f_s = tensile stress in the steel bar
A_s = nominal cross-sectional area
Σ_o = nominal bar perimeter
L = embedded bar (bond) length
$P = f_s A_s$ = force acting on the bar

Slabs, beams, walls, and other structural members properly designed for flexure and tension generally are considered satisfactory in bond. Occasionally, however, bond is checked for compliance with maximum allowable values. American Association of State Highway and Transportation Officials (AASHTO), for example, stipulates that the maximum bond stress should not exceed the value given by the following expression

$$u_{max} = \frac{4.8\sqrt{f_c'}}{d} \tag{4-4}$$

but with an upper limit 500 psi (3.45 N/mm^2). According to Eq. (4-4), the allowable bond stress is proportional to the square root of the ultimate concrete strength f_c' and inversely proportioned to the nominal bar diameter d. For $f_c' = 4350$ psi or 30 N/mm^2, which is normal grout strength, and $d = 1$ in, the allowable bond stress from Eq. (4-4) is 317 psi or 2.2 N/mm^2.

In anchor work, the development of grout–tendon bond is often considered adequate for the load transfer, as emphasis is placed on the ground–grout interface. Relating tendon–grout response to conventional bond mechanisms, it is often assumed that any embedment length sufficient for the ground–grout interface load transfer also ensures sufficient bond length at the grout–tendon contact. However, the mode of failure of a tendon bonded into grout of in situ ground anchors is markedly different from the bond failure caused in pullout tests for conventional concrete work. Since in most ground anchors the grout is usually in tension, as is the steel, the mechanism

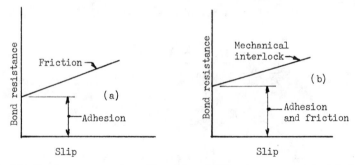

Fig. 4-3 Idealized representation of bond in reinforced concrete; (a) plain bars; (b) deformed bars.

of bond action greatly depends on the respective elastic properties of the steel and grout.

Guidelines and standards for permissible grout–tendon bond are, therefore, based on experimental data that relate specifically to known field conditions of anchorage installations. For preliminary analyses and rule of thumb estimates, recommended allowable bond stresses under proof load should not exceed the following values, based on grout strength 30 N/mm^2 (Littlejohn, 1980).

1.0 N/mm^2 (145 psi) for clean plain wire or plain bar tendon

1.5 N/mm^2 (220 psi) for clean crimped wire tendon

2.0 N/mm^2 (290 psi) for clean strand or deformed bar

These recommendations apply also to parallel multiple-unit tendons with clear spacing not less than 5 mm. For noded tendons that mobilize mechanical interlock or grout shear strength, the minimum spacing criterion does not apply. More guidelines are given in subsequent sections and in Table 4-3.

The Mechanism of Bond between Cement Grout and Tendon, Experimental Investigation

Figure 4-4 shows an enlarged view of grout–tendon interface. The surface irregularities are shown magnified, but in reality they are assumed large enough to cause all three bond interactions: adhesion, friction, and interlock.

Adhesion. This is developed as physical attraction that attaches the microscopically rough steel surface and the surrounding grout. The two media stick together, and the process is further promoted by molecular association. Breakdown of adhesion is assumed to occur over those portions of the tendon where bar slip corresponding to steel indentations dislodges the two mate-

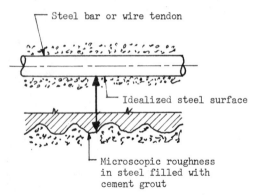

Fig. 4-4 Magnified view of grout–tendon interface showing normal microregularities on the steel surface.

rials and causes their detachment. This, however, occurs at a microscopic scale.

Friction. This mechanism depends on the lateral confining stress, the surface roughness of the steel, and the magnitude of slip. Friction decreases with increasing lateral tension, whereas the longitudinal tendon pull has no influence on the magnitude of friction. Longitudinal pull is, however, necessary for the condition of bar slip. There is also evidence that dilatancy and wedge action contribute to increased frictional resistance.

Mechanical Interlock. Deformed bar tendons resist pullout as the same bars in concrete. With plain tendons (bar, wire, or strand) mechanical interlock is manifested at all major irregularities such as ribs, twists, and couplings. Bearing failure against these projections can lead to splitting of the grout column or local crushing of the grout.

The three components of bond are shown graphically in Fig. 4-5 as function of slip. In this context, bond is a mechanism progressively compounded. For short embedment lengths and very small tendon slip adhesion prevails, but with increasing embedment and larger slip all three modes will be present.

Distribution of Bond

Figure 4-6 shows the progressive course of bond distribution at successive stages of pullout test (Gilkey et al., 1956). The curves in (a) represent bond stress intensities between steel and concrete. The curves in (b) show stresses in the bar at successive points along the bond length; hence they are stress distribution curves. Stress transfer is possible only by bond, which is present only in the region of changing stresses in the steel or the grout.

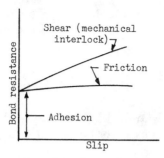

Fig. 4-5 Idealized representation of the three components of bond at grout–tendon interface.

The distribution of bond is developed under the following mechanism:

1. Bond resistance is first developed as cohesion at the proximal end. Only as slight slip occurs are tension and bond stresses progressively transferred distally.
2. As the pull increases, the bulk of bond resistance begins to move toward the distal end. Between the proximal end and the region of maximum bond concentration there is a fairly uniform friction drag of moderate intensity.

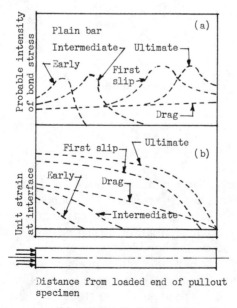

Fig. 4-6 Progressive course of bond distribution during pullout test: (a) variation of bond stress and migration with increased load; (b) variation of total tensile stress. (From Gilkey et al., 1956.)

3. "First slip" occurs only after the bulk of bond resistance has migrated along almost the entire bond length and is near the distal end.
4. After appreciable slip the primary adhesion disappears, and the force is counteracted solely by friction or drag resistance, which is roughly one half the ultimate resistance obtained (see Fig. 4.6).

Research work by Hawkes and Evans (1951) and later by Phillips (1970) has provided a theoretical basis for bond distribution expressed by the exponential function

$$u_x = u_o e^{-\frac{Ax}{d}} \qquad (4\text{-}5)$$

where u_x = bond stress at distance x from the proximal end
u_o = bond stress at proximal end
d = nominal bar diameter
A = empirical constant relating axial stress to bond stress in the anchorage material

Philips (1970) has shown that if P is the total tension force (equal to the sum of the total bond stresses acting on the bond area of the tendon), then

$$P = \frac{\pi d^2 u_o}{A} \qquad (4\text{-}6)$$

Combining Eqs. (4-5) and (4-6) we obtain

$$\frac{u_x}{P}(\pi d^2) = A e^{-\frac{Ax}{d}} \qquad (4\text{-}7)$$

Equations (4-5) and (4-7) are plotted in Figs. 4-7 and 4-8, respectively, which show the variation of shear stresses along the bond zone and its dependence on the constant A. For greater values of A, the stress concentration is at the free or proximal end, whereas for smaller values bond stresses are more evenly distributed along the bond length.

Values of A have been estimated for steel anchorages in concrete (A = 0.28) by Hawkes and Evans (1951), but more meaningful and relevant values for A should be forthcoming applicable to ground anchors. The factor A may relate to the physical tendon characteristics and vary with the tendon type (bar, wire, or strand).

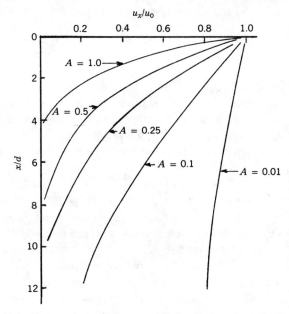

Fig. 4-7 Theoretical bond stress distribution along fixed anchor length under pullout load. (From Hawkes and Evans, 1951.)

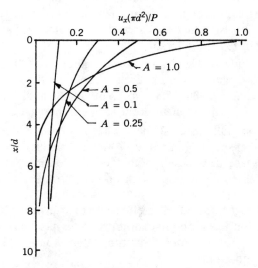

Fig. 4-8 Load distribution along fixed anchor length for various values of empirical constant A; valid for relatively large AL/d ratios. (From Phillips, 1970.)

Factors Affecting Magnitude of Bond

Littlejohn and Bruce (1977) offer the following comments on bond and the factors affecting its magnitude, applicable mainly to bar tendons.

1. Bond resistance, contrary to accepted belief, is not linearly proportional to the compressive strength of grout. Bond continues to increase with increasing grout strength, but in the upper grout strength range this increase occurs at smaller ratio. Specifically, for grout strength less than 21 N/mm² (or 3000 psi) bond appears to increase linearly with grout strength. The rate of increase is reduced and is almost zero at high values of grout strength. Thus, in the upper grout strength range (UCS > 42 N/mm² (or 6000 psi) no increase in bond should be accepted for additional grout strength.

2. If a given embedment is increased, the resulting additional bond is not proportional to the additional length. The average unit bond stress that can be sustained by a plain bar tendon is greater for shorter bond length. Hence, double bond length should not be expected to double bond resistance, but the increase may justify the extra length.

3. Variations in age and curing conditions alter bond resistance to a lesser degree compared to the same effects on compressive grout strength.

4. Information on the effect of bar spacing is limited, but for clear spacing between 1d and 3d the difference in bond is insignificant.

For wire and strand, bond is affected by the following factors:

1. The helical arrangement of exterior wires causes strand to rotate while undergoing slip through a grout channel, but the corresponding bond increase is not significant. A strand rotation of 15° has been observed after pullout tests (Anderson et al., 1964).

2. The magnitude of bond increases by about 10 percent for every 7-N/mm² (1000-psi) increase in grout strength in the strength range from 16 to 52 N/mm² (2300–7500 psi). Wire and strand, therefore, respond more favorably than bar to grout strength increase.

3. Results from pullout tests where wire or strand tendon was subjected to lateral pressure from zero to 17 kN/m², mechanically applied, show linear increase in bond strength. Tendon confinement in the grout is therefore essential, and grout shrinkage detrimental.

Effect of Rust and Coating. The effect of steel surface condition on bond has been studied in tests, and useful information has been disseminated by Kemp et al. (1968) for bars and Armstrong (1948), Base (1961), and Hanson (1969) for wire and strand.

Tendons are exposed first to the aggressivity of the air environment while they are stored prior to construction, and then to the aggressivity of the ground and its effects on the steel–grout interaction.

The appearance of deep flakey rust on bars, occurring usually after 6–8 months of exposure, blocks the contact surface and inhibits bond development to the same degree as in the unrusted tendon. This problem is prevented if all loose rust is wiped off from the tendon surface prior to homing. However, slightly rusted bars with exposure up to three months can develop bond resistance better than unrusted or wiped rusted bars. Loose powdery rust that appears on bars the first few weeks of ordinary exposure should cause no concern since it has no significant effect on bond.

Tendons should be kept clean, but caution is necessary in selecting cleaning methods since certain protective waxes and films from degreasing agents have deleterious effect on bond. A light, thin uniform surface rusting provides useful visual quality control.

Epoxy coated bars develop strength essentially similar to uncoated bars for film thickness less than 10 mils. Both liquid and powder epoxies perform equally well, and the application method does not significantly affect bond surface. Where bentonite slurry is used as the flushing medium, its effect on bond strength should be explicitly understood. This can be more serious for tendons where adhesion is the predominant bond mechanism. Bentonite can also be trapped around and against bar projections and irregularities with a subsequent effect on bond development.

Corrosion (see also subsequent sections), as it relates to cracking, can become prime deterent to the transfer of load. Corrosion is likely to start where the bar intersects a crack, and in the short term it will have significant influence on crack width. If cracking occurs around a ribbed deformed bar, at the end the force is primarily transferred from steel to concrete by mechanical action of the ribs as adhesion is largely lost.

Allowable Bond and Transmission Length

The allowable bond values and stresses recommended in this section will provide safe and conservative transmission lengths. A transmission length is the minimum length required to transfer initial prestressing to the grout column by bond.

More refined procedures allow estimation of embedment lengths in terms of bar diameter, but experimental evidence suggests that for small diameter ordinary strand the transmission length is not proportional to strand diameter. This is evident from Table 4-2, which gives transmission lengths for strand for loads 70 percent ultimate and for grout strength of 35–48 N/mm^2 (Littlejohn and Bruce, 1977).

Tests on Dyform compact strand embedded in grout with strength of 41–48 N/mm^2 have established average transmission lengths of 30–36 diameters at 70 percent ultimate load. Further variance in transmission length for strand is also shown in FIP (1974b) survey of codes and specifications worldwide. According to this survey, compact strand has transmission lengths 25 percent

TABLE 4-2 Transmission Lengths for Small Diameter Strand

Diameter of Strand (mm)	Transmission Length[a]	
	(mm)	(Diameters)
9.3	200 ($\pm$25)	19–24
12.5	330 ($\pm$25)	25–28
18.0	500 ($\pm$50)	25–31

[a] Range of results given in parentheses.

From Littlejohn and Bruce (1977).

higher than the normal 7-wire strand, and a sudden release of load will increase the transmission length by an additional 25 percent.

For single plain and deformed bars used as tendons, maximum allowable bond stresses are shown in Table 4-3, based on British standards. For a group of bars in the same unit a reduction factor is applied but without regard to group geometry and minimum spacing. This factor is 0.8, 0.6, and 0.4 for 2, 3, and 4 bars in the same group, respectively. From Table 4-3 it is seen that the allowable bond stress for deformed bars and grout strength of 30 N/mm^2 is 2.2 N/mm^2, which is the same value estimated from Eq. (4-4) for concrete of the same strength and bar diameter 1 in.

Bright or rusted, plain, or indented wire with small offset crimp (0.3 mm) may require transmission length of 100 diameters and for grout strength not less than 35 N/mm^2. With larger crimp (1.0 mm), the bond length may be reduced to 60 diameters. In the same context, about 80 percent the maximum stress is developed in lengths of 70 and 54 diameters for the small and large crimp wire, respectively. Galvanized treatment can reduce bond to less than half. Furthermore, tests on 5-mm-dia. wires (Morris and Garret, 1956) have shown that the minimum necessary embedment is slightly over 1 m (about 3.5 ft).

Although the grout–strand bond is higher than the bond on single wires due to spiral interlock, the bond drops rapidly if the embedment is less than 0.6 m (2 ft). Tests on Freyssinet anchors with spacers have shown that each

TABLE 4-3 Maximum Allowable Bond Stress for Bar Tendons, Stipulated by the British Code, for Neat Cement Grout

Type of Bar	Characteric Strength of Grout (f'_{cu} N/mm^2)			
	20	25	30	40+
	Maximum Bond Stress (N/mm^2)			
Plain	1.2	1.4	1.5	1.9
Deformed	1.7	1.9	2.2	2.6

strand withstands from 156 to 178 kN (35–39 kips) at 0.6-m embedment (strand capacity 178–270 kN). From this, investigators have concluded that strand anchors do not need embedment lengths greater than 1.5 m (Golder Brawner Assocs., 1973).

In general, higher uniformity is evident for strand–grout working bond values than for bar and wire. The average bond developed by bars, particularly deformed types, is significantly higher than strand and wire bond values. However, under the guidelines and recommendations of most current codes, the actual factor of safety against bond failure is well in excess of 2.

More useful data on bond values for the three types of tendon are given by Littlejohn and Bruce (1977).

Selection of Tendon Bond Length. For cement grout anchors, the minimum tendon embedment lengths recommended by a majority of codes are (a) for tendons homed and bonded in situ, bond length = 3 m (10 ft); and (b) for tendons bonded under factory-controlled conditions, bond length = 2 m (6.5 ft).

For shorter bond lengths, or where doubt exists regarding the actual factor of safety, bond should be confirmed by full scale tests. A minimum grout cover 20 mm ($\frac{3}{4}$ in) should be maintained, usually with the use of centralizers, and the cross-sectional area of the steel tendon should not exceed 15 percent of the borehole area (Littlejohn et al., 1978).

Debonding

Debonding occurs mostly with high-capacity anchors (those with design loads in excess of 200 tons metric (or 450 kips), and is manifested as the ductile tendon transfers shear to the brittle cement grout over a relatively long transmission length. As the load increases, cracking of the grout at the interface begins to dominate, and causes loss of adhesion and friction over a critical portion of the tendon embedment. Along this length tendon extension appears erratic, and tends to complicate interpretation of load–extension data, especially where acceptance criteria are based on extension limits. Cracking of the grout also creates zones of unprotected tendon that can initiate corrosion attack. Tendon density within the borehole appears to affect the degree of debonding; hence it dictates the limits in tendon cross-sectional area specified in the foregoing section.

Tests conducted by Muller (1966) on BBRV anchors stressed to 220 tons (metric) indicate an unusual pattern of bond distribution along the fixed length of the anchor (8 m long). This is shown in Fig. 4-9, and evidently the stress distribution along the fixed zone is essentially nonuniform. At a load of 50 tons, the force is by large transmitted over 75 percent the proximal section, about 5.5 m, and almost dissipates at the distal quarter section, giving average bond value 0.22 N/mm². At 185 tons most of the load is recorded at the lower half (4 m) of the fixed length, and debonding of the

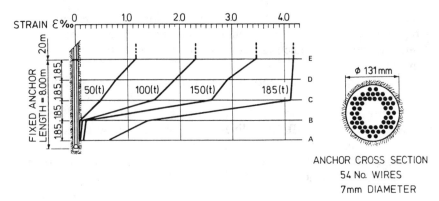

Fig. 4-9 Stress–strain distribution along tendon in fixed anchor zone; anchor stressed to 220 tons. (From Muller, 1966.)

tendon is apparent along the upper half. About 30 tons are resisted at the bottom, with a corresponding stress concentration there. Between points A and C, the average bond stress is 0.98 N/mm^2, and bond is inactive above this zone. At 280 tons (not shown in Fig. 4-9), complete debonding of the tendon occurred, and the entire load was resisted by the wedge at the end of the anchor. Based on uniform distribution, the average bond stress would be 0.65 N/mm^2, or well below the allowable and the actual at 180 tons.

Decoupling equivalent to an extension of 2 m in the free length has been reported by Eberhardt and Veltrop (1965) during stressing of a test anchor installed in basalt, with capacity 1300 tons, fixed anchor length 11.5 m, and diameter 406 mm. In some instances, spacers can cause decoupling, but these effects are uncertain and cannot be quantified.

4-4   FAILURE OF GROUND–GROUT BOND

Basic Assumptions and Considerations

For most anchored structures the fixed anchor length is located beyond the zone of influence of the potentially unstable area. Anchors are long and deep enough so that bond at the anchor–ground interface is likely to fail before shear failure occurs in the ground mass. For straight-shaft, cylindrically shaped anchors, a convenient assumption is that shear resistance is mobilized at the interface of the borehole, and is uniformly distributed along the fixed length. The entire load is resisted in this fashion, whereas the free length is not engaged in this interaction. Under these conditions, the total shear resistance developed at the interface is a function of fixed anchor dimensions and applied load. Where the anchorage diameter is larger than the shaft diameter, the bearing capacity of the transfer area is added to the total ultimate capacity, which is thus derived from a compatibility of displacements necessary to mobilize these reactions.

However, experimental and theoretical work has shown that shear resistance between ground and grout column (whether soil or rock) is more complex than the foregoing idealized model, and this complexity gives rise to essentially nonuniform bond distribution. Whereas in many instances an assumed mechanism of load transfer may result in designs that are grossly inaccurate, it is reassuring to know that where certain sections of the fixed zone are overloaded and shear failure is imminent, other sections begin to receive and resist load so that equilibrium is reestablished. It is conceivable, for example, that as a high-capacity anchor is loaded to its limit, the bond stress at the proximal end will be extremely high and possibly approach the condition of failure whereas the most distal parts may remain inactive and in effect redundant.

For relatively shallow anchorages failure originating in the soil mass before the anchor itself fails is quite likely. However, this occurrence becomes very remote as the fixed length becomes embedded in deeper strata since in this case much higher load will be required to cause failure in the soil mass. Eventually the bond between ground and grout may be exceeded along the fixed length, and the anchor may pull out but without failure of the ground mass. If this failure is limited to a slip only, the regroutable anchor discussed in the foregoing sections can restore anchor function and balance initial load capacity lost in this fashion.

Bond resistance (adhesion or friction) is also known to depend on the soil properties. An increase in the relative density of sand generally increases the angle of internal friction, which in turn increases the frictional resistance at the interface. For cohesive soils, increase in stiffness or decrease in plasticity usually implies higher shear strength, with a corresponding improvement in bond capacity.

Other factors that have considerable effect on bond resistance relate to field operations. For example, the use of rotary percussive hammers to advance a casing in sand increases the normal stress, and this improves friction. In cohesive soil drilling without casing or with casing using flushing water tends to have softening effects and thus reduce shear. Postgrouting generally improves load capacity in proportion to the magnitude of postgrouting pressure (see also Section 4-9).

A Simple Theoretical Expression for Bond

For straight-shaft fixed length, the average (in this case uniform) shear (also referred to as "bond") stress τ along the ground–grout interface can be related to the applied load P by the simple expression

$$P = \pi D L \tau \qquad (4\text{-}8)$$

where D is the effective grout column diameter and L is the fixed anchor length. This simplified approach is generally acceptable by many codes of practice and investigators (Fargeot, 1972; Mascardi, 1973; Coates, 1970; White, 1973). It is valid under the following assumptions:

1. The transfer of load from the grout to the ground occurs uniformly over the fixed length.
2. The borehole and the fixed length have the same diameter.
3. Failure occurs by sliding at the ground–grout interface for a smooth borehole, or by shearing along a zone adjacent to the interface for a rough borehole (failure is thus manifested along the weaker shear zone, which may be at the interface or away from it).
4. Debonding does not occur at the ground–grout contact area.
5. There are no discontinuities or weak planes that can alter the process of failure.

The total shear resistance, or bond, at the interface is, consistent with Eq. (4-8), the summation of two components: adhesion and friction. Thus, the shear stress τ is expressed as

$$\tau = c_a + \sigma_n \tan \delta \qquad (4\text{-}9)$$

where c_a = adhesion between ground and grout
σ_n = normal effective stress on the anchor zone
δ = friction angle between the soil and the grout

Where shear strength tests are carried out on representative rock and soil samples, all the factors in Eq. (4-9) are lumped into a single parameter. In this case, the allowable bond stress is estimated from shear strength test values with an appropriate factor of safety, normally not less than 2. In most instances, however, the magnitude and actual distribution of bond is more complex than the simple expression of Eq. (4-9). This subject warrants consideration of all factors affecting bond, reviewed in the following sections for various types of soil and rock.

4-5   FAILURE OF ANCHORS IN ROCK—STRAIGHT SHAFT

Among the four anchor types shown in Fig. 2-2, the straight-shaft tremie-grouted type A is considered most suitable and applicable to rock because of low cost and simplicity of construction. A straight-shaft anchor generally is adequate for the transfer of load, and because of its frequent use it has provided the basis for most theoretical and experimental work on rock anchorages.

For preliminary estimates and feasibility studies of proposed anchorages in rock, the bond length can be established from direct pullout tests. This procedure consists of placing a core sample vertical in the center of a steel form and filling the annular space with cement grout. After curing, the sample is pressed out and the bond strength is estimated from the pullout force over the contact area.

Ultimate bond strength values from various sources are available for normal conditions of loading and various types of rock. Table 4-4 gives such a summary within typical ranges for the materials indicated. An excellent survey on rock–grout bond values recommended for design or employed in practice has also been compiled by Littlejohn and Bruce (1977).

Theoretical Data

Stress distribution around a cylindrical anchorage in triaxial stress field has been investigated by Coates and Yu (1970) using finite-element methods. Parts (a) and (b) of Fig. 4-10 show the basic geometry and finite-element mesh, respectively, for the anchor model, which allowed load in tension or in compression with stresses calculated for either loading.

The results show dependence of the shear (bond) stress distribution on the ratio of the elastic moduli of the anchor material E_a and the rock E_r. This variation is plotted in Fig. 4-11 for a fixed anchor length six times the radius, and for E_a/E_r ratios 0.1, 1.0, and 10. The smaller this ratio (meaning stronger rock and weaker grout) the larger the stress concentration at the proximal end, whereas higher ratios (meaning softer rock and stronger grout) result in bond that is more evenly distributed. For E_a/E_r close to or exceeding 10, which means anchorages in soft rock, uniform stress distribution is acceptable for all practical purposes. In this case anchor design can assume bond resistance equal to the shear strength of the weaker medium (rock).

Tension in rock anchors will cause shear stresses at the interface, which will in turn induce tensile stresses in the rock, with maximum value at the

TABLE 4-4 Ultimate Bond Strength Values for Various Types of Rock

| | Ultimate Bond Strength | |
Rock Type	(psi)	(N/mm²)
Granite and basalt	250–450	1.72–3.10
Dolomite limestone	200–300	1.38–2.07
Soft limestone	150–200	1.03–1.38
Slates and hard shales	120–200	0.83–1.38
Soft shales	30–120	0.21–0.83
Sandstone	120–150	0.83–1.03
Weathered marl	25–36	0.17–0.25

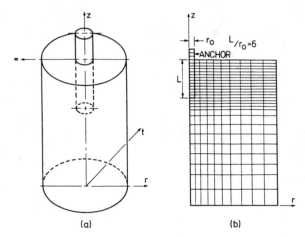

Fig. 4-10 Stress distribution around cylindrical anchorage in triaxial stress field: (a) geometry of rock anchor; (b) finite-element mesh. (From Coates and Yu, 1970.)

proximal end. These stresses follow rapid dissipation radially at the distal end as shown in Fig. 4-12. For a load of 1500 kN (337 kips) in a 75-mm-dia. hole, the estimated average maximum tensile stress at the proximal end is 145 N/mm², whereas at the distal end this stress is 48 N/mm². It is inconceivable, however, that rock will sustain these high-tensile stresses; hence

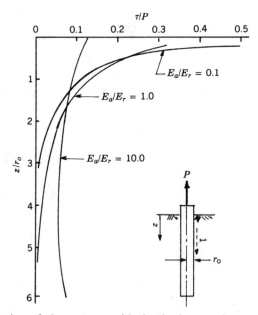

Fig. 4-11 Variation of shear stress with depth along rock–grout interface; model study. (From Coates and Yu, 1970.)

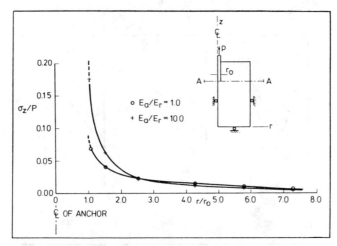

Fig. 4-12 Variation of tensile stress σ_z in rock adjacent to the end of tension anchor. (From Coates and Yu, 1970.)

cracking is likely to occur in the direction of the anchor and will dissipate radially outwards reaching an equilibrium position. These cracks possibly account for anchor creep often observed for some time after stressing.

Littlejohn (1982) quotes E_a values for normal grout in the range 20–70 kN/mm^2, whereas other investigators quote values from 10 to 20 kN/mm^2 for cement composition and water/cement (W/C) ratios commonly used in practice. Accepting average E_a values for grout of about 15 kN/mm^2, or 1.5×10^4 N/mm^2, it follows that the rock must have E_r values below 0.15 $\times 10^4$ N/mm^2 before uniform bond distribution can be assumed. Relating rock compressive strength and elastic modulus as proposed by Judd and Huber (1961), whereby UCS $= E_r/350$, it appears that the rock strength in this case should be less than 1500/350 or 4.5 N/mm^2 (about 650 psi), which excludes the majority of anchors installed in rock. Indeed, the usual E_a/E_r ratio is between 0.1 and 1.0, and for this range bond distribution is markedly nonuniform. In this case, stress concentration at the proximal end is very likely, with stress values probably 5–10 times the average stress level.

It appears that the mechanics of anchors in strong rock is not as yet fully explained, although their construction and performance becomes less of a problem because of the adequacy of the installation procedures available in current practice. However, with high-capacity anchors, the subject of high stresses at the proximal end as well as the effect of debonding on stress distribution deserves further analytical treatment. As an initial approach to better understanding, Phillips (1970) suggests the following pattern (see Fig. 4-13):

1. Following debonding, the restraint imposed by the rock on the uneven rock–grout interface causes dilation. Additional anchor movement is

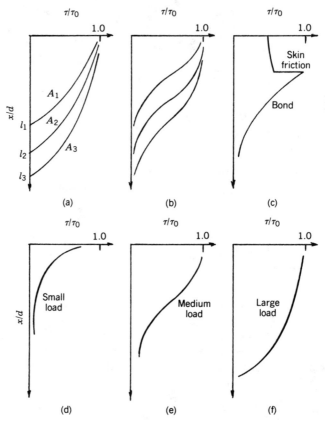

Fig. 4-13 Hypothetical stress distribution around partly debonded fixed anchor length in rock. (From Phillips, 1970.)

possible only through further shear failure of the grout, with a possible stress distribution as shown in Fig. 4-13(a) and (b).

2. Residual bond stresses not affected by dilation will essentially depend on the magnitude of normal pressure acting on the interface. This pressure will vary along the debonded length, and if it is less than the grout shear strength, the stress distribution will be as shown in Fig. 4-13(c). If it is greater than the grout shear strength, the stress distribution will revert to that of parts (a) and (b).

3. Under applied load, the stress distribution diagrams are likely to be as shown in parts (d), (e), and (f) of Fig. 4-13. At large loads, the entire anchor is practically debonded. At this stage the stress is distributed according to the amount of relative movement and the degree of dilation or frictional shear strength mobilized, represented by part (f).

Empirical Record on Bond Distribution

Tests by Berardi (1967) confirm the theoretical predictions of Coates and Yu discussed in the foregoing section. These tests suggest an active zone independent of the total fixed length, and dependent on anchor diameter and strength characteristics of the supporting rock, especially its modulus of elasticity.

The tests involved two anchors installed in marly limestone, with modulus of elasticity 3×10^4 N/mm², or about twice the E_a value of normal grout, and uniaxial compressive strength 100 N/mm² (about three times the strength of normal grout). Both anchors had straight shaft 120 mm in diameter.

The diagrams of Fig. 4-14 are plotted using data from strain gauge measurements, and clearly show uneven bond distribution. Further results show that bond tends to be uniform for high E_a/E_r ratios, and becomes markedly nonuniform for low ratios, that is, for rock with high strength.

Selection of Bond Values and Fixed Anchor Length

Where shear strength tests are performed on representative rock samples, Littlejohn (1980) recommends maximum average working bond stress (assuming uniform bond distribution) not greater than the minimum shear strength divided by the factor of safety (≥ 2 and probably ≈ 3). This approach is valid for soft rocks with uniaxial compressive strength less than 7 N/mm² (1000 psi), and where the holes have been drilled using rotary–percussive techniques. If shear strength data and pullout tests are not available, the ultimate bond stress is often taken as one-tenth the uniaxial compressive strength of massive rock (100 percent recovery) up to a maximum value 4.2

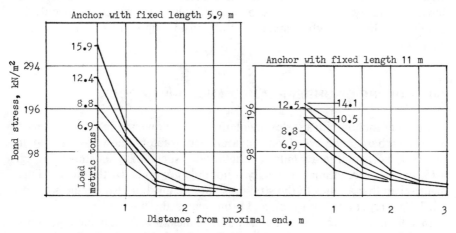

Fig. 4-14 Distribution of bond along fixed anchor length in rock. Data obtained from load tests. (From Berardi, 1967.)

N/mm^2 (600 psi). For rocks classified as granular weathered variations the one-tenth value may be unnecessarily low, and in this case the ultimate bond stress can be taken as 20–35 percent UCS.

Recommended design bond values are summarized in Table 4-5 for a wide range of igneous, metamorphic, and sedimentary rocks. Where shown, the factor of safety correlates ultimate and working bond stress based on uniform bond distribution. The average working bond stress ranges from 0.35 to 1.4 N/mm^2 (50–200 psi).

Fixed anchor lengths used or recommended in practice for cement grouted rock anchors are presented in Table 4-6, and cover Europe, North America, and South Africa. Fixed lengths longer than those indicated are justified where nonhomogeneous conditions prevail and the design cannot be confirmed by tests. However, there is no evidence to indicate that fixed anchor lengths greater than 10 m (30 ft) can have any real effect on anchor capacity; hence this value is accepted as the upper limit. Fixed lengths shorter than those indicated may be sufficient, especially under a generous factor of safety and in certain controllable conditions. For short anchors, however, a sudden change in rock quality or discontinuities along the fixed length combined with construction inefficiency can amplify the effect on anchor capacity. Thus, the specified minimun length for the fixed anchor zone is 3 m (10 ft).

In general, estimation of fixed anchor length should be supplemented by relevant geologic and geotechnical data. Among these, groundwater conditions are important, especially water table, rate of flow, pressure, and aggressivity. Site investigations (discussed in detail in subsequent sections) should also address the potential of construction difficulties.

In soft rocks, weatherability and durability should be assessed from samples obtained at the depth of the proposed fixed zone. The degree of rock weathering will affect not only ultimate bond but also load-deflection behavior. A fairly accurate value of the modulus of elasticity can only be obtained at high cost, but is very useful considering its effect on bond distribution.

4-6   FAILURE OF UNDERREAMED ANCHORS IN ROCK

The type D anchor with a series of enlargements shown in Fig. 2-2 is common in stiff to hard cohesive deposits. Many contractors, however, use it in relatively soft rock in a similar configuration, specifically as shown in Fig. 4-15. In this case, ultimate capacity is not only governed by the shear failure mechanism that applies to the straight shaft anchor, but also depends on the crushing strength of both rock and grout material.

In solid massive rock with few bedding planes, it is reasonable to assume that each underream will exert forces on the rock mass that tend to be dissipated at an angle of 45° from the direction of the pull. The conical stress fields thus produced will overlap and form one large cone expanding upwards

TABLE 4-5 Rock–Grout Bond Values Recommended for Design

Rock Type	Working Bond (N/mm²)	Ultimate Bond (N/mm²)	Factor of Safety	Source
		Igneous		
Medium hard basalt		5.73	3–4	India—Rao (1964)
Weathered granite		1.50–2.50		Japan—Suzuki et al (1972)
Basalt	1.21–1.38	3.86	2.8–3.2	Britain—Wycliffe-Jones (1974)
Granite	1.38–1.55	4.83	3.1–3.5	Britain—Wycliffe-Jones (1974)
Serpentine	0.45–0.59	1.55	2.6–3.5	Britain—Wycliffe-Jones (1974)
Granite and basalt		1.72–3.10	1.5–2.5	USA—PCI (1974)
		Metamorphic		
Manhattan schist	0.70	2.80	4.0	USA—White (1973)
Slate and hard shale		0.83–1.38	1.5–2.5	USA—PCI (1974)
		Calcareous Sediments		
Limestone	1.00	2.83	2.8 (temporary)	Switzerland—Losinger (1966)
Chalk—Grades I–III (N = SPT in blows/0.3 m)	0.005N	0.22–1.07 0.01N	2.0 (temporary) 3.0–4.0 (permanent)	Britain—Littlejohn (1970)
Tertiary limestone	0.83–0.97	2.76	2.9–3.3	Britain—Wycliffe-Jones (1974)
Chalk limestone	0.86–1.00	2.76	2.8–3.2	Britain—Wycliffe-Jones (1974)
Soft limestone		1.03–1.52	1.5–2.5	USA—PCI (1974)
Dolomitic limestone		1.38–2.07	1.5–2.5	USA—PCI (1974)

TABLE 4-5 (Continued)

Arenaceous Sediments				
Hard coarse-grained sandstone	2.45		1.75	Canada—Coates (1970)
Weathered sandstone		0.69–0.85	3.0	New Zealand—Irwin (1971)
Well-cemented mudstones		0.69	2.0–2.5	New Zealand—Irwin (1971)
Bunter sandstone	0.40		3.0	Britain—Littlejohn (1973)
Bunter sandstone (UCS>2.0 N/mm²)	0.60		3.0	Britain—Littlejohn (1973)
Hard fine sandstone		2.24	2.7–3.3	Britain—Wycliffe-Jones (1974)
Sandstone	0.69–0.83	0.83–1.73	1.5–2.5	USA—PCI (1974)
Argillaceous Sediments				
Keuper marl		0.17–0.25 (0.45 c_u)	3.0	Britain—Littlejohn (1970) c_u = undrained cohesion
Weak shale	0.10–0.14	0.35		Canada—Golder Brawner (1973)
Soft sandstone and shale		0.37	2.7–3.7	Britain—Wycliffe-Jones (1974)
Soft shale		0.21–0.83	1.5–2.5	USA—PCI (1974)
General				
Competent rock (where UCS > 20 N/mm²)	Uniaxial compressive strength—30 (≤1.4 N/mm²)	Uniaxial compressive strength—10 (≤4.2 N/mm²)	3	Britain—Littlejohn (1972)

Weak rock	0.35–0.70			Australia—Koch (1972)
Medium rock	0.70–1.05			
Strong rock	1.05–1.40			
Wide variety of igneous and metamorphic rocks	1.05		2	Australia—Standard CA35 (1973)
Wide variety of rocks	0.98			France—Fargeot (1972)
	0.50			Switzerland—Walther (1959)
	0.70			Switzerland—Comte (1965)
		1.20–2.50		Switzerland—Comte (1971)
	0.70		2–2.5 (temporary) 3 (permanent)	Italy—Mascardi (1973)
	0.69	2.76	4	Canada—Golder Brawner (1973)
	1.4	4.2	3	USA—White (1973)
		15–20% of grout crushing strength	3	Australia—Longworth (1971)
Concrete		1.38–2.76	1.5–2.5	USA—PCI (1974)

TABLE 4-6 Fixed Anchor Lengths for Cement Grouted Rock Anchors Recommended in Practice

Fixed Anchor Length (meters)		
Minimum	Range	Source
3.0		Sweden—Nordin (1968)
3.0		Italy—Berardi (1967)
	4.0–6.5	Canada—Hanna and Seeton (1967)
3.0	3.0–10.0	Britain—Littlejohn (1972)
	3.0–10.0	France—Fenoux and Portier (1972)
	3.0–8.0	Italy—Conti (1972)
4.0		South Africa—Code of Practice (1972)
(very hard rock)		
6.0		South Africa—Code of Practice (1972)
(soft rock)		
5.0		France—Bureau Securitas (1972)
5.0		USA—White (1973)
3.0	3.0–6.0	Germany—Stocker (1973)
3.0		Italy—Mascardi (1973)
3.0		Britain—Universal Anchorage Co. Ltd. (1972)
3.0		Britain—Ground Anchors Ltd. (1974)
3.5		Britain—Associated Tunnelling Co. Ltd. (1973)
(chalk)		

From Littlejohn (1980).

in the direction of the top of the borehole, as shown in Fig. 4-15. Pullout resistance is provided by this large inverted cone, so that the deeper the penetration of the anchor in rock material the larger the ultimate load that can be resisted by the anchor–rock interaction. In bedded or fractured rock, cone size and shape will vary with the distribution of bedding and cleavage planes and also with the grout take in the fissures, whereas in badly decomposed and broken material the stress distribution pattern will approach an underreamed anchor in clay.

It is documented from field experience that the mode of failure of shallow anchors in rock (total length <10 ft) follows the conical form described, although there is difference of opinion regarding the angle of the cone (see also subsequent sections). High-capacity anchors should not be constructed with the fixed zone in close proximity to the surface, since their capacity is governed by cone failure.

An important factor relevant to underreamed anchors is the mechanism of load transfer from the steel tendon to the grout and then to the rock. Stress distribution should be as even as possible, and local concentration should not occur. In this context, a plain smooth bar terminating in end-bearing plate is unsuitable since it will cause premature failure of the grout or rock by crushing in the end zone due to intense stresses concentrated at this location.

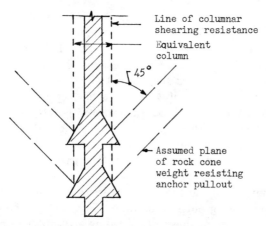

Line of columnar
shearing resistance

Equivalent
column

$45°$

Assumed plane
of rock cone
weight resisting
anchor pullout

Fig. 4-15 Diagrammatic presentation of underreamed anchor in rock. Assumed failure plane along equivalent column.

For rock anchors with average borehole diameter about 115 mm (or $4\frac{1}{2}$ in), underreams are usually formed with 12-in (300-mm) diameter. Either two or four underreams are provided per anchor. The advantages of underreams diminish as rock strength increases, and with hard massive rock underreaming is not necessary. Provided the tendon layout is designed to ensure uniform bond distribution (e.g., basketed and bonded strand tendon), the cone effect is approximated by the following empirical expression relating fixed length L, applied load P, underream diameter D_u, and working (allowable) shear stress τ_w (Nicholson et al., 1982):

$$ L = \frac{P}{\frac{2}{3}\pi D_u \tau_w} \tag{4-10} $$

Equation (4-10) replaces the cone effect by an equivalent columnar shear resistance. The working (allowable) stress τ_w is derived from the ultimate shear resistance of the grout or rock (whichever is smaller) with an appropriate factor of safety, not less than 2 and preferably 3.

4-7 FAILURE OF COMPRESSED TUBE ANCHORS IN ROCK

The response of compressed tube anchors to load was discussed in Section 3-2 in conjunction with the behavior shown in Fig. 3-7. The transfer of force from the tendon to the anchorage medium through a tube subjected to compression is accomplished under increased bond resistance along the interface as the walls of the tube undergo bulging. This process also eliminates the disking normally observed around a grouted bar or strand in tension. The shifting of the shear resistance zone along the tendon shown in Fig. 4-

6 for tension-type anchors as the tensile force increases is essentially absent in compressed tubes. Indeed, results of tests carried out by Ivering (1981) confirm that there is no significant movement of the bond wave along a grouted steel tube subjected to compression until the axial compressive stress in the tube exceeds the elastic limit.

This conclusion is consistent with test results reported by Schwarz (1972) showing that in compressed tube anchorages the effects of friction drag, commonly manifested with grouted bars in tension, do not appear. These anchors were loaded first to the plastic limit and then beyond this limit, but the resulting extension of the reaction length caused no appreciable increase in load-carrying capacity.

Lateral expansion of the tube in compression reduces slip between the tube and surrounding grout considerably. Figure 4-16 shows stress/strain diagram obtained from loading and unloading a 10-m-long tube anchor. The system consists of a steel tendon 32 mm in diameter, and a steel tube, 52 mm OD. The stress–strain characteristics lead to the conclusion that friction and slip at the end of the compressed tube anchorage can be disregarded (Weber, 1966).

Problems in the Transfer of Load. Improved and modified versions of compressed tube anchors have appeared on the market. For example, the new type of Stump Duplex anchor shown in Fig. 3-8 incorporates a seal that prevents entry of grout into the annulus between the tendon and the inside wall of the tube, and also includes a mechanism for destressing and removing the tendon from the tube.

However, irrespective of improvements, certain problems associated with the transfer of load have been experienced and reported by several investigators, and these problems are yet to be solved. For example, several anchor failures have been documented, and it is interesting to note that most of them actually occurred while the compression on the tube was very low, in some instances of the order of 10 percent the design load. These surprising failure incidents are clearly inconsistent with the expected ultimate load, and in most cases they occurred suddenly. Ivering (1981) reports similar failures in laboratory tests on compressed tubes. Inasmuch as the space inside the tube was not grouted, failure was caused by lack of sufficient bond between the outside face of the tube and the surrounding grout rather than along the rock–grout interface. The implicit conclusion is that in a tube of relatively smooth outside surface, compressive stresses of considerable intensity must be developed and cause the tube to bulge as shown in Fig. 3-7(a) before adequate bond resistance becomes available by the process of lateral expansion and radial compression. Until the latter becomes effective, failure in early loading stages is always likely but can be prevented by increasing the initial grip of the compressed tube. This can be accomplished by the following means: (a) increasing the length of tube, (b) increasing

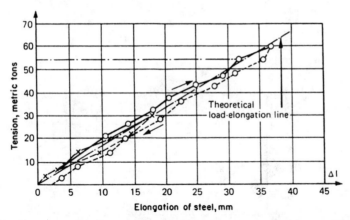

Fig. 4-16 Stress–strain diagram for 10-m-long compressed tube anchor. (From Weber, 1966.)

surface roughness, (c) welding steel spacers to the tube, and (d) forming the tube in the shape of wedge.

Results from Pushout Tests. Table 4-7 shows results from pushout tests on compressed tubes carried out at the New South Wales Institute of Technology. The tests involved four steel tube specimens embedded in concrete blocks. The tubes were mild steel, with outside and inside diameters of 40 and 25 mm, respectively.

The higher ultimate load for the shorter tubes is explained by the increased surface roughness on these specimens (indentation depth 0.05–0.10 mm). These tubes also incurred greater slip at both the loaded and the free ends. Data from a short tube specimen with 90 mm OD and 900-mm embedded length are plotted in Fig. 4-17. One of the most significant results is that load capacity in terms of bond resistance reaches its maximum when the

TABLE 4-7   Results on Pushout Tests on Compressed Tube Anchors (Outside Diameter 40 mm; Inside Diameter 25 mm)

Specimen No.	Length (mm)	Ultimate Load (kN)	Slip at Loaded End (mm)	Slip at Free End (mm)
1	1100	250	0.290	0.0005
2	1100	350	0.738	0.004
3	900	275	0.745	0.005
4	900	350	1.495	0.009

From Ivering (1981).

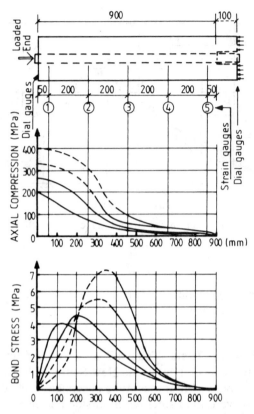

Fig. 4-17   Axial compression and bond stresses. Compressed tube specimen: 90 mm outside diameter; 900 mm embedded length. (From Ivering, 1981.)

compressive stress at the point of load application exceeds the elastic limit. Thus, with compressive stress 400 MPa (clearly beyond the elastic range), the bond stress reaches a maximum 7 MPa on the bond resistance wave. With compressive stress 260 MPa (in this case the yield stress), the maximum bond stress on the wave curve is 4 MPa. Under maximum compression (applied load 308 kN) the calculated average bond stress over the embedded length is 3.17 MPa, which is well in excess of the average grout–tendon bond stress for tension tendons.

Contrary to the foregoing experimental results, elastic analysis shows only a minor effect of developed lateral stresses and induced radial compression. Inasmuch as this discrepancy is yet to be explained, at present the design of compressed tube anchorages should be approached with caution in weak foundation media where developed lateral compression is limited.

4-8   FAILURE OF ANCHORS IN SAND

Basic Principles

In many instances, anchors in cohesionless soil can sustain loads in excess of 300 kips over a fixed length of 4–8 m (13–26 ft) and with shaft diameter 10–15 cm. These reported loads cannot be explained by the classic laws and theories of soil mechanics. However, this load capacity has been explained by backanalysis of field tests data whereby the effect of soil conditions, anchor dimensions, construction techniques, and miscellaneous factors that are not numerically assessed is quantitatively measured and included in the analysis. Thus, experience demonstrates that ultimate load of anchors in sand depends on the following: (a) relative density and degree of uniformity of the soil; (b) fixed anchor geometry and dimensions (mainly the length and to a lesser degree the diameter); (c) method of grout injection and grout pressure used; (d) dilatancy in the soil, which can result in higher normal stresses, hence greater friction at the grout–soil interface; and (e) to a lesser degree, the drilling methods and equipment.

In the context of design priorities, before the loads and fixed anchor zones are established it is necessary to obtain accurate soil data, including, sieve analysis, grading curves, angles of internal friction, and sand strata thickness, all discussed in detail in subsequent sections.

Anchors in sand can have their fixed length formed in the configuration of type *A, B* or *C* in Fig. 2-2. Each type is suitable and feasible for a given set of ground conditions, namely, soil density, porosity, and particle size. Furthermore, the use of pressure grouting techniques implies that increased anchor capacity can be attained since greater penetration of cement grout into surrounding soil medium causes consolidation and densification, whereas a residual "locked-in" grout pressure remains in the fixed zone after completion of pressure grouting.

In relatively fine but dense compact sand, cement grout tremied into a borehole will most likely produce a straight shaft in the fixed zone. Cement grout injected under pressure of 10–40 psi (0.069–0.28 N/mm^2) in sand with permeability between 10^{-1} and 10^{-2} cm/s will penetrate the soil pores and most likely produce type *B* anchor zone, normally of nonuniform diameter. Likewise, in coarse sand with relatively nonuniform characteristics, grout injection under pressure is likely to produce type *C* fixed zone. Examination of grouted bodies extracted from fixed anchor zones confirms the effect of soil density and grouting pressure on anchor load capacity and configuration. The zone immediately adjacent to the borehole is recompressed, and the accompanying densification results in increased skin friction.

From the foregoing it follows that several interaction modes are possible. For a straight shaft to be produced, the cement particles of the grout mix must be blocked at the interface although some water from the mix is filtered toward the soil. Enlarged cylinder (type B), with either uniform or variable diameter, is obtained where grout mix penetrates into soil pores through deep filtration extended beyond the borehole interface under higher pressure. The irregular form of type C is created in fairly nonuniform soil varieties and in conjunction with high injection pressure whereby fissures and grout roots are formed erratically until further penetration is restrained by the shear strength of the cement grout.

Empirical Estimation of Load Capacity (Straight Shaft and Enlarged Cylinder)

Enlarged Cylinder. For low pressure grouted type B anchor in sand, field trials have provided the following empirical rule for estimating ultimate load:

$$T_f = LN' \tan \phi \tag{4-11}$$

where T_f = ultimate load capacity (kN)
$\quad\ L$ = fixed anchor length (m)
$\quad N'$ = a constant factor
$\quad \phi$ = angle of internal friction of sand

The constant parameter N' includes the effect of overburden pressure, fixed anchor diameter, in situ stress field, and dilation characteristics. The drilling techniques assume rotary–percussive tools with water flushing. Grouting pressures are up to 1 N/mm^2 (145 psi).

Where the fixed anchor length is quoted in meters (m), N' values range from 400 to 600 kN/m. If the fixed anchor length is expressed in feet (ft), N' values range from 27 to 41 kips/ft. Equation (4-11) takes into account soil permeability from 10^{-1} to 10^{-2} cm/s, overburden depth to the top of the fixed anchor 20–45 ft (6–14 m), effective average diameter of the enlarged cylinder in the fixed zone 15–24 in (380–610 mm), and range of anchor lengths within the limits normally employed in practice.

Straight Shaft. Equation (4-11) is also valid for straight-shaft fixed anchor zone. In this case, soil permeability is likely to be from 10^{-2} to 10^{-4} cm/s, which will inhibit permeation of sand by cement grout so that an essentially smooth and straight shaft will result. The N' values range from 130 to 165 kN/m if the fixed length is quoted in meters, and from 9 to 12 kips/ft if this length is given in feet. Likewise, Eq. (4-11) is valid for overburden depth to the top of the fixed anchor 18–30 ft (5.5–9 m), and effective average shaft diameter 7–8 in (280–310 mm).

Equation (4-11) provides simple but crude estimation of fixed anchor lengths and load capacity with probably upper and lower limits, but tends to be conservative since it does not consider important ground parameters. Deviation between estimated and actual load capacity is more significant if this rule is applied to dense overconsolidated alluvium where N' values were initially established in normally consolidated materials (Littlejohn, 1980).

Effect of Pressure Grouting. This is known to increase anchor load capacity since greater penetration of cement grout into soil is achieved. As a rule of thumb, permeability of the order 10^{-1} to 10^{-2} cm/s will allow soil infiltration by the grout, and permeability of the order 10^{-3} to 10^{-4} cm/s will cause soil densification local to the borehole. In conjunction with pressure grouting, a further empirical formula is derived from field trials, and relates ultimate load as follows:

$$T_f = p'\pi DL \tan \phi \qquad (4\text{-}12)$$

where D and L are again effective anchor diameter and fixed length, respectively, and p' is the grout pressure taken as 2 psi (0.014 N/mm²) for every foot of overburden above the top of fixed anchor. This value of p' is used as average over the fixed length. The intent of Eq (4-12) is to express the increase in grout–soil friction due to pressure grouting, but it also relates to known dimensions. It is interesting to note that actual grout pressures different from p' as given above have been used to check anchor capacity determined in this manner, but deviations have been found to be subject to variability in site conditions (Nicholson et al., 1982). For example, the insertion into Eq. (4-12) of the very high postgrouting pressure associated with the tube-a-manchette technique will give results unrealistic and unsafe. Thus, the foregoing empirical rule is recommended if used in conjunction with actual grout pressures which are compatible with the definition of the p' parameter.

Theoretical Analysis (Straight Shaft and Enlarged Cylinder)

The ultimate load capacity T_f in sand is derived in a more general form from bearing capacity theories as follows:

$$T_f = A\sigma'_v\pi DL \tan \phi + B\gamma h \frac{\pi}{4} (D^2 - d^2) \qquad (4\text{-}13)$$

where A = ratio of constant pressure at the fixed grout–soil interface to the average effective overburden stress
γ = unit weight of soil (effective weight below the water table)
h = depth of overburden to top of fixed anchor
L = fixed anchor length
σ'_v = average effective overburden pressure adjacent to fixed length (taken at midpoint)
D = effective fixed anchor diameter

ϕ = angle of internal friction
B = bearing capacity factor
d = nominal anchor diameter above fixed zone

Equation (4-13) includes the effect of side shear and end bearing, and is applied under the following guidelines:

1. The dimension D usually is estimated from grout intake and in conjunction with ground porosity. In coarse sand and gravel and for borehole nominal diameter 100–150 mm (4–6 in), values of D attained range from 400 to 500 mm (16–20 in), or from $3d$ to $4d$. Grout pressure is relatively low and less than 1 N/mm² (145 psi).
2. In medium dense sand grout permeation is limited and can cause local compaction only. For the same borehole diameters (100–150 mm) and grouting pressure up to 1 N/mm², D values may be from 200 to 250 mm or $1.5d$ to $2d$.
3. For very dense sand D is further reduced, in the range 180–200 mm or $1.2d–1.5d$.

The parameter B is generally smaller than N_q (conventional bearing capacity factor). For the slenderness ratio of ground anchors N_q/B = 1.3–1.4 (Trofimenkov and Mariupolskii, 1965). The factor N_q is related to ϕ as shown in Fig. 4-18 for h/D = 25. The influence of slenderness ratio h/D on N_q is shown in Table 4-8, and clearly this effect diminishes for h/D > 25 so that N_q is reasonably estimated from Fig. 4-18. For compact sandy gravel (ϕ = 40°) and compact sand (ϕ = 35°), in situ measured values of B are 101 and

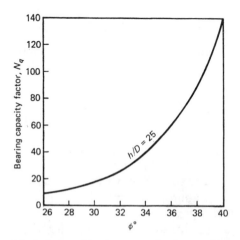

Fig. 4-18 Relationship between bearing capacity factor N_q and angle of internal friction ϕ° for slenderness ratio 25. (From Littlejohn, 1980.)

TABLE 4-8   Approximate Relationship between Bearing Capacity Factor N_q and Slenderness Ratio of the Anchor

	ϕ				
h/D	26°	30°	34°	37°	40°
15	11	20	43	75	143
20	9	19	41	74	140
25	8	18	40	73	139

From Littlejohn (1980).

31, respectively (Littlejohn, 1980). For the same friction angles, values derived from Fig. 4-18 are about 100 and 35, respectively, and in good agreement with in situ values.

Values of A for enlarged cylinder (type B anchor) and for grout pressures of up to 1 N/mm² (145 psi) have been measured in situ. For compact sandy gravel with $\phi = 40°$, $A = 1.7$. For compact sand with $\phi = 35°$, $A = 1.4$.

Considerable doubt exists with respect to the real value of the end-bearing component of Eq. (4-13) in anchor design. Because side shear is mobilized at smaller axial displacement than end bearing, the entire shaft resistance must be developed before any load can be transferred by end bearing. For very dense compact soil the end bearing surface is small, hence end bearing is a very small component of the total resistance. In loose material, on the other hand, end-bearing surface is large, but before any bearing resistance is mobilized there anchor yielding may be excessive and equivalent to failure.

Ignoring end bearing, Eq. (4-13) reduces to the following

$$T_f = K_1 \pi D L \sigma_v' \tan \phi \qquad (4\text{-}14)$$

where all parameters are the same, and K_1 is an appropriate coefficient of earth pressure. Equation (4-14) has been used regionally for grouted bar anchors in medium to dense sandy gravel with cobbles ($\phi = 35°–42°$). Probable K_1 values are shown in Table 4-9. The value 1.4 for dense sand was obtained from field tests in Boston for the Bauer anchor (Oosterbaan and Gifford, 1972).

Where the effective fixed anchor diameter D is in question, Eq. (4-14) may yield misleading results. In such cases, the following expression is suggested, relating ultimate load to effective borehole diameter d

$$T_f = K_2 \pi d L \sigma_v' \tan \phi \qquad (4\text{-}15)$$

where K_2 is empirically estimated to range from 4 in coarse silt and fine sand, to 9 for dense sand and gravel, with an average value of 6 for injection pressures 0.3–0.6 N/mm². Comparison of K_1 and K_2 values, and also noting the approximate relation between D and d, suggests that the products $K_1 D$

TABLE 4-9   Recommended Values for Earth Pressure Coefficient K_1 for Use in Eq. (4-14)

K_1	Soil Type	Injection Pressure
0.5–1.0	Fine sand and silt for low and high relative density, respectively	Low grout pressure
1.4	Dense sand	Low pressure
1.4–2.3	Medium to dense sandy gravel with cobbles	No grout injection pressure

and $K_2 d$ are largely equivalent; hence Eq. (4-14) and (4-15) involve the same uncertainties and approximations.

Effect of Grouting Pressure. If the soil is not compacted or displaced during casing installation, and if no residual grout pressure is left at the fixed length grout–soil interface on completion of the injection phase, the factor A in Eq. (4-13), estimated in the range 1–2, may reduce to the K_0 value. This reduction is, however, unnecessarily severe, and even if the grout is tremied, A is not likely to be less than 1. For most soils, especially with fine-grained materials, A depends on the residual grout pressure at the fixed zone, which is considered function of the injection pressure. During injection, cement particles form a filter cake at the interface through which only water escapes; hence part of the injection pressure is transmitted to the soil. When grouting is completed, the shear strength of the cement mix combines with ground restraint to cause a residual pressure to be locked into the system. This mechanism provided the basis for developing Eq. (4-12), which relates ultimate load to injection pressure.

In this context, several contractors use a modified version of Eq. (4-12) as follows

$$T_f = p_g \pi DL \tan \phi \qquad (4\text{-}12a)$$

where p_g is the grout injection pressure. Equation (4-12a), however, tends to overestimate pullout capacity, even if $\frac{2}{3}$ of p_g is actually used in estimating T_f (Littlejohn, 1980). As the in situ permeability increases, filter cake formation is replaced by rheological blocking whereby more grout escapes, hence more injection pressure tends to dissipate until further flow of grout is restrained by its own plastic stiffening. It is not surprising, therefore, that in soil with higher permeability such as gravelly sand a lower injection pressure remains locked in. This important difference should be recognized in interpreting ultimate capacity values, for example the data from Fig. 2-18. Essentially, such data are valid for type C anchors where erratic permeation of the grout into soil occurs under injection pressures normally exceeding 2 N/mm² (290 psi).

Load Capacity of Type *C* Anchors

In view of the random geometry of the fixed zone in type *C* anchors, theoretical predictions of load capacity are not reliable. Ultimate load estimates usually are based on design curves derived from field tests for a particular range of soil and ground parameters.

In alluvium, limited data are available from tests in medium sand, and also in variable deposits of sand and gravel, for effective borehole diameters of 100–150 mm. Estimated ultimate load capacity at the fixed anchor length is 90–130 kN/m for injection pressure 1 N/mm^2, and 190–240 kN/m for injection pressure 2.5 N/mm^2 (Littlejohn, 1980).

Tests on Anchors in Sand

Comprehensive tests on anchors in sand have been carried out by Werner (1972), Ostermayer (1974), Wernick (1977), Ostermayer and Scheele (1978), Fujita et al. (1978), Shields et al. (1978), and Somerville (1981). These tests, in addition to details on the mode of load transfer, have also provided empirical design rules for type *C* anchors.

Figure 4-19(a) shows the effect of soil type, density, and fixed anchor length on ultimate load. The graphs represent results of tests on 30 anchors loaded to failure. Fixed anchor length is seen to vary from 2 to 10 m, and the injection pressure has an average value of 0.5 N/mm^2.

The results confirm the conclusions earlier reported by Ostermayer (1974). For a given soil, ultimate load capacity increases rapidly with higher soil density, and for the same relative density the load capacity increases with higher coefficient of uniformity. The increase in load-carrying capacity observed with length tapers off and diminishes for fixed length exceeding 10 m. The load capacity appears to increase with fixed anchor diameter up to 10 cm dia., but in the range 10–15 cm this effect becomes rather uncertain, as shown in Fig. 2-19(b). Compared with these factors, increase in grouting pressure over the range 0.5–5.0 N/mm^2 was found to have a less marked effect on pullout capacity.

The calculated skin friction from these tests is as high as 500 kN/m^2 [10 ksf (kilopounds per square foot)] for sand, and almost 1000 kN/m^2 (20 ksf) for sandy gravel. Inasmuch as these values are considerably higher than those predicted by conventional soil mechanics, a possible explanation is the locking-in and wedging effect caused by dilatancy as the anchor is given a pull. Soil dilatancy is manifested by the relocation of soil particles within a soil continuum under stress, the relocation causing expansion of the soil. If it is restricted laterally, the result can be increase of the in situ active stress by as much as 2–10 times the effective overburden stress. In situ density measurements by standard or dynamic penetration tests are related to ultimate load capacity in Fig. 4-20 for the three fixed anchor lengths used in the tests—namely, 3, 6, and 9 m. The capacities shown include the effect

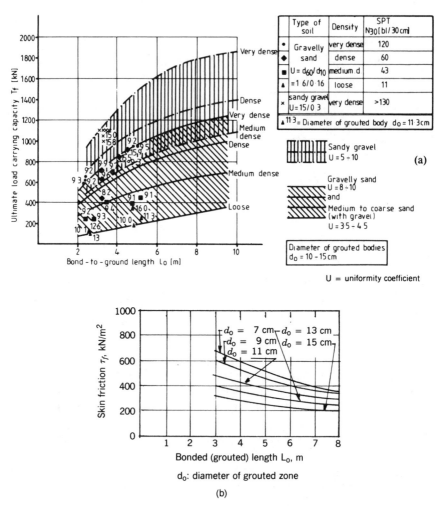

Fig. 4-19 (a) Ultimate load capacity of anchors in sandy gravel and gravelly sand showing the effect of soil type, density, and fixed anchor length; (b) effect of diameter and fixed length on skin friction. (From Ostermayer and Scheele, 1978.)

of dilatancy. Fluctuations in test results are thus possible due to soil inhomogeneity. Direct correlation between maximum skin friction and N values from standard penetration tests is given in Fig. 4-21 (Fujita et al., 1978), derived by an analytical model in conjunction with field tests.

Dilatancy has been theoretically investigated by Somerville (1981) assuming that the soil above an anchor behaves in a manner appropriate to solid material, and the soil continuum is extensive in the lateral direction. For a given anchor inclination and depth, the stress conditions during anchor pullout are expressed in terms of total overburden stress by an appropriate factor M, which is the soil dilatancy factor. It is then possible to correlate

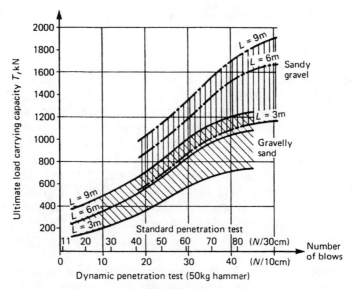

Fig. 4-20 Relationship between ultimate load capacity, fixed anchor length, and dynamic penetration resistance: sandy gravel and gravelly sand. (From Ostermayer and Scheele, 1978.)

ultimate load and fixed anchor length from an analysis of the failure mode, whereas results therefrom are verified and modified by backanalysis of in situ tests. The soil dilatancy factor is directly related to the soil relative density at the point of anchor failure. Values of M between 3.9 and 4.6 are theoretically possible with soil of high relative density, especially when the anchor is at a critical inclination (estimated between 40° and 44° from the horizontal). Dilatancy tests on various soils and relative densities have been carried out by Rowe (1962).

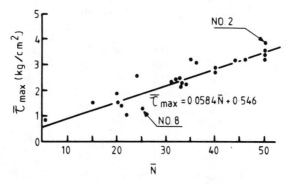

Fig. 4-21 Correlation of maximum skin friction and average (mean) value. (From Fujita et al., 1978.)

Somerville (1981) has produced Table 4-10, which shows values for various soil relative densities, anchor inclinations, and pullout loads. Two fixed anchor diameters have been used in the backanalysis, with the larger applied to pressure grouting.

Unusually high magnitude of load transfer was experienced with anchors in very dense fine to coarse gravelly sand in Washington, D.C. (Shields et al., 1978). Since the diameter of the extracted grouted body was only slightly greater than the borehole, these investigators concluded that at ultimate load as much as 25 ksf (1200 kN/m^2) skin friction was mobilized at the interface, corresponding to a radial stress 36 ksf, which cannot be accounted for by classic theories. Dilatancy was again suggested, in conjunction with increased density and uniformity coefficient. This effect was assumed to have been achieved by the construction characteristics, namely the driving of closed-end casing followed by pressure grouting, which served to increase relative density.

Evaluation of Test Results. Estimation of ultimate load capacity based on the assumption of average uniform friction still remains a good practical choice, although actual field values are rare since contractors are often reluctant to report anchor failures. Actual distribution of friction from instrumental field tests is shown in Fig. 4-22 for fixed anchor lengths from 2 to 4.5 m, and for loose, medium dense, and very dense sand.

The apparent effect of soil density is clearly demonstrated by the maximum friction values 150, 300, and 800 kN/m^2 for loose, medium dense, and very dense gravelly sand, respectively. The 4.5-m-long anchor in loose and medium dense gravelly sand has near uniform friction distribution along the ground–grout interface. However, for dense and very dense materials, max-

TABLE 4-10   Correlation between Dilatency Factor _M_ and Pertinent Soil–Anchor Parameters[a]

Anchor No.	Inclination (β°)	Fixed Length (m)	Force _T_ (kN)	φ'	RD (%)	M D = 1.33d = 152 mm	M D = d = 114 mm
1	35	7.0	765	38°	88	2.37	3.16
2	35	7.0	730	38°	88	2.26	3.01
3	35	7.0	510	38°	88	1.59	2.12
4	35	7.0	612	38°	88	1.90	2.53
5	35	7.0	635	38°	88	1.97	2.63
6	30	7.0	384	36°	85	1.52	2.03
7	30	7.0	720	36°	85	2.83	3.80
8	32	7.0	720	37°	85	2.56	3.41

Key to notation: M = dilatancy factor; RD = relative density; D = diameter of grouted body; d = casing diameter; β = anchor inclination with horizontal; φ' = effective friction angle.
From Somerville (1981).

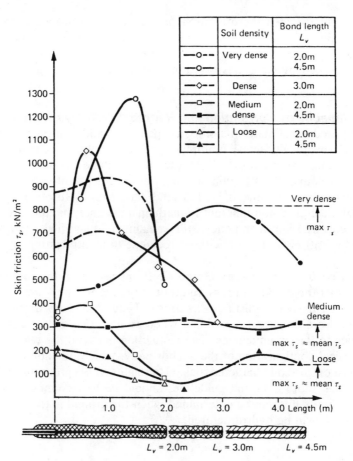

Fig. 4-22 Distribution of long-term friction τ_s at ultimate load in relation to fixed anchor length and soil density; anchor diameter 91–126 mm in gravelly sand. (From Ostermayer and Scheele, 1978.)

imum friction values occur along a relatively short length, which begins at the proximal end and shifts distally as the load increases. It is interesting to note that displacements of the order of 2–3 mm only are sufficient to mobilize high load resistance values.

Pressure grouting may restress the soil medium and thus restore it to its initial in situ relative density, but it appears from backanalysis results that the extent of this effect is not always certain. A disturbed soil continuum should not be expected to be replaced in its particle interlocking by pressure grouting once this interlocking has been disturbed by the installation process.

If the limit value or maximum friction can be assumed to be the same for different fixed anchor lengths, it follows that the mean value of friction for long anchors is smaller than for short anchors, and this is apparent in Fig. 4-19(a). Stated otherwise, there is a critical limit to the effective fixed anchor length beyond which there is no significant increase in load capacity. As shown in Fig. 4-23, for dense sand the load capacity increases by insignificant amounts for fixed anchor lengths greater than 6.7 m. The 6–7-m fixed-length limit suggested by Ostermayer is thus meaningful and economically practical.

Load Capacity by Correlation and Regression Analysis. Kramer (1978) has proposed to estimate anchor load capacity in sand using statistical methods, and specifically linear multiple regression analysis (Littlejohn, 1980). This procedure represents a refined attempt to include all relevant regression constants, boundaries of grain size distribution, depth of overburden, and other pertinent fixed anchor and ground parameters. However, even with the analytical precision inherent in mathematical analysis, the effect of different construction techniques still remains unquantified, and thus no method is entirely complete without recourse to in situ testing.

Distinction between Types B and C. It appears from the foregoing discussion that the design of pressure-grouted type *B* or type *C* anchors is best approached from two distinct directions. Type *B* anchors are reasonably assessed from empirical equations, whereas type *C* is better analyzed by reference to design envelopes. The distinction between the two types is in the physical configuration of the anchor body, which, in addition to its dependence on the soil characteristics, also relates to the grout injection pressure. Thus, more data and guidelines should be forthcoming on injection pressure limits if it can be determined when the ground is to be permeated and when it will be hydrofractured.

4-9 FAILURE OF ANCHORS IN CLAY

General Considerations

Load capacity of anchors in clay generally is low, unless it can be improved by special procedures, because of the low adhesion. More problems in the load transfer will arise if long-term creep occurs and if the anchor hole is allowed to soften. Load capacity can be improved by (a) injecting irregular gravel into the augered hole over the fixed anchor length, together with the cement grout; (b) using high-pressure grouting; and (c) using bells or underreams in the fixed anchor zone. Each of these types responds to load differently.

Tremie-Grouted Straight Shaft

As a first approximation, pullout capacity for tremie-grouted straight shaft anchor is derived theoretically as

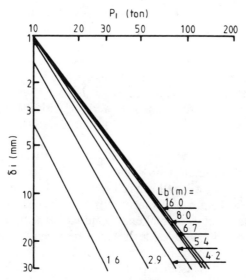

Fig. 4-23 Effect of fixed anchor length on load (P_1)–displacement (δ_1) relationship. (From Fujita et al., 1978.)

$$T_f = \pi DLas_u \qquad (4\text{-}16)$$

where s_u = average undrained shear strength over fixed length
$\quad a$ = adhesion factor

For stiff London clay ($s_u > 90$ kN/m^2) values of a range from 0.30 to 0.35, which is also consistent with the dilute nature of cement grout. This range is considered conservative compared to a values close to 0.45 often used by piling contractors. Values of a close to 0.45 have been confirmed for stiff clayey silt ($s_u > 95$ kN/m^2 or 2 ksf) in South Africa. However, a values for straight-shaft anchors in stiff overconsolidated clay in Taranta, Southern Italy, were found to be between 0.28 and 0.36. In general, the adhesion factor a decreases with increasing shear strength of clay.

Results from tests carried out by Ostermayer (1974) show that the boring technique has decisive effect on load capacity. Boring without casing or with casing combined with flushing water will probably lower the ultimate load, since it tends to have a softening action along the walls of the hole. The use of flushwater is however, economically expedient, hence after boring is completed the tendon should be expediently inserted and grouted. The interaction between clay and grout during setting of the latter is uncertain and incompletely documented. Tests indicate that during this process a complex migration of fluids across the interface takes place (Tanaka, 1980), and in stiff clays this has resulted in reduced average shear bond during quick loading in the order of one-third the undrained shear strength (Evangelista and Sapio, 1978).

The effect of fixed length and diameter is yet to be systematically investigated. Results thus far indicate that average bond resistance (adhesion) is independent of measured diameter in the range 80–160 mm; in other words, within this range total load capacity increases linearly with diameter. Bond resistance is also independent of fixed length up to a value 100 kN/m² (about 2 ksf). For higher values, slight decrease is likely with increase in bond length. Bond at the interface tends to increase with increasing consistency and decreasing plasticity of clay. Useful data demonstrating the foregoing effects on bond are shown in Fig. 4-24.

Ostermayer (1974) also reports tests carried out to investigate the feasibility of increasing ultimate load capacity in adverse soil conditions by increasing the fixed anchor length. In relevant tests, bond resistance was measured over a fixed length of 18 m in stiff to very stiff highly plastic clay. Except for peaks at the proximal end, bond resistance was nearly constant along the fixed length, with values ranging between 40 and 80 kN/m². An interesting anchor behavior in these soil conditions is that tension-type anchors had lower bond resistance than did the equivalent compression type.

The load-transfer characteristics between tension and compression-type anchors have been studied in full scale tests by Mastrantuono and Tomiolo (1977). Figure 4-25 shows layout and load distribution diagrams for two anchors. One is the TPT, a compression-type anchor, whereas the IRP is a conventional tension anchor. The load distribution in both anchors along their fixed length was measured with the use of strain gauges. For the compression anchor load distribution begins at the distal end, where the load transfer is greater, and dissipates toward the proximal end. For the tension anchor, the load distribution begins at the proximal end and is absorbed almost entirely before it reaches the distal end. For the tension anchor, as the load increases its transfer peaks at half-point along the fixed length, and dissipates toward the distal end.

Distribution of Bond. The behavior of anchors in stiff clay is characterized by nonlinear variation and distribution of shear bond at the clay–grout interface both at low stress levels and at failure, which is also the observed performance of anchors in strong rock and in dense sand.

This is clearly confirmed in Fig. 4-26, which shows the distribution of bond stresses in stiff overconsolidated clay for two anchors loaded to failure (Evangelista and Sapio, 1978). The soil in this case has an estimated $E = 6.9 \times 10^4$ kN/m², and average undrained shear strength 270 kN/m². Noting that E values for normal grout mixes are from 1 to 2×10^7 kN/m², it is evident that nonuniformity in stress distribution in stiff clay exists even at modular ratios exceeding 100. The mechanism of load transfer in relation to anchor and soil parameters that normally affect load distribution warrants therefore further studies and investigation. At present, the optimum upper limit 10 m in fixed anchor length appears practical and reasonable for design purposes.

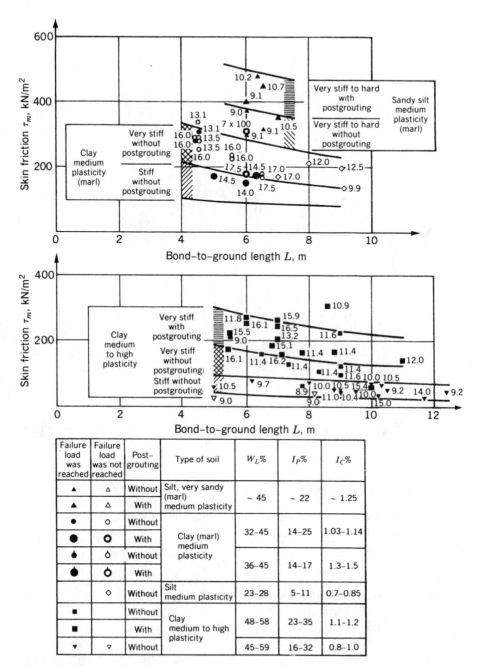

Fig. 4-24   Bond resistance in cohesive soil for various fixed anchor lengths with or without postgrouting. (From Ostermayer, 1974.)

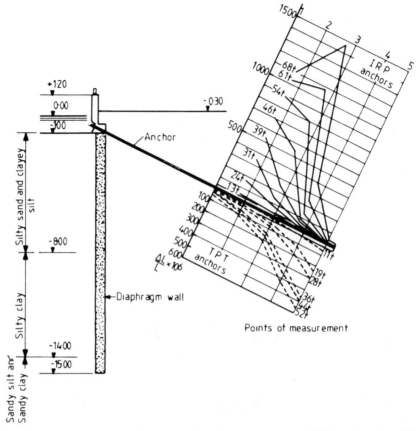

Fig. 4-25   Load distribution of anchors in clay: compression (TPT) and tension (IRP) anchor. (From Mastrantuono and Tomiolo, 1977.)

Underreamed Anchors in Clay

For the configuration of underreamed anchor shown in Fig. 4-27, ultimate load capacity is

$$T_f = T_u + T_e + T_s \tag{4-17}$$

where T_u = side shear in underream length
T_e = end bearing on clay
T_s = side shear along shaft length

Assuming that failure occurs in an ideal elastic medium, T_u, T_e, and T_s are as follows:

$$T_u = \pi D L s_u a_u \tag{4-18a}$$

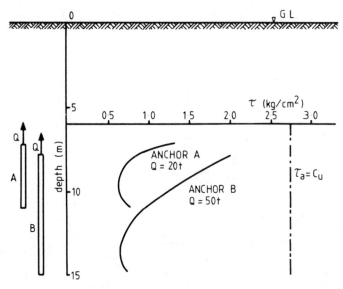

Fig. 4-26 Estimated bond stresses at failure for two anchors of different fixed length. Field tests in overconsolidated clay. (From Evangelista and Sapio, 1978.)

$$T_e = \frac{\pi}{4}(D^2 - d^2)N_c s_u \qquad (4\text{-}18b)$$

$$T_s = \pi d l s_u a_s \qquad (4\text{-}18c)$$

where D = diameter of underream
   L = length of underreamed section
   s_u = average undrained shear strength of clay
   a_u = efficiency coefficient, usually in the range 0.75–0.95 reflecting soil disturbance
   d = diameter of shaft
   N_c = bearing capacity factor
   l = shaft length (part of fixed length)
   a_s = shaft adhesion factor

Bearing capacity N_c values for stiff to hard clays range from 6 to 13, but a value close to 9 is often used, particularly in London clay, and confirmed in practice. Typical range for underream diameter D is 350–400 mm, although much larger values have been used, of the order of 550 mm. Typical values for d are 130–150 mm, but larger values, of the order of 175 mm, are common and in conjunction with larger underream diameters. The underream length L usually is from 3 to 7.5 m, with common values 6 m. The shaft length l ranges from 1.5 to 3 m. The shaft adhesion factor a_s can be assumed to have the same values as for straight-shaft anchors. Further reduction in the ef-

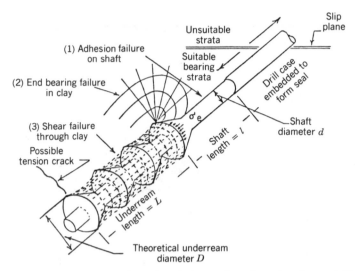

Fig. 4-27   Schematic presentation of underreamed anchor showing mobilization of anchor resistance at failure, specifically, shear and direct bearing.

ficiency factors for both underream shear and end-bearing components is recommended where there is strong possibility of soil disturbance and softening during construction, or where the clay adjacent to the fixed anchor zone contains open or sand-filled fissures.

Optimum underream spacing is achieved when shear plug and end bearing failure occur simultaneously. For this condition to exist $T_u = T_e$ or

$$\pi D L s_u a_u = \frac{\pi}{4} (D^2 - d^2) N_c s_u . \qquad (4\text{-}19)$$

Rearranging Eq. (4-19) we obtain the following expression, giving the maximum underream spacing ΔL necessary for plug failure to prevail

$$\Delta L < \frac{(D^2 - d^2)}{4 D a_u} N_c \qquad (4\text{-}20)$$

or

$$\frac{\Delta L}{D} < \frac{(D^2 - d^2)}{4 D^2 a_u} N_c \qquad (4\text{-}20\text{a})$$

Assuming that $(D^2 - d^2)/D^2 \approx 1$, and using $a_u = 0.75$ and $N_c = 9$, Eq. (4-20a) is reduced to $\Delta L < 3D$; thus, for optimum design underream spacing should be less than three times the underream diameter.

The foregoing analysis combines the recommendations introduced in two independent design approaches by Littlejohn (1970b) and Bassett (1970). For the analysis of anchors in London clay at Oxford Ness, Bassett (1970) also suggested to include the effective stress normal to the proximal end in the end bearing component (See also Nicholson et al., 1982).

Experimental Investigations. Direct comparison between straight shaft and underreamed anchor in the same soil conditions is shown in Fig. 4-28(a) (Wroth, 1975). Both types have 150 mm diameter. The straight shaft with fixed anchor length 10.7 m failed by yielding at 1000 kN (225 kips), whereas the underreamed anchor with underream length 3 m withstood load close to 1500 kN without indication of failure. For this example, Truman Davies (1977) has quantified the associated advantages by field measurements in London clay and backanalysis showing overall improvement more than five times the efficiency of straight shaft anchors.

Bastable (1974) has evaluated results of tests on underreamed anchors used in the construction of a highway underpass in North London, and installed in clay with undrained shear strength 175 kN/m² (3.7 ksf). The underreams are spaced at 1.15 m center to center, and have diameter 540 mm. Several anchors were tested, each with different number of underreams from 2 to 7 in order to study the effect on ultimate load. After completing the loading cycle to the test load, the anchors were loaded to ultimate failure, defined as the load at which increase in strain does not produce an increase in load. Both test and ultimate loads are shown in Table 4-11. Overall anchor movement is between 7 and 15 mm at test load, and 40–50 mm at failure. Figure 4-28(b) shows the ultimate load as function of the number of underreams. By extrapolating an ultimate load 325 kN to one underream and in conjunction with Eq. (4-18b), N_c has a value of 9. On the other hand, con-

TABLE 4-11   Ultimate Loads for Test Anchors in London Clay
(Undrained Shear Strength = 175 kN/m²)

Test Anchor	Test Load (kN)	Underream No.	Ultimate Load (kN)
1	445	2	650
2	840	4	1160
3	1310	6	1540
4	415	2	650
5	1050	5	1300
6	1550	7	—
7	675	3	—
8	630	3	790
9	940	5	—
10	1510	7	—

From Bastable (1974).

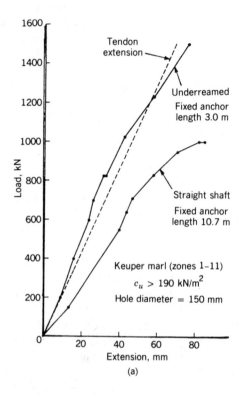

(a)

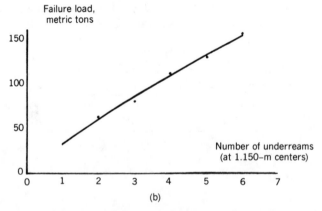

(b)

Fig. 4-28 (a) Comparison of load extension curves for straight-shaft and under-reamed anchors in clay. (From Wroth, 1975.) (b) Diagrammatic presentation of failure load as function of number of underreams. Anchors of Table 4-11. (From Bastable, 1974.)

sidering the slope of the curve between one and two underreams in conjunction with Eq. (4-18a), the value of a_u is computed as 0.75. The diagram
shows clearly an increase in ultimate load-carrying capacity with increasing
number of underreams, but with distinct degradation of one of the load
components at ultimate load compared with its peak value reached at lower
movement. This agrees with plate loading tests in London clay showing
reduction in bearing capacity as much as 10–20 percent following full mobilization of end-bearing resistance.

This complex behavior has been also investigated by Bassett (1977). Beginning with the total stress approach of Eq. (4-17), the theoretical predictions are compared with results from laboratory tests. Theoretical and test
data are shown in graphical form in Fig. 4-29. Part (a) represents the changes
in load capacity as function of underream spacing for an anchor with three
underreams, and evidently the most efficient performance is achieved with
underream spacing about $3D$. For a typical underream spacing $1.5D$, the
effect of the number of underreams on ultimate load capacity is shown in
part (b), and this load is clearly seen to increase with increasing number of
underreams. This increase is essentially linear, but reverts to a nonlinear
form if consistent displacement criteria are imposed on the definition of
failure load. The value of a_u is estimated at 0.63–0.79.

The observed rapid migration of pore water pressure around the underreams and the associated softening of clay has been investigated by Tanaka
(1980) using finite-difference and finite-element methods applied to a single
underream model. The study concentrated on the pore water pressure during
undrained loading, and on the subsequent consolidation. Theoretical predictions were correlated with laboratory tests, which show good agreement
with the consolidation rate. The consolidation mechanism for a load increment is diagrammatically presented in Fig. 4-30. An important observation
is that large pore pressures develop above the underream with negative
values below, resulting in large hydraulic gradients at the edge of the underream. This manifestation becomes quite significant when extended to an
anchor with several underreams.

Contrary to the experience gained in Germany where underreamed anchors did not produce the expected results probably because of large plastic
deformations and creep displacement, the underreaming method can provide
safe working loads from 500 to 1000 kN (110–225 kips) in stiff to hard clay.
This represents 50–80 percent improvement over the capacity attained by
straight shaft anchors, and reflects an average load safety factor of 3. In
view of the limited data available on softening effects around the underreams,
careful assessment of the proposed construction techniques and drilling
methods is necessary during design. Of equal importance is the time taken
for drilling, underreaming and grouting, and although practical limits cannot
be imposed this time should be kept to a minimum. According to current
experience, underreaming is ideally feasible in clays with undrained shear
strength greater than 90 kN/mm² (2 ksf). Difficulties have been encountered

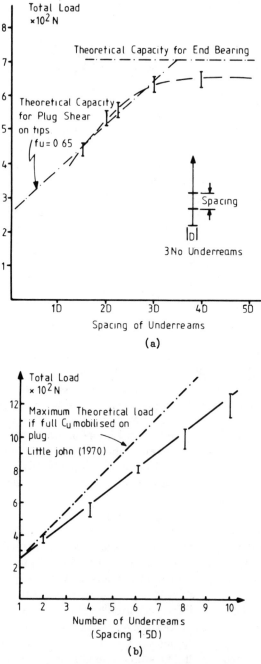

Fig. 4-29   Load transfer in anchors with underreams: (a) load capacity versus underream spacing; (b) ultimate load as function of underreams shown for theoretical and experimental procedures. (From Bassett, 1977.)

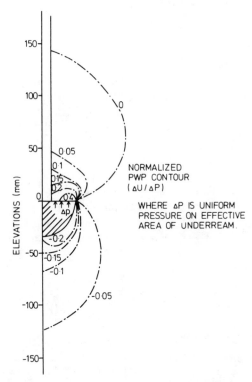

Fig. 4-30 Predicted pore water pressure around underreamed anchor in clay. (From Tanaka, 1980.)

in the form of local collapse or breakdown of the neck portion between underreams where s_u values are in the range 60–70 kN/m². Underreaming is considered impractical and should not be attempted if s_u is less that 50 kN/m², or 1 ksf.

Anchor Load Resistance with Postgrouting

Postgrouting was first tried in West Germany, where considerable work has been carried out and reported by Ostermayer (1974). With this method, high injection pressure is used to cause hydraulic fracture in the clay locally and simultaneously lock in the normal effective stresses acting on the anchor. Reported successful applications show increase in shear resistance along the interface from 120 kN/m² to almost 300 kN/m² for stiff clay of medium to high plasticity.

The data shown in Fig. 4-24 represent a relatively large number of tests in a variety of soil conditions; hence they are useful as design aid for borehole diameters from 80 to 160 mm. The lowest values of shear resistance are 30–80 kN/m², recorded in clay with medium to high plasticity ($I_c = 0.8$–1.0)

without postgrouting. The highest values (>400 kN/m^2) are shown in sandy silts of medium plasticity and very stiff to hard consistency ($I_c = 1.25$) with postgrouting. The method has also been shown to be effective in increasing shear resistance in very stiff clays by about 25–50 percent, but better results are claimed in stiff clays of medium to high plasticity.

Figure 4-31 shows the effect of postgrouting in quantified form for clays of medium to high plasticity. The diagrams indicate linear increase in shear resistance with increasing postgrouting pressure until the latter attains values of 3 N/mm^2 (about 450 psi). Postgrouting pressures should be well below the values at which bursting of the grout can occur.

Whereas the distinction between postgrouting and regrouting is not entirely clear, both mechanisms appear to have the same effect on the development of shear resistance. An example of load transfer by regroutable anchor, in this case the IRP system described in Section 3-4, is from trial tests at Bagnolet east of Paris carried out to assess anchorage feasibility in

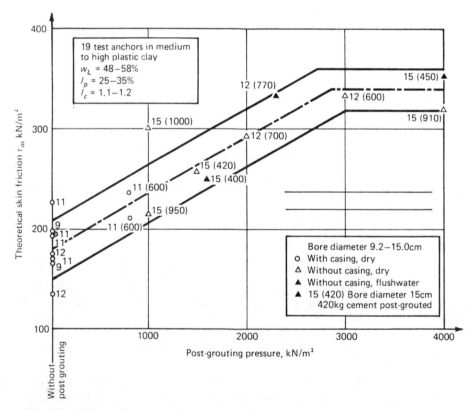

Fig. 4-31 Effect of postgrouting pressure on shear bond for anchors in clay. (From Ostermayer, 1974.)

gypsum bearing marl formations (Jorge, 1969). The in situ undrained shear strength of marls ranges from 27 to 76 kN/m². The fixed anchor zone length was 13, 20, and 26 ft, in borehole diameters 114 and 122 mm. Initially, the holes were grouted at a relatively low pressure (70–130 psi), and thereafter the anchors were tensioned to failure (excessive displacement). The regrouting and tensioning-to-failure step was repeated twice and each time under much higher pressure. Ultimate load capacity, fixed anchor length and grout pressure range are correlated in Fig. 4-32. The ultimate load appears to be linearly proportional to the fixed anchor length, and this linear dependence is true for all three grouting stages. However, the ultimate load capacity increases rapidly with each regrouting stage, and for the 20-ft fixed length the same load at the third grouting stage is almost three times larger than the ultimate load at first grouting.

For the same in situ conditions Jorge (1969) investigated also the effect of the drilling fluid on ultimate load capacity in view of the softening effect that flushwater can have on cohesive soil. The results are shown in Fig. 4-33, which correlates ultimate load capacity and grout pressure in conjunction with two different drilling fluids, namely, water and cement slurry. In the low range of grouting pressure (100 psi or 0.7 N/mm²) both fluids perform in the same manner, probably because the associated soil disturbance is inconsequential with respect to soil strength. In the upper pressure range, however, the diagrams show almost 30 percent increase in ultimate load capacity when the cement slurry is the drilling fluid. This is probably explained by the lesser extent of soil disturbance and softening during high-pressure grouting due to the previously afforded better protection of the borehole by the cement slurry.

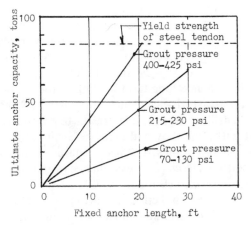

Fig. 4-32 Correlation of ultimate load, fixed anchor length, and grouting pressure in regroutable anchor. (From Jorge, 1969.)

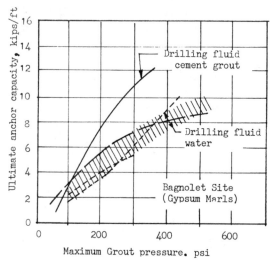

Fig. 4-33 Ultimate load capacity as function of grout pressure and drilling fluid. (From Jorge, 1969.)

The Rationale of Analysis

The effect of soil disturbance during anchor hole drilling is widely accepted, yet incompletely understood. Short term behavior, in particular, is affected by mechanical disturbance and soil wetting around the borehole. Flushing with water and grouting with liquid cement mix are further sources of moisture changes prompting pore fluid migration across the anchor–soil interface. Most of this disturbance appears to be confined to a relatively narrow zone around the drilled hole, although its exact extent depends on construction conditions.

Several investigators have carried out triaxial tests on undisturbed, softened, remolded, and reconsolidated samples of London clay (Chandler, 1968). In these tests, the effective angle of friction ϕ' remained almost unchanged, but remolding reduced the effective cohesion c'. Although this evidence is limited, it may suggest that the clay will revert to its original properties with time; hence long-term loading in the anchor is not affected by initial construction disturbance. Thus, the design hypothesis emerges that during anchor hole drilling the in situ lateral earth pressures around the shaft are relieved, but with time they will increase again although they may not reach the original in situ values. In these conditions, ultimate load predictions can be based on effective stress analysis, and several investigators have followed this approach (Burland, 1973).

Assuming that the horizontal effective earth pressures on the anchor shaft will increase following construction and initial disturbance, the ultimate load capacity will also increase under quick or slow loading in a process known as "ageing." Since the rate of ageing is important in determining the duration of a slow anchor test, it inhibits the investigation of the increase of ultimate anchor load if it is carried out under slow loading. Whitaker and Cooke (1966) report a 12 percent increase in short-term uplift capacity of straight tension piles in London clay for the period from 15 to 62 weeks.

Ultimate Anchor Resistance. In the analysis presented in the foregoing sections, shaft resistance is correlated to undrained shear strength s_u by an adhesion factor a. Agreement with measured values generally is good, but the scattered range of individual results suggests inherent empirical limitations and sensitivity to variable construction factors.

Interpreting test results by total stress analysis is meaningful if the test duration is short so that significant drainage of the clay around the stressed zone does not take place. However, this claim is not accepted in its entirety, and it is instead suggested that because the zone of distortion around an anchor shaft is thin some drainage will always occur during anchor testing (Burland, 1973), The same investigator proposed a relationship involving the coefficient β, which is the ratio of mean effective overburden pressure to the mean mobilized shaft resistance. This approach resulted in good agreement between theory and test results in London clay and narrowed the scattered range of individual results, although it did not amplify the correlation of analysis with construction effects.

A third approach supports the view that the normal testing procedures of anchorage installations produce neither an entirely drained nor entirely undrained loading condition (Rice and Hanna, 1981). It is interesting to note that variations in ultimate pullout capacity of large-diameter shallow anchor footings have been reported when the test duration was changed. According to this view, therefore, a factor that limits the relevance of test results to actual field loading conditions (usually long sustained loads) is the relatively short duration of load tests. Conclusions are drawn from short-term load tests when the anchor is expected to be stressed by long-term loads and after the ageing process has ended.

Anchor Displacement. From the foregoing sections, it appears that anchor displacement is not a critical factor where resistance to pullout is primarily derived from shear bond at the interface. Indeed, a relative displacement always occurs, and is of such magnitude that shear resistance will be mobilized at the interface. However, when a bearing capacity factor is included in pullout resistance anchor displacement must be considered, since much longer movement is necessary to develop the bearing load component. This is the case with underreamed anchors. In these conditions the development of load resistance must be considered from the compatibility of displacement necessary to mobilize each component load.

Experimental Documentation. The adequacy of load-transfer theories is yet to be tested and confirmed. There are further interrelated factors that must be isolated and analyzed, including pore water pressure changes, comparison of ultimate capacity for quick and slow loading after completion of the ageing process, comparison of anchor displacement under quick and slow loading, and effect of clay stress history.

Tests to analyze and study these factors have been carried out by Rice and Hanna (1981) for straight-shaft and underreamed anchors under field and laboratory conditions. These investigators have concluded the following:

1. Considerable increase in shaft resistance under quick loading occurs during the first months after construction, but the rate of increase is slower with larger shaft. If this observation is extrapolated to large-diameter shafts (such as bored piles), it shows the significance of ageing process which may continue for many years after construction.

2. Loss of strength due to remolding may not be observed during the initial load tests. In particular, a shaft was tested several times before such a loss was recorded.

3. Ultimate shaft resistance measured at great age for quick loading showed poor agreement with predictions from total and effective stress analysis.

4. Significant negative pore pressures occur around the anchor during quick tests lasting a few hours. Effective stress analysis is, therefore, not applicable where the soil is subjected to load variations over short time periods.

5. Ultimate shaft resistance during slow tests at great age has values equivalent to an earth pressure coefficient of about unity, but significant load loss occurs in a quick load test.

6. For underreamed anchors the bearing capacity factor was less than the theoretical values. This is probably because as the test proceeds tension cracks in the clay develop near the underream.

4-10 CREEP AND LONG-TERM LOADING

"Creep" is the time-dependent effect of static loading on the anchor. Long term static loading can cause displacement of the anchor, which changes with time. This cumulative effect represents creep in (a) the soil and (b) the anchor components—namely, creep of the grout, steel relaxation, partial debonding of the steel–grout interface, and creep of the tendon connections with the wall and the anchorage. Anchors, therefore, must be designed and installed considering these effects so that creep displacements during the service life of structure will not constitute objectionable movement.

Creep in Soil

Before a failure load is reached, large creep displacements under constant load can take place in cohesive soils and also in uniform grained cohesionless soils, implying a time-dependent stability. In such soils the average size of the interconnecting pores is very small so that pore water flow is retarded by viscosity action. Resistance to flow is measured in terms of flow rate and expressed as soil permeability. The latter is a significant quantitative difference that distinguishes soil behavior under constant load. For example, sand and normally consolidated clay can conceivably have similar effective stress shear strength parameters, but the permeability of the clay is much lower. This difference makes the stability of clay time dependent under static load, whereas the sand responds to loading changes almost immediately.

When saturated clay is loaded, the effective stress undergoes only minor immediate changes, and the pore water takes most of the load. With time, however, the excess pore water pressure is dissipated by drainage away from the zone of increased pressure and into an adjacent zone of lower pressure. The net result is increase in effective stress and a time dependent reduction of soil volume within the zone of influence, a process known as "consolidation." The soil structure stiffens with an associated decrease in settlement and higher strength.

With short term loading the stressed clay does not undergo quick changes in water content or in volume, but the load increment generally will cause some distortion of the stressed zone. With time, significant changes in the effective stress together with changes in soil configuration no longer constitute stable conditions, and the clay enters a state of plastic flow.

Thus, time dependent soils can experience creep displacements under constant load and before structural failure of the anchor occurs. Therefore, in the design of permanent anchors the creep behavior of the soil must be considered and information should be obtained about creep displacement as a function of time. Current theories express the relationship between displacement and time in the form of an exponential mathematical function; thus, a straight line is obtained when this function is plotted to a semilogarithmic scale (Ostermayer, 1974). The slope of this line is defined as the creep coefficient, and it increases for every subsequent loading step. The ultimate load is assumed to have been reached when the displacement does not decrease with time but continuous under load. This condition provides the definition of anchor failure and serves as basis for establishing working loads.

Ostermayer (1974) and the German Code recommend that the creep coefficient K_Δ can be calculated from the expression

$$K_\Delta = \frac{\Delta_2 - \Delta_1}{\log(t_2/t_1)} \tag{4-21}$$

where all symbols correspond to the notation of Fig. 4-34. Values of K_Δ can be estimated at different stages of loading and then recorded as shown. In the above expression, the contribution to creep associated with the components of the anchor is not separated. Creep displacement contributed by partial debonding in the steel–grout interface, creep of cement grout and

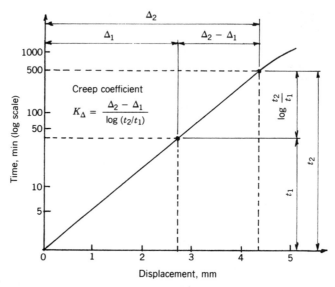

Fig. 4-34 Time–displacement curve for anchor in clay, plotted on logarithmic scale.

relaxation of the steel tendon can amount to a creep coefficient 0.4 mm. Increase beyond this value is the result of creep in the grout–soil interface. Creep in the anchor itself is discussed in the following section.

Figure 4-35 shows creep coefficients plotted as function of the mobilized carrying capacity (ratio of test load to failure load).
Creep values at the beginning of load application are relatively small but increase rapidly (creep coefficient >1) as follows:

- For medium to highly plastic clay of stiff consistency at 40 percent of the failure load.
- For medium to highly plastic clay of stiff to very stiff consistency at 55 percent the failure load.
- For medium to highly plastic clay of very stiff to hard consistency at 80 percent the failure load.
- For uniform grained sand at 80 percent the failure load.

It is evident that these limits indicate the beginning of plastic flow around the grouted shaft, and this stage should be avoided in permanent anchors. Furthermore, these results are for straight shaft anchors only. Ostermayer (1974) reports larger creep displacements with underreamed anchors, probably associated with local stress concentrations resulting in consolidation and plastic deformation.

Inasmuch as the creep phenomenon and the associated anchor response are not fully understood, construction of permanent anchors is not recommended where a soil has large organic content, where it consists of loose

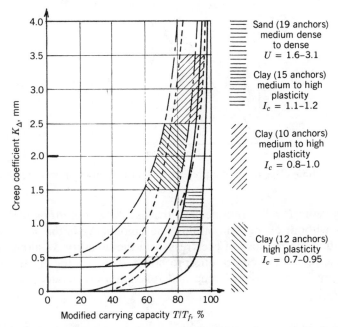

Fig. 4-35 Creep coefficient in relation to mobilized carrying capacity (ratio of test load to failure load) and maximum working load. Results from 56 tests on permanent anchors. (From Ostermayer, 1974.)

sand, or where it consists of cohesive materials with consistency smaller than 0.9 or liquid limit less than 50 percent.

Creep in Anchor Components

Creep in the Grout. In certain cases, anchorage grouts may be susceptible to creep, especially where admixtures are added to improve the antibleeding characteristics. Most cement-base grouts do not experience any significant creep under sustained loads. However, some chemical grouts will introduce time-dependent deformations, but these are used primarily for stabilization and strengthening of sand deposits.

Relaxation of Steel Tendon. Relaxation and creep represent the behavior of steel with time, and result in approximately equal loss of prestress. These mechanical characteristics are generally assessed under controlled laboratory test conditions. Properly devised and performed tests relate the type of steel tendon to both relaxation and creep in a precise manner, and allow the prediction of prestress loss during service. Useful data on relaxation were presented in Section 2-5.

Relaxation behavior is shown diagrammatically in Fig. 4-36. With increasing stress, the strain also increases (elastic deformation). At some point

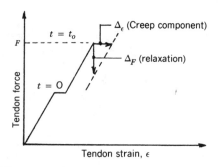

Fig. 4-36   Diagrammatic presentation of relaxation in steel tendon.

of stress value F relaxation occurs; that is, the stress drops with a loss of load but without changes in strain. The amount Δ_F is the steel relaxation representing the net loss of prestress, and the quantity $\Delta\epsilon$ along the strain axis is the relaxation contribution to creep. Thus, in the elastic range and under identical conditions creep and relaxation are related by the simple form

$$\sigma_r = E c_r \qquad (4\text{-}22)$$

where σ_r = relaxation rate
c_r = creep rate of the steel tendon
E = elastic Modulus of steel

Littlejohn and Bruce (1977) draw the following important conclusions on relaxation:

1. The initial conception that relaxation values peak at 1000 h approaching ultimate values is grossly misleading (see also Section 2-5).
2. The use of stabilized wire can reduce prestress loss from 5 to 10 percent to about 1.5 percent at 75 percent f_{pu}.
3. Relaxation rate varies with initial stress, and is a function of the type of steel. For initial stress up to 50 percent f_{pu}, relaxation is very small. For initial stress greater than $0.55\ f_{pu}$, relaxation can be estimated from

$$\frac{f_t}{f_i} = 1 - \frac{\log t}{10}\left(\frac{f_i}{f_y} - 0.55\right) \qquad (4\text{-}23)$$

where f_t = residual stress at time t
f_i = initial stress
f_y = 0.1 percent proof stress at working conditions and temperature
t = time in hours after application of initial stress

4. With initial stress at 70 percent f_{pu}, restressing at 1000 h reduces total relaxation to almost one-quarter of its value. For initial stress at 80 percent f_{pu}, the reduction is one-half. A high degree of accuracy in estimating relaxation loss is not warranted in practice since the significant parameter is the residual stress in the tendon.

5. Important practical feature is the effect of strand jacks on the relaxation of prestressed strand, in relation to the tendency of the latter to unwind. The presence of a torsional component in the jacking force can contribute markedly to relaxation loss.

Creep in the Steel. Creep in the steel, like creep in the soil, is difficult to rationalize theoretically or measure experimentally. Present understanding of the subject is based mainly on the work carried out by Fenoux and Portier (1972).

A diagrammatic presentation of creep in steel is shown in Fig. 4-37(a), where creep strain rate is plotted versus sustained load. Apparently creep starts from the lowest stress values. The creep rate c_r increases over the range 0–30 percent f_{pu}, it becomes constant up to the proportional limit (in this case 0.68 f_{pu}), and thereafter it increases rapidly with higher loads. Continuous strain increase appears beyond the 0.2 percent elastic limit (proof stress). Fenoux and Portier (1972) have reported that the creep for a test stress near the proportional limit in 2 min is 0.2 mm/m of free anchor length. It appears that creep does not terminate with time in spite of an apparent stabilization, although no accurate practical indication can be given with respect to the time after which creep becomes insignificant.

Creep rates can be assumed to be independent of the steel type for stresses below the proportional limit. In this range, creep can be estimated from the following:

$$\text{Creep at time } t \text{ after lockoff} = c_r \log t \qquad (4\text{-}24)$$

The linear proportionality between creep and relaxation rates in the lower stress range is also shown in Fig. 4-37(b). The curve is essentially similar to that of Fig. 4-37(a) in form and shape. Likewise, the relaxation rate increases rapidly beyond the proportional limit.

Littlejohn and Bruce (1977) have assembled examples of documented case histories in which loss of load in the tendon was involved. These data illustrate the importance of the causes of creep and relaxation, and provide a useful reference source.

4-11 REPETITIVE LOADING

Repetitive loading exerts cyclic shear stresses on the soil surrounding the anchor. In sand, after a number of cycles the soil begins to densify because

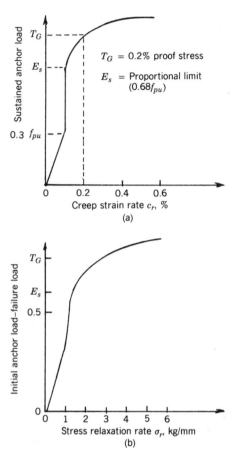

Fig. 4-37 (a) Diagrammatic presentation of creep in steel: creep strain rate plotted versus sustained anchor load; (b) correlation between creep and stress relaxation. (From Fenoux and Portier, 1972.)

of reorientation of particles. Without a corresponding change in volume the normal stress must decrease. Moussa (1976) has carried out cyclic simple shear tests on dry sand showing gradual decrease in the vertical stress as the number of cycles increases. In similar conditions, Youd and Craven (1975) maintained the vertical stresses constant, which resulted in increased density with increasing number of cycles. Under cyclic loading the gradual decrease in the vertical stress around the fixed anchor zone is followed by a corresponding reduction in the bond friction of the anchor until the member fails. In clays, the cyclic shear stresses are accompanied by gradual cumulative increase in pore pressure, which reduces the effective stress and results in reduced bond. Cyclic shear stresses can also cause remolding of the clay with loss of shear strength.

Repetitive load tests on anchors have been limited, and show a paucity of results. However, from the few field and model tests reported to date it appears that repetitive loading can cause anchor failure at load levels below the ultimate static pullout capacity. On the other hand, situations involving unavoidable or regular repetitive load changes represent limited if not sporadic cases in anchor applications. The most likely causes of fluctuating loads are tidal changes, temperature variations, extreme wind loads, and occasional wave action. Dynamic loads such as earthquake effects or impact from heavy moving loads where inertia forces are significant compared to static load intensities are not considered in this review.

Repetitive tests on soil anchors carried out by Soletanche (Pfister et al., 1982) for a sea wall project indicated that if the peak cyclic load were less than 63 percent the ultimate static load the net anchor movement would diminish after five cycles. For larger cyclic loads anchor displacement continued to increase at constant or increased rates. In these tests the number of cycles was limited to 50.

Certain conclusions can be drawn from work on field dead and prestressed anchors in sand, and cylindrical-shaped anchors in sand under laboratory conditions. Carr (1971) reports that load cycling increases anchor displacement, and unloading to zero instead to one-half the peak load increases anchor movement per cycle further. Abu Taleb (1974) performed repeated load tests on prestressed anchors, and concluded that repetitive loading reduces the prestress force. The number of cycles necessary to remove the entire prestress decreases with increasing cyclic load amplitude. In the same test group higher prestress load resulted in smaller anchor displacement per cycle.

Andreadis et al. (1978) carried out a model study on plate anchors in saturated sand. Likewise, the conclusion was that the number of cycles to produce failure decreases with increasing cyclic load amplitude. For a cyclic load level 20 percent the ultimate static capacity, significant increase in strain occurred after 5000 cycles. Furthermore, the absolute magnitude of the peak cyclic load influences the number of cycles to failure but to a lesser extent.

Figure 4-38(a) shows the cumulative effect of load cycles on anchor displacement. These data were obtained from repetitive load tests on 38 mm-dia. plate-shaped anchors in dry medium dense sand (Al-Mosawe, 1979). The load was applied at the rate of one cycle per minute. The graphs represent a composite condition, and show anchor displacement as function of maximum applied load and cyclic load amplitude. The most severe condition results when the load during each cycle is completely removed, whereas if part of the load is maintained, the function of the anchor for a particular displacement range is prolonged. Part (b) of Fig. 4-38 shows the change in the displacement rate per cycle as function of the number of cycles. It is interesting to note that in this study anchor displacement continues to decrease infinitely but never ceases; hence anchor pullout did not occur. Com-

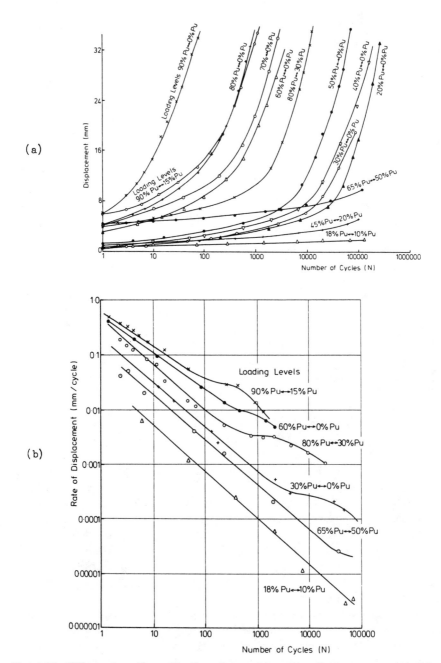

Fig. 4-38 Effect of number of load cycles on (a) anchor displacement—repetitive loading with variable load amplitude; (b) rate of anchor displacement. (From Al-Mosawe, 1979.)

paring repetitive loading effects on dead and prestressed plate anchors, Al-Mosawe (1979) concluded that prestressing can improve longevity markedly.

The effect of alternating loads (tension to compression) on plate anchors has been investigated by Hanna et al. (1978). The conclusion is that stress reversal associated with alternating loads can have more severe effects than repetitive loading. The number of cycles to failure varies considerably with the amplitude of the alternating loads.

REFERENCES

Abu Taleb, G. M. A., 1974: "The Behavior of Anchors in Sand," Thesis, Sheffield Univ.

Adams, J. I., and T. W. Klym, 1972: "A Study of Anchorages for Transmission Tower Foundations," *Can. Geotech. J.* **9** (89).

Al-Mosawe, M. J., 1979: "The Effect of Repeated and Alternating Loads on the Behavior of Dead and Prestressed Anchors in Sand," Thesis, University of Sheffield, England.

American Concrete Institute, 1966: "Bond Stress, The State of the Art," ACI Committee 408.

Anderson, G. F., J. H. Rider, and M. A. Sozen, 1964: "Characteristics of Prestressing Strand." Prog. Report 13, Eng. Exp. Sta., Univ. Illinois Urbana.

Andreadis, A., R. C. Harvey, and E. Burley, 1978: "Embedment Anchors Subjected to Repeated Loading," *J. Geotech. Eng. Div., ASCE,* **104**, (GT7, July).

Anonymous, 1977: "World-Wide Interest in Ground Anchors," *Int. Constr.,* **16** (11, Nov.).

Anonymous, 1970: *Ground Anchors, Cons. Eng.* (Special Supplement) (May), p. 15.

Armstrong, W. E. I., 1948: "Bond in Prestress Concrete," *J. Inst. Civ. Eng.,* **33**, 19–40.

Associated Tunnelling Ltd., 1973: "Report on Ground Anchor Tests-Thames Flood Prevention," Unpublished (14 pp.), Lowton St. Mary's near Warrington, England.

Banks, J. A., 1955: "The Employment of Prestressed Techniques on Allt-na-Lairige Dam," *Proc. 5th Int. Cong. on Large Dams,* Paris, pp. 341–357.

Base, G. D., 1961: "An Investigation of the Use of Strand in Pretensioned Prestressed Concrete Beams," Research Report No. 11, 12 pp., Cement and Concrete Assoc., Lond.

Bassett, R. H., 1970: "Discussion to Paper on Soil Anchors," *Proc. Conf. on Ground Eng..* Institution of Civ. Eng., London, pp. 89–94.

Bassett, R. H., 1977: "Underreamed Ground Anchors," Specialty Session No. 4, *9th Int. Conf. on Soil Mech. and Found. Eng.,* Tokyo, pp. 11–17.

Bastable, A. D., 1974: "Multibell Ground Anchors in London Clay," *Proc. 7th FIP Cong. Tech. Session on Prest. Concrete Found. and Ground Anchors,* pp. 33–37, New York.

Berardi, G., 1967: "Sul Comportamento Deglic Ancoraggi Immersi in Terreni Diversi," Univ. Genoa, Inst. Constr. Sc. Series III, No. 60, 18 pp.

Boyne, D. M., 1972: "Use of Skin Friction Values in Rock Anchor Design," BSc (Honors) Thesis, Eng. Dept., Univ. Aberdeen.

Burland, J. B., 1973: "Shaft Friction of Piles in Clay—A Simple Fundamental Approach," BRE Current Paper 33/73.

Buro, M., 1972: "Rock Anchoring at Libby Dam," *West. Const.* (March), 42, 48, 66.

Carr, R. W., 1971: "An Experimental Investigation of Plate Anchors in Sand," Thesis, Sheffield Univ.

Chandler, R. J., 1968: "The Shaft Friction of Piles in Cohesive Soils in Terms of Effective Stress," *Civ. Eng. Public Works Rev.*, No. 63.

CIRIA, 1967: "The Effect of Bentonite on the Bond Between Steel Reinforcement and Concrete," Constr. Ind. Res. Int. Assoc. Interim Report 9, London.

Clifton, J. R., H. F. Beeghly, and R. G. Mathew, 1975: "Nonmetallic Coatings for Concrete Reinforcing Bars," Natl. Bureau Stand., Dept. of Commerce, Report No. NBS-BSS/65, Washington, D.C.

Coates, D. F., 1970: "Rock Mechanics Principles," Dept. Energy. Mines and Resources. Mines Monograph No. 874. Ottawa.

Coates, D. F. and Y. S. Yu, 1970: "Three Dimensional Stress Distributions Around a Cylindrical Hole and Anchor," *Proc. 2nd Int. Conf. on Rock Mech.*, Belgrade, pp. 175–182.

Comte, C., 1971: "Technologie des Tirants," Inst. Research Found. Kolibrunner/Rodio, Zurich, 119 pp.

Conti, N., 1972: "Reply to FIP Questionnaire."

Costa Nunes, A. J., 1971: "Metodas de Ancoragens," Parts 1 and 2, Associacao dos Antigos Alumos da Politecnica, 50 pp.

Drossel, M. R., 1970: "Corrosion Proofed Rock Ties Save Building from Earth Pressures," *Constr. Meth.* (May), 78–81.

Eberhardt, A., and J. A. Veltrop, 1965: "Prestressed Anchorage for Large Trainter Gate," *Proc. ASCE. J. Struct. Div.*, **90** (ST6), 123–148.

Evangelista, A., and G. Sapio, 1978: "Behavior of Ground Anchors in Stiff Clays." Specialty Session No. 4, *9th Int. Conf. Soil Mech. Found Eng.*, Tokyo, pp. 39–47.

Fargeot, M., 1972: "Reply to FIP Questionnaire."

Fenoux, G. Y., and J. L. Portier, 1972: "La Mise en Precontrainte des Tirants," *Travaux*, **54** (449–450), 43–53.

FIP, 1972: "Draft of the Recommendations and Replies to FIP Questionnaire" (1971). FIP Subcommittee on Prestressed Ground Anchors.

FIP, 1974b: "Questionnaire on Choice of Tendon," FIP Notes 50, pp. 4–8, Cement and Concrete Assoc., 52 Grosvenor Gardens, London.

Fujita, K., K. Ueda, and M. Kusabuka, 1978: "A Method to Predict the Load Displacement Relationship of Ground Anchors," *Revue Francaise de Geotechnique* (3), 58–62.

Gilkey, H. J., S. J. Chamberlin, and R. W. Beal, 1956: "Bond between Concrete and Steel," Eng. Rep. No. 26, Iowa Eng. Exp. Sta., Iowa State College, Ames.

Golder Brawner Assocs., 1973: "Government Pit Slopes Project III: Use of Artificial Support for Rock Slope Stabilization," Parts 3.1–3.6, Unpublished, Vancouver, Canada.

Gosschalk, E. M., and R. W. Taylor, 1970: "Strengthening of Muda Dam Foundation Using Cable Anchors," *Proc. 2nd Cong. Int. Soc. Rock Mech.*, Belgrade, pp. 205–210.

Ground Anchors Ltd., 1974: "The Ground Anchor System" (28 pp.) Reigate, Surrey.

Hanna, T. H., and J. E. Seeton, 1967: "Observations on a Tied-Back Soldier-Pile and Timber-Lagging Wall," *Ontario Hydro. Res. Qtly.*, **19** (2), 22–27.

Hanna, T. H., E. Sivapalon, and A. Senturk, 1978: "The Behavior of Dead Anchors Subjected to Repeated and Alternating Loads," *Ground Eng.*, **11** (3, April).

Hanson, N. W., 1969: "Influence of Surface Roughness of Prestressing Strand on Bond Performance," *J. Prestr. Conc. Inst.*, **14** (1), 32–45.

Hawkes, J. M., and R. H. Evans, 1951: "Bond Stresses in Reinforced Concrete Columns and Beams," *Struct. Eng.*, **29** (12), 323–327.

Irwin, R., 1971: "Reply to FIP Questionnaire."

Ivering, J. W., 1981: "Developments in the Concept of Compression Tube Anchors," *Ground Eng.*, **14** (2, March), 31–34, London.

Jackson, F. S., 1970: "Ground Anchors—The Main Contractor's Experience," *Suppl. Ground Anchors, Cons. Eng.* (May).

Jennings, J. E., and D. J. Henkel, 1949: "The Use of Underreamed Pile Foundations in Expansive Soils in South Africa," CSIR Research Report No. 32, pp. 9–15, Pretoria.

Jorge, G. R., 1969: "The Re-groutable IRP Anchorage for Soft Soils, Low Capacity or Karstic Rocks," *Proc. 7th Int. Conf. on Soil Mech. and Found. Eng.*, Specialty Session No. 15, pp. 159–163, Mexico City.

Judd, W. R., and C. Huber, 1961: "Correlation of Rock Properties by Statistical Methods," Int. Symp. on Mining Research.

Kemp, E. L., F. S. Brezny, and J. A. Unterspan, 1968: "Effect of Rust and Scale on the Bond Characteristics of Deformed Reinforcing Bars," *J. Am. Conc. Inst.*, **65** (9), 743–756.

Koch, J., 1972: "Reply to FIP Questionnaire."

Koziakin, N., 1970: "Foundations for U.S. Steel Corp. Building in Pittsburg, Pa.," Civ. Eng. Pub. Works Rev. (Sept.), 1029–1031.

Kramer, H., 1978: "Determination of the Carrying Capacity of Ground Anchors with the Correlation and Regression Analysis," *Revue Francaise de Geotechnique* (3), 76–81.

Larson, M. L., W. R. Willette, H. C. Hall, and J. P. Gnaedinger, 1972: "A Case Study of a Soil Anchor Tie-Back System," *Proc. Spec. Conf. on Performance of Earth and Earth Supported Structures, Purdue Univ., Ind.*, **1** (2), 1341–1366.

Littlejohn, G. S., 1970a: "Anchorages in Soils—Some Empirical Design Rules," *Suppl. Ground Anchors, Cons. Eng.* (May).

Littlejohn, G. S., 1970b: "Soil Anchors," *ICE Conf. on Ground Eng.*, London, pp. 33–44, and discussion, pp. 115–120.

Littlejohn G. S., 1972: Some Empirical Design Methods Employed in Britain, "Part of Questionnaire on Rock Anchor Design," Geotechnic Research Group, Dept. Eng., Univ. Aberdeen (Unpubl. Tech. Note).

Littlejohn, G. S., 1973: "Ground Anchors Today—A Foreword," *Ground Eng.* 6 (6), 20–23.

Littlejohn, G. S., 1980: "Design Estimation of the Ultimate Load-Holding Capacity of Ground Anchors," *Ground Eng., Found. Publ.* (Nov.), Essex, England.

Littlejohn, G. S., 1982: "Design of Cement Based Grouts," *Proc. Grouting in Geotechnical Eng. ASCE*, New Orleans.

Littlejohn, G. S., and D. A. Bruce, 1977: "Rock Anchors—State of the Art," Found. Publ., Essex, England.

Littlejohn, G. S., D. A. Bruce, and W. Deppner, 1978: "Anchor Field Tests in Carboniferous Strata," *Revue Francaise de Geotechnique* (3), 82–86.

Littlejohn, G. S., B. Jack, and Z. Sliwinski, 1971–72: "Anchored Diaphragm Walls in Sand," *Ground Eng.* (Sept. 4, 1971), 14–17; (Nov. 1971), 18–21; (Jan. 5, 1972), 12–17.

Longworth, C., 1971: "The Use of Prestressed Anchors in Open Excavations and Surface Structures," Australian Inst. Mining and Metallurgy (Illwarra Branch) Symposium on Rock Bolting, Feb. 17–19, Paper No. 8, 17 pp.

Losinger and Co., 1966: *Prestressed VSL Rock and Alluvium Anchors*, Techn. Brochure, Bern, 15 pp.

Lundahl and Adding, 1966: "Dragforankringer i flytbenagen Mo under Grundvattenytan," *Byggmastaren*, **44**, 145–152.

Lutz, L., and P. Gergelen, 1967: "Mechanics of Bond and Slip of Deformed Bars in Concrete," *ACI J.*, **64** (11), 711–721.

Mascardi, C., 1973: "Reply to Aberdeen Questionnaire" (1972).

Mastrantuono, C., and A. Tomiolo, 1977: "First Application of a Totally Protected Anchorage," *Proc. 9th Int. Conf. on Soil Mech. and Found. Eng.*, Tokyo. Specialty Session, pp. 107–112.

Mathey, R. G., and D. Watstein, 1961: "Investigation of Bond in Beam and Pullout Specimens with High-Yield Strength Deformed Bars," *ACI J.*, **32**, 1071.

Mitchell, J. M., 1974: "Ground Anchors," DIC Dissertation, Dept. Civ. Eng., Imperial College of Science and Tech., Jan.

Mohan, D., V. N. S. Murthy, and G. S. Jain, 1969: "Design and Construction of Multi-Underreamed Piles," *Proc. 7th Int. Conf. on Soil Mech. and Found. Eng.*, pp. 183–186, Mexico City.

Morris, S. S., and W. S. Garrett, 1956: "The Raising and Strengthening of the Steenbras Dam (and Discussion)," *Proc. ICE*, Part 1, **5** (1), 23–55.

Moschler, E., and P. Matt, 1972: "Felsanker und Kraftnessanlage in der Kaverne Waldeck II," *Schweizerische Bauzeitung*, **90** (31), 737–740.

Moussa, A. A., 1976: "Equivalent Drained–Undrained Shearing Resistance of Sand to Simple Shear Loading," *Geotechnic*, Vol. 25, No. 3, pp. 485–494.

Muller, H., 1966: "Erfahnungenmit Verankerungen System BBRV in Fels-und Lockergesteinen," *Schweizerische Bauzeitung*, **84** (4), 77–82.

Neely, W. J., and M. Montague-Jones, 1974: "Pull-Out Capacity of Straight-Shafted and Underreamed Ground Anchors," *Die Siviele Ingenieur in Suid-Africa Jaargang* **16** (4), 131–134.

Nicholson, P. J., D. D. Uranowski, and P. T. Wycliffe-Jones, 1982: "Permanent Ground Anchors—Nicholson Design Criteria," Fed. Highwy. Adm. Office of Research and Development, U.S. Dept. Transp., Report No. FHWA/RD-81/151, Washington, D.C.

Nordin, P. O., 1968: "In situ Anchoring," *Rock Mech. Eng. Geol.*, **4**, 25–36.

Oosterbaan, M. D., and D. G. Gifford, 1972: "A Case Study of the Bauer Anchor," *Proc. Spec. Conf. on Performance of Earth and Earth Supported Structures*, **1** (2), 1391–1400.

Ostermayer, H., 1974: "Construction Carrying Behavior and Creep Characteristics of Ground Anchors," *ICE Conf. on Diaphragm Walls and Anchorages*, London, pp. 141–151.

Ostermayer, H. and F. Scheele, 1978: "Research on Ground Anchors in Non-Cohesive Soils," *Revue Franciase de Geotechnique* (3), 92–97.

PCI Post-Tensioning Committee, 1974: "Tentative Recommendations for Prestressed Rock and Soil Anchors," PCI, Chicago, 32 pp.

Pfister, P., G. Evers, M. Guillaud, and R. Davidson, 1982: "Permanent Ground Anchors, Soletanche Design Criteria," Fed. Hwy. Admin., U.S. Dept. Transp., Washington, D.C.

Phillips, S. H. E., 1970: "Factors Affecting Design of Anchorages in Rock," Cementation Research Report R48/70, Cementation Research Ltd., London.

Radhakrishna, H. S., and J. I. Adams, 1973: "Long term Uplift Capacity of Augered Footings in Fissured Clay," *Can. Geotech. J.* **10** (4), 647–651.

Rao, R. M., 1964: "The Use of Prestressing Techniques in the Construction of Dams," *Indian Concrete J.* (Aug.), 297–308.

Rawlings, G., 1968: "Stabilization of Potential Rockslides in Folded Quartzite in Northwestern Tasmania," *Eng. Geol.* **2** (5), 283–292.

Rice, S. M., and T. H. Hanna, 1981: "Tests on Pull-Scale Vertical Anchors in Stiff Glacial Till Soil," *Ground Eng.*, **14** (2, March), 16–28.

Robinson, K. E., 1969: "Grouted Rod and Multi-Helix Anchors," *Proc. 7th Int. Conf. Soil Mech. Found. Eng.*, Specialty Session No. 15, Mexico City, pp. 126–130.

Rowe, P. W., 1962: "The Stress–Dilatancy Relation for Static Equilibrium of an Assembly of Particles in Contact," *Proc. Royal Soc. A*, **269**, 500–527.

Schousboe, J., 1974: "Prestressing in Foundation Construction," *Proc. Tech. Session on Prest. Concrete Found. and Ground Anchors*, pp. 75–81, 7th FIP Cong., New York.

Schwarz, H., 1972: "Permanent Anchorage of 30-m-High Retaining Wall," *Die Bautechnik*, **49** (9, Sept.).

Shchetinin, V. V., 1974: "Investigation of Different Types of Flexible Anchor Tendons (Rock Bolts) for Stabilization of Rock Masses," *Gidroteknicheskoe Stroitel Stuo*, **4** (April), 19–23.

Shields, D. R., H. Schnabel, and D. E. Weatherby, 1978: "Load Transfer in Pressure Injected Anchors," *Proc. ASCE*, **104** (GT9), 1183–1196.

Somerville, M. A., 1981: "A Design Equation for Inclined Ground Anchor Fixed Length in Cohesionless Soil, *Ground Eng.*, **14** (2, March), 26–28.

Sommer P., and F. Thurnherr, 1974: "Unusual Application of VSL Rock Anchors at Tarbela Dam, Pakistan," *Proc. Tech. Session on Prestr. Concrete Found. and Ground Anchors*, pp. 65–66, 7th FIP Cong., New York.

Standard Association of Australia, 1973: "Prestressed Concrete Code CA35–1973," Section 5—"Ground Anchorages," pp. 50–53.

Stocker, M. A., and M. Sozen, 1964: "Investigation of Prestressed Reinforced Concrete Bridges," Part V, "Bond Characteristics of Prestressing Strand," Univ. Illinois Eng. Res. Sta. Bull. 503.

Suzuki, I., T. Hirakawa, K. Morii, and K. Kanenko, 1972: "Developments Nouveaux dans les Fondations de Pylons pour Lignes de Transport THT du Japan," Conf. Int. des Grande Reseaux Electriques a Haute Tension, Paper 21-01, 13 pp.

Tanaka, Y., 1980: "Consolidation Behaviour of a Single Underreamed Anchor in Clay," Ph.D. Thesis, Univ. Sheffield, England.

Togrol, E., and A. Saglamer, 1978: "Short-Term Capacity of Ground Anchors," *Bull. Techn. Univ. Istanbul*, **31** (1), 13 pp.

Trofimenkov, J. G., and L. G. Mariupolskii, 1965: "Screw Piles Used for Mast and Tower Found.," *Proc. 6th Int. Conf. Soil Mech. & Found. Eng.*, **2**, 328–332, Montreal.

Truman Davies, C., 1977: "Report on Discussion to Session IV. A Review of Diaphragm Walls," ICE, London.

VSL–Losinger, 1978: "Soil and Rock Anchors," VSL Intern., Bern.

Walther, R., 1959: "Vorgespannte Felsanker," *Schweizerische Bauzeitung*, **77** (47), 773–777.

Weber, E., 1966: "Injection Anchor, the Stump Bohr A. G. System," *Schweizerische Bauzeitung*, **84** (6, Feb.), 11.

Werner, H. U., 1972: "Die Tragkraft langzylindrischer Erdanker zur Verankerung von Stutzwanden unter besonderer Berucksichtigung der Lagerungsdichte Kohasionsloser Boden," Dissertation, Reichsuniversitat, Ghent, Belgium.

Wernick, R., 1977: "Stresses and Strains on the Surfaces of Anchors," Specialty Session No. 4, *Proc. 9th Intern. Conf. on Soil Mech. and Found. Eng.*, Tokyo, pp. 113–119.

Whitaker, T., and R. W. Cooke, 1966: "An Investigation of the Shaft and Base Resistance of Large Bored Piles in London Clay," Symp. on Large Bored Piles, ICE, London, pp. 7–49.

White, R. E., 1973: "Reply to Aberdeen Questionnaire."

Wosser, T., and M. Darragh, 1970: "Tie-Backs for Bank of America Bldg. Excavation Wall," *Civ. Eng.* (N.Y.), **40** (3), 65–67.

Wycliffe-Jones, P. J., 1974: Personal Communications.

Wroth, C. P., 1975: Report on Discussion on Papers 18–21, page 166. ICE Conf. on Diaphragm Walls and Anchorages, London.

Xanthakos, P. P., 1979: *Slurry Walls*, McGraw-Hill, New York.

Youd, T. L., and T. N. Craven, 1975: "Lateral Stress in Sand During Cyclic Loading," *J. Geotech. Eng. Div. ASCE*, **101** (GT2).

XANTHAKOS, Petros

Ground anchors and
anchored structures

0471 52500

£63·00   £73

willy

PL

CHAPTER 5

DESIGN CONSIDERATIONS

5-1 GROUND AND SITE INVESTIGATIONS

General

Ground and site investigations usually are carried out to confirm the feasibility of a project as a whole, but adequacy in this respect does not necessarily ensure the feasibility of a proposed anchorage installation. This section, therefore, deals with the special topics that are relevant to the design, construction, testing, protection, and monitoring of anchor systems. Whereas adequate data may be available to indicate the feasibility and advantages of anchors for a given project, these data nonetheless may be insufficient to permit their economic design and construction, and this demonstrates the importance of a detailed knowledge of the ground.

In general, an adequate investigation program will include the following stages: (a) initial site reconnaissance and field survey; (b) main field and laboratory geotechnical investigations, which also includes chemical analysis; and (c) investigation during construction.

The exact time of program initiation will depend on the nature and extent of data which are available when ground anchors are considered including institutional and legal aspects. Likewise, the work to be carried out in any one stage will depend on the overall scope of the project. For example, a major retaining wall for a deep basement in a built-up area will require extensive field and laboratory investigations, whereas a simple rock bolt system can be predetermined by visual field observation and mapping. In the same context, the data required for the safe design of temporary anchors are essentially similar to the data necessary for permanent work, hence the scope of investigation may be the same except for corrosion protection and monitoring. The adoption of a stage program of investigation is also advantageous where the main constraints of the project relate to economy rather than time. Each stage can thus be undertaken following consideration of available data, and after a clear commitment to proceed with the planning.

Site Reconnaissance and Field Survey

This initial stage is not intended to determine the need for a ground anchorage as solution to an engineering problem. Rather, it is undertaken with the intent to demonstrate that further investigations are warranted, although useful technical data may also be assembled. This program involves also a so-called desk study that considers information available of the site (ground and physical condition) in documentary form, and is assembled in the following four groups.

Site Topography. This includes the collection and analysis of various maps and aerial photographs. Site topography is useful in obtaining a preliminary plan and profile of the proposed structure and its anchorage in relation to the main features at the site.

Site Geology. Initial studies of geologic and soil survey maps are undertaken to determine the general ground conditions (soil or rock) expected to be encountered. Previous experience with the local geology can be useful in achieving a high level of geologic information pertinent to the site.

Groundwater Conditions. General information on ground water conditions is essential and can be obtained by preliminary tests at the site, whereas information on tides, meteorology, and hydrology can be from other available sources. Observation of surface water runoff patterns, seepage, and vegetation growth is useful in assessing potential drainage problems. Obvious environmental features will indicate the potential aggressivity of groundwater as it may relate to corrosion problems.

Site History. This includes details and records of previous past development, and data or intentions for developing new sites in the area, particularly if this involves extensive underground work. For example, contemplated future tunneling or shaft construction will have significant effects on in situ soil properties related to groundwater changes and ground disturbance.

Where the anchorage zone is formed under other structures or buildings, it is essential at this stage to determine the nature, condition, and location of any existing foundation elements, basements, and substructures. Legal aspects (discussed in other sections) should be recognized and adequately addressed.

In the context of the general construction requirements, site inspection at the initial conception of a project will ensure that anchorage installation and the associated operations can be scheduled with regard to traffic conditions, preservation or relocation of existing utilities, and interaction with existing buildings.

Main Field Investigation

Extent and Intensity of Investigation. The function, geometry, and operational characteristics of anchors relate to the ground conditions, especially around the fixed zone. Minor variations in ground conditions must be given greater attention because of the higher sensitivity of anchors to ground changes compared with more conventional foundation elements. For example, the recording of stratification is important where thin layers of silt and sand are intermixed within a clayey soil intended for the fixed zone of underreamed anchors. In this case the effect will be as shown in Fig. 5-1, and will severely limit the load capacity of the underream.

Since anchors are installed horizontally as commonly as they are vertically, lateral variations in ground properties must be investigated and determined in the same detailed fashion. Investigation methods normally consist of vertical probes, from which it follows that horizontal anchorages will require more boreholes than in other underground work. If the fixed anchor length is underneath an existing structure, drilling for inclined boreholes is still possible, but practical difficulties preclude the boring of inclined investigation boreholes in soil. In this case it will be necessary to resort to supplementary investigations ahead of anchorage construction during the contract period.

The number and locations investigated by borings, probes, or in situ tests and the depth to which they must be extended will be determined with regard to the project type, site shape and dimensions, and data available from previous investigations. For deep anchorages it can be assumed that soils of

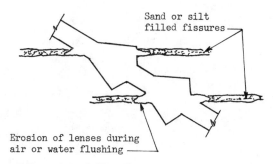

Fig. 5-1 Effect of sand-filled fissures on underream configuration.

greater uniform bearing value will be encountered in deeper than in upper layers. Deeper layers mobilize the supporting reaction whereas the upper strata generate the active pressures; hence fewer probes will probably be required to predict ground resistance than to establish ground active action.

A typical boring layout for anchorage installation is shown in Fig. 5-2 (Otta et al., 1982), and can be used as guide for boring locations and depth. In this case, main borings along the wall alignment are drilled to a depth equal to twice the difference between ground surface level and the depth of a known geologic stratum. Main boring spacing is 50 m (or 150 ft). Intermediate borings of first order are drilled after the results of the main borings are known to a depth equal to twice the difference between ground level and the level of a uniform soil layer determined from the main borings. Intermediate borings of second order are drilled only if a considerable change is recorded in the upper layers, and are spaced as the main borings. Their depth depends on results already obtained.

Additional test borings should be drilled where sloping ground exists, or with potential landslide. Where very long anchors are expected, several test sites should be selected to investigate drilling conditions above the fixed anchor zone.

Sampling. Available sampling techniques are well documented. Samples are taken by standard tube penetrometer, Shelby tube, or NX rock coring to obtain material for identification and testing, and for determining rock quality RQD index.

For anchorage work, samples should be taken from each stratum at max-

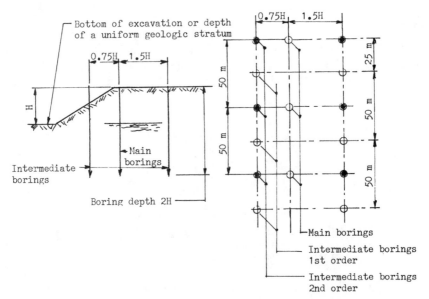

Fig. 5-2 Typical boring layout for an anchorage installation.

imum intervals 1.5 m (5 ft) in thick strata. Intermediate disturbed samples suitable for simple classification tests should also be obtained. In variable ground formations continuous undisturbed sampling may be necessary in the probable region of the fixed zone. In granular soils investigation of density by Dutch Cone Penetrometer or SPT will be justified. Field vane or Dutch Cone tests in cohesive soils will give good undrained shear strength values. For good anchor design it is essential to obtain the in situ permeability in the region of bond resistance. Furthermore, groundwater level should be monitored to allow determination of drilling and grouting pressure as well as effective stresses during service life.

If rock is encountered, discontinuity frequency and orientation data together with information on joint continuity and roughness are important for determining the size and shape of rock mass liable to fail. Data obtained from rock exposures, borehole interface observations, and RQD parameters will be useful in backanalysis of water test data to determine pregrouting requirements. Maximum continuous core recovery should be obtained, which implies core diameters not less than 75 mm. Weak rocks, which are difficult to core, are investigated with the SPT, which provides relative measure of in situ quality.

Groundwater. Determination of ground water conditions are equally essential for the overall design and construction of the project, particularly where deep excavation is contemplated. It must be emphasized that the speed of boring, combined with the addition of water to stabilize the borehole or as circulation fluid during drilling, usually precludes the measurement of equilibrium groundwater conditions during the investigation period. However, all observations of water conditions during boring or drilling should be recorded in the investigation process. For example, the percentage of circulation water return during rotary drilling may help for the initial assessment of groundwater level in rock.

Long term groundwater conditions can be better measured with the use of standpipe piezometers. Where groundwater is contained in several aquifers separated by impervious strata, piezometers must be installed at different levels to record the head at each aquifer.

Additional Requirements. In relatively soft rocks, weatherability and durability should be assessed, particularly at the depth of the proposed fixed zone. Stress–strain characteristics, that is, determination of the modulus of elasticity, are important because of their effect on the pattern and magnitude of bond stress distribution. Radial stress–strain characteristics of the ground mass can be obtained in both soil and soft rock by a pressuremeter test within a borehole. For strong rock the Goodman jack is useful, but if results are difficult to determine, deformability measurements from cores should be carried out.

The extent of field investigation relates to the importance of the project and the associated risks in its execution. If necessary, in situ anchor tests are often carried out to clarify design proposals. In addition, however, to the prior knowledge of certain geologic and geotechnical data, the site investigation enhances the design if it is followed by a construction stage (discussed in the following sections) where drill logs, penetration rates, grout consumptions, and check pullout tests are monitored in order to detect difficult or changed conditions. These terms must be adequately defined in view of the obvious legal implications where doubt exists about anchor competence.

Laboratory Investigation

Soil Properties. Soil properties relevant to anchor design are (a) unit weight in natural condition, (b) angle of internal friction, (c) cohesion, (d) particle size (in cohesionless and mixed soils), (e) in situ density, (f) permeability, (g) liquid and plastic limits, and (h) unconfined compressive strength.

Rock Properties. For permanent rock anchors or for soil anchors having their fixed zone partially extended into rock, the following properties are relevant to design: (a) modulus of elasticity (Young's modulus E); (b) uniaxial compressive strength; (c) existing interfaces between various strata; and (d) the presence of water in the joints, its quality, quantity, and pressure conditions.

For general classification of soils, tests should be carried out to determine the grading of granular materials and the liquid and plastic limits for cohesive soils. Grading also provides empirical range of permeability and an indication of grout penetration under pressure.

Angle of internal friction can be approximately determined from Standard Penetration Test SPT results, but should be confirmed by laboratory shear box tests for free-draining materials. For granular soils of mixed grading peak shear strength of samples can be obtained from direct shear tests for densities from loose to dense.

For cohesive soils the shear strength should be obtained from triaxial compression tests on representative samples. The type of test, specifically, drained or undrained, will depend on the design method (total or effective stress analysis), the mass permeability of the soil, and the expected stressing rate of the anchor. Where the ground displays high mass permeability, that is, silts, clays with permeable fabric, chalk, or marl, both undrained and effective shear parameters are relevant to anchor design. If the anchor is expected to induce a high stress level in the clay between the fixed zone and the structure, the compressibility characteristics should be determined, since they can give guidance on possible loss of prestress through case history comparisons (Littlejohn, 1980). Where soils of high plasticity or compressibility are encountered, their long-term consolidation characteristics should be determined since these data may indicate the probability of consolidation or creep.

Particle size distribution is of great significance, and enables the soil to be described according to the shape of the distribution curve and permeability. In this context, the groutability of the anchor can be assessed, and predictions can be made about the shape and configuration of the fixed zone.

In rocks, index tests such as "point load strength" are suitable and have reasonable cost. Alternatively, it is possible to determine directly the uniaxial compressive strength, and occasionally the tensile strength, on specially prepared test cylinders. A shear box test can be used to determine shear strength of intact material or an existing discontinuity. These tests are intended to provide assessment of rock mass stability and the shear bond in the fixed anchor zone. For bond distribution, however, rock deformability is the main parameter and should be obtained from stress–strain relationships established from uniaxial compression tests or, preferably, from in situ pressuremeter tests.

The susceptibility of rock to weathering can be assessed by the "slake durability test." This allows the sensitivity of rock to flushing water to be examined together with possible mineral reaction with grout or groundwater. In this respect swelling tests are also relevant.

Chemical Analysis. Suitable tests are usually carried out on a routine basis to assess the overall corrosion hazard. Where an aggressive or corrosive environment exists, a comprehensive chemical analysis is mandatory to determine (a) the aggressivity of groundwater with respect to cement and (b) the aggressivity of the soil with respect to metals. This program is discussed in detail in the Chapter on corrosion.

Investigation during Construction

Where the initial investigation suggests that ground conditions may be prone to random variations, data obtained during anchor drilling should be recorded on a daily basis. This record should include variations in strata levels, ground types, and conditions that may require design changes and different installation procedures. The performance of test anchors should also be within the scope of ground investigation, and analyzed with regard to field and laboratory data. These aspects affect the long term anchor performance, and are discussed in detail in the sections on anchor testing and serviceability requirements.

5-2 LEGAL CONSIDERATIONS

Statutory Obligations

Pertinent codes define the role of persons engaged in construction work with respect to health and safety. These guidelines may vary in a wide spectrum of judicial and labor systems, and they may dominate construction markets regionally. In the United States OSHA (Occupational Safety and Health Act) defines the safety requirements for underground construction including excavations, trenching and shoring, tunnels and shafts, caissons, cofferdams, and compressed air. Similar codes are applicable in other countries, combined with national and local statutory requirements.

Anchorage installations generally encounter only few unintentional interruptions because of accidents or due to safety precautions. The occurrence of hazards is confined to three areas.

Protection of the Work Site. Anchorages generally are installed during general excavation, hence they involve a reduced risk of hazards associated with gases, fumes, mist, and oxygen deficiency, except for some dust problems during earth moving. Open excavations, however, present the risk of falls. Occasionally, falsework, shoring, and platforms are required to prestress or distress and remove tiebacks, with some hazards.

Materials Transporting and Handling. The usual materials are tendons, anchor heads, bearing plates, and equipment for drilling, homing, grouting, stressing, and testing. Hoisting and lifting is confined to incidental operations, and the most important handling problem is the transportation and installation of tendons.

Safeguards from Sections of Building Codes. These apply mainly to the shoring and bracing of excavations, and especially the requirements for underpinning existing foundations. In some instances, however, local regulatory authority may be exercized in a manner affecting the installation procedures of an anchorage.

Encroachment under Adjacent Property

Ground anchorages usually encroach beneath adjacent property, and consent of its owner is therefore necessary. This owner may be a private entity or a public agency. If consent should be forthcoming, it should be cautioned that this may be a time-consuming process and therefore the planning of the project can benefit from early action on this matter.

A wayleave or license is sufficient, but the agreement is binding only on the parties concerned and is not a right on the land. This may not be satisfactory for permanent installations where (a) consequences and effects on adjacent land must be considered, (b) ground stability problems may arise, (c) the installation can cause heave or transmission of high direct or indirect pressure on the supported wall, (d) further contractual agreement will be necessary regarding stressed or unstressed anchor parts, and (e) permanent anchors must be monitored and recorded.

In many instances an easement may be necessary, in the form of an agreement in perpetuity. In any case, in exchanging a consent the engineer or contractor should indemnify the adjacent owner against damage or claims by other parties. If consent is withheld, the structure and its support must be redesigned.

Public Liability, Pollution Aspects, and Civil Remedies

The contracting parties are liable for any damage to underground services, utilities, foundations and private structures in the course of drilling, or for damage and injury arising from accidents occurring during prestressing of tendons or dismantling of obsolete anchorages, or for damage to private property associated with excessive ground movement and settlement during and after construction.

Environmental protection agencies and pollution control acts govern the pollution aspects for this type of work, including the effects of grouting, resins and chemical additives, and the discharge of trade affluent into any. land, lake, pond, waterway, and public stream. Noise control codes establish the allowable noise level during anchorage installation.

Civil remedies available to owners of adjacent property whose rights are infringed upon, violated or threatened, include steps to pursue abatement, injunction and damages, and often criminal responsibility. In this respect, engineers are cautioned to seek competent legal advice and become familiar with pertinent statutory provisions and regulations before undertaking the planning of an anchorage project.

5-3 STABILITY OF A MASS OF GROUND

The analysis of anchor load capacity and modes of failure discussed in Chapter 4 should be supplemented, and often preceded, by an analysis of stability of the ground mass above the anchor, although in most cases individual anchors are installed in sufficient depth so that failure of the soil or rock above is quite unlikely. The methods presented in this section apply mainly to anchors constructed in vertical or steeply inclined in a downward direction.

Uplift Capacity of Rock Anchors

For rock, estimation of uplift capacity is based on crude cone or wedge mechanisms whereby the system is equated to the weight of a specified rock cone. Where this cone is situated below the water table, submerged weight is used. This arrangement is shown in Fig. 5-3(a) and (b), and is invariably conservative since it is based primarily on weight ignoring any tensile or shear strength in the rock. Where the rock mass displays heterogeneity or discontinuities, this method is not entirely applicable and necessitates the modified versions shown in Fig. 5-3(c)–(e).

The generalized cone approach described above is further refined and detailed as shown in Fig. 5-4. Part (a) distinguishes cone failure between a straight-bond anchor and a plate anchor, whereas part (b) shows the effect of a group of anchors on the production of a flat vertical plane at the interface of adjoining cones. As anchor spacing for a single line is further reduced, failure assumes a simple continuous rock wedge.

The shape and configuration of the failure volume is generally accepted, but opinion is divided with respect to position along the profile of the anchor. This is shown in Table 5-1, and evidently both the cone angle and its apex vary within relatively wide limits. In the same context, the rock cone method yields variable data on the factors of safety against wedge failure. Factors of safety 3 and 2 have been used by Schmidt (1956) and Rawlings (1968), respectively, whereas Littlejohn and Truman-Davies (1974) report a factor of safety 1.3 and 1.6 for anchors at the Devonport Nuclear Complex. It is not uncommon in current practice to reduce the factor of safety further (often close to unity) in view of the fact that other parameters contributing to resistance are ignored, for example, the shear strength of rock. However, this extra contribution is offset when anchors are installed in highly fissured loose rock strata, especially if they contain interstitial material or high pore water pressures.

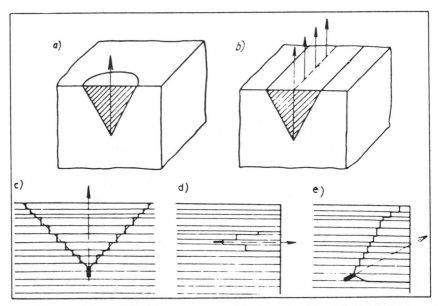

Fig. 5-3 Configuration of rock mass assumed to be mobilized at failure: (a) single anchor in isotropic medium; (b) line of anchors in isotropic medium; (c) perpendicular to planes of discontinuity; (d) parallel to planes of discontinuity; and (e) at acute angle to planes of discontinuity. (From Hobst and Zajic, 1977.)

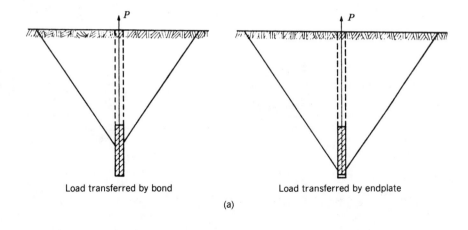

Load transferred by bond   Load transferred by endplate

(a)

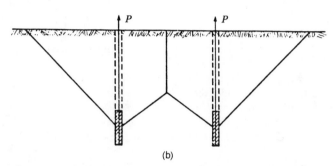

(b)

Fig. 5-4 (a) Geometry of cone, assumed to be mobilized when failure occurs in a homogeneous rock mass; (b) interaction of inverted cones in a general stability analysis.

All these factors have been considered by Hobst (1965) in presenting the empirical expressions shown in Table 5-2, where

τ = shear strength of rock (tons/m^2)
F = factor of safety (usually 2–3)
s = anchor spacing (m)
ϕ = angle of friction across fractures in rock mass
γ = specific gravity of rock
P = anchor load

From Table 5-2, it is evident that the shear strength is relevant to failure for homogeneous rock, whereas weight is the dominant parameter for fissured rock masses. Roch shear strength contributes a major component to the ultimate pullout resistance associated with cone failure, and this demon-

TABLE 5-1   Geometry of Rock Failure Cone and Its Relative Position

Geometry of Inverted Cone		Source
Included Angle	Position of Apex	
60°	Base of anchor	Canada—Saliman and Schaefer (1968)
60°	Base of anchor	USA—Hilf (1973)
90°	Base of anchor	Britain—Banks (1955)
90°	Base of anchor	Britain—Parker (1958)
90°	Base of anchor	Czechoslovakia—Hobst (1965)
90°	Base of anchor	USA—Wolf et al. (1965)
90°	Base of anchor	Canada—Brown (1970)
90°	Base of anchor	Australia—Longworth (1971)
90°	Base of rock bolt	USA—Lang (1972)
90°	Base of anchor	USA—White (1973)
90°	Base of anchor where load is transferred by endplate or wedges	Germany—Stocker (1973)
90°	Middle of grouted fixed anchor where load is transfered by bond	Germany—Stocker (1973)
90°	Middle of anchor	Britain—Morris and Garrett (1956)
90°	Middle of anchor	India—Rao (1964)
90°	Middle of anchor	USA—Eberhardt and Veltrop (1965)
90°	Top of fixed anchor	Australia—Rawlings (1968)
90°	Top of fixed anchor	Austria—Rescher (1968)
90°	Top of fixed anchor	Canada—Golder Brawner (1973)
60°[a]–90°	Middle of fixed anchor where load is transferred by bond	Britain—Littlejohn (1972)
60°[a]–90°	Base of anchor where load is transferred by endplate or wedges	
90°	Top of fixed anchor	Australia—Standard CA35 (1973)
60°	or base of anchor	

[a] 60° employed primarily in soft, heavily fissured, or weathered rock mass.
From Littlejohn and Bruce (1977).

TABLE 5-2 Suggested Depth of Anchor for Overall Cone Stability

Rock Type	Formula for Depth of Cone	
	One Anchor	Group of Anchors
"Sound" homogeneous rock	$\sqrt{\dfrac{FP}{4.44\tau}}$	$\dfrac{FP}{2.83\tau s}$
Irregular fissured rock	$\sqrt{\dfrac{3F(P)}{\gamma\pi\tan\phi}}$	$\sqrt{\dfrac{FP}{\gamma s\tan\phi}}$
Irregular submerged fissured rock	$\sqrt{\dfrac{3F(P)}{(\gamma-1)\pi\tan\phi}}$	$\sqrt{\dfrac{FP}{(\gamma-1)s\tan\phi}}$

From Hobst (1965).

strates the importance of obtaining quantitative data or rock fracture geometry and shear strength prior to design. Brown (1970) suggests that in homogeneous massive rock, pullout resistance depends on the shear strength and the surface area of the cone, which for a 90° angle is proportional to $4\pi h^2$, where h is the depth of embedment. Suggested maximum allowable shear stresses acting on the cone surface vary from 0.034 N/mm² (Saliman and Schaefer, 1968) to 0.024 N/mm² used by Hilf (1973) in conjunction with a factor of safety 2 on a test load displacement not exceeding 12 mm.

Experimental Evidence. There is an impressive scarcity of data on anchor failure in the rock mass, hence documentation of stability theories is not readily available. Saliman and Schaefer (1968) present possible failure modes based on test results at Trinity Clear Creek, as shown in Fig. 5-5. The results represent four tests on deformed steel bars grouted into 70-mm-dia. holes, 1.5 m deep in shale sediments. In all cases failure occurred when a block of grout and rock pulled out. The propagation of cracks to the surface provides indication of the cone of influence. Using rock bulk density 2 Mg/m³, backanalysis of the failure loads gives the following factors of safety: between 7 and 23 if the apex of the 90° cone is assumed at midpoint, and between 0.9 and 2.9 for a cone with its apex at the base.

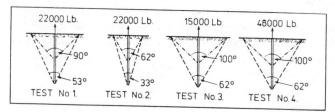

Fig. 5-5 Possible failure modes for anchors in rock mass, based on results at Trinity Clear Creek. (From Saliman and Schaefer, 1968.)

Littlejohn and Bruce (1977) report tests in laminated dolomite carried out by Brown (1970) involving shallow anchors. In this case, the shape of the pullout zone could not be observed, but the broad area over which the rock surface was uplifted suggested failure along a horizontal bedding plane.

Recent comprehensive tests have been reported by Littlejohn and Bruce (1977) and involved 57 anchors installed in the upper carboniterous sediments of the Millstone Grit series. In these tests, failure in the rock mass occurred for embedment depth up to 1.5 m (5 ft), and bond failure occurred at greater depth. The former failure was manifested by the structural configuration of the rock mass. These investigators proposed the following empirical relationship for the ultimate pullout resistance

$$P_f = 600 \, d^2 \qquad (5\text{-}1)$$

where d is the depth of embedment in meters and P_f is given in kilonewtons.

Effect on Anchor Spacing. In anchorage practice, the general trend is toward larger anchor loadings in conjunction with improved utilization of weaker ground. Although rock failure of this type is normally restricted to shallow anchorages, the occurrence of laminar failure or excessive fixed anchor movement should not be excluded. Whereas classification of rock masses with particular reference to fracture geometry will assist design, for closely spaced anchors it is often advisable to consider staggered anchor layout whereby some anchors are longer than others. If bedding planes occur normal to the anchor axis, the staggered lengths will reduce the stress intensity across such planes at the level of the fixed zone. Besides the cone failure, however, the possibility of laminar failure should also be considered in choosing anchor spacing and depth. Useful data and guidelines on this subject are given by Littlejohn and Bruce (1977).

Uplift Capacity of Soil Anchors

The expanding use of vertical anchors in soils warrants the importance of investigating the probability of failure in the soil mass. According to one method of analysis, failure is assumed to be manifested along an expanding conic plug with increasing diameter from the top of the fixed anchor zone as shown in Fig. 5-6. In this case, the weight of soil mass is considered together with shear resistance along the assumed failure surface.

Experience, however, shows that generalized shear failure in a soil mass is associated with relatively shallow installations, and is more common where the top of the fixed anchor zone is less than 3–4 m (10–13 ft) from the ground surface. With deeper installations, anchor failure is confined to the fixed anchor zone and does not propagate to the ground surface.

A different approach with reference to the configuration and volume of soil mass engaged at failure is presented by the French Code, and it is shown

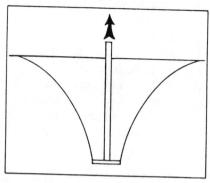

Fig. 5-6 Geometry of soil mass assumed to be mobilized in pullout failure.

in Fig. 5-7. The probable volume of soil involved in uplift failure of the soil mass has the configuration shown in part (a), which resembles a rough cylindrical shape ending with a cone at its lower end. In practice, this cylinder is substituted by a regular cone as shown in (b) for uniform cohesionless soil, or by a combination of cone and cylinder as shown in (c) if the upper soil layer consists of cohesive materials. Point O is the apex of the cone, and is taken as the half-point in the fixed anchor length. This analysis gives a convenient and practical method for predicting failure in a soil mass, and is valid for vertex angle $\beta \leq \frac{2}{3}\phi$. In this case, maximum uplift resistance of the anchor is equated to the weight of the soil mass contained in the cone, and does not include shear resistance along the failure surface.

The equivalent cone method presented above can be extended to consider stratified soil, or soil with groundwater.

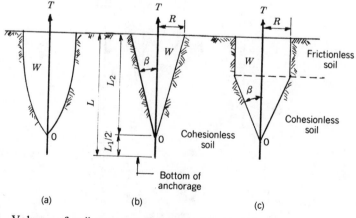

Fig. 5-7 Volume of soil engaged at failure: (a) theoretical cylindrical shape; (b) practical configuration for cohesionless soil; (c) practical configuration for frictionless soil. Point O is half-point in the fixed anchor length and is the vertex of the cone.

Uplift with Overlapping Anchors. For anchors with overlapping cones, the stability of the ground is analyzed as shown in Fig. 5-8. Part (a) shows an isolated anchor, and part (b) shows the overlapping in the zone of influence of the cones of two adjacent anchors. The reduction in uplift capacity for each anchor represents the volume defined by the common chord.

If T is the uplift capacity of the individual anchor, T' is the uplift capacity of the anchor in a group of anchors with overlapping cones, a is the anchor spacing, and R is the cone radius as shown in Fig. 5-8, then the adjusted uplift anchor capacity can be estimated from the relation

$$T' = \psi'T \tag{5-2}$$

where ψ' is a function of the ratio a/R, and is estimated with the help of the diagram and the table shown in Fig. 5-9.

Soil Stability with nearly Horizontal Anchors. A frequent practical case involves shallow individual anchors installed along a profile nearly horizontal or with a small inclination as shown in Fig. 5-10. This problem can arise with the uppermost row of tiebacks in a retaining wall before the next excavation level is reached or where unavailability of a deep anchorage zone restricts the anchor profile to a shallow depth. Failure is characterized by pullout of the anchorage and a mass of soil in front of the installation. As the anchor is stressed, the force is transmitted to the surrounding soil mass, which begins to yield in front of the anchorage. Under more load the shear surface in the soil mass manifests a failure plane, followed by pullout of the

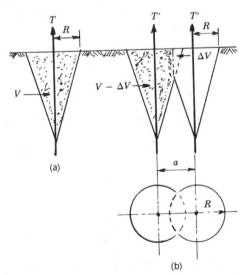

Fig. 5-8 Geometry of soil mass assumed to be mobilized at failure; anchors with overlapping cones of influence.

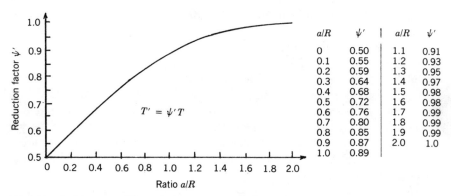

a/R	ψ'		a/R	ψ'
0	0.50		1.1	0.91
0.1	0.55		1.2	0.93
0.2	0.59		1.3	0.95
0.3	0.64		1.4	0.97
0.4	0.68		1.5	0.98
0.5	0.72		1.6	0.98
0.6	0.76		1.7	0.99
0.7	0.80		1.8	0.99
0.8	0.85		1.9	0.99
0.9	0.87		2.0	1.0
1.0	0.89			

Fig. 5-9 Adjusted uplift anchor capacity calculated from the volume of cone influence as function of the ratio a/R.

anchor. The failure mechanism in the soil mass is thus similar to the general shear failure of shallow footings.

The stability conditions can be analyzed in two ways. In the first, stability is checked assuming that failure occurs as shown in Fig. 5-10(a), whereby passive resistance and active pressures are developed in the passive and active zones as shown provided sufficient movement in the ground occurs. Resistance of the soil mass to this failure is equal to the difference $P_P - P_A$. Furthermore, the shear resistance mobilized along the plane that coincides with the axis of the anchor must be compared to the force applied to the anchor, as shown in Fig. 5-10(b). Both types of failure must be

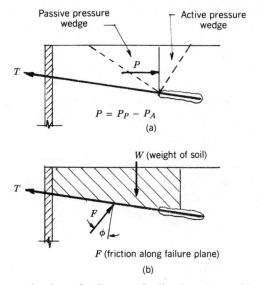

Fig. 5-10 Failure mechanism of soil mass. Inclined anchors with shallow depth.

checked, and a zone of influence in the horizontal profile of the anchorage must be defined geometrically for each anchor.

Practical experience shows that as long as the embedment depth is at least 4 m (13 ft), failure of the soil mass is unlikely even in soft or loose soil. Stability with more rows of anchors is discussed in subsequent sections.

The foregoing practical methods of analysis highlight the complexity of the anchor problem associated with close anchor spacing and anchor prestressing. Pressure grouting also appears to affect stress changes in the soil adjacent to the anchor axis. At present, group action through anchor interaction is thought to be minimized by introducing a lower limit to anchor spacing.

5-4 SELECTION OF FIXED ANCHOR LOCATION

Whether anchors are to be placed in soil or in rock, the design of an anchorage usually begins with the identification and selection of a suitable fixed anchor zone. Several interrelated factors influence this decision, and they are grouped in the following categories according to their relative significance.

Geologic and Geotechnical Data. Initially suitable strata or ground layers must be identified, possessing sufficient strength for the transfer of load and reasonably accessible for the construction operations. These data are also analyzed in relation to optimum anchor capacity.

Structural Requirements of the Wall System. This is based on relevant structural analysis to establish compatibility in the wall–anchor–soil system. It will include consideration of wall stiffness, multiple-tier requirements, vertical anchor component and effect on wall stability at the base, and determination of slip planes and critical zones of the ground mass at failure.

Interaction with Site Conditions. The main concern in this case is with existing services and structures adjacent to the site, or with those to be planned in the future.

Construction Considerations. The selected anchorage zone should accommodate the expected conditions during construction and service, namely, site access, installation sequence, drilling and grouting, time restraints, probable anchor removal or detensioning, and long service monitoring.

Suitable Ground Strata

The presence of suitable ground formations at reasonable depth should be established from the initial soil investigation. On the other hand, initial analysis of project requirements will determine a probable range of anchor loads, which is compared with the load capacities likely to be attained in the designated fixed anchor location. The limits of anchor lengths stipulated for soil and rock should be followed. Fixed anchor zones should not terminate in one line in order to avoid crack formation between the anchor body and the ground behind it. A stable anchor zone must be at least 4–5 m below the ground surface; otherwise the danger of local ground rupture will remain.

The stability of the wall–anchor–soil system is improved if the fixed anchor zone is located in the lower vicinity of the wall, but this criterion must be applied in conjunction with the optimum anchor inclination. If the ground consists of layers with variable shear strength, in the interest of expediency and economy anchor slope and length should be selected so that the transfer of load is accomplished in the layer with the most favorable strength characteristics. This will also provide greater ultimate load capacity thus improving efficiency.

Ground types considered suitable for fixed anchor zones are discussed in Chapter 4. In general, nonplastic soils or dense soils of medium to low plasticity together with granular materials such as silty sand or coarse sand and gravel will provide acceptable fixed anchor zones. Plastic clays and silts, backfill materials, and soils of high organic content should be rejected. A rule of thumb is to check the vertical distance from the anchor entry point to the first suitable ground layer. If it exceeds 100 ft (30 m), the feasibility of an economical anchor design begins to diminish. As it becomes considerably greater, anchorage construction approaches the range of high costs and other alternatives should be considered.

Stability of the Structure–Anchor–Soil System

A fixed anchor zone should be located well beyond the ground mass expected to interact with the wall–anchor system under load. Besides the failure mechanism shown in Fig. 5-10, the location of the fixed zone must inhibit development of similar limiting conditions. These are approached as the soil–structure system undergoes qualitative changes rendering it inoperative.

Figure 5-11 shows potential limiting conditions for a wall supported at the top by a single row of anchors and at the bottom by sufficient embedment below excavation level. In part (a) the wall structure is overloaded beyond its structural capacity, and is on the verge of failure, which may also pull the anchor out of the fixed zone.

With insufficient anchor length beyond the slip plane and also with insufficient embedment below excavation depth, the wall may shift by rotating as shown in (b). Passive earth resistance, which provides the supporting reaction of the wall below excavation level, is manifested in connection with the slip plane of the ground mass and is perceptible through measurement of wall deformation. Equilibrium is established by increasing wall embedment, which will also result in longer anchor zone.

In (c) the wall has sufficient embedment below excavation level and is

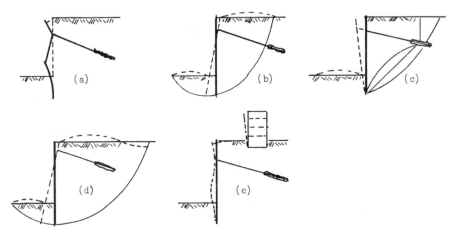

Fig. 5-11   Limiting condition for anchored walls as they relate to fixed anchor location.

stable in this zone, but it tilts forward as shown because the anchor is too short and its fixed zone is within the area which slips. In this case the stability of the ground mass must be analyzed after the fixed anchor zone is moved beyond the potential slip plane. Measureable ground deformations occur before this condition is reached.

The condition shown in (d) involves slipping of the ground mass and rotation of the wall. It occurs because of two unstable factors: insufficient wall embedment and fixed anchor zone within a ground mass prone to failure. Ground deformation measurements can be used to monitor this condition and provide indication of pending danger.

The condition shown in (e) involves stable structure–anchor–soil interaction, but excessive ground deformation associated with large horizontal wall displacements results in unstable foundation conditions for the existing structure. Lateral wall measurements will signal the beginning of this situation. Incidents of this nature may be associated with excessive anchor yielding, ground movement above and below excavation level, and walls that are too flexible and deformable.

Figure 5-12 illustrates a problem common with deep vertical cuts. The lateral earth stresses are considerable because of the depth, and the resulting vertical component likewise has considerable magnitude. With steeply inclined anchors, normally chosen to moderate anchor length necessary to reach the fixed anchor zone, this vertical component increases further and may result in shear failure as shown. The stability of the ground, in this case bedrock, is not only a matter of strength but also depends on the presence or absence of fissures, clay-filled seams, weak joints, and cracks. The same problem can likewise arise with conventional earth-supporting walls where the vertically induced component of the anchor load exceeds the bearing capacity of the wall at its base.

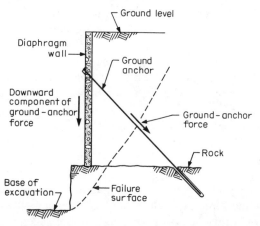

Fig. 5-12 Vertical loads from anchor tension and chiseling into rock, causing shear failure in the rock mass below the base of the wall.

The potential of vertical drawdown of the wall under excessive anchor loads, demonstrated in both soil and rock, is avoided by decreasing the vertical anchor component, by increasing the bearing capacity of the wall at the base, or both. If a suitable fixed anchor zone location exists close to the surface, anchor inclination can be as flat as 15°, which reduces the vertical anchor component considerably. If the depth of the anchor zone requires much steeper angles (45°–50°), the walls must be designed to provide a bearing capacity corresponding to the expected vertical loads.

5-5 SELECTION OF ANCHOR SPACING AND INCLINATION

In general these are factors depending on site access conditions and sequence of construction, but also relate to the optimum number of anchors and optimum anchor load. Structural analysis usually dictates the optimum number of rows and the optimum vertical spacing of anchor levels. This scheme must, in turn, be verified by the expected construction sequence and coordinated with other operations. For example, the most favorable excavation sequence may not provide excavation stage levels corresponding to the selected anchor tier location. Alternatively, extra surcharge loads from heavy construction equipment or incidental materials will impose the need for extra lateral supports. If access to the site is restricted, it will inhibit the use of large drilling equipment leaving the choice to small diameter rotary and percussive tools.

Anchor Spacing. Optimum anchor spacing, both vertical and horizontal, is attained in conjunction with the analysis of the structure or wall to be supported. Since anchors at their heads constitute point supports or reaction supports for continuous beam bracing, their inception and final location follows the analysis of the structure under the expected loading conditions. For example, a diaphragm wall has normal minimum thickness 2 ft (0.6 m), and this provides sufficient structural stiffness for the wall to be designed as two-way slab. Vertical and horizontal anchor spacing is in this case coordinated to provide a balanced distribution of bending moments and shears in either direction.

Secondly, anchor spacing may be determined by the horizontal or vertical structural continuity of the wall, illustrated in Fig. 5-13. The wall types shown have distinct structural and physical characteristics. They may be physically continuous, or consist of vertical and horizontal panels through construction or open joints. Anchor spacing in this case must accommodate the modular configuration of each panel and the constituent reinforcing cages. Irrespective of this analysis, however, anchor spacing should be consistent with anchor loads determined from the available capacity of the fixed zone and potential anchor types.

Anchor spacing may also be dictated by the expected deviation of hole alignment or by possible interference between fixed anchor zones, especially where pressure grouting is involved. If general design considerations dictate close anchor spacing and it appears that the zones of stressed soil or rock will overlap, staggered layouts or lengths are used as shown in Fig. 5-14. Staggered arrangement reduces stress overlapping, especially across planes of ground weakness.

The potential of anchor interference because of close spacing can be assessed once the anchor length, anchor hole tolerance, and probable size or diameter of the fixed zone are determined. The zone of stressed ground can be assumed to be three times the radius of the effective fixed anchor. In this context and taking into account possible alignment deviation, spacing should be selected to separate bond lengths by 6–9 ft (2–3 m), and this guideline is applied to vertical as well as horizontal spacing. Where entry points are closer, the separation of the bond zone can be as shown in Fig. 5-14.

Anchor Inclination. Most anchors are inclined to facilitate anchor hole drilling, homing, and grouting. Furthermore, anchors must be inclined to avoid adjacent foundations and buried structures, or to reach a suitable ground layer. Alternatively, variation in the inclination is chosen to achieve bond length separation.

An angle of 15° with the horizontal is considered by many contractors the minimum practical inclination that can accommodate proper grouting procedures. Furthermore, small anchor inclination implies a lack of overburden depth in the fixed zone, which limits anchor capacity. Within the range of moderate depths most soil anchors are installed at an angle of 15°–30°. Where a suitable ground for anchorage is relatively deep (in excess of 10 m or 35 ft), a steeper angle (usually 45°) may be selected as compromise between anchor length and the associated decrease in horizontal component for a

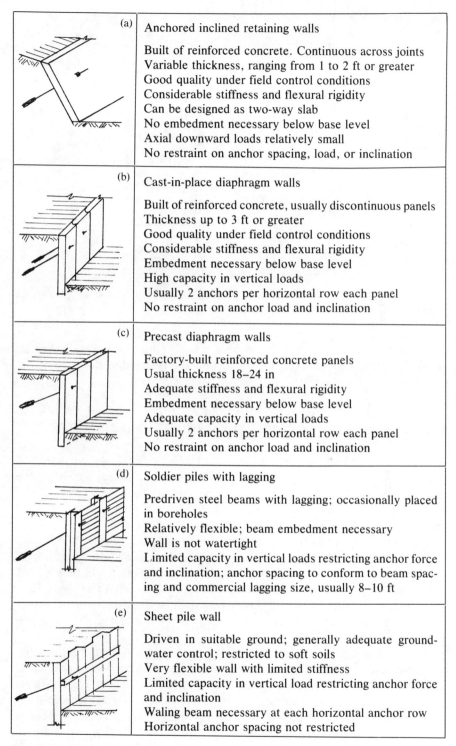

(a)	**Anchored inclined retaining walls** Built of reinforced concrete. Continuous across joints Variable thickness, ranging from 1 to 2 ft or greater Good quality under field control conditions Considerable stiffness and flexural rigidity Can be designed as two-way slab No embedment necessary below base level Axial downward loads relatively small No restraint on anchor spacing, load, or inclination
(b)	**Cast-in-place diaphragm walls** Built of reinforced concrete, usually discontinuous panels Thickness up to 3 ft or greater Good quality under field control conditions Considerable stiffness and flexural rigidity Embedment necessary below base level High capacity in vertical loads Usually 2 anchors per horizontal row each panel No restraint on anchor load and inclination
(c)	**Precast diaphragm walls** Factory-built reinforced concrete panels Usual thickness 18–24 in Adequate stiffness and flexural rigidity Embedment necessary below base level Adequate capacity in vertical loads Usually 2 anchors per horizontal row each panel No restraint on anchor load and inclination
(d)	**Soldier piles with lagging** Predriven steel beams with lagging; occasionally placed in boreholes Relatively flexible; beam embedment necessary Wall is not watertight Limited capacity in vertical loads restricting anchor force and inclination; anchor spacing to conform to beam spacing and commercial lagging size, usually 8–10 ft
(e)	**Sheet pile wall** Driven in suitable ground; generally adequate groundwater control; restricted to soft soils Very flexible wall with limited stiffness Limited capacity in vertical load restricting anchor force and inclination Waling beam necessary at each horizontal anchor row Horizontal anchor spacing not restricted

Fig. 5-13 Various wall types; structural and physical characteristics affecting anchor spacing, load, and inclination.

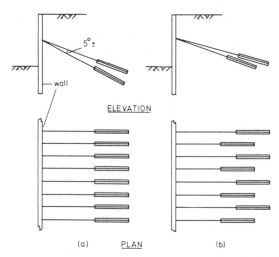

Fig. 5-14 Staggered anchors with close spacing: (a) varying inclination; (b) varying length.

given axial anchor capacity. Steeper anchor inclination also means increased vertical load component, part of which must be resisted at the base of the wall.

5-6 SELECTION OF ANCHOR TYPE, LENGTH, AND DIAMETER

Overall Anchor Length. Given the anchor entry points, the inclination of the installation, and the depth of the suitable ground strata for fixing the anchor, the overall length must satisfy the geometry of the system. Total anchor length obtained in this manner should be assessed in the context of anchor cost, and compared with other possible solutions.

For normal anchor installations, especially those associated with retaining walls, overall anchor length of 40–70 ft (12.5–21 m) is quite common, with a minimum fixed length of 20 ft (6 m). In this range, the economy of the system is well documented. Where the overall anchor length exceeds 125–150 ft, the economic advantages should be scrutinized and subjected to direct comparison with other alternatives.

Fixed Anchor Length. Reference to the appropriate procedures, graphs, and tables presented in Chapter 4 for rock and soil anchors will enable initial determination of fixed anchor length in conjunction with the grouting pressure and the predicted anchor zone type. Field tests will check and confirm the adequacy of this length, with provisions to modify the remaining anchors if the initially selected fixed length is not satisfactory. Regrouting is also a possible remedy where anchor capacity must be increased.

For anchor fixed zones in rock type A is normally assumed, with a fixed anchor length not less than 3 m (10 ft), and not more than 10 m (33 ft). Under certain conditions it is conceivable that a shorter fixed length may be sufficient, however a sudden drop in rock quality along the fixed zone compounded with construction inefficiencies can deprive the installation of its pullout capacity.

For rock cavities (tunnels and caverns) anchor forces and length cannot be determined independently of each other. Both parameters must be selected beforehand and then tested in successive converging steps until their compatibility can ensure the stability of rock around the cavity as it relates to the optimum deformation of the rock mass and mobilization of rock strength.

For rock anchors stabilizing foundation mats or retention systems, the empirical record presented in Chapter 4 can be supplemented by the data shown in Table 5-3.

For anchors in sand, estimation of fixed anchor length may be based on theoretical or semiempirical relations for type B anchor, whereas design curves established from field experience on a range of soils are probably the best procedure for estimating fixed length for type C anchors. Subject to field confirmation, a broad range of anchor capacity and fixed anchor length is shown in Table 5-4.

The selection of fixed length for anchors in clay should recognize the relative validity of undrained shear strength and effective stress analysis. Further difficulties arise in estimating the bearing capacity factor in under-reamed anchors, the reduction coefficients applied to the side shear components to include the effect of soil disturbance and softening during construction, the utilization of higher injection pressure with and without postgrouting, and the upper time limit specified for drilling, underreaming, and grouting. The soil conditions adjacent to the fixed zone will also have significant effects on load-carrying capacity. For example, where the clay adjacent to the fixed zone contains open or sand-filled fissures, a 50 percent reduction in side shear and bearing components is not unlikely.

Anchor Hole Diameter. This depends mainly on anchor size and type, corrosion protection requirements, drilling procedures, and ground conditions. It is interesting to note, however, that a common range of drilled hole diameters is 3–6 in (75–150 mm). Since the vast majority of soil anchors are drilled in cased hole, the weight of the casing and its handling appear to impose an upper limit in hole size, so that at present the 6-in hole is common and practical in this respect. Typical hole sizes are shown in Figs. 2-20 and 2-21 for anchors in rock using percussive methods. The effect of corrosion protection on drilled hole diameter is obvious. For single corrosion protection, the table in Fig. 2-20 shows that 27 strands can be accommodated in a 6-in hole, but for double corrosion protection the same number of strands will require an 8-in hole, as is evident from the table in Fig. 2-21.

TABLE 5-3   **Maximum Load-Bearing Capacity and Fixed Anchor Length in Rock**

Type of Rock	Load-Bearing Capacity (Maximum) (kN)		Fixed Anchor Length L (m)
	High Degree of Fissuring	Low Degree of Fissuring	
Granite, gneiss, basalt, hard limestones, and hard dolomites	≤2000	≤4000	4–6
Soft limestones, soft dolomites, hard sandstones	≤1200	≤2000	4–6

Selection of Tendon Type. Section 2-4 provides a comparative description of tendon types together with relevant mechanical characteristics, and the associated advantages and disadvantages. The choice of tendon type is further dictated by (a) the working life of the system and the requirements of corrosion protection, (b) the load capacity, and (c) the drilling methods and homing conditions at the site. An important consideration dictating the choice of tendon type is the actual cost of installation after loss of prestress, which must include the cost of testing and anchor monitoring. A final factor to be considered is regional preference and previous experience with each type.

5-7   DESIGN OF ANCHOR HEAD

Anchor head details, function, and design requirements were discussed in Section 2-6. In general, anchorage components are detailed and standardized by the tendon supplier or manufacturer, and relevant data include such com-

TABLE 5-4   **Maximum Load Capacity and Fixed Anchor Length for Anchors in Sand**

Type of Ground	Load-Bearing Capacity (Maximum) (kN)[a]		Fixed Anchor Length L (m)[b]
	Unconsolidated Deposit	Consolidated Deposit	
Sandy gravel	≤600	≤1000	4–6
Silty sand	≤400	≤600	4–6

[a] 1 kN = 0.2248 kips (force).
[b] 1 m = 3.279 ft.

ponents as standard anchorages, template and custom-designed anchorages, devices to measure force and deformation, anchorage recess, and stressing clearance. Anchor heads for permanent installations are additionally provided with corrosion protection on the basis that external conditions are aggressive, although this is not identified at the time of installation.

Distribution Plate. The distribution bearing plate transfers the anchor load into the main structure, and is located directly under the anchor head. Its location represents a zone of maximum shear, for which the structure must be designed accordingly.

Where the distribution plate is bedded on to concrete, the pad should not exceed 10 percent of the plate width or 10 cm (4 in) in thickness. The allowable bearing stress on the concrete pad should be as stipulated by the applicable concrete codes.

Concrete Blocks. Where the anchor inclination is severe and the prestressing force considerable, the load of transfer is better accomplished by the concrete support shown in Fig. 5-15. The concrete block is cast after the main wall, but is designed as a reinforced concrete member. A shear-friction concept can be applied assuming that failure in the connection area can occur in the most undesirable manner. Structural continuity and resistance to shear is provided by extending the steel bars into the main wall as shown.

Steel Grillage Support. If anchors transfer their loads to a steel structure, specifically, a sheet pile wall or soldier beams, they will probably apply this load to a waling beam. The connection usually involves a steel bearing plate or a steel bracket, designed and detailed according to applicable steel codes.

If a group of anchors support a structure, the design of the structure or parts therefrom should recognize a loading stage where one anchor in the

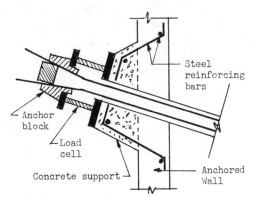

Fig. 5-15 Concrete block used as support to transfer load from the anchor to the structure.

group fails whereas the remaining units continue their function and remain active under load. The elimination of one support will cause the structure and its members to be overstressed, which is acceptable by the design rules and standards established in this section provided the overstressing does not exceed the allowable working stress in bending or in shear by more than 50 percent. The load from the anchor that became inoperative will be redistributed to the neighboring units and their components. Any overstressing thus produced should not exceed the working anchor load by more than 25 percent while remedial measures are under way to replace the failed unit.

5-8   ESTIMATION OF LOCKOFF OR TRANSFER LOAD

The lockoff load is the sum of the design (working) load plus a load allowance to compensate for seating loss and long-term time-dependent loss (creep in the soil, creep in anchor components, and steel relaxation). After the design load has been verified by testing, the anchor is prestressed to a level that includes the adjustments due to the expected loss, and this load is locked off. Seating loss is instantaneous and involves slippage in the holding parts and devices. Usually, it is quoted in the range $\frac{1}{8}$–$\frac{3}{8}$ in (3.2–9.5 mm). Long-term loss due to steel relaxation, creep, temperature effects, and soil deformation can be estimated under the guidelines discussed in Chapter 4 and in conjunction with technical data supplied by the tendon manufacturer. A frequent range is 10–15 percent the transfer load. The following example illustrates how the lockoff load is estimated.

An anchor unit consisting of seven strands (see also Fig. 2-21) has a confirmed working (design) load of 145 kips (645 kN). Noting that each strand has 41.3 kips (184 kN) f_{pu}, the design working load is 50 percent f_{pu} (total $f_{pu} = 41.3 \times 7 = 289$ kips), which is the final prestressing force that must remain in the tendon after all losses occur. The total cross-sectional area of the tendon is computed at 1.07 in², and the quoted modulus of elasticity for the steel is 28×10^6 psi. Further data are the expected seating loss quoted at $\frac{1}{4}$ in, the long-term loss assumed 8 percent the design load, and the free anchor length 40 ft.

Initially, the elongation of the steel tendon ΔL is computed for a load 145 kips as

$$\Delta L = \frac{PL}{AE} \qquad (5\text{-}3)$$

or

$$\Delta L = \frac{145 \times 1000 \times 40 \times 12}{1.07 \times 28 \times 10^6} = 2.3 \text{ in (58 mm)}$$

If the seating loss is $\frac{1}{4}$ in and the expected long-term loss 8 percent the design load, the combined tendon elongation during stressing is

$$2.3 \times 1.08 + 0.25 = 2.48 + 0.25 = 2.73 \text{ in (69 mm)}$$

This total elongation 2.73 in is entered now into Eq. (5-3) with P as the unknown factor, or

$$P = \frac{2.73 \times 1.07 \times 28}{4 \times 12} = 170 \text{ kips (756 kN)}$$

Under these conditions and assumptions the tendon will be prestressed to 170 kips (or 59 percent f_{pu}) so that the final effective prestress after losses will be the design working load of 145 kips.

The foregoing method of analysis is quick and convenient because of its practical value as long as it is understood that an anchored structure does not react as a prestressed structure does. Basic to the evaluation of these effects is the establishment of time behavior of the medium within which the prestress loss is expected to occur, as well as changes in actual earth pressures and loads expected to be manifested in the final configuration of construction. Equally valid in determining lockoff load is the concept of stiffness and deformability of the ground mass, which resists the prestress application. This subject is discussed in subsequent sections.

5-9 LOADS ACTING ON ANCHORS

The concept of loading discussed in Section 4-1 identifies two types of loads: those acting on the anchored structure, and special static loads deliberately induced by pull during testing or at lockoff stage. The second group of loads is intentionally manifested by prestressing the anchors to the desired level. These prestress loads may persist with time or may revert to the loads acting on the structures during the structure–soil interaction.

Loads acting on the the anchored structure include the following:

Lateral Loads. These consist of (a) lateral earth stresses, which are generally dependent on the magnitude of strains developed in the ground; (b) lateral pressures caused by surcharge loads acting at ground surface; (c) lateral stresses induced by concentrated loads, such as footings, acting within a mass of soil; and (d) water pressure.

Vertical Loads. These include the weight of the anchored structure and reactions from interacting loads reaching the anchors indirectly. Besides the forces and loads transmitted from above, a structure may be subjected to the action of upward forces caused by ground reaction, heave, and uplift.

Construction Loads. These are created in two ways: (a) by changing the earth stresses existing at some limiting state and (b) by inducing loads associated with construction operations and equipment.

Dynamic Loads. These may include vibratory effects from earthquake activity or impact from nearby heavy moving loads, and are of such intensity that they must be included in the design.

Load intensity, effects, and distribution are treated in detail in subsequent sections.

5-10   FACTORS OF SAFETY

Steel Tendon. Section 2-5 presented a general review of allowable stresses and factors of safety for bar, wire, and strand, used here and abroad for permanent anchors. This review was documented by the data shown in Tables 2-3, 2-4, and 2-5.

A more general summary is shown in Table 5-5, which covers recommendations contained in codes of practice as well as suggestions by anchor contractors and engineers. This summary shows a definite trend to increase the measured and ultimate factor of safety to 1.5 and 2.0, respectively.

The recommended factor of safety for steel tendons, irrespective of type, by this author is merely derived as an extension of the working (design) stresses and loads expressed by Eqs. (4.1) and (4.2) for permanent and temporary anchors, respectively. Accordingly, the following guidelines are introduced.

For permanent anchors:
Working stress = 50 percent f_{pu}
Ultimate factor of safety = 2.0
Measured factor of safety = 1.5
For temporary anchors:
Working stress = 62.5 percent f_{pu}
Ultimate factor of safety = 1.6
Measured factor of safety = 1.25

The measured factor of safety is the test (proof) load divided by the working load. It implies that the test (proof) load must be at least 1.5 and 1.25 times the design (working) anchor load for permanent and temporary anchors, respectively.

According to Section 2-2, temporary anchors will remain in service less than 6 months. The same classification may include installations expected to remain in service up to 18 months provided a monitoring program is included.

Relating the margin of safety against structural failure to the degree to which public safety is involved is a philosophy to which this author has not subscribed. The underlying doctrine is that structural failure should always be prevented with a factor of safety that statistically, and also in the context of engineering analysis, is satisfactory, and this irrespective of public liability and order. Interestingly, material strength and stiffness for anchor tendons should be as expected, since by definition the margins of error in their quoted values should not exceed 5 percent. If the estimated (calculated) loads and forces acting on anchors are close to the actual developed during service, the anchorage should perform satisfactorily within acceptable deformations. Hence, if the intent of introducing contingencies is to avoid a growing situation whereby engineers will tend to specify a higher factor of safety, it follows that the most logical and reliable approach is to scrutinize the magnitude of loads and combinations therefrom expected to act on the anchorage.

The foregoing factors of safety should be applied to all anchor components for which precise mechanical and strength characteristics are available, hence they include anchor head and its parts.

Ground–Grout and Grout–Tendon Interface. The literature contains a conspicuous shortage of specific recommendations and guidelines pertinent to the factor of safety that should be applied to these two media. Most codes are reluctant to stipulate a specific procedure, citing the uncertainty of data relevant to these media, and the engineer therefore must judge on the basis of the best information available what safety factors are prudent. This reluctancy appears to originate from two sources. Initially, practical experience documents the variability of factors affecting the transfer of load, and points to the need to improve the quantitative and qualitative level of this record. Secondly, this lack of uniformity is assumed to be compensated by the generous bond lengths normally provided at these interfaces, to be followed by a recourse to trial tests. All present variables taken into account, a minimum factor of safety at least 2.5 and preferably 3.0 for the ultimate static load should be applied to the ground–grout or grout–tendon interface, unless full-scale field tests confirm that a lower value is satisfactory.

Interestingly, static loads are not necessarily the governing criterion. Where excessive movement cannot be tolerated, which is often the case in urban excavations, load–displacement relationships appear to merit attention. In this case the soil–structure system may reach a point where it no longer satisfies the requirements for which it was designed, hence a limit state is approached. Overall stability is still provided, but the design must also consider a serviceability criterion that may give rise to a limiting condition related to permissible displacement rather than ultimate carrying ca-

TABLE 5-5 Allowable Stresses and Factors of Safety Recommended for Anchor Tendons

Working Stress (%)	Test Stress (%)	Measured Safety Factor	Ultimate Safety Factor	Remarks	Source
50	75	1.5	2	With respect to (w.r.t.) characteristic tensile strength	Britain—Littlejohn (1973)
<50	70	1.5	>2	w.r.t. characteristic tensile strength	Britain—Mitchell (1974)
50	75–80	1.5–1.6	2	w.r.t. characteristic tensile strength	Britain—Ground Anchors Ltd. (1974)
	80				Britain—CP 110 (1972)
<60	<90	1.5	1.75	w.r.t. yield strength	Germany—DIN 4125 (1972)
70	77	1.1	1.43	"Swissboring SA" BBRV anchors	Switzerland—Descoeudres (1969)
<69	≤90	>1.3	>1.45	w.r.t. yield strength	Switzerland—Draft recommendation (1973)
<70	<95	1.36	1.43	w.r.t. 0.1% residual elongation	France—Fargeot (1972)
60				w.r.t. elastic limit	France—Adam (1972)
53–66	80	1.2–1.5	1.5–1.9	w.r.t. ultimate tensile strength	France—Fenoux and Portier (1972)
≤60		1.3	2	w.r.t. elastic limit	France—Bureau Securitas (1972)
			1.5–2	w.r.t. elastic limit	Italy—Mascardi (1972)
65 or 85			1.54	w.r.t. ultimate tensile strength	Finland—Laurikainen (1972)
				w.r.t. elastic limit	Finland—Laurikainen (1972)
59 or 71			1.69	w.r.t. ultimate tensile strength	Czechoslovakia—Voves (1972)

Reference	w.r.t.	Measured safety factor	Ultimate safety factor	(%)
Czechoslovakia—Voves (1972)	w.r.t. elastic limit	>1.2	>1.75	<57, <69
Czechoslovakia—Draft Standard (1974)	w.r.t. ultimate tensile strength			60
Canada—Golder Brawner Assocs. (1973)	w.r.t. ultimate tensile strength	1.33	1.67	80
USA—White (1973)	w.r.t. ultimate tensile strength	1.5	2	50, 75
Brazil—de Costa Nunes (1971)	w.r.t. ultimate tensile strength	1.1	1.7–1.9	55–60
Brazil—da Costa Nunes (1971)	w.r.t. elastic limit	1.1		90
South Africa—Perry-Davies (1968)	w.r.t. ultimate tensile strength		1.54	65 or 80
South Africa—Perry-Davies (1968)	w.r.t. yield strength			
South Africa—Johannesburgh (1968)	w.r.t. ultimate tensile strength	>1.2	≥1.43	≤70
South Africa—Code (1972)	Wires w.r.t. ultimate tensile strength		≥1.43	≤70
South Africa—Code (1972)	Bars w.r.t. ultimate tensile strength		≥1.49	≤67 or 80
South Africa—Code (1972)	Bars w.r.t. 0.2% proof strength	1.25	≥1.49	
Australia—Koch (1972)	w.r.t. ultimate tensile strength			75
Australia—Code CA 35 (1973)	w.r.t. ultimate tensile strength		≥1.67	<60
New Zealand—Irwin (1972)	w.r.t. ultimate tensile strength	1.5	2	50, 80

Measured safety factor = test load/working load

Ultimate safety factor = failure load/working load

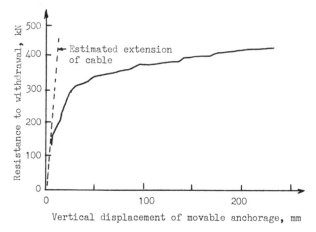

Fig. 5-16   Load–displacement curve for compact fine to medium sand (ϕ = 35°).

pacity. It thus becomes prudent to establish a yield load producing unacceptable movement and correlate it to the ultimate static load. This approach becomes more significant as less discretion is exercised in choosing suitable ground for anchorages or where such ground simply is not available. Typical examples are situations where an increased factory of safety (>2) against static load results after the structural ground–anchor system is proportioned to deal with (a) large displacement for a given load increment and (b) creep in plastic soil.

An example of load–displacement behavior is shown in Fig. 5-16, and in this case it constitutes a serviceability criterion. As the load increases beyond 200 kN, the anchor is seen to undergo vertical displacement at rapidly increasing rates, and this behavior establishes a limit value in the design load. In the same context, the concept of serviceability factor enters the analysis of soils susceptible to creep. If the criterion of suitability and acceptance tests is a creep coefficient K_Δ less than 2 mm under a load 1.5 times the working load, this would clearly imply a value of the latter related to ultimate static load by a factor of safety less than 2.0 (see also Fig. 4-35). Since this is not the intent of the serviceability approach, this observation prompted Ostermayer (1974) to recommend a K_Δ value not more than 1 mm for a load 1.5 times the working load. All independently, but clearly arbitrarily, other investigators recommend a factor of safety at the ground–grout interface not less than 3.5 for installation in ground liable to creep.

Movement detrimental to surroundings and procedures to reduce it are discussed in subsequent sections. It appears, however, that as anchor design recognizes the effects of limiting conditions, an appropriate factor of safety S_L will be introduced to express protection against a yield condition, and then correlated to the factor of safety S_u for the ultimate load capacity. These two different levels of anchor assessment, depending on whether field tests or calculated ultimate loads are used, will become essential, and with reference to urban excavations they may give rise to a statistical treatment.

The factors of safety introduced in this section are thus intended to correlate the design with ultimate static loads. As a matter of consistency, they are higher for the ground–grout and the grout–tendon interface in order to reflect the much higher margin of error in establishing appropriate soil strength parameters and bond distribution patterns.

Ground Mass. This mode of failure usually is checked by approximate methods, briefly reviewed in Section 5-2 for both rock and soil anchors. This situation will arise more frequently as the types and classes of tiedown structures and foundations subjected to uplift are expanded. Ground mass failure is also conceivable with the uppermost (top) row of tiebacks in excavations, especially with shallow anchor inclination. This condition is best analyzed by considering all the parameters contributing to resistance within the assumed failure zone. Prevention of failure should be ascertained with a factor of safety not less than 2.5 and preferably 3.0.

5-11 SUGGESTED DESIGN PROCEDURE

Typical objectives in anchor design include (a) determination of working load, (b) estimation of fixed length and overall anchor length, (c) selection of tendon type, and (d) selection of stressing load. The latter not only confirms the safe working load, but also becomes mandatory where it is necessary to control the behavior of the anchored structure within a broad range of excavation stages and loading conditions.

As an essential supplement, the analysis must also consider (a) potential ground movement, magnitude, and distribution; (b) rigidity of the structure and the stiffness of the ground; (c) compatibility of the ground–structure–anchor system; (d) the behavior of clayey ground during anchor prestressing; and (e) in case of rock anchors, anisotropy, inhomogeneity, and rock mass discontinuities.

Two additional considerations will enhance the results: (a) the design procedure and step-by-step approach should be simple and state the underlying assumptions and variables clearly and completely and (b) anchor design cannot be separated from the analysis and design of the anchored structure. Thus, the following design procedure is recommended.

Step 1. From available soil data, establish the location of suitable fixed anchor zones. From a consideration of relevant soil and rock strength parameters, establish probable range of anchor load capacity (static conditions) and probable range of fixed anchor lengths. Knowing the entry point and the distal end, estalbish overall anchor length, assuming anchor angle 30° with the horizontal. Establish corrosive tendencies of the environment.

Step 2. For the conditions established in step 1, obtain probable anchor cost data. At this time it is necessary to provide the initial justification of anchor choice as opposed to other alternatives. When anchor costs have been checked and static load requirements are satisfied, the engineer should consider the overall function of the project and how it relates to serviceability aspects.

Step 3. Establish the geometry and dimensions of the project. Assuming that some form of excavation is involved, identify the details of the ground support. Given the excavation depth and sequence, structure rigidity, soil stiffness, intended prestress level, and expected ground movement, establish an appropriate lateral earth stress pattern. This may have as basis a given state (active, passive, at rest, etc.), or it may be based on a stress–strain model assuming elastic–plastic behavior. Identify and quantify other loads expected to act on the system.

Step. 4. Establish a logical anchor pattern and layout in terms of spacing, number of anchors per wall or structural unit, and number and spacing of horizontal rows. Consider the factors discussed in Section 5-5, and also excavation levels and sequence, anchor load, and support requirements for the structure.

Step 5. Carry out a structural analysis to determine the horizontal loads on anchor units (point reactions), bending moments and shears on the structure, and probable movement at each excavation level. The structure and the anchors should be included in a single analysis for optimum results, but several iterations may be necessary before this is achieved. Simultaneously, for each excavation stage, carry out a stability analysis for the structure–ground–anchor system, and establish probable slip planes. Step 5 is based on numerical techniques, and requires appropriate computer programs.

Step 6. With the horizontal loads on each anchor known, select anchor inclination as suggested in Section 5-5. With the anchor inclination known, establish the axial anchor load, and determine its vertical component.

Step 7. With the vertical load on the system estimated from step 6, check the stability at the base of the structure. If conditions are not stable, consider a modified anchor inclination, or increase the bearing capacity by extending the structure deeper or by special foundation features.

Step 8. Compare actual design anchor loads with the load carrying capacity per anchor unit assigned to the fixed zone in step 1, for normal anchor types and configuration. Factors to be considered are higher grouting pressure, postgrouting, underreams, and features contributing to increased load capacity. Note that a smaller anchor inclination results in smaller axial load and vertical component. If fixed anchor capacity is grossly exceeded, more anchors should be provided and the system redesigned.

Step 9. Establish location and length of fixed anchor zone, compatible with the intended anchor type (*A, B, C,* or *D*). Locate the proximal point (upper end) at least 5 ft (1.5 m) beyond the slip failure plane. Select a suitable tendon type (bar, wire, or strand) including anchor head. Establish the requirements for corrosion protection, and specify a protective system.

Step 10. Establish a suitable load testing program and type and number of tests (see also Chapter 7). Establish the requirements for long-term monitoring. Establish anchor prestress requirements and lockoff load. Check to confirm that these are compatible with the lateral earth loads used in design.

Step 11. Prepare anchor plans and specifications. In general, these should include (a) general anchor location plan, with general notes; (b) anchor details, including data on prestressing steel, installation procedure, testing, and stressing procedure; (c) anchor-head details, fixed anchor length, proof load, and jacking force; (d) proof load testing frame details; and (e) corrosion protection details. Samples of anchor plans and specifications are included in subsequent sections, together with design examples of anchored structures.

REFERENCES

Adam, M., 1972: "Reply to FIP Questionnaire."

Banks, J. A., 1955: "The Employment of Prestressed Techniques on Allt-na-Lairige Dam," 5th Int. Cong. on Large Dams, Paris, pp. 341–357.

Brown, D. G., 1970: "Uplift Capacity of Grouted Rock Anchors," *Ontario Hydro. Res. Q.*, **22** (4), 18–24.

Bureau Securitas, 1972: "Recommendations Regarding the Design, Calculation, Installation and Inspection of Ground Anchors," Editions Eyrolles, Paris.

Da Costa Nunes, A. J., 1971: "Metodes de Ancoragens," Parts 1 and 2, Assoc. dos Antigos Alumos da Politecnica, 50 pp.

Descoeudres, J., 1969: "Permanent Anchorages in Rock and Soils," 7th Int. Conf. Int. Soc. Soil Mech. and Found. Eng. Mexico, Paper 15–17. pp. 195–197.

Eberhardt, A., and J. A. Veltrop, 1965: "Prestressed Anchorage for Large Trainter Gate," *Proc. ASCE, J. Struct. Div.*, **90** (ST6), 123–148.

Fargeot, M., 1972: "Reply to FIP Questionnaire."

Fenoux, G. Y., and J. L. Portier, 1972: "Lamise en precontrainte des Tirants," *Travaux*, **54** (449–450), 33–43.

Golder Brawner Assocs., 1973: "Government Pit Slopes Project III; Use of Artificial Support for Rock Slope Stabilization," Parts 3.1–3.6. Unpublished, Vancouver, Canada.

Ground Anchors Ltd., 1974: *The Ground Anchor System* (28 pp.), Reigate, Surrey.

Hilf, J. W., 1973: "Reply to Aberdeen Questionnaire," (1972), unpublished.

Hill, J. W., 1973: "Reply to Aberdeen Questionnaire," (1972), unpublished.

Hobst, L., (1965): "Vizepitmenyek Kihorgonyzasa," *Vizugi Koziemenyek*, **4**, 475–515.

Hobst, L., and J. Zajic, 1977: "Anchoring in Rock," *Development in Geotechnical Engineering*, Vol. 13, Elsevier Scient. Publ., Amsterdam.

Johannesburg, City Engineer's Dept., 1968: "Cable Anchors, Design Procedures," Unpublished Memorandum (4 pp.).

Koch, J., 1972: "Reply to FIP Questionnaire."

Lang, T. A., 1972: "Theory and Practice of Rock Bolting," Trans. Am. Inst. Mlg. Eng., pp. 333–348.

Laurikainen, 1972: Reply to FIP Questionnaire.

Littlejohn, G. S., 1972: "Some Empirical Design Methods Employed in Britain. Part of Questionnaire on Rock Anchor Design," Geotechnics Research Group. Dept. Eng., Univ. Aberdeen (Unpublished Technical Note).

Littlejohn, G. S., 1973: "Ground Anchors Today—A Foreword," *Ground. Eng.*, **6** (6), 20–23.

Littlejohn, G. S., 1980: "Ground Anchors, State of the Art," *Civ. Eng. Surveyor*, (Aug.), London.

Littlejohn, G. S., and D. A. Bruce, 1977: *Rock Anchors—State of the Art*, Found. Publ., Ltd., Essex, England.

Littlejohn, G. S., and C. Truman-Davies, 1974: "Ground Anchors at Devonport Nuclear Complex," *Ground Eng.*, **7** (6), 19–24.

Longworth, C., 1971: "The Use of Prestressed Anchors in Open Excavations and Surface Structures," Australian Inst. Mining and Metallurgy (Illwarra Branch), Symp. on Rock Bolting, Feb. 17–19, Paper No. 8, 17 pp.

Mascardi, C., 1972: "Reply to Aberdeen Questionnaire."

Mitchell, J. M., 1974: "Some Experiences with Ground Anchors in London," ICE Conf. on Diaphragm Walls and Anchors, Lond. (Sept.), Paper No. 17.

Morris, S. S., and W. S. Garrett, 1956: "The Raising and Strengthening of the Steenbras Dam (and Discussion)," *Proc. ICE*, Part 1, (1), 23–55.

Ostermayer, H., 1974: "Construction Carrying Behavior and Creep Characteristics of Ground Anchors," *Proc. Diaphragm Walls and Anchorages*, Inst. Civ. Eng., London.

Otta, L., M. Pantucek, and P. R. Goughnour, 1982: "Permanent Ground Anchors, Stump Design Criteria," U.S. Dept. Transp., Fed. Hwy. Admin. Washington, D.C.

Parker, P. I., 1958: "The Raising of Dams with Particular Reference to the Use of Stressed Cables," *Proc. 6th Cong. on Large Dams*, New York, Question 20, 22 pp.

Perry-Davies, R., 1968: "The Use of Rock Anchors in Deep Basements," Ground Eng. Ltd., Johannesburgh, South Africa, unpublished paper.

Rao, R. M., 1964: "The Use of Prestressing Technique in the Construction of Dams," *Indian Concrete J.* (Aug.), 297–308.

Rawlings, G., 1968: "Stabilization of Potential Rockslides in folded Quartzite in Northwestern Tasmania," *Eng. Geol.*, **2** (5), 283–292.

Rescher, O. J., 1968: "Amenagement Hongrin-Leman-Soutenement de la Centrale en Cavernes de Veytaux par Tirants en Rocher et Beton Projete," *Bull. Tech. de la Suisse Romande*, **18** (Sept. 7), 249–260.

Rawlings, G., 1968: "Stabilization of Potential Rockslides in Folded Quartzite in Northeastern Tasmania," *Eng. Geol.* **2** (5), 283–292.

Saliman, R., and R. Schaefer, 1968: "Anchored Footings for Transmission Towers," ASCE Annual Meeting and National Meeting on Structural Engineering, Pittsburgh, Pa., Sept. 30–Oct. 4, Preprint 753, 28 pp.

Schmidt, A., 1956: "Rock Anchors, Hold TV Tower on Mt. Wilson," *Civil Eng.*, **56**, 24–26.

Standards Association of Australia, 1973: "Prestressed Concrete Code CA35," Section 5—"Ground Anchorages," pp. 50–53.

Stocker, M. F., 1973: "Reply to Aberdeen questionnaire."

Voves, B., 1972: "Reply to FIP Questionnaire."

White, R. E., 1973: "Reply to Aberdeen Questionnaire."

Wolf, W., G. Brown, and E. Morgan, 1965: Morron Point Underground Power Plant, Source Wrunsing.

CHAPTER 6

CORROSION AND CORROSION PROTECTION

6-1 GENERAL REQUIREMENTS

The protection of steel tendons from corrosion attack in principle involves the same problems encountered in ordinary prestressed structures, except that with anchorage these conditions are much more severe. The environment in which anchors are set generally is more aggressive with high degree of humidity or water seepage and the usual presence of salt solutions. Where natural soil conditions are heterogeneous and the ground strata have relatively unknown characteristics, incidents of corrosion are very liable to occur if anchors are left unprotected. Thus, only in case of confirmed non-aggressive environment or temporary installations may corrosion protection be omitted.

Interestingly, applications of permanent anchors are in many instances contracted under the condition of long-term responsibility, and this trend is likely to continue with an expanding market. It appears, therefore, that a standardized corrosion protection philosophy is essential, although the choice and the degree of protection is left with the engineer.

Invariably, in current practice steel tendons are chosen from high-strength steels used in prestressed concrete. Variations in the manufacturing process are unavoidable and affect material properties including the mechanism of anchor resistance to corrosion. Whereas no general claim can be made that some steels or tendon types resist corrosion attack better than others, a general requirement is that all steel tendons should be effectively protected over their entire length, including the stressing anchorage and coupling devices.

6-2 DEFINITION AND MECHANISM OF CORROSION

Basic Concepts

Most metals used in the anchor industry are generally obtained by extraction from their oxides in a process requiring input of energy. In their final form the refined metals are less stable than in their natural form, and under appropriate conditions they tend to revert to oxides; that is, corrosion is initiated. If a constraint is not present to inhibit this tendency, the metal will react with oxygen and water from its environment to form oxides and/or hydroxides. Schrier (1976) gives the following general form of this reaction

$$M + O_2 \xrightarrow{\text{H}_2\text{O}} M(OH)_2 \quad \text{or} \quad M_xO_y$$

where M represents the metal. This essentially electrochemical transformation involves metal dissolution and simultaneous conversion of oxygen and water to hydroxyl ions. During the process electrons are transferred from the metal to form the hydroxyl ions, and metal ions migrate into the aqueous electrolyte. The sites where dissolution of the metal occurs (corrosion sites) are anodic, whereas sites where oxygen and water are converted into hydroxyl ions are cathodic.

As anodic and cathodic action occur simultaneously, the corrosion process is initiated only when both sites are available and active; an example where this case is manifested is surface inhomogeneities on a single metal. Electrochemical corrosion is thus promoted as the potential difference between the anodic and cathodic sites, and its rate is directly proportional to the magnitude of the current flowing between the electrodes and their respective areas. Furthermore, the potential developed by each electrode depends on the chemical influence of the environment.

Development of Cells. Where either two electrodes of the same metal are placed in different ionically conducting and connecting media, or electrodes of different metals are placed in the same conductive environment and are connected electronically, a cell will be set up. A metal immersed in solution will develop electrical potential, the nature and value of which will depend on the thermodynamics of the system and influenced by the chemistry of the solutions. A reversible or standard potential of a metal (E_0) is defined as that developed by the metal in contact with a solution of its own ions at unit activity for a given temperature and pressure [usually 25°C and one atmosphere (1 atm)]. Standard electrode potentials have been established for most commonly used metals in their pure state with reference to a standard hydrogen electrode. However, conditions usually encountered in practice involve the potential of a metal (in a solution) appearing in an indefinite and irreversible form, hence they are not so readily defined as E_0 values. It is because such potentials are fundamental to the corrosion process that corrosion mechanisms are not readily controllable.

Galvanized Microcell and Bimetallic Action

Two metals in contact both electronically and ionically will give rise to an electrochemical cell due to the difference in the respective metal potentials. The metal with the more noble potential will function as the cathode of the cell thereby causing the other metal to act as the anode, and under favorable conditions causing it to corrode. The metal with the higher E_0 is in this case the more noble element, or potentially less reactive. Ideally, the potential of the cell would be determined by the difference in the respective E_0 values, but in general the cell current (or corrosion rate) is controlled by electrochemical influences and polarization of one or both metals forming the cells, assisted by environmental conditions. The smaller the anodic area the more severe the attack as metal loss is central to corrosion current and rate (King, 1977), although in reality there is no quantitative relationship between weight loss and the difference in E_0 values. In ground anchors galvanic cells are set up by the contact of different metals constituting the same unit.

Whereas bimetallic corrosion is manifested by the formation of galvanic microcells between differing metals, galvanic microcells can also develop on a single piece of metal or alloy under appropriate conditions in ionically conductive environment between regions of varying composition and therefore different electrode potentials. Grain boundaries, in particular, are less noble than the interior of crystal grains as are lattice defects within crystals due to differences in composition and increased lattice energy. Inclusions in the metal surface can stimulate formation of microcells where the latter acts as the cathode of the cell.

Active–Passive Cells. If a metal or alloy passivates (becomes less reactive) by forming an integral protective oxide film on its surface, active–passive cells can develop at defects in the oxide film as result of inhomogeneities in metal composition or film fracture due to stresses in the metal. The remaining oxide film acts as the cathode stimulating corrosion at defective sites or anode areas under favorable environmental conditions. Likewise, the relative area of anode to cathode is essential in determining the severity of corrosion. Interestingly, the initial potential difference available from the active–passive cell is not sufficient for corrosion to continue; hence the process becomes dependent on factors such as availability of oxygen to maintain the cathodic process. In this manner, an active–passive cell can develop into a differential aeration cell.

Differential Aeration Cells

In ground anchorages, development of corrosion is particularly sensitive to these cells, and can occur as tendons pass between zones of different porosity or from disturbed to undisturbed layers. Differential aeration cells are formed where a metal experiences an ionic environment with dissolved oxygen at different concentrations. Metal areas surrounded by higher oxygen concentration form cathodic zones, whereas those with the low concentration are the anodic areas. Significant corrosion rate is reached only if an appreciable amount of oxygen is present; hence attack is usually observed close to the boundary of the oxygenated cathodic regions. An important example of differential aeration is crevice corrosion, whereby the metal is attacked in a crevice formed at the contact with another metal according to a mechanism causing rapid oxygen consumption. A similar mechanism is associated with pitting corrosion, although in this case the presence of differential aeration cells is not essential and the process requires merely the localized breakdown of the passive oxide film. These cells are usually microcells.

Differential Concentration Cells

These can form when the metal is in environment of varying ionic strength, but the presence of oxygen is likewise important in determining corrosion rates. The nature of ionic species is essential to corrosion sites. Species with different pH values can also interact to start corrosion, since the pH is measure of the hydrogen ion concentration. The effect of pH alone depends, however, considerably on the thermodynamics of the system. Interestingly, pH values are not always reliable indicators of ground agressivity, and chemical composition together with nature and resistivity of the environment are also important factors (Palmer, 1974).

Differential Embedment. In this case differential cells are developed on a larger scale where the metal is embedded in two or more types of environment capable of producing ionic action and are integrally connected. The same factors discussed in the foregoing paragraph are likewise essential, and the foregoing considerations still apply. Differential aeration cells can for example be set up on cables or tubes buried in soil at the passage from aerated to nonaerated soil (clay for instance).

6-3 TYPES OF CORROSION

With respect to the causes of corrosion and the resulting effects, the types of corrosion can be grouped into three main categories: (a) generalized attack, (b) localized attack (shallow or deep pitting), and (c) cracking (due to either hydrogen embrittlement or stress corrosion). These three categories are illustrated in Fig. 6-1.

Generalized Attack

Corrosion atack in this case is approximately uniform and covers the surface of the metal as shown in Fig. 6-1(a), where discrete anodic and cathodic

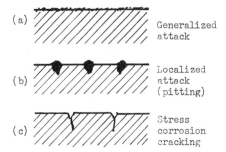

Fig. 6-1(a) Generalized attack

Fig. 6-1(b) Localized attack (pitting)

Fig. 6-1(c) Stress corrosion cracking

Fig. 6-1 Main types of corrosion.

sites do not exist as such or fluctuate over the surface. For this form to occur, the anode and cathode areas must be equal as are anodic and cathodic polarization, and both processes control corrosion rate equally. At some point, it is possible for the corrosion product to form a continuous film that thereafter may act as protective layer and inhibit further attack.

In essense this may be termed chemical corrosion since it corresponds to attack by acid in a laboratory. The metal is gradually transformed into ferrous ions, uniformly from the outer surface inwards. The accompanying reduction in cross section is basically uniform, whereas the center of the metal remains intact and sound. The rust thus formed has no cohesion and is therefore easily displaced by circulation or infiltration, after which attack can resume with increasing acuity.

Localized Attack

This may be termed electrochemical corrosion, and is manifested as deep or shallow pitting as shown in Fig. 6-1(b). The formation of holes causes local stress concentration and eventually premature failure. With conventional prestressing steel, pitting has been observed in the presence of salts used for thawing ice, or near seawater.

Where separate corrosion cells are present on the metal surface, localized corrosion can occur. Separate corrosion cells are distinguished by variation in the electrode potential over the metal surface. The process becomes more localized as the ratio of cathodic to anodic area increases under chemical and/or physical inhomogeneities in the metal or electrolyte, whereas one of the electrical reactions has overall control over the corrosion rate which is unpredictable in practice. Ionization occurs at the anode or cathode, constituting a bimetallic cell as shown in Fig. 6-2(a).

Localized attack is associated with the presence of a protective oxide film on the metal or alloy. Pitting or crevice corrosion will occur in the presence of aggressive ions such as chloride. Pitting can have severe consequences, yet the overall metal loss is small. Whereas the distinction between localized corrosion and pitting is not exact, the latter is often defined in terms of pit

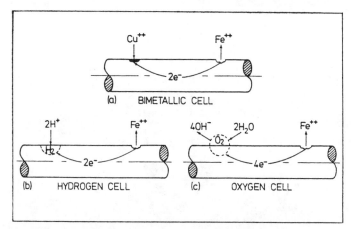

Fig. 6-2 Idealized representation of three major modes of corrosion; (a) corrosion by pitting; (b) corrosion by hydrogen embrittlement; (c) corrosion involving oxygen. (From Longbottom and Mallett, 1973.)

geometry. Thus, it has been suggested that the transition from pitting to localized attack occurs when the ratio of average pit width to depth is 4 or less (Champion, 1962), although a ratio of 1 is widely accepted as the definition of pit (Schrier, 1976). The mechanism of corrosion by pitting is shown in Fig. 6-3.

Stress Corrosion–Hydrogen Embrittlement (Cracking)

This is a form of corrosion where physical causes predominate, although stress corrosion cracking (SCC) is produced by the combined action of static tensile stress on the steel and localized corrosion. SCC is more commonly encountered with alloys where a passivating oxide film is present with appropriate corrosion media. Its precise mechanism is not completely understood. It appears, however, that localized action of corrosion produces a narrow pit, which allows the tension forces to concentrate at the tip of the pit, resulting in the formation of fresh metal surfaces where further dissolution can occur. With this combined action propagation occurs causing cracking either along grain boundaries or along slip planes within the crystal lattice. The accompanying reduction in cross-sectional area leads to failure by plastic yielding.

Unlike other forms of corrosion, SCC depends greatly on the stress condition of the metal, and increases with increasing tensile stress. Certain steels appear to be more liable to this type of corrosion, but a precise classification is yet to be made. Uhlig (1971) indicates that high-strength steels with yield strength higher than 1241 N/mm^2 (180 ksi) or a Rockwell C hardness value greater than 40, are susceptible to SCC. If sulfides are present, Phelps (1967)

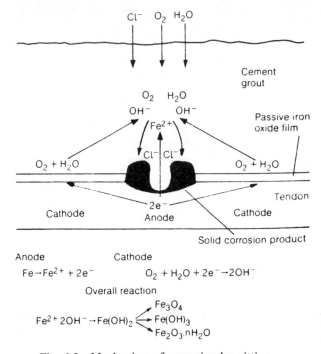

Fig. 6-3 Mechanism of corrosion by pitting.

suggests to reduce the equivalent Rockwell C number to 22. Interestingly, tensile tests do not reveal any weakening of the so-called susceptible steels (Portier, 1974). The stress in this case must be maintained for some time before the steel breaks suddenly under stresses that sometimes are one-half those at instantaneous failure.

Among the causes observed, the best known is hydrogen embrittlement, mainly affecting highly stressed carbon steels. This involves the migration of atomic hydrogen into the metal lattice where molecules are formed and produce internal pressure in the metal. The atomic hydrogen may have been formed by corrosion of the metal or by corrosion of the baser metal in direct contact with the former. Cracking of the metal can then occur as direct result of tensile stresses developed by the hydrogen itself, but also where a certain critical tensile stress is externally applied (sometimes called hydrogen cracking). The process is more severe in steel if bismuth, lead, sulfur, tellerium, salenium or arsenic are present (FIP, 1986). Atomic hydrogen may enter the metal over an extended period, and failures of this type have been reported years after installation.

Experiments with doped distilled water have shown that as hydrogen infiltrates the metal the process is helped by deformations produced by tensile stresses causing lack of cohesion of the crystals. This effect however may not be instantaneous, and the hydrogen needs time to penetrate the material. It is conceivable that the surface state, which is different between rolled steel and drawn steel, may favor this infiltration. From a survey of reports on hydrogen embrittlement it appears that oil quenched and tempered steels are more susceptible to this process than drawn types (Littlejohn and Bruce, 1977). However, there is no unanimous opinion about the susceptibility of prestressing steel to hydrogen embrittlement in highly alkaline grout.

An idealized representation of this process is shown in Fig. 6-2(b).

Other Forms of Corrosion

Bacterial Attack. A common form of bacterial attack is in association with the metabolic processes of sulfate-reducing bacteria (SRB) utilizing sulfate in anaerobic conditions. These are found in spaces isolated from atmospheric oxygen, particularly in sulfate-bearing clays or organic soils below the water table. SRBs are most active at pH values 6.2–7.8 (Hadley, 1939) when cathodic reaction in a corrosion cell releases hydrogen. If the corrosion cell is located in more open environment than clay, allowing some ingress of atmospheric oxygen to the boundaries of the anaerobic region, the sulfur cycle can be completed by the metabolism of sulfur-oxidizing bacteria (SOB) producing sulfuric acid, which in turn can dissolve metal and grout. Bacterial attack can either be localized (i.e., pitting), or more generalized depending on the nature of soil, depth of embedment, and the presence of any protective coating or oxide film on the metal surface.

Corrosion Fatigue. This results from the combined action of corrosion and cyclic stresses. Unlike SCC, corrosion fatigue occurs in most aqueous media and is not connected with special combinations of aggressive ion and metal. The prevailing mechanism involves exposure of oxide-free, cold-worked metal that becomes anodic and corrodes, and under cyclic stressing transgranular cracks gradually develop.

Corrosion Involving Oxygen. The presence of oxygen accelerates corrosion at the cathode as shown in Fig. 6-2(c), but this reaction is typically favored by alkaline conditions. At the anode oxygen concentration leads to formation of a protective, passivating layer of rust. The highly alkaline grout can protect the steel through rust formation, but if the grout has cracks water will remove the rust, eventually dissolving the metal.

6-4 AGGRESSIVITY OF ENVIRONMENTS

Guidelines including quantitative limits on aggressivity of environments have been prepared and introduced by various institutions, for example, Bureau Securitas (1972), FIP (1973), and more recently FIP (1986). With the increasing realization that failure of highly stressed steels under the influence of corrosion is quite complex, and as yet it is impossible to specify the condition giving rise to this process, it appears that two safe principles are available in dealing with the problem. The first is a comprehensive analysis of the various environments to which the steel will be exposed to determine the aggressivity level, and the second is to adopt a design philosophy oriented toward ensuring complete protection of the prestressing steel.

Ground Aggressivity

Aggressivity toward Metals. Water content, aggressive ion content (presence of chloride and sulfate ions), and permeability of the ground are prime factors affecting corrosion. However, a generalized measure of redox potential and soil resistivity is now accepted as index of risk and can be applied to the assessment of potential corrosiveness (King, 1977).

Table 6-1 shows data correlating soil corrosiveness to resistivity and redox potential, valid for soils of single composition. Where the anchors pass through layers of different composition, these data should be interpreted with caution to avoid development of differential embedment cells. ASTM report STP 741 (1979) gives useful information on the measurements and processes required. If such data are not available, soil and groundwater samples should be taken for chemical analysis, discussed in subsequent sections.

Aggressivity toward Grout. The most serious conditions affecting grout durability are (a) sulfate-bearing ground and groundwater and (b) acid-bearing ground and groundwater.

Aqueous solutions of sulfates attack the set grout according to a chemical reaction depending upon the type of sulfate and cement. The rate of attack is influenced by the permeability of grout. The process involves also the following factors: (a) normal groundwater table and seasonal fluctuations

TABLE 6-1   Corrosiveness of Soils Related to Values of Resistivity and Redox Potential

Corrosiveness[a]	Resistivity (Ω/cm)	Redox Potential, (mV) (corrected to pH = 7); Normal Hydrogen Electrode
Very corrosive	<700	<100
Corrosive	700–2000	100–200
Moderately corrosive	2000–5000	200–400
Mildly corrosive or noncorrosive	>5000	>430 if clay soil

[a] In the absence of the above tests, ground and groundwater samples should be taken for detailed chemical analysis in order to judge aggressivity.

and (b) form of construction. If sulfate conditions around the set grout are unavoidable, the best defense against attack is to control the type and quality of grout, that is, by the use of dense grout with low permeability. Relevant concrete codes are, by extension, recommended for guidelines on the requirements for grout exposed to sulfate attack with reference to the type of cement, minimum cement content and maximum water/cement (W/C) ratio.

Conditions involving the presence of acids can be critical because of the vulnerability of cement to acid attack, but for most uses below ground level little erosion of carefully prepared grouts should be expected when the pH is about 5.5 and the water is stagnant. More critical conditions are discussed in the following section.

Aggressivity and Chemistry of Groundwater. In unfamiliar regions it is customary to determine the chemical properties of groundwater and the associated effects, particularly with reference to cement attack. In general, chemical analysis of groundwater of natural origin involves the following tests:

1. pH values
2. Smell
3. Potassium permanganate in mg per liter
4. Total hardness in mval/liter or O_d
5. Carbonate and noncarbonate hardness
6. Magnesium content in mg per liter
7. Ammonium in mg/liter
8. Sulfate content in mg/liter
9. Chloride in mg/liter
10. Lime-dissolving carbonic acid in mg/liter, determined using Heyer's marble test

Pure water may be termed aggressive if the CaO concentration is less than 300 mg/liter. Such waters display the tendency to dissolve the free lime and hydrolyse the silicates and aluminates in the cement.

Acid waters with pH between 5.5 and 4.5 are considered aggressive since they can attack the lime in the cement. They may be found with dissolved carbon dioxide or contain humic acids, but a frequent source is industrial waste. If the pH is in this range, a dense grout should be used, and the addition of pulverized-fuel ash may be beneficial (Gutt and Harrison, 1977). Values of pH below 4.0 are unusual, and since water aggressivity in this case is very strong, alternative grouts (other than those based on Portlant cement) should be considered.

Waters with high sulfate content react with tricalcium aluminate present in cement to form salts that disarrange the cement by swelling. Among these

TABLE 6-2   Parameter Limits for Assessing Aggressivity of Groundwater

	Aggressivity		
Test	Weak	Strong	Very Strong
1. pH value	6.5–5.5	5.5–4.5	<4.5
2. Lime-dissolving carbonic acid (CO_2) in mg/liter determined by Heyer's marble test	15–30	30–60	>60
3. Ammonium (NH_4^+) in mg/liter	15–30	30–60	>60
4. Magnesium (Mg^{2+}) in mg/liter	100–300	300–1500	>1500
5. Sulfate (SO_4^{2-}) in mg/liter	200–600	600–3000	>3000

are selenious water and magnesian water. With the presence of these salts, the conditions become very aggressive when the salt concentration exceeds 0.5 g/liter for selenious water and 0.25 g/liter for magnesian water. These values refer to stagnant water, and for flowing water the equivalent limits are 40 percent these values.

The aggressivity of groundwater can be judged by quick reference to Table 6-2. These limits apply to stagnant or weakly flowing water of ample supply. This is assumed to attack the anchor immediately and the effects are not diminished by reaction with the grout. The highest degree of aggressivity is assigned to groundwater even if it is obtained in only one of the five classes shown. If the values in two or more classes lie in the upper quarter of a range, the degree of aggressivity is increased by one grade.

Higher aggressivity must be accepted for higher temperatures and pressures or where the grout is subjected to mechanical abrasion due to swiftly flowing or agitated water. The degree of aggressivity decreases at lower temperatures if small amount of water is present and the water is still, and where the aggressive constituents can be reactivated slowly, for example in low permeability ground ($k < 10^{-3}$ cm/s).

Aggressivity of Grout toward Steel

The concensus of opinion is that steel tendons can be adequately protected if they are surrounded by an alkaline environment with a pH range of 10–13. Hydrated cement has normal pH value of 12.6, which inhibits the presence of aggressive ions. At this pH a passive film forms on the steel surface that reduces the rate of further corrosion to inconsequential levels. Cement grout cover can provide therefore chemical as well as physical protection to the steel. It is, however, normal to accept a long-term loss of alkalinity owing to the permeability and porosity of grout. This can occur by reaction with acidic gases in the atmosphere or by leaching water from the surface. These acidic gases react with the alkali and neutralize them by forming carbonates and sulfates, at the same time reducing the pH. If the carbonated front can penetrate sufficiently into the grout, it may intercept the steel tendon, causing it to corrode if oxygen and moisture are available. The defense in this case is to provide an adequate grout cover sufficiently thicker than the carbonated front (normally a few millimeters) so that the steel will continue to remain in alkaline environment.

As mentioned in previous sections, cracks in the grout are formed following high prestressing, shrinkage, or other factors. Crack formation will provide ingress of the atmosphere and aggressive ions, and designate an entry for the carbonation front. If a crack is initially formed and gradually reaches the steel, considerable protection can be lost especially with the tendency of the unit for debonding, which exists under tensile loading, as shown in Fig. 6-4. If this occurs, it will disrupt the direct contact of the steel with the alkaline grout environment and destroy protection in the vicinity of debonding. Subsequent corrosion will depend upon several factors such as size of crack, loading conditions (constant or fluctuating), degree of exposure, and environmental effects. In some instances, the cracks may be closed by products of carbonation reactions, ingress of debris or combinations therefrom, all restricting the supply of oxygen and moisture and inhibiting further corrosion. However, if the cracks are not closed or if they propagate to other areas because of fluctuating loads, oxygen and moisture will gain access to unprotected steel surfaces, and corrosion should be expected to continue at an unpredictable rate.

Influence of Cracking. Beeby (1978) has shown that, with regard to corrosion of steel in concrete, corrosion is likely to start first where a bar intersects a crack. In the short term (in this case 2 years), the influence of crack width on the amount of corrosion found near the crack is significant. In the long term, however (10 years or longer), the influence of crack width on the amount of corrosion is negligible, and this conclusion is based on observations of 0.05–1.5-mm cracks. Furthermore, the smaller the crack,

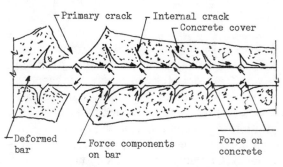

Longitudinal section of axially loaded specimen

Fig. 6-4 Crack formation and conditions approaching debonding. (From Goto, 1971.)

the lower the corrosion risk. Where cement grout is used to provide the protective cover in permanent anchorages, the use of crack-inhibiting steel is now considered desirable. Although there is limited field evidence as to what crack widths are acceptable in a cementitious protective barrier, an upper limiting crack width 0.1 mm is proposed as guideline, based on work by Houston et al. (1972), Ryell and Richardson (1972), Schiessl (1975), Beeby (1978), Naus (1979) and O'Neill (1980).

The idealized representation of cracking around a ribbed deformed bar shown in Fig. 6-4 is proposed by Goto (1971). At this stage the force is primarily transferred from steel to concrete by the mechanical action of the ribs, as adhesion between the bar surface and the concrete is largely lost. As the steel rusts, the corrosion products generally occupy a volume sometimes two to three times the volume of lost metal. As a result, the corrosion products from a small reduction in the cross-sectional area of the tendon will produce internal stresses sufficient to disrupt the surrounding grout. The ability of grout to resist these stresses is dependent upon the location of the tendon unit, the tensile strength of the mix, and the cover thus provided. These considerations impose the maximum acceptable ratio of steel area to grout cross-sectional area discussed in the foregoing sections.

Stability of Alkaline Conditions. The passivity associated with alkaline conditions can be reversed in the presence of chloride ions, despite the remaining high level of alkalinity in the grout. The chloride ions can depassivate the metal locally and promote active metal dissolution. The reaction takes place with calcium aluminates and ferrites in the grout to form calcium chloroaluminate and chloroferrite compounds, which are solid and prevent the chloride from reacting. These remain stable only as they are in chemical equilibrium with a small amount of chloride in the aqueous phase of the grout. The chloride in the solution is the prime agent that is free to initiate corrosion in the steel. At low levels of chloride ion in the aqueous phase corrosion will be limited, but as the concentration increases, the risk is significantly higher. Thus, the amount of chlorides in the grout as well as the amount of free chloride in the aqueous phase will by large determine the risk of corrosion. In this context, the total chloride content of the grout derived from all sources should not exceed 0.1 percent by weight of cement.

In marine environment, changes occur between hydrated cement and sea water, some of which are harmful and others are beneficial. It is possible that low-permeability concrete can become watertight in seawater. Although the exact cause is not known, one explanation is the blocking of pore spaces by the crystallization or precipitation of chemical products created by interaction between seawater ions and hydrated cement (FIP, 1986). The free corrosion potential on an embedded steel tendon may also fall and stabilize at very low values, specifically, after 6–18 months in submerged conditions (Fidjestol and Nilsen, 1980).

6-5 RISK OF CORROSION AND AGGRESSIVE CIRCUMSTANCES FOR ANCHORAGES

The Corrosion Problem in Conventional Prestressing

Although from the statistical point of view corrosion instances are normally kept secret, quite often it has become necessary to repair the sheaths of existing conventional prestressing systems by injecting resins. In most cases the principal causes of corrosion risk of steel, despite the effective protection of the basic medium with concrete, relate to poor workmanship and construction techniques.

The first questionable practice is the "do-it-yourself" trend promoted by several manufacturers of jacking devices and sealing products for an application which is highly specialized, and where prestressing methods and the residual tension of some cables require competent judgment and utmost attention. Injection, in particular, is a markedly sensitive phase, and should be performed only by skilled specialists.

Furthermore, in conventional prestressing many cables have horizontal alignment with high and low points, and this means that pockets may be created during injection allowing bleeding to occur if the grout is not stable. Since the water content in a cement grout is usually kept low to ensure its stability, it follows that the lower the water/cement ratio the less injectable the grout, hence the higher the risk of blockage. Interestingly, an unsatisfactory injection has little chance of being detected probably for 10 years after it is completed.

Another cause of problem relates to the type of sleeves used in posttensioning. Sleeves are often made in the form of spirals so that they can easily bend and assume the curved configuration of the prestressing system. These sleeves, however, are not always watertight.

Under these circumstances it is possible, especially if there is a leak because the stressing cable has been improperly stressed or the concrete has cracked, for runoff water charged with pollutants to circulate along the prestressing steel in such a manner and at such a rate that it is not passivated by the cement. A last contributing factor is contact between cables and the sleeves. There is strong indication that the line of contact is oxidized, either as a result of water taking a preferred path or because of creation of microbatteries.

Most of the foregoing risks are absent with anchorages. Ground anchors are generally inclined or vertical so that setting of any unstable cement will occur downwards, and in practice it is sufficient to fill the top section under pressure after the lower cement has set. Since the borehole is always straight, plastic sleeves can be used for the free length without the risk of being torn during tensioning because of the curvature of the tendon. These sleeves are sufficiently watertight laterally, and this inhibits formation of microbatteries along the contact surface. Although the presence of aggressive water in a subsoil must generally be accepted, it is encountered in a milder form as the depth of the anchorage increases with the exception of marine environments.

Aggressive Circumstances and Risks Specific to Ground Anchors

Risk Due to Uplift Pressure. Ground anchors often are used to stabilize foundation rafts and tiedown structures. These are generally located below the groundwater table, and hence are subject to uplift. In this case, the slightest orifice will serve as drain cock, and water may then flow along the tendon. The problem is more serious with strand anchors, and in the past it has been remedied by injecting resin to cover the affected areas. More recently, the use of an epoxy pitch is claimed to penetrate the tendon core and ensure complete watertightness.

Overall Sealing. According to one practice favored by French contractors (Portier, 1974) complete sealing of the tendon can be achieved by multiple injections. According to this claim, if grout is injected without pressure, two results are possible: (a) the weight of the steel tendon may bring the steel into contact with soil, even with the use of centralizers and spacers; and (b) shearing of the sealing compound may occur along the contact line with the soil. Thereafter, the cement follows the deformation of the steel and cracks in the same locations. At this stage, the risk of corrosion exists.

However, when injection is carried out in several stages, a recentering of the tendon in the borehole follows. Each new injection breaks the existing grout, flows around it, and lodges between the anchor and the soil as the anchor bulb thickens.

In soils that are permeable but not injectable, the open radial fissures of type *C* simply do not form, since shearing cannot occur in contact with the soil. A critical condition is therefore reached with grouts which shear longitudinally in contact with the steel tendon, but without radial cracking. This sheath of grout, which is part of the soil structure, becomes compressed at some stage of tensioning or at the final steel tension, a release causing precompression of the sealing.

Free Length. A major problem is to prevent longitudinal paths by which water can flow along the axis of the sleeve, due to uplift pressure. Equally important is to ensure material performance of the injection process. If the injection is carried out after tensioning, a grout sleeve tube situated at the base of the free end will allow cement to be injected under pressure upwards.

Anchor Head. This zone deserves the same attention because it can be vulnerable for many reasons: (a) any possible setting is likely to occur at this point; (b) leaks emerge at this location; (c) mechanical and heat stresses can create electric couples out of proportion with those of the sealing; (d) it is the zone exposed to a much more corrosive environment, namely, the atmosphere; and (e) because it is readily accessible, it is often the practice to neglect its full protection until corrosion is documented, at which time remedial measures are too costly, if at all possible.

Aggressive Circumstances. Protection of the steel tendon against corrosion can be chosen only in conjunction with the potential corrosion hazards. The main factors affecting the selection of suitable protective systems relate markedly to ground aggressivity. Thus, the following conditions will demand consideration of complete corrosion protection (FIP, 1986).

1. Anchorage sites exposed to seawater, containing chlorides and sulfates.
2. Anchorages in saturated clays with low oxygen content and high sulfate content.
3. Anchorages in evaporite rocks containing chlorides (i.e., salt-lake deposits).
4. Anchorages in soils near the vicinity of chemical factories producing corrossive effluents, or which are subjected to corrosive atmosphere.
5. Anchorages passing through ground layers with fluctuating water levels.
6. Anchorages passing through partially saturated soils.
7. Anchorages passing through strata of differing nature with regard to chemical composition and difference in water or gas content.
8. Anchorages under cyclic loading causing cyclic stress changes within the tendon.

In general, corrosion is enhanced by exposure to the combined action of oxygen and chlorides, anaerobic conditions in the presence of sulfates, or severe fluctuations of load and high stress levels. For initial assessment of these conditions, reference can be made to Tables 6-1 and 6-2 for ground and groundwater aggressivity, respectively. Anchorages in hard rock of low permeability should be considered being in essentially nonaggressive conditions.

Sensitive decisions must often be made regarding temporary anchorages in ground with some corrosion history and potential. In this case, examinations of buried metals in the same vicinity can establish the corrosion history and provide useful data as to the degree of protection required. Special consideration should be given to construction effects on groundwater flow and seepage, especially where it is possible to cause water diversion through regions containing aggressive chemicals.

6-6 OBJECTIVES OF CORROSION PROTECTION

Unacceptable Corrosion Conditions

The design philosophy of protecting an anchorage against corrosion is to ensure that, during the effective life of the system, the probability of unacceptable corrosion is acceptably small. Thus, a rational approach to design is to satisfy the following criterion (Beeby, 1978):

$$\text{Design life} \leq t_0 + t_1 \qquad (6\text{-}1)$$

where t_0 = time from construction to the initiation of corrosion (depassivation)

t_1 = time from initiation of corrosion to the occurrence of unacceptable corrosion

The time t_0 is the time taken for the aggressive front to penetrate the grout system and reach the steel. This time will depend on whether the grout has cracked, the crack width, the grout cover, and the nature and conditions of the environment. For cracked grout in marine environment it is reasonable to assume $t_0 = 0$. In determining the factors that influence the time t_1, the first requirement is to define unacceptable corrosion. For reinforced concrete structures this stage is reached with the onset of spalling. For anchorages, the occurrence of prestress loss in excess of 15 percent should indicate unacceptable corrosion and cause concern. If the sole cause of this loss is indeed corrosion, it is likely that loss of mechanical interlock might occur in advance of spalling.

The time taken to produce loss of mechanical interlock will depend on the corrosion rate, and the amount of corrosion necessary to disrupt the grout. These variables will, in turn, depend on the grout cover, density and watertightness, alkalinity, tendon size and configuration, and loading condition.

Presently, the available theoretical and empirical data do not warrant a safe estimation of t_0 or t_1, hence the longevity of anchorages must be inferred essentially from statistical data and past performance of installations under similar conditions of aggressiveness and protection.

Reducing the Corrosion Risk

Water. Regardless of the type of corrosion, this process can occur only in ionic medium, and under natural conditions water is the most widespread agent. The renewal of water will increase the risk, and humidity is even more dangerous. Factors that are closely interdependent include the supply of oxygen, the promotion of the microcell effect by the formation of a cathode at the water–air interface, and the action of hydrogen embrittlement.

Electrochemical Potential. Figure 6-5 shows three regions of attack. Region I is characterized by the formation of ferrous ions and generalized dissolution. Hence, in order to avoid corrosion, it may be sufficient to choose the pH range 10–13, that is, remains in the region created by grouts. However, as discussed in the foregoing sections, this protection is entirely inadequate, and despite the passivating action of grout, there may be corrosion by pitting (region II) under the influence of chloride ions present in the cement. Furthermore, region III, corrosion with crack formation, may occur.

From these remarks, it appears that the risk of corrosion can greatly be reduced by the following measures:

1. Ensuring a pH environment of 9–12 in the grout to guard against the risk of dissolution. Chloride, sulfide, sulfate, and carbonate ions tend to lower the pH of the grout and thus promote electrolytic activity.
2. Avoiding the presence of harmful ions in contact with the steel, that is, providing a barrier that envelopes the steel completely to inhibit corrosion by pitting.
3. Selecting steels that are less liable to corrosion under tension, and eliminating from the mixes anions that favor the passage of hydrogen.
4. Preventing to the extent possible the circulation of water to resist the renewal of corrosive circumstances.

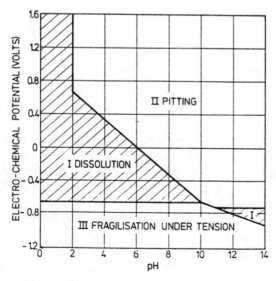

Fig. 6-5 Relation of types of corrosion to pH and electrochemical potential (From Caron, 1972.)

5. Avoiding porous grout by selecting a suitable type of cement and a proper water/cement ratio.

6. Reducing the level of prestressing since this accelerates corrosion.

7. Using cold-drawn carbon steel instead of quenched and tempered varieties, which are more susceptible to stress corrosion.

8. Avoiding repetitive loading (cyclic) that destroys protective rust. With reference to measure 5, it has been found that in certain cases the protective grout cover of 25 mm (1 in) is not sufficient. Therefore, more relevant to grout quality is grout porosity rather than cover thickness.

Prestressing the steel may accelerate the rate of intensity of corrosion, nonetheless the elastic and strength properties of nonstressed steel are likewise affected. Certain types of steel are more liable to corrode than others, as mentioned in measure 7 (above). Stress corrosion is more accute than ordinary corrosion because (a) stressing and releasing, if repeated, gradually destroys the protective oxide film; (b) stressing enhances the formation of microfissures; and (c) prestressing steel is, to begin with, more susceptible than plain steel.

It is clearly redundant to try to prevent the presence of water in the vicinity of anchorages, in view of the considerable pressures involved that cause it to infiltrate the very substance of the anchor system through the smallest aperture (valves, screw, threads, etc). Furthermore, several anticorrosive systems are based on the anchor being immersed in water of suitable pH value.

6-7 REQUIREMENTS OF CORROSION PROTECTION

Degree of Protection against Corrosion

Whereas most engineers will agree with the definition and the signs of unacceptable corrosion, the time taken to produce it still remains dependent on many uncertain and unquantified parameters. The protection of permanent anchors presents an important task, although not crucial. Since underground construction work, where anchorages are normally useful, is implemented in urban or in industrial sites, it is possible that with ongoing and future industrial development the natural environment of the subsoil will become much more hostile.

The condition of a tendon unit during manufacturing and storing should be carefully checked for defects such as longitudinal depressions in wire or bar that do not impair the physical and mechanical properties. Currently, these units are acceptable provided such defects are less than 4 percent the nominal diameter of the tendon components or up to 1 mm maximum depth of depression (FIP, 1986), determined from statistical assessment of tendon samples. For strand, this criterion applies to individual wires within the strand.

A film of rust on the tendon at the time of its delivery is not necessarily harmful and may improve bond, but tendons showing signs of pitting or transverse defects should be rejected under any circumstances. Adequate protection should be afforded to the tendon and its parts by the manufacturer to avoid corrosion and mechanical damage before delivery. At the site, care should be exercised during storage to ensure protection of the tendon prior to its homing (see also Section 2-9).

Methods and criteria used to determine the protection level reflect the following factors: (a) the intended effective (economic) life of the anchorage and the anchored structure, (b) the aggressivity of the environment, and (c) the consequence of failure caused by corrosion. It is erroneous to consider corrosion protection only where these consequences involve endangered public safety, since the cost of structural damage alone is likely to exceed the cost of protection. Hence, the decision whether the rate of corrosion merits the expense of protection is irrelevant and academic, irrespective of public safety and human injury.

Temporary Anchors. By definition, these will remain in service up to six months. If left unprotected, they will probably corrode in time. It is however reasonable to assume that the cement grout will protect the fixed anchor length, and the specified minimum cover is normally provided for most fixed anchor types. In general, a grout cover not less than 25 mm (1 in) should be specified. The need to provide some form of protection over the free length is sometimes recommended but not always enforced. Exceptions are extremely aggressive conditions such as marine environment. In these cases, a combination of grease and tape in the free length is good practice.

Semipermanent Anchors. By definition, these will remain in service from 6 to 18 months. Corrosion rate will vary with the environment and working conditions. If the environment exhibits corrosive tendencies or where there is risk of local damage or corrosion by pitting, the results will justify the expense of protection. If monitoring is included in the program and the anchorages are not at immediate risk within their working life, protection can be provided and graded according to the severity of the service conditions.

In this case, the minimum cover as well as the watertightness of the grout becomes more critical. A minimum cover 30 mm should be virtually guaranteed. Equally important is the structural integrity of the grout since reliance must be placed on the condition that the grout is not cracked. If these guarantees are not provided, some additional protection should be considered. Protection of the free length may still be based on a single protective system.

Permanent Anchors. As a general rule, permanent anchorages should be protected, preferably with double protection. The design solution may be based on the assumption that an aggressive environment will at some time exist, and that environmental changes during the service life cannot be predicted but they are likely to occur so that exposure to aggressive conditions cannot be excluded. The protective system should not adversely affect the handling of the tendon or the mechanism of bond and its effect on the transfer of load.

In some instances, such as with low-capacity rock bolts used solely as secondary reinforcement, a simple grout cover may suffice. For high-capacity anchorages installed in low-permeability rock, it would be prudent to insist on at least one physical barrier, although satisfactory performance has been documented where an alkaline environment is the only protection against corrosion. An unprotected bar anchor is shown in Fig. 6-6.

The choice of class of protection should ordinarily be left with the engineer after considering the many variable parameters involved. In this regard, however, basic guidelines may be useful and provide a rational approach for designing an effective protective system.

Single and Double Protection. By definition, "single protection" constitutes one physical barrier between the steel tendon and the corrosive front. "Double protection" means that two such barriers are provided, and the main purpose of the outer barrier is to protect the inner barrier against damage during tendon handling and homing. In this context, the outer barrier may be considered sacrificial and redundant. Double protection has evolved as mandatory anchorage requirement in most European countries and North America, but it is not universally accepted as necessary.

Requirements of Protective Systems

Under a growing need to establish standards of protective system, the following requirements have been identified and apply to all anchors. Accordingly, a protective system should satisfy the following:

1. The system should have an effective life at least equal to the anticipated service life of the anchorage.
2. It should not interact with the environment, and should not have adverse effects on the efficiency of the protected anchor and its parts.
3. It should not restrict movement of the free length, especially where anchors must be restressed or with reference to the functions and requirements discussed in Section 2-3.
4. It should consist of materials that are mutually compatible with the deformability and permanence of the anchorage, and will also inhibit potential induction of corrosive conditions.

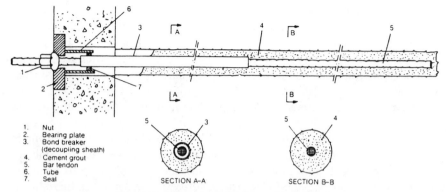

1. Nut
2. Bearing plate
3. Bond breaker
   (decoupling sheath)
4. Cement grout
5. Bar tendon
6. Tube
7. Seal

SECTION A-A   SECTION B-B

Fig. 6-6 Unprotected bar anchor in rock.

5. The use of these materials should involve a single treatment, since protective systems (with few exceptions) can be neither replaced nor maintained.

6. The system should neither fail during stressing to proof load, nor disrupt the tendon–grout interaction, especially at junctions between components.

7. It should be flexible but strong enough to withstand handling stresses and distortions during manufacture, transport, and installation.

8. It should be delivered and packed in a manner allowing easy inspection before installation.

Materials and Principles of Protection

The foregoing requirements are satisfied if a protective system can exclude moist gaseous atmosphere around the metal by completely enclosing it within an impervious covering or sheath. The effectiveness of this treatment depends on maintaining the continuity of the covering; on external fluid pressure gradients across coatings and joints; on content and cleanliness of the atmosphere during application of the coating; on junction details, especially at the fixed anchor zone and anchor head; and on the electrochemical potential at the metal surface.

The industry offers a variety of protective coatings or coverings. It appears that the principles of protection are essentially the same for all parts of the anchor system, but different details apply to the fixed length, the free length, and the anchor head. Among the protective materials available are one or more coatings applied during manufacture of the tendon and basically attached to it, as well as materials introduced as fluids within the coating. Grout injected in situ to form the anchorage zone and bond the tendon to the ground is not considered part of the protective system unless its quality and integrity is assured. However, its passivating alkalinity should be recognized.

Single protection has only one of these barriers, whereas double protection includes a sheath and a coating, together with materials injected externally after homing or internally during manufacture. Double protection includes both a physical barrier to corrosion and an electrochemical barrier associated with a fluid material, which may harden for additional physical protection.

Ideally, fluid materials should remain ductile in their final configuration to provide a physical barrier that does not crack or debond in service as the system undergoes differential strains. When a tendon in cement grout is stressed (with tension-type anchors), cracks within the fixed length typically occur at about 50–100-mm intervals (2–5 in), and of widths up to 1 mm or more (Graber, 1980; Meyer, 1977). The severity of corrosion can be reduced if the crack width is limited to 0.1 mm under the influence of the alkaline environment of the cement grout. Unlike a cementitious material, cracked resin does not provide protection since the resin is inert (FIP, 1986), although certain resins exhibit stress–strain characteristics ensuring their performance without severe risk of cracking. Crack width in cements can be controlled and reduced by the use of spiral steel cages or meshes within the grout, which determine crack spacing, but more data should be forthcoming to substantiate this concept in practice.

Nonhardening fluid materials, such as greases, have been found to have limitations as corrosion protection agents. The reasons are:

1. Fluids are susceptible to drying out, which is followed by shrinkage and change in chemical properties.
2. Fluids are liable to leakage if slight damage is sustained by their containment sheaths.
3. Having no shear strength, they are easily displaced.
4. Their long-term chemical stability, which determines susceptibility to oxidation, is not known with certainty.

Because of these reasons, nonhardening materials must themselves be protected or properly contained within means that must also be resistant to corrosion. However, media such as grease provide an essential function in the protective system: they act as fillers to exclude the atmosphere from the surface of the steel tendon, balance the correct electrochemical environment, and reduce friction in the free length. Nonetheless, they are not considered a permanent physical barrier to corrosion. In this context, a layer of grease is not an acceptable physical barrier in the decoupled free length, although it is acceptable in a restressable anchor head where it can be replaced or replenished. Greases are discussed in the following sections.

Typically, for effective protection the entire tendon must be protected. Partial protection will only induce more severe corrosion on the unprotected part. Thus the extent of treatment for the fixed length, the free length, and the anchor head must be similar in principle. The use of a larger-size tendon, that is, a thicker metal section, with a provisional sacrificial area in lieu of the physical barrier, should not be expected to give more effective protection. Corrosion is seldom uniform and extends rapidly and preferentially toward localized pits and surface irregularities. The presence of such corrosion pits cannot be reversed by brushing or covering, and once pitting is observed in high-tensile steel it should be rejected for ground anchors.

Noncorrodable metals have been considered for anchorage components, and this category includes nonferrous metals and stainless steel. These may be found suitable after checking their electrochemical behavior relative to other components and stress corrosion characteristics. Nonmetallic fibres may also be used after investigation of their effective life in stressed conditions in potentially aggressive environments that may be different from those aggressive to steel.

6-8 PROTECTIVE SYSTEMS OF THE FREE LENGTH

This zone of the anchor is usually protected by the injection of setting fluids (i.e. cementitious grouts) to surround and enclose the tendon, by preapplied coatings, or by combinations thereform.

Figures 6-7 and 6-8 show typical examples of protection. The anchor of Fig. 6-7 consists of strand tendon and has double corrosion protection in the free length consisting of individually greased and sheathed units enclosed in a plastic tube. The anchor system of Fig. 6-8 consists of a bar tendon with single protection along its free length, in this case a plastic tube. Interestingly, the grout cover is not considered part of the protection, although it provides a physical barrier. In both instances, the protective system permits uninhibited extension of the tendon during stressing and thereafter.

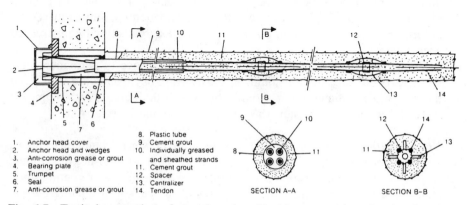

1. Anchor head cover
2. Anchor head and wedges
3. Anti-corrosion grease or grout
4. Bearing plate
5. Trumpet
6. Seal
7. Anti-corrosion grease or grout
8. Plastic tube
9. Cement grout
10. Individually greased and sheathed strands
11. Cement grout
12. Spacer
13. Centralizer
14. Tendon

SECTION A–A　　SECTION B–B

Fig. 6-7 Typical protection of strand anchor (double-protected anchor head and free length, unprotected fixed tendon length).

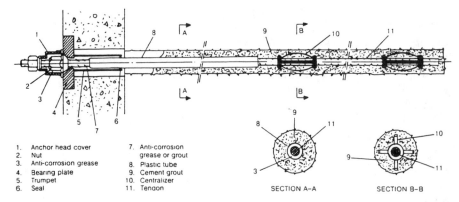

1. Anchor head cover
2. Nut
3. Anti-corrosion grease
4. Bearing plate
5. Trumpet
6. Seal
7. Anti-corrosion grease or grout
8. Plastic tube
9. Cement grout
10. Centralizer
11. Tendon

SECTION A–A   SECTION B–B

Fig. 6-8 Typical protection of bar anchor (double-protected anchor head, single protected free length, unprotected fixed tendon length).

Injected Materials

Solidifying Fluids or Suspensions. These materials, usually cement-base grout, are injected either within protective coatings or after stressing the anchorage. They also may be combined with a sheath where a higher degree of protection is needed, or where the tendon must remain restressable. The material is tremied from the bottom of the hole displacing air and water as it rises. Centralizers should be used to ensure uniform cover of injected materials. Packers may be necessary to retain external grout placed in situ in holes inclined upward. A small injection pressure (0.25 N/mm^2 or 40 psi) will ensure full penetration of the grout and exclusion of fluids from the hole.

The cement must be essentially free of sulfides and other aggressive elements. The total sulfate and chloride content of the grout should not exceed 4 percent and 0.1 percent (by weight) of cement, respectively, and for this the mixing water should not contain more than 500 mg of chloride ions per liter. If admixtures are used, they should not contain more than 0.1 percent of chlorides, sulfates, or nitrates. Recommended minimum grout cover around the tendon is 10 mm, but where doubt exists it should be increased accordingly.

Where the cement grout is used as an internal stage of permanent protection, specifically, annulus filling inside a sheath or capsule, the total bleed during simple sedimentation should not exceed 0.5 percent by volume. If the bleed tendency is higher, the bleed water should be removed.

Viscoelastic Fluids. These are primarily bitumen fillers. Mandatory checks are made to ensure that bitumen solvents do not contain chlorides or sulfides. If used for double protection in combination with a sheath or coating, they should not adversely affect the properties of the sheath or coating by dissolution, chemical attack, or at temperatures required to maintain fluidity for injection. The bitumens should remain sufficiently solid but flexible in their cool state in order to remain in their position even in pervious ground, and some form of checking is prudent to that effect.

Liquids and Gels to Control pH. These include lime suspensions, sodium silicates, and silica gels. If uncoated tendons must remain in the borehole for some time (a month or longer) under aggressive conditions, it may be prudent to protect the tendon temporarily with a suitable liquid in order to maintain a pH between 9 and 12. This fluid can be displaced when the final filler is injected. In this case, the liquid must be contained either by impervious ground or by closed impervious sheath.

Greases

An appropriate choice of grease can be made only with regard to the requirements and service conditions of an anchorage. Greases normally should be compounded in order to provide corrosion prevention characteristics inherent with stressed high-tensile steel tendons.

In view of the limited guidelines that are available at present on the boundary conditions of the physical and chemical properties of appropriate greases, their suitability should be documented where possible by reference to similar applications. The general requirements are summarized as follows:

1. Greases should not contain substances that can enhance corrosion, specifically, unsaturated fatty acids and water, and individual contents of sulfides, nitrates and chlorides should not exceed 5 ppm. Anticorrosion compounds should be described.
2. Greases should be stable against water and oxygen and should not separate into soap and oil.
3. Greases should satisfy the standards set forth in this section with regard to bacterial and microbiological degradation resistance, low moisture vapor transmission, and high electrical resistivity.
4. Greases should be compatible with wrapping or sheathing materials which may be applied after greasing, and should not affect barrier properties of coverings.

The most suitable grease for long-term protection is obviously the type exhibiting the optimum combination of properties, although absolute standards cannot be established. Thus, if documentary evidence shows that a grease has been successfully used in providing long-term protection to anchorages, its basic properties should be quantified with the aid of infrared (IR) spectroscopy. Greases used for nuclear reactor vessels or naval marine applications are good choices for consideration.

Petroleum jelly, either pure or enriched, with dissolved inhibitors and lithium-based greases, has been used successfully in West Germany, the United Kingdom, and the United States.

Table 6-3 shows results of parallel tests on three grease types. Among all the properties checked, the following are considered satisfactory and acceptable: penetration (worked and unworked), water content, sulfur content, ash content, foreign particles, and total acidity (British Standards Institution, 1982).

The drop point of 62°C given in Table 6-3 for grease 3 is considered satisfactory (FIP, 1986), provided operating temperatures do not exceed 42°C, but this safety margin of 20°C is the absolute minimum in order to avoid the risk of separation with possible loss of properties. Evaporation losses for greases 2 and 3 are higher than the normal limits, and could lead to hardening with time.

Oxidation stability results indicate that the process of breakdown is accelerated after 300 h under test conditions at 99°C. If the rate of reaction

TABLE 6-3   Data and Results of Three Greases Employed in Anchorage Practice to Protect Steel Tendons

Test	Method	Grease 1	Grease 2	Grease 3
Drop point, °C	BS 2000: Part 132	126	93	62
Penetration,				
Unworked at 77°F	IP 50	167	169	102
Worked at 77°F		350	260	216
Water content, % (by weight)	IP 74	0.10	0.03	0.01
Sulfur content, % (by weight)	BS 2000: Part 61	2.24	1.47	2.99
Evaporation loss, % (by weight)	ASTM D972	0.1	0.9	0.8
Ash content, % (by weight)	IP 4	0.01	2.82	0.01
Foreign particles, per cm³	IP 134			
≥25 μm		200	Too dark	600
≥75 μm		Nil	to count	Nil
≥125 μm		Nil	particles	Nil
Oxidation stability,	BS 2000: Part 142			
Pressure drop, lb/in², after				
100 h		10	11	10
200 h		11	17	16
300 h		13	25	24
400 h		17	43	39
Total acidity, mg KOH/g	IP 1(B)	0.79	3.91	0.27

From British Standards Inst. (1982).

TABLE 6-4 Specified Properties of Grease

Porperty	Test Specification[a]		Acceptance Criterion
	Petroleum-Based Grease	Calcium Lithium Soap-Based Grease	
Base no.	ASTM-D-974 (modified)	—	15 min
Contents of chlorides	ASTM-D-512	ASTM-D-512	5 ppm max.
Contents of nitrates	ASTM-D-992	ASTM-D-992	5 ppm max.
Contents of sulfides	APHA no. 428	APHA no. 428	5 ppm max.
Oxidation stability			Maximum loss:
100 h	ASTM-D-942	ASTM-D-942	70 kPa
400 h			140 kPa
1000 h			210 kPa
Corrosion resistance, 14 days at 25°C and 100% relative humidity	ASTM-D-1743	ASTM-D-1743	Incipient corrosion no more than three spots of a size sufficient to be visible to the naked eye; max. rating = 2
Drop point	ASTM-D-566	ASTM-D-566	Minimum 60°C
Cone penetration: worked at 25°C	ASTM-D-937	ASTM-D-217	Minimum 250 units (0.1 mm = 1 unit) max. 350 units
Flash point	ASTM-D-93	ASTM-D-93	Minimum 150°C
Effects of salt spray testing, 1-mm-thick grease layer, 500 h	ASTM-B-117	ASTM-B-117	No corrosion
Oil separation test, % by weight	—	App. "G" of BS 3223 (1960) or IP 121/57	Maximum 3%
Evaporation loss, % (by weight)	ASTM-D-1972	ASTM-D-972	Maximum 0.5%

[a] Unless otherwise specified, the latest issue of referenced documents applies.

From Brian-Boys and Howells (1984).

increases by a factor of 2–3 (average 2–5) for every 10°C rise in temperature, it follows that the times equivalent to 300 h for ambient temperatures of 25 and 10°C would be as follows:

$$\text{Ambient temperature 25°C, time} = 300 \times 2.5^{[(99-25)/10]h} = 30 \text{ years}$$

$$\text{Ambient temperature 10°C, time} = 300 \times 2.5^{[(99-10)/10]h} = 120 \text{ years}$$

Breakdown due to oxidation will increase acidity and fluidity.

From the results of Table 6-3 it appears that grease 1 is the most suitable of the three types tested. This grease has the following advantages: (a) a higher drop point, affording better adhesion to the tendon at elevated temperatures; (b) higher penetration for easier pumping and better void-filling properties; (c) lower evaporation loss, diminishing the risk of hardening; and (d) much higher oxidation stability.

Useful comparative data are shown in Table 6-4, and include basic properties of greases according to the recommendations of the Geotechnical Control Office of the Government of Hong Kong (Brian-Boys and Howells, 1984).

Tendon Coatings

Tendon coatings should be applied under factory conditions, either by the manufacturer, or on site in special workshops where air-drying and clean conditions are provided.

Bonded Metallic Coatings. This process includes galvanizing, zinc spraying, and electroplating, all producing an absorbed surface coating. They should be applied only in the factory by the tendon steel manufacturer. Sacrificial metallic coatings should be confined to temporary anchorages. A coated tendon is always liable to damage; hence special care is necessary during handling. They should be chosen under sufficient information and competent advice, since there is some reservation about the effectiveness of a thin surface film under highly stressed cyclic loading, and possible flaws in surface treatment may enhance corrosion by creation of bimetallic cells.

Bituminous and Metallic Paints. Most authorities consider these fairly unreliable for strands owing to difficulties in obtaining uniform coating, and because they are also subject to damage during handling. They are suitable for tendon protection during storage and before use.

Tapes. In this group are polypropylene or grease-impregnated fabric tapes, generally considered effective for temporary anchorages. Tapes should be applied by wrapping with minimum 50 percent laps. During wrapping contact with the tendon should be maintained, hence the latter must be greased before wrapping to exclude atmosphere and give the tendon flexibility to move within the coating.

Plastic Sheaths. Continuous-diffusion impervious polypropylene or polyethylene sheaths applied under factory conditions are used for temporary and permanent anchorages. Their minimum thickness should be 1 mm, but with 1.5 mm nominal thickness. Plastics susceptible to ultraviolet (UV) light are suitable, provided that carbon black or UV inhibitors are incorporated to resist degradation. With reference to potential exposure to fire when corrosion-promoting chlorides are released, plastic sheathing is considered safe since this hazard is extremely unlikely.

Sheaths are effective as long as the internal annular space is filled during manufacture with an appropriate resin, cementitious material or grease to exclude atmosphere and create an appropriate electrochemical environment. A heat-shrinkable tube with a preapplied sealant may be used, but is not considered acceptable as double protection system.

When close-fitting sheaths are used together with grease or a sealant, it is essential to ensure that the coating clearance around the tendon is sufficient so that the tendon can be stressed without frictional resistance. If setting fluids are used in combination with a sheath giving substantial clearance, provisions should likewise be made to ensure that the tendon can extend without restrictions; thus, an additional sheath or tube can be used to act as bond-breaker.

Metal Sheaths. Light corrugated metal sheaths should not be considered for protection, since they can be easily perforated by corrosion. Where metals are chosen, their electrochemical characteristics must be compatible with the tendon metal in order to avoid induction of corrosion potentials between differing metals.

Sheath Joints. Bars used for tendons are not transported in rolls, and their sections must be therefore effectively connected in situ. Sheath or coating joints should not interfere with the continuity of the protective system along the entire tendon length, with respect to physical and electrochemical effects.

Reliable joints can be obtained by overlapping at least 25 mm (1 in), combined with liberal use of solvent glues appropriate for the sheathing material. Loose sleaves should have overlaps at least 50 mm (2 in) and fit easily over the basic coating with clearance allowing injection or extrusion of the bonding agent.

Heat-shrinkable tubing is suitable for connecting sheaths, provided the components are of the same quality approved for tendon protection. Normal overlap should be 350 mm (14 in) minimum for butt joints without solvent.

Any voids at joints within the sheath should be completely filled to exclude atmosphere. Joint details should accommodate injection of cementitious material or greases and similar sealing compounds with simultaneous displacement of air. Excess filler must be extruded during tightening of screwed connections or during injection of the joint voids between the tendons or sheath.

A typical sheathed joint detail for bar tendon is shown in Fig. 6-9.

6-9   PROTECTIVE SYSTEMS OF THE FIXED LENGTH

The fixed anchor length must receive the same degree of protection as the free length. Furthermore, materials as well as their structural configurations

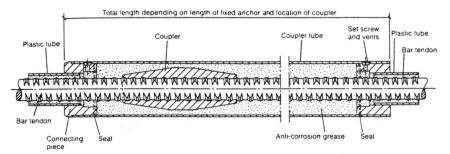

Fig. 6-9   Coupler detail in the free length of bar tendon.

must be capable of receiving and transferring tensile loads from one medium to another and eventually to the ground, and the only mechanism available is by bond. Accordingly, strength and deformability features must be ascertained in terms of their structural behavior. From the study of bond distribution discussed in the foregoing sections, it may be assumed that stresses are gradually transmitted to the ground along the entire fixed length until at some point or near the end the stress remaining in the tendon is practically zero. Where the corrosion problem is critical, it is prudent to ensure that the distal end of the tendon is redundant and therefore unstressed but enclosed.

The deformation or associated distortion of individual elements of the corrosion protective system should not be allowed to reach creep stage, nor expose the steel tendon through cracking. In practice, however, both cracking and creep of individual components is always likely, but the corresponding requirements of each mode are opposite, and few materials are available which can satisfy both particularly under the stress intensity involved in anchor testing and stressing.

Cement Grout

This material is invariably the agent used to transmit the fixed anchor load to the ground. This is not a reliable electrochemical barrier, although its alkalinity must be recognized. The minimum cover specified for this zone should always be provided. Cement grout is brittle and either bonds to or encases the tendon; hence it will invariably crack following extension of the tendon when preloading the anchor.

Caution is necessary when using washers at suitable intervals to induce local compression in front of the plate, since these systems are not proof against cracking. In this case, the entire fixed length does not necessarily function in compression. Furthermore, these devices can cause decoupling of the tendon from the grout at the washer location; otherwise the grout will bond to the tendon in such a way that the load transfer will begin at the proximal end thereby reverting the grout to tension.

Epoxy

Certain nonmetallic coatings, especially epoxies, exhibit physicochemical durabilities as well as strength and protective qualities that make them suitable for corrosion protection. The coating materials included for consideration in this section are restricted to organic formulations, the most important criteria being inertness toward the constituents of cement paste and also chloride ions, favorable creep characteristics, film integrity and protective qualities, and bond to steel. The abrasion resistance of suitable epoxy coating is also acceptable. However, a large variation has been observed between the relative flexibilities of epoxy coatings, with the powder systems giving better flexibilities than the liquid ones. Polyvinyl chloride coating, for example, has excellent flexibilities even in film thickness of 35 mils. Epoxies are tough materials and therefore should be more resistant to abuse.

The effect of coated bars on structural integrity has been favorably assessed by pullout and creep tests. Epoxy coated bars with average film thickness of 5–11 mils have shown acceptable bond strengths to concrete (and by extension to grout) as measured in pullout tests. Most epoxies have also shown acceptable creep rates, that is, comparable to those of uncoated bars. However, polyvinyl chloride-coated bars have unacceptable bond and creep characteristics.

Epoxy and polyester resins may be substituted for cementitious grouts, but generally are more expensive. When used alone as bonding agent between the tendon and the ground resins can be formulated to deform without cracking and are thus suitable for corrosion protection without the necessity of sheathing (this is possible with rock bolts). A comprehensive report and study on epoxies is given by Clifton et al. (1975), following research sponsored by the U.S. Department of Transportation. Considering flexibility, bond strength, creep characteristics, and corrosion protection requirements, this report concluded that the optimum film thickness of epoxy coatings on steel bars is 7 mils, with acceptable deviation of 2 mils.

When epoxy and polyester resins are used to encapsulate fixed lengths of tendon in combination with sheaths, compatibility of elastic properties of all components of the anchor must be considered and ensured in order to avoid decoupling of the resin from the sheath or sheath from grout when stressed.

Details

Figure 6-10 shows a bar anchor protected by a double protective system. The tendon, in this case, is a ribbed bar. Cement grout cracks adjacent to the ribs are calculated to be less than 0.1 mm wide. Therefore, the grout and plastic corrugated sheath provide two physical barriers to corrosion. Longitudinal cracking should be given consideration, and depends on the lateral restraint. If uncontrolled longitudinal cracking is possible, the system reverts to single corrosion protection.

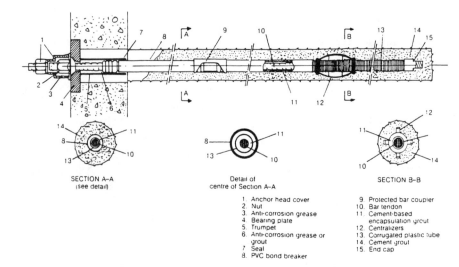

1. Anchor head cover
2. Nut
3. Anti-corrosion grease
4. Bearing plate
5. Trumpet
6. Anti-corrosion grease or grout
7. Seal
8. PVC bond breaker
9. Protected bar coupler
10. Bar tendon
11. Cement-based encapsulation grout
12. Centralizers
13. Corrugated plastic tube
14. Cement grout
15. End cap

Fig. 6-10 Double corrosion protection for a bar anchor.

Figure 6-11 shows corrosion protection for a multiwire tendon with monobar jacking head. The protective system in the fixed (bond) length is classified as double protection if cement cracks are controlled and shown to be less than 0.1 mm wide and uncontrolled longitudinal cracking can be excluded. In the free length the system is classified as single protection since the anticorrosive paste is not considered a physical barrier.

Figure 6-12 shows double corrosion protection for the fixed (bond) length of strand tendon. If, however, the grout within the corrugated sheath is cement-based, the tendon bond length protection reverts to a single system.

Nonstressed elements of the tendon, that is, the threaded length of bars protruding beyond the nuts, should always be enclosed within the protected system. In instances where protection is not specified, cement grout cover over the fixed length may be considered adequate for temporary anchors only, and on the explicit understanding that more detailed protection is not necessary.

Corrugated Ducts

In addition to protection, the plastic sheath forming the primary element of protection must also transmit stresses from the filler to the external grout without displacement, distortion, or distress. Effective shear transfer is accomplished if the sheaths are corrugated. Established guidelines specify a corrugation pitch 6–12 times the sheath wall thickness, and corrugation amplitude not less than three times the wall thickness, which should be 1 mm minimum. Duct materials must be impervious, and common types are polypropylene, polyethylene and plastic. Duct joints may be screwed, and al-

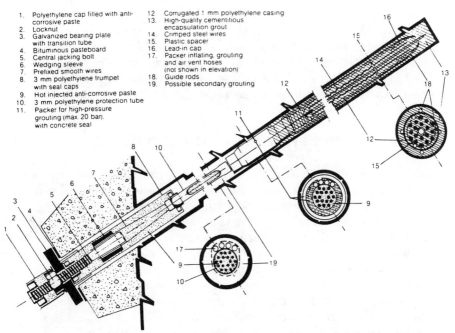

1. Polyethylene cap filled with anti-corrosive paste
2. Locknut
3. Galvanized bearing plate with transition tube
4. Bituminous pasteboard
5. Central jacking bolt
6. Wedging sleeve
7. Prefixed smooth wires
8. 3 mm polyethylene trumpet with seal caps
9. Hot injected anti-corrosive paste
10. 3 mm polyethylene protection tube
11. Packer for high-pressure grouting (max. 20 bar) with concrete seal
12. Corrugated 1 mm polyethylene casing
13. High-quality cementitious encapsulation grout
14. Crimped steel wires
15. Plastic spacer
16. Lead-in cap
17. Packer inflating, grouting and air vent hoses (not shown in elevation)
18. Guide rods
19. Possible secondary grouting

Fig. 6-11 Single protection for multiwire tendon with monobar jacking head.

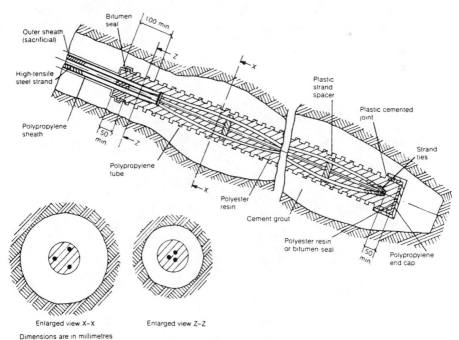

Bitumen seal
100 min.
Outer sheath (sacrificial)
High-tensile steel strand
Polypropylene sheath
50 min.
Polypropylene tube
Plastic strand spacer
Plastic cemented joint
Strand ties
Polyester resin
Cement grout
Polyester resin or bitumen seal
50 min.
Polypropylene end cap

Enlarged view X–X   Enlarged view Z–Z

Dimensions are in millimetres

Fig. 6-12 Double protection of fixed (bond) length for strand tendon.

ways cemented to preclude ingress to fluids. Where the anchor length permits, unjointed ducts are preferred.

Where metal ducts, either plain or corrugated, are considered, they must be compatible with the tendon metal, and appropriate certification from the manufacturer should be requested. A double sheath protection of free and fixed anchor length of strand tendon, shown in Fig. 6-13, consists of two concentric high-strength plastic corrugated ducts.

6-10 PROTECTIVE SYSTEMS OF THE ANCHOR HEAD

Certain anchor-head details may be left with the manufacturer, or they may be standardized. In either case, the system consists of a bearing plate, the main anchor head, a trumpet, and a protective cover. Custom-designed anchor heads are frequently specified. In this context, they are not entirely prefabricated. Furthermore, because of tendon elongation associated with prestressing, friction grips for strand and locking nuts on bars cannot lock the system in a fixed position until the entire extension has been achieved. Locking devices and arrangements require exposed wire, strand, or bar on which to grip, and any preformed corrosion protection at this end must be removed. This exposes tendon metal in two locations, above and below the bearing plate, which must be protected separately, as is the bearing plate and other exposed anchor head accessories.

In aggressive circumstances early anchor head protection is indicated for both temporary and permanent anchorages. The essence of protection in

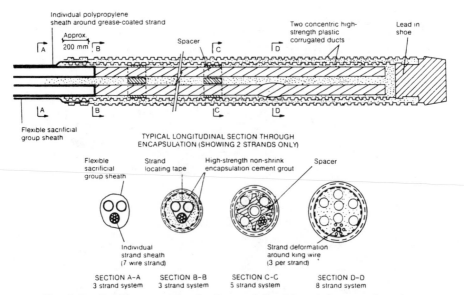

Fig. 6-13 Double protection for free and fixed length for strand tendon.

this case is to enclose the exposed metal parts and accessories while imparting to the system freedom to extend under prestressing and thereafter under the working stresses. Typical examples of anchor head protection are shown in Figs. 6-14, 6-15, and 6-16, and also in the illustration of corrosion protection for the free and the fixed length.

Inner Head. Protection at this location is to ensure effective overlap with the free-length protective system, protect the short tendon section below the bearing plate, and isolate the short section passing through the plate. These requirements may be satisfied with a telescopic section of sheathing and, after tensioning, fillers that will displace any water and injected both within and outside the telescopic sheath.

Cement grout is not suitable for inner head protection, and primary grout should not be in contact with the structure, since it may crack during movement of the anchor against the structure. This area, therefore, must be protected with deformable ductile materials impervious to water, and these may be preplaced or injected but fully contained within surrounding ducts with an end seal.

In saturated or damp conditions, it may be impracticable or difficult to exlude every vestige of water during the protection application; hence the design of the telescopic sheath must provide a full enclosure around the

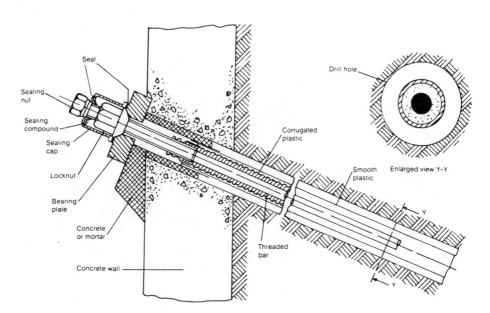

Fig. 6-14 Typical corrosion protection detail for anchor head; double protection of bar tendon.

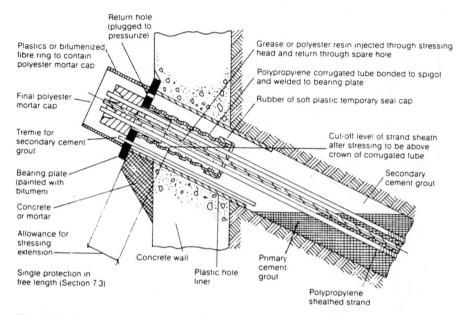

Return hole
(plugged to
pressurize)

Plastics or bitumenized
fibre ring to contain
polyester mortar cap

Final polyester
mortar cap

Tremie for
secondary cement
grout

Bearing plate
(painted with
bitumen)

Concrete
or mortar

Allowance for
stressing
extension

Single protection in
free length (Section 7.3)

Concrete wall

Plastic hole
liner

Grease or polyester resin injected through stressing
head and return through spare hole

Polypropylene corrugated tube bonded to spigot
and welded to bearing plate

Rubber of soft plastic temporary seal cap

Cut-off level of strand sheath
after stressing to be above
crown of corrugated tube

Secondary
cement grout

Primary
cement
grout

Polypropylene
sheathed strand

Fig. 6-15 Typical corrosion protection detail for anchor head; double protection of strand tendon.

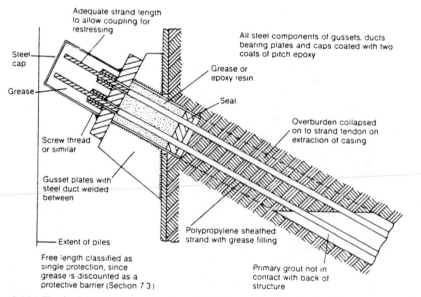

Adequate strand length
to allow coupling for
restressing

Steel
cap

Grease

Screw thread
or similar

Gusset plates with
steel duct welded
between

Extent of piles

Free length classified as
single protection, since
grease is discounted as a
protective barrier (Section 7.3)

All steel components of gussets, ducts
bearing plates and caps coated with two
coats of pitch epoxy

Grease or
epoxy resin

Seal

Overburden collapsed
on to strand tendon on
extraction of casing

Polypropylene sheathed
strand with grease filling

Primary grout not in
contact with back of
structure

Fig. 6-16 Typical corrosion protection detail for anchor head; double protection of restressable strand tendon.

exposed metal to ensure against the possibility of water flow through the cell during service.

Where the protective materials are injected, a lower injection pipe and an upper vent pipe should be used for complete filling of the void and full expulsion of water and air (FIP, 1986). Injected materials should be conveyed by tremie to the lowest part of the duct, thereby displacing fluids upward and toward the vent. If the space is restricted, simple grease gun techniques are a good alternative.

Frequently, the duct through the anchored structure is exposed to wet conditions, and in this case a brittle grout is fairly unreliable in providing a water seal outside the tube. Experience shows that the grout will probably be squeezed, displaced, moved, or fractured as the structure deforms during service.

Where standard anchor head details are applicable, some elements of corrosion protection may be prefabricated. These include a rigid plastic sheath, resin-bonded to a metal spigot welded in turn to the back of the bearing plate. During installation of the latter, the plastic tube is slid externally and telescopically over the tendon coatings, but adequate tolerance must be provided.

Outer Head. With restressable anchors, or with anchors subject to load monitoring, protection of parts above the bearing plate (bare tendon, friction grips, and locking nuts) should recognize the requirement to remove both the anchor head cap and the contents to allow tendon access for restressing. The protective system will depend on the details of the stressing and locking method and equipment. Generally, however, grease is used with plastic or steel caps. Additionally, a suitable seal and mechanical coupling between the cap and the bearing plate should be included.

If the tendon is not the restressable type, the cap and its contents may be fixed. Resins and other setting sealants are suitable, and the mechanical coupling between the cap and the bearing plate may be omitted.

If the design calls for the anchor head to be completely enclosed by the structure (e.g., concrete blocks), the overhead components can be encased in good concrete, and rely on adequate cover if the environment is not aggressive and ingress to water and atmosphere is restrained.

Bearing Plate. Current practice calls for the bearing plate and other steel accessories to be painted with bitumastic or similar protective materials. Prior to painting, all steel surfaces should be cleaned and all rust and deleterious matter removed by sand blasting or acid pickling. The selected coatings should be compatible with the materials used for corrosion protection of the anchor head. The side of the bearing plate against the structure and other inaccessible parts should be treated before installation. It is quite acceptable for bearing plates on concrete structures to be set directly on a concrete pad or in a seating formed by epoxy or polyester mortar.

6-11   CATHODIC PROTECTION

Cathodic protection is an acceptable method for underground pipelines and marine structures. It consists of passing an artificial electric current through the ground with the intent to polarize the metal cathodically. A diagrammatic representation of the process is shown in Fig. 6-17.

Cathodic protection may be considered for anchorages, but caution is necessary in assessing its effectiveness and reliability because of the following disadvantages:

1. The protection must be extended along the entire length of the anchor; otherwise intermediate corrosion centers may appear.
2. For best results, the cathodic protection needs virtually complete saturation in electrolytes. The system must be used therefore in combination with protected coverings where partially saturated soils are to be traversed.
3. Determination of the electric current required to maintain protection during service is still empirical and rather uncertain; hence it cannot be predicted as well as the response of other protective systems.
4. Full protection cannot be ensured, since there is the potential of damage by underwater corrosion to adjacent buried metals.
5. An additional tendon must be inserted and fit in a system that is already congested, and without interfering with the load-carrying capacity.
6. The need to maintain and occasionally renew sacrificial anodes implies a continuing cost. Furthermore, accessibility to replace such anodes is difficult for anchorages, and sometimes even impossible.

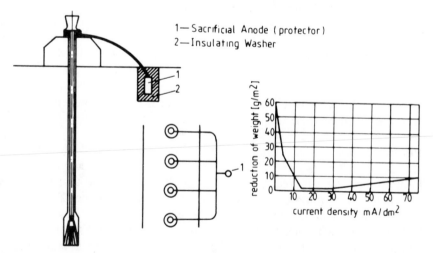

Fig. 6-17   Diagrammatic representation of cathodic protection for an anchor. (From Hobst and Zajic, 1977.)

7. Stress corrosion in high-tensile steels cannot be precluded with the same certainty as in the case of coatings.

Portier (1974) mentions anchorage projects in zones where the difference of potential of the soil attained 100 V, particularly in the vicinity of railway stations In these cases cathodic protection might appear appropriate, but after careful consideration it was excluded because of the risk of inducing electroosmosis or electrodrainage in the soil, and because the cost of energy and supervision of the installation was estimated to be too high.

Further reluctance to use this method relates to the risk of producing lines of current that may enter and leave the steel in relatively pervious environment, thereby creating microbatteries. Certain specialists also fear that this operation may in the long term create a more serious corrosion problem under tension than other forms of protection.

6-12 PREPROTECTED BOND LENGTH ANCHORS

An example of preprotected bond anchor (factory applied) is shown in Fig. 6-18. With this type, the steel tendon is protected over the entire bond length by a zibbed plastic sheath, with the annular space filled with epoxy pitch or cement mix at the manufacturing stage.

For this operation, the injection sleeve pipe is located beside the tendon steel and maintained in position by spacers. The protective plastic ribbed sheath that covers the fixed length is elastic, rotproof, waterproof, and non-corrodable. It is also sufficiently strong to resist and transfer high bond stresses during service. An epoxy pitch or a cement–bentonite mix is injected into the annular space between the ribbed sheath and the steel tendon at the manufacturing stage and under factory-controlled conditions.

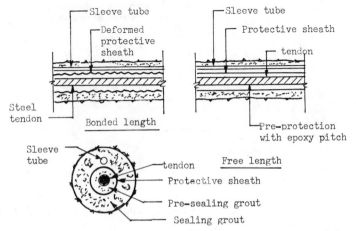

Fig. 6-18 Preprotected bond length anchor; schematic section. (Soletanche.)

6-13   A BRIEF SURVEY OF CORROSION INCIDENTS

The permanence of ground anchorages and long-term effectiveness of corrosion protection systems is assessed in this section and also in Section 6-14 in terms of documented case histories.

Cheurfas Dam. This structure was mentioned in Section 1-2, and a typical cross section is shown in Fig. 1-2. Approximately 30 years after the 1000-ton (metric) anchors were installed (between 1931 and 1935), a survey indicated that they were subject to considerable corrosion, despite the elaborate protective system shown in Fig. 6-19. These conclusions were based on three main observations: (a) a nearly steady average reduction of 5 percent the initial tension force on most anchors occurred between 1938 and 1969; (b) a very substantial cumulative loss of tension was observed on certain anchors, which were retensioned to 1000 tons (metric), attributed to long-term effects; and (c) a virtually complete failure was manifested at the anchor head of two units (Portier, 1974).

Since most anchors were of the restressable type, their top part was projected beyond the bearing plate and was protected by a tarpaulin sheet

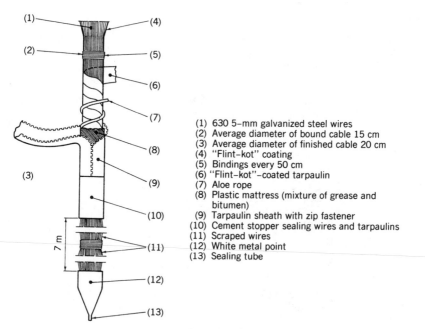

(1) 630 5-mm galvanized steel wires
(2) Average diameter of bound cable 15 cm
(3) Average diameter of finished cable 20 cm
(4) "Flint-kot" coating
(5) Bindings every 50 cm
(6) "Flint-kot"-coated tarpaulin
(7) Aloe rope
(8) Plastic mattress (mixture of grease and bitumen)
(9) Tarpaulin sheath with zip fastener
(10) Cement stopper sealing wires and tarpaulins
(11) Scraped wires
(12) White metal point
(13) Sealing tube

Fig. 6-19   Corrosion protection for the 1000-ton anchor in Cheurfas Dam, Algeria. (From Cambefort, 1966.)

and bitumen in order to fascilitate their individual retensioning at any time. Floodwaters and repeated restressing rapidly destroyed the fabric, and the sun melted the bitumen. The heads were left in this condition for many years, and some corroded to failure.

The 5 percent load loss observed over a period of 30 years is by current standards nominal and reflects relaxation and long-term loss in the steel tendon, irrespective of any corrosion effects. The very high time yield observed with certain anchors might have been the combined result of improper handling during tensioning and corrosion attack in the presence of swelling marls and water in the body of the dam.

Portier (1974) suggests that the problems encountered with the anchor heads can be avoided if they are not placed in spillways, and if their detensioning is not performed at the expense of protection.

Joux-Tarare Dam. This incident was among the first documented cases involving stress corrosion cracking (SCC), discussed in Section 6-3, and exemplifies the combined action of high static tensile stress and localized corrosion.

Between June and October 1952, eight 1300-ton anchors of the Cheurfas type were installed and tensioned in the upper works of the Joux dam in France. The tendons were stressed to 67 percent f_{pu}. Controls performed several months after installation indicated that the residual tension had been reduced from $0.67f_{pu}$ to practically zero. On exposing the tendon, broken strands were noticed, (see Fig. 6-20), apparently the result of SCC initiated by high tensioning, groundwater aggressivity, and poor storage conditions.

Recognizing the effect of high tension on the steel tendon in a corrosive environment, a direct recommendation that followed the study of this incident was to reduce the stress level in the steel by limiting the working load of permanent anchors to 55 percent the failure strength.

World Trade Center. In this example, slurry (diaphragm) walls 70 ft deep were temporarily braced by six rows of tiebacks, stressed to 100 percent their design (working) load, and installed at 45°. The walls are keyed into underlying bedrock, mainly for the transfer of the considerable vertical load imposed on the structure during the service of the temporary bracing. The final permanent lateral support is provided by the underground floor system, and the tiebacks were eventually distressed.

Since the tiebacks were temporary, corrosion protection was omitted, although monitoring was specified and carried out. During service the tiebacks showed certain corrosion effects, and an expensive system of cathodic protection was installed (Feld and White, 1974). Although zinc loss was considerable, elimination of corrosion was not accomplished. A view of the excavation and the anchorage system is shown in Fig. 1-14.

Fig. 6-20 Joux Dam (1952); (a) view of dam crest during floor period; (b) failure of tendons by stress corrosion cracking (SCC). (From Portier, 1974.)

6-14 DOCUMENTED PERFORMANCE OF ANCHORAGES

Interestingly, the subject of corrosion protection may never be placed in the right perspective, since corrosion failures are either not reported or seldom well documented.

To date, however, about 35 case histories of failure by tendon corrosion have been collected and fairly documented. One led to collapse of the complete structure–anchor–ground system, but under the combined effect of installation procedures and corrosion protection, neither of which is acceptable by current standards (FIP, 1986).

Of these cases, 24 related to permanent installations (protected or unprotected), and 11 were temporary anchorages with no protection specified other than cement grout cover along the fixed length and occasionally a decoupling sheath slong the free length. Included in this group are the Cheurfas and Joux dams discussed in Section 6-13.

Various pertinent data for these cases are summarized in Table 6-5 based on reported case histories of tendon corrosion by Portier (1974), Herbst (1978), Nurnburger (1980), Weatherby (1982), and personal communications by FIP (1986).

Relevance of Tendon Type and Location. It appears that corrosion is localized. No tendon type is exempt from the process, and no special system has immunity to corrosion. Nine incidents involved bar, 19 involved wire, and 7 involved strand. For each tendon type, the service period before failure extended from several weeks to many years. Failures occurring a few weeks after installation have been caused by stress corrosion cracking or hydrogen embrittlement.

The case histories confirm that quenched and tempered plain carbon steels and high-strength alloy steels are more susceptible to hydrogen embrittlement than other varieties, hence these steels should be selected with extreme caution where environmental conditions are known to be dangerous and aggressive.

Interestingly, in this survey corrosion incidents are associated with certain anchor components more frequently than with others. Thus 19 incidents occurred at or within 1 m (3.5 ft) of the anchor head, 21 incidents involved the free length, and only two occurred in the fixed length. In terms of cause there is no specific pattern, and these incidents are fairly random with possible exception of the choice of steel.

Corrosion Time. The duration of service at failure is extremely variable, ranging from a few weeks to 31 years. The following observations are appropriate:

- Nine incidents occurred within six months after installation, namely, cases 2, 8, 11, 15, 18, 23, 24, 30, and 33. Of these, four were permanent anchorages with some or full protection.

TABLE 6-5   Case Histories of Anchorage Corrosion and Their Relevant Conditions

Case No.	Date of Installation	Time in Service at Failure	Geographic Location	Type of Structure	General Environment	Ground Conditions	Type of Tendon	Working Load or Stress Level
1	1934	31 years	Algeria	Anchored dam	Dry air	Masonry overlying sandstone	630 wires, 5 mm dia. (1100–1300 N/mm^2)	10,000 kN (65% UTS)
2	1952	A few months	France	Anchored dam	Temperate climate	Concrete overlying rock	Multiwire	13,000 kN (67% UTS)
3	1955	16 years	Czechoslovakia	Prestressed concrete dam	Humid air	Concrete overlying rock	4.5-mm-dia. smooth patented wires	4 MN
4	1955	26 years	Sweden	Underground power station crane beam	Humid air	Rock underlying concrete with reported water leakages (pH = 7–8)	26-mm-dia. bar (80/105)	300 kN
5	1959	10 months	West Germany	Underground power station	Temperate climate	Rock; below water table; water contained very little chlorides, i.e., fairly nonaggressive	15 oval ribbed wires (1570 N/mm^2)	(74% of elastic limit)
6	1961	2 years	U.S.A.	Anchored cofferdam	—	Soil overlying rock in the presence of saltwater	35-mm-dia. bar	—
7	1963	A few years	West Germany	Anchored retaining wall adjacent to river	Temperate climate	Soil; below water table; water-contained industrial pollutants and high chloride ion content	22 8-mm-dia. wires (ST 135/150)	—
8	1964	A few weeks up to 1 year	Algeria	Anchored dam	Dry air	Masonry overlying sandstone	54 7-mm-dia. cold-drawn wires (1265/1432 N/mm^2)	1960 kN (75% of elastic limit)
9	1965	7 years, 9 years	U.K.	Rock face stabilization	Temperate climate	Limestone	44 high-tensile steel wires	2000 kN
10	1960s	8 years	West Germany	Anchored retaining wall	Temperate climate	—	5.2-mm-dia. wires (alloy steel)	—

Anchorage Category	Corrosion Protection	Corrosion No.	Failure Location	Remarks and/or Diagnosis
Permanent	Coated tarpaulin covered by mixture of grease and bitumen with outer tarpaulin sheath over free length; cement grout cover in fixed length	4 anchorages	Beneath the anchor head	Floodwaters and repeated tensioning tore the tarpaulin fabric and the exposed internal bitumen cover melted under high ambient temperatures, thereby removing protection; localized corrosion
Permanent	Coated tarpaulin covered by mixture of grease and bitumen with outer tarpaulin sheath over free length; cement grout cover in fixed length	Wires in 2 anchorages	Beneath the anchor head	Corrosion failure under tension linked to type of steel; decision taken to limit working stresses to 55% UTS thereafter, but to increase proof loading up to 1.5 times working load on occasions
Permanent	Grease-impregnated glass fiber bandage and outer asphalt wrapping over free length; cement grout cover in fixed length	4 anchorages	Beneath the anchor head	Fully corroded wires exposed in spite of protective wrapping; by contrast, on the same site and in the same location, steel tendons comprising 37 bundles of 19 wires of 2.9 mm dia. were undamaged—here the internal spaces of the ropes were filled in the factory with red lead sealing compound
Permanent	Bitumen coating of anchor head; cement grout cover over tendon length	One anchorage	In fully bonded length 2.5 m up from crane beam anchor head	Virtually no trace of grout cover; significant pitting and typical reduction in cross-sectional area was 6.8%; depth of deepest crack 1.3 mm; failure attributed to intergranular stress corrosion; steel judged to be sensitive to cracking
Permanent	No protection at anchor head; jute wrapping-impregnated bitumen in free length	17 anchorages	5 in anchor head; 12 in free length of which 5 were within 0.5 m of anchor head	Deep localized corrosion where bitumen protection was missing (differential aeration postulated); this protection could not withstand damage during installation or environmental attack; steel judged to be sensitive to corrosion
Temporary	No protection over free length; cement grout cover in fixed length	A few anchorages	Free length	Brittle failure; groundwater was corrosive due to presence of sulfuric acid formed from cinders falling for many years from steel locomotives; brine may also have contributed
Permanent	Tendon encased in cement grout	3 anchorages	85% of ruptured wires failed in vicinity of concrete deadman–tendon interface	Surface corrosion and pitting observed in tendons; insufficient grout cover and presence of chlorides noted; stress corrosion and cracking also located; tendon bending due to ground movement
Permanent	Road oil loaded with red lead but anchor head waiting several weeks before protective filling placed	Several individual wires	Button headed wires at anchor head	Brittle failure under tension; button heads were cold forged on site
Permanent	Bituminous infilling as a surround for the free length (piped in hot); cement grout cover in fixed length	24 wires in 3 anchorages; a further 13 wires in same anchorages 2 years later	0.6–1 m beneath anchor head	Stress corrosion due to an aqueous environment
Permanent	Tendon painted with bitumen over free length; cement grout cover in fixed length	3 anchorages	Free length	Although no corrosion-producing elements found, stress corrosion postulated where bitumen protection had broken down; surface corrosion and heavy pitting on wires; some pits had small fissures

(continued)

TABLE 6-5 (Continued)

Case No.	Date of Installation	Time in Service at Failure	Geographic Location	Type of Structure	General Environment	Ground Conditions	Type of Tendon	Working Load or Stress Level
11	1960s	3 months	U.K.	Anchored floor of dry dock	Temperate climate and saline atmosphere	Rock	Quenched and tempered low-alloy bars (1500 N/mm^2)	(67% UTS)
12	1967–68	6–18 months	West Germany	Underground pumped storage scheme	Temperate climate	Rock	18 8-mm-dia. wires (1500–1700 N/mm^2)	900–1700 kN
13	1968	Within 3 years	Switzerland	Underground power station	Temperate climate	Rock	12 8-mm-dia. wires	650 kN
14	1968	11 years	France	Anchored foundation blocks	Temperate climate	Coal mine waste fill; above water table	8–12-mm-dia. oval ribbed wires (1450/1600 N/mm^2)	1720 kN (67% of elastic limit)
15	Before 1969	A few days and 100 days	West Germany	Anchored retaining wall supporting a rail track	Temperate climate	Soil fill	6 12.2-mm-dia. wires	—
16	1968–69	—	U.S.A.	Anchored retaining wall	—	Landfill with high organic content overlying mica schist; brackish groundwater	12.7-mm-dia. strand of 4.2-mm wire (270 K grade)	≤2640 kN
17	1969	10 years	U.S.A.	Anchored retaining wall	Acidic; adjacent to waste acid neutralization plant	Fill (clays and silts)	32-mm-dia. high-strength bars (1033 N/mm^2 ultimate)	636 kN
18	1969	A few weeks	France	Anchored retaining wall	Temperate climate	Above water table; chlorides and sulfates in water from sewer leakages	8 12-mm-dia. ribbed wires (1450/1600 N/mm^2)	1030 kN (63% of elastic limit)
19	1969	5 years	Malaysia	Rock strengthening	Humid	Rock	36 7-mm-dia. wires	700 kN
20	1970	28 months	New Zealand	Anchored retaining wall	—	Clay overlying sandstone	42 7-mm-dia. wires	UTS = 2570 kN; initial tensioning to 48% UTS; tendon designed to work at up

Anchorage Category	Corrosion Protection	Corrosion No.	Failure Location	Remarks and/or Diagnosis
Permanent	Bare bar, ungrouted over free length; cement grout cover in fixed length	2 anchorages	Free length	Corrosion pitting leading to hydrogen-induced stress corrosion cracking at failure; free-length grouting actioned in 1977, since when no corrosion failures have been observed
Permanent	A chemical filler (oil-based unsaturated fatty acid polymer) surrounding the free length; cement grout cover in fixed length	Majority of the 133 anchorages installed	Free length	Stress corrosion due to leaching out of nitrate ions from chemical protective filler
Test	Tendon installed in borehole with anchor heads at either end; no protection by design for test purposes	2 anchorages (11 wires and 10 wires)	Stressed length between anchor heads	Presence of sulfides caused embrittlement of the steel
Permanent	Ordinary Portland cement and polyethylene outer tube over free length	4 anchorages	0.2–2.5 m beneath anchor head	Brittle failure under tension, initiated at surface oxidized local defects
Temporary	Tendon unprotected over free length; cement grout cover in fixed length	3 anchorages	Free length	Tendons not heavily corroded; failure judged to be due to corrosion fatigue as a result of bending due to fluctuating loads from railway being transmitted through frozen ground; cracks in steel noted at failure location
Temporary	Bentonite–cement grout cover plus outer steel pipe in free length; in addition, a sacrificial zinc ribbon anode was installed with each tendon; cement grout cover in fixed length	—	Beneath the anchor head	Brittle corrosion failure of tendon where bentonite–ceement grout cover had dropped 1–1.2 m; hydrogen sulfide was present in the soil and the sacrificial anode was consumed near the anchor head of the failed tendons
Permanent	No protection of anchor head; in free-length grease, paper wrapping and plastic sleeve embedded in cement grout; cement grout cover in fixed length	8 anchorages	Beneath the anchor head	Heavy pitting leading to brittle failure of unprotected tendon
Permanent	Ordinary Portland cement grout and mild steel outer tube in free length; cement grout cover in fixed length	6 anchorages	0.1–0.5 m beneath anchor head	Brittle failure under tension; decarboned steel at wire perimeter; incomplete filling of protective grout beneath anchor head
Permanent	Anchor outer head protected by sealing cap infilled with grease and injected under pressure; polypropylene sheathed wires surrounded by bitumen placed in situ over free length; polypropylene sheathed wires with stainless steel end barrels surrounded by cement grout in fixed length	1 anchorage comprising 36 wires, of which 33 were broken	Underside of anchor head and at bare section of wires immediately above plastic	Stress corrosion cracking of wires; inadequate filling of inner head region with bitumen; exposed bare wires subject to wetting and drying cycles; groundwater of low pH suspected
Permanent	Polypropylene extruded sheathing of individual wires with outer plastic tube infilled with a mastic sealant; ribbed alkathene tube and epoxy resin cover in fixed length	5 wires	In free length 1–8 m below anchor head	Surface corrosion cracking; mastic filler found to be hygroscopic and it was suspected that the mineral oil softened the polypropylene sheathing; also speculated that the polypropylene sheathing may have been damaged during trans-

(*continued*)

TABLE 6-5 (*Continued*)

Case No.	Date of Installation	Time in Service at Failure	Geographic Location	Type of Structure	General Environment	Ground Conditions	Type of Tendon	Working Load or Stress Level
								to 66% UTS
21	Before 1971	—	West Germany	Anchored retaining wall	Temperate climate	—	15 5.2-mm-dia. wires (alloy steel)	—
22	Before 1971	Within a year	West Germany	Anchored retaining wall	Temperate climate	—	5.2-mm-dia. wires (alloy steel)	—
23	1971	6 weeks	U.S.A.	Anchored retaining wall	—	Acidic soil embankment comprising mainly blast furnace slag; soil moist adjacent to tendon	32-mm-dia. bar hot-rolled, drawn, and stress-relieved (1100 N/mm^2 ultimate)	—
24	1971	4 weeks	U.S.A.	Anchored retaining wall	—	Moist soil with low pH	35-mm-dia. bar, hot-rolled, drawn, and stress-relieved	—
25	1972	2 years	South Africa	Restraint for cantilevered grandstand	Seasonal wetting and drying	Fill	5 12.2-mm-dia. strands	450 kN
26	1972–73	In the early stages of contract	New Zealand	Anchored retaining wall	—	Clays and silts overlying sandstone	Multiwire tendon	490–1050 kN (50% UTS)
27	1973	11 years	U.K.	Anchorage restraint of abutment which was yielding initially	Temperate climate	Fill overlying clay and weak rock	4–5 15.2-mm-dia. strands	350 kN (50% UTS max.)
28	1974	—	New Zealand	Anchored bridge abutment	—	Rock	34 7-mm-dia. wires	1000 kN (50% UTS)
29	1974	5 years	Algeria	Concrete dam raising	Dry air	Concrete	36 15.2-mm-dia. strands	—

Anchorage Category	Corrosion Protection	Corrosion No.	Failure Location	Remarks and/or Diagnosis
				port and installation; 1 m of polypropylene sheathing stripped off below anchor head before tendon installation and stressing
Temporary	Tendon unprotected over free length; cement grout cover in fixed length	2 anchorages	Free length	Heavy pitting and occasional cracking of wires noted; chemical analysis of corrosion products indicated 0.25% sulfur content but no chlorides
Temporary	Tendon encased in cement grout	5 anchorages	Free length	Heavy corrosion and pitting in certain zones where there was no adhering cement grout; other sections of tendon that were completely grout-free displayed general corrosion; no corrosion where tendon still bonded to grout; brittle failure recorded; tendon bending and overstressing also induced by ground deformations; analysis of corrosion products indicated 0.63% sulfates but no chlorides or sulfides
Temporary	Tendon unprotected over free length; cement grout cover in fixed length	4 anchorages	Free length	Stress corrosion cracking postulated
Temporary	Tendon unprotected over free length; cement grout cover in fixed length	—	Free length	Stress corrosion cracking postulated
Permanent	Polypropylene sheathed and greased strands in free length; cement grout cover in fixed length	No failure, but one tendon located with unacceptable corrosion, i.e., pitting, and all 9 anchorages condemned	Fixed anchor zone	Some doubts were expressed over the efficacy of the grouting of the fixed anchor length where no special precautions had been taken; when one anchorage was excavated, grout cover in fixed zone ranged from nil to 6 mm, and pitting up to 1 mm in depth was measured
Temporary	Unprotected in free length; cement grout cover in fixed length	—	—	Ground movements created severe overloading of tendons in certain locations; corrosion of tendon
Permanent	No anchor head protection; greased and sheathed strands over free length; cement grout cover in fixed length	One strand in each of 2 anchorages	Beneath the anchor head	Failure due to stress corrosion
Permanent	Polypropylene sheathing over wires with a secondary protection of outer tube and mastic infilling in free length; corrugated tube–grout encapsulation over fixed length	—	—	Protective ducting in free length damaged during transportation, permitting leakage of mastic filler that had softened at the high ambient temperature; protected tendons stored several months on site before installation
Permanent	Free-length annulus grouted with acrylamide chemical; cement grout cover in fixed length	—	Beneath anchor head	Where duct had not been filled properly with acrylamide grout, a tar epoxy was poured in to fill upper 0.5 m; failure occurred at the base of the tar epoxy

(continued)

TABLE 6-5 (*Continued*)

Case No.	Date of Installation	Time in Service at Failure	Geographic Location	Type of Structure	General Environment	Ground Conditions	Type of Tendon	Working Load or Stress Level
30	1975	6 months	France	Anchored retaining wall	Temperate climate	Above water table; nothing suspicious	32-mm-dia. ribbed bars (1079–1225 N/mm^2)	640 kN (74% of elastic)
31	1976	5 years	Switzerland	Anchored abutments for pipeline bridge	Temperate climate	Fill overlying sands and gravels overlying rock	10 12.7-mm-dia. steel strands	1130–1150kN
32	1977	Within 3 years	Hong Kong	Anchored retaining wall	Humid and slightly saline	Nonaggressive fill overlying completely weathered granite that improves with depth to moderately strong granite	7 12.9-mm-dia. Supa strands	1050 kN
33	1977	4 months	West Germany	Anchored retaining wall	Temperate climate	Fill consisting of slag and ash; sulfate content = 200 mg/liter	32 mm dia. hot-rolled and threaded bars (1100 N/mm^2 UTS)	—
34	1978	4 years	South Africa	Slope stabilization	Humid	Weathered sedimentary rock	4 or 6 15.2 mm dia. strands	590 kN 890 kN (60% UTS)
35	1980	1–3 years	Hong Kong	Stabilization of rock	Humid and slightly saline	Rock	High-tensile steel bars	500–650 kN

From FIP (1986).

Anchorage Category	Corrosion Protection	Corrosion No.	Failure Location	Remarks and/or Diagnosis
Temporary	Polyethylene outer tube over free length; cement grout cover in fixed length	2 anchorages	3 m and 8 m beneath anchor head	Brittle failure under tension
Permanent	Polyethylene sheathed strands in free length with asphalt filling; cement grout cover in fixed length	3 anchorages	In fixed anchor length within 500 mm of free length	Bridge collapse due to failure of anchored abutment; severe corrosion of strands in proximal zone of fixed anchor length that was only partially grouted; tendon exposed to aggressive groundwater containing sulfides and chlorides in fill and sandy gravel; poor construction practice and lack of quality controls, such as water testing, led to inadequate grouting; fixed anchor straddled permeable soil and rock
Permanent	Anchor head encased in concrete; grease and plastic sheathing over free length; cement grout cover in fixed length	1 anchorage (2 strands)	Beneath the anchor head and in the free length	No corrosion protection provided immediately beneath anchor head; considerable delays experienced between stressing and concrete encasement of anchor head; metallographic examination of tendon wires in 45 anchorages showed up to 2.7% and 12% loss of diameter for delay periods of 1–8 months and 16–36, months respectively; it was also speculated that strands had been stored on site for some time (allowing corrosion to develop) before greasing and sheathing of free length
Temporary	No protection at anchor head, polythene tube over free length; cement grout cover in fixed length	2 anchorages	50 mm beneath anchor head; middle of free length	Failure adjacent to anchor head due to brittle fracture at a deep pit; second failure attributed to hydrogen embrittlement; ground deformations also present, leading to bending and overstressing; lack of protection and use of corrosion-susceptible steel highlighted overall; sulfur compounds present as corrosion products
Permanent	Grease-filled or cement-grouted outer anchor head; PVC sheathed and greased strands in free length; cement grout cover and epoxy resin coating over fixed length	2 anchorages	Underside of anchor head	Ground movement after service increased tendon loads by up to 20%; grease filling and capping of anchor head inadequate to stop infiltration of surface water to inner head; stray currents from adjacent electrified rail line (15–20-m distance) identified; sulfate-reducing bacteria located in annulus between strand and PVC sheathing in some cases
Permanent	Cement grout plus sheath over free length and tendon bond length; grease at bar couplers	10 anchorages	Up to 20 m beneath anchor head but always adjacent to a coupling joint	All fractures occurred over a small area where neither grout nor grease was in contact with the bar—this small air void resulted from the method of encapsulation; metallurgical examination showed pitting corrosion and hydrogen embrittlement; traces of chloride salts were present on the bar after assembly, which probably initiated pitting

- Five corrosion failures occurred within the period 6–18 months, namely, cases 5, 12, 16, 22, and 26. Two of these were permanent installations.
- The majority of failures (21 incidents) occurred during the period from 18 months to 31 years, which is the upper time limit for the cases studied. Although limited statistical value can be inferred from these figures, it appears that the 18-month or longer service period used to define permanent anchors is not inconsistent with this record. However, the relative large number of failures recorded within 6 months after anchorage installation raises the question of serviceability of unprotected anchors, even those of very short duration.

Fixed Length. The two incidents involving the fixed anchor length were caused by inadequate grouting of the tendon bond length. In one case, this lack of protection exposed 3 m (10 ft) of tendon to aggressive groundwater containing chlorides and sulfides. This incident (case 31 in Table 6-5) involved the failure of three rock anchors bracing an abutment, and occurred after 5 years in service. This failure caused the collapse of a pipeline bridge. The following conditions during construction were recorded: (a) no borehole was drilled at this location, and the rock stratum was inferred from a borehole 25 m (80 ft) away; (b) drilling for the anchors was poorly supervised, and drill records were not kept; (c) water or pregrouting tests were not carried out prior to tendon homing; and (d) grout injection procedures were not monitored, and instead a fixed quantity of grout was preplaced sufficient only for the tendon bond.

This problem could have been detected by water or pregrouting tests, and prevented if one protective sheath had been applied over the tendon bond length.

Free Length. The relatively high number of incidents in the free length, compared to only few at the fixed length, suggests more aggressive or combined causes augmenting intensified failure of the anchor system. In this survey, free length failures were caused by the following reasons: (a) tendon overstressing due to ground movement initiating pitting corrosion or corrosion fatigue; (b) absence of cement grout or inadequate grout cover in tendons exposed to chlorides in industrial waste fills or organic materials; (c) disruption of bitumen cover because of lack of elasticity; (d) poor choice of protective materials, incompatible with the anchor system and its components; (e) poor storage conditions on site and for periods long enough to cause initial corrosion damage; and (f) poor execution of the protection system and its details.

Anchor Head. Documented causes of anchor-head failure are (a) lack of protection (extended even for only a few weeks in aggressive conditions);

(b) incomplete protection, such as inadequate cover due to improper filling; and (c) damage to the protective filler during service.

In case 32 of Table 6-5, there was considerable delay between tendon stressing and concrete capping of the anchor head. For a delay between 16 and 36 months, a loss of wire diameter 12 percent was recorded.

Exposure of the anchor head to the atmosphere contributes to the corrosion risk and increases the corrosion potential. This simple fact suggests that the anchor head should be protected with at least the same standards that are applied to the free and fixed anchor length. Noting that 19 failures occurred within 18 months after installation, early protection of the anchor head is always indicated, and at best it should be applied after grouting, irrespective of the service life. Where a delay is unavoidable, the anchor head should be protected temporarily with the use of plastic paint, grease-impregnated tape, or other suitable cover.

REFERENCES

American Society for Testing and Materials, 1979: "Underground Corrosion," ASTM Symp. on Corrosion of Metals, Williamsburg, Virginia.

Arup, H., 1979: "A Recording Instrument for Measuring Corrosivity in Offshore Seawater." Offshore Tech. Conf., Houston, Paper OTC 3602, 2129–2134.

Beeby, A. W., 1978: "Corrosion of Reinforcing Steel in Concrete and Its Relations to Cracking," *Struct. Eng.*, **54** (3), 77–81.

Bird, C. E., and F. J. Strauss, 1967: "Metallic Coatings for Reinforcing Steel," *Materials Protection*, **6**, 48.

Brian-Boys, K. C., and D. J. Howells, 1984: "Model Specification for Prestressed Ground Anchors," Geotech. Control Office. Hong Kong, GCO Publ. (3/84).

British Standards Institution, 1982: "Recommendations for Ground Anchorages," Draft for Development DD81, BSI, Lond.

Burdekin, F. M., and G. P. Rothwell, 1981: "Survey of Corrosion and Stress Corrosion in Prestressing Components Used in Concrete Structures with Particular Reference to Offshore Applications," Cement & Concrete Assoc., Slough.

Bureau Securitas, 1972: "Recommendations Regarding the Design, Calculation, Installation and Inspection of Ground Anchors," Editions Eyrolles, 51 Boulevard Saint-Germain, Paris (Ref. TA 72).

Cambefort, H., 1966: "The Ground Anchoring of Structures," *Travaux*, **46** (April–May), 15 pp.

Caron, C., 1972: "Corrosion et Protection des Ancrages Definitives," *Construction*, (Feb.), pp. 52–56.

Champion, F. A., 1962: *Corrosion Testing Procedures*, Chapman & Hall, London.

Clifton, J. R., H. F. Beeghly, and R. G. Mathley, 1975: "Nonmetallic Coatings for Concrete Reinforcing Bars," U.S. Dept. of Commerce, National Bureau of Standards, Washington, D.C.

Comte, C., 1971: "Tech. des Tirants," Inst. Research Found. Kolibrunner/Rodio, 119 pp. Zurich.

Cornet, I., and B. Bresler, 1966: "Corrosion of Steel and Galvanized Steel in Concrete," *Materials Protection*, **5**.

Coyne, A., 1930: "Perfectionnement aux barrages-poids par l'adjonction de tirants en acire," (Genie Civil, Aout).

Duffaut, Duhoux, et Heuze, 1973: "Corrosion des aciers dans le beton arme. Essais realises dans l'estuaire de la Rance de 1959 a 1971," *Annales ITBTP* (May).

Environmental Degradation by De-Icing Chemicals and Effective Countermeasurements, 1973: page 25. Highway Research Record No. 425.

Federation Internationale de la Precontrainte, 1976: "Report of Prestressing Steel, 1. Types and Properties" (FIP/5/3), Slough.

Feld, J., and R. E. White, 1974: "Prestressed Tendons in Foundation Construction," *Prestr. Concrete Found. and Ground Anchors*. 7 FIP. Cong., pp. 25–32, New York.

Fidjestol, P., and N. Nilsen, 1980: *Reinforcement Corrosion in Concrete*, Veritas, Bergen.

FIP, 1986: "Corrosion and Corrosion Protection of Prestressed Ground Anchorages."

FIP, 1972: "Draft of the Recommendations and Replies to FIP Questionnaire," (1971). FIP Subcommittee on Prestressed Ground Anchors.

FIP, 1973: "Final Draft of Recommendations FIP Subcommittee on Prestressed Ground Anchors."

Frazier, K. S., 1965: "Value of Galvanized Reinforcing in Concrete Structures," *Materials Protection* **4**, 53.

Goto, Y., 1971: "Cracks Formed in Concrete Around Deformed Tension Bars," *J. Am. Conc. Inst.*, **68** (4), 244.

Graber, F., 1980: "Excavation of a VSL Rock Anchor at Tarbela," VSL Silver Jubilee Symp., Losinger Ltd., Bern, unpubl. work.

Gutt, W. H., and W. H. Harrison, 1977: "Chemical Resistance of Concrete," BRE Current Paper 23/77, Bldg. Research Establishment, Garston.

Hadley, R. F., 1939: "Microbiological Anaerobic Corrosion of Steel Pipe Lines," *Oil Gas J.*, **38** (19), 32.

Hamner, N. E., 1970: "Coatings for Corrosion Protection," Chapter 14 in *NACE Basic Corrosion Course*, A. des Brasunas and N. E. Hamner, eds., National Assoc. Corrosion Engineers. Houston, Tex.

Hausman, D. A., 1967: "Steel Corrosion in Concrete," *Materials Protection*, **6**, 19.

Herbst, T. F., 1978: "Safety and Reliability in Manufacture of Rock Anchors," Int. Symp. on Rock Mech. Related to Dam Foundations, Rio de Janeiro.

Hilf, J. W., 1973: "Reply to Aberdeen Questionnaire," (1972), Unpublished.

Hobst, L., and J. Zajic, 1977: "Anchoring in Rock," Development in Geotechnical Engineering, Vol. 13, Elsevier Scient. Publ., Amsterdam.

Houston, J. T., et al., 1972: "Corrosion of Reinforcing Steel Embedded in Structural Concrete," Report No. CFHR-3-5-68-112-I, Centre of Hwy. Research. Univ. Texas.

King, R. A., 1977: "A Review of Soil Corrosiveness with Particular Reference to Reinforced Earth," TRRL Supplementary Report 316. Transport and Road Research Laboratory, Crowthorne.

Koch, J., 1972: "Reply to FIP Questionnaire," Unpublished.

Larson, T. D., P. D. Cady, and J. C. Theisen, 1969: "Durability of Bridge Deck Concrete," Report No. 7, College of Engineering, Pennsylvania State Univ., April.

Lee, H., and K. Neville, 1967: *Handbook of Epoxy Resins*, McGraw-Hill, New York, pp. 6-45 to 6-52.

Littlejohn, G. S., 1973: "Report on Tendon Corrosion of Ground Anchorages Adjacent to Bridge No. 5, Clear Water Bay Road, Hong Kong," Unpublished.

Longbottom, K. N., and G. P. Mallett, 1973: "Prestressing Steels," The Structural Engineer, **51**, (12), pp. 455-471.

Littlejohn, G. S., and D. A. Bruce, 1977: *Rock Anchors—State of the Art*, Found. Publ. Ltd., Essex, England.

Mayne, J. E. O., J. W. Menter, and M. J. Pryor, 1960: "The Mechanism and Inhibitions of the Corrosion of Iron by Sodium Chloride Solution," Part I, *J. Chem. Soc.*, 3229.

Meyer, A., 1977: "Report on Discussion to Session VI by J. M. Mitchell. A Review of Diaphragm Walls," Thomas Telford Ltd., London.

Mitchell, J. M., 1974: "Some Experiences with Ground Anchors in London," *ICE Conf. on Diaphragm Walls and Anchorages*, London, Sept., pp. 129-133.

"Modern Electrical Methods for Determining Corrosion," 1973: "Rates," NACE Tech. Unit Comm. T-3D on Instruments for Measuring Corrosion, NACE Publ. 3D170.

Mozer, J. D., A. C. Bianchini, and C. E. Kelser, 1965: "Corrosion of Reinforcing Bars in Concrete," *J. Am. Concrete Inst. Proc*, **62**, 909.

Naus, D. J., 1979: An evaluation of the effectiveness of selected corrosion inhibitors for protection of prestressing steels in PCPVs. Oak Ridge National Laboratory, Tennessee.

O'Neill, E. F., 1980: Study of reinforced concrete beams exposed to marine environments. Performance of concrete in marine environments, Report SP-65, American Concrete Institute.

Nurnberger, U., 1980: "Analysis and Evaluation of Failures in Prestressed Steel," *Forschung. Strabenbau und Strabenverkehrstechnik*, **308**, 1–95.

Ostermayer, H., 1974: "Construction, Carrying Behavior and Creep Characteristics of Ground Anchors," *Proc. Diaphragm Walls and Anchorage Conf.*, Inst. Civil Eng., London, pp. 141–151.

Palmer, J. D., 1974: "Soil Resistivity—Measurement and Analysis," *Materials Performance* (Jan.), 41–46.

Pascoe, W. R., 1968: *Plastic Coatings for Metals, Modern Plastics Encyclopedia Issue*, McGraw-Hill, New York.

PCI Post-Tensioning Committee, 1974: "Tentative Recommendations for Prestressed Rock and Soil Anchors," PCI, Chicago, 32 pp.

Pender, E., A. Hosking, and B. Mattner, 1963: "Grouted Rock Bolts for Permanent Support of Major Underground Works," *J. Inst. Eng. Austral.*, **35**, 129–150.

Phelps, E. H., 1967: "A Review of the Stress–Corrosion Behaviour of Steels with High Yield Strength," Conf. on Fundamental Aspects of Stress Corrosion Cracking, Ohio State Univ., Columbus, Sept. 11–15.

Portier, J., 1974: "Protection of Tie-Backs Against Corrosion," *Prestressed Concrete Found. and Ground Anchors*, 7th FIP Congress, pp. 39–53, New York.

Pourbaix, M., 1966: *Atlas of Electrochemical Equilibria in Aqueous Solutions*, Pergamon Press, New York, pp. 409–410.

Rehm, G., 1968: "Corrosion of Prestressing Steel," *Proc. Symp. on Prestressing*, Madrid (June).

Robinson, R. C., 1972: "Design of Reinforced Concrete Structures for Corrosive Environments," *Materials Performance*, **11**, 15.

Ryell, J., and B. S. Richardson, 1972: "Cracks in Concrete Bridge Decks and Their Contribution to Corrosion of Reinforcing Steel and Prestressing Cables," Report IR51, Ontario Ministry of Transport and Communication.

Schiessl, P., 1975: "Admissible Crack Width in Reinforced Concrete Structures," Contribution II, 3–17, Inter-Assoc. Colloqu. on the Behavior of In-Service Concrete Structures, Liege.

Schrier, L. L., 1976: *Corrosion*, Newnes-Butterworths, London.

Soletanche Co., Ltd., 1970: "Other Types of Anchor," *Suppl. Ground Anchors, Cons. Eng.* (May), 13, 15.

Spellman, D. L., and R. F. Stratfull, 1969: "Chlorides and Bridge Deck Deterioration," Research Report No. M&R 635116-4, Division of Highways, State of Calif.

Stern, M., 1958: "A Method for Determining Corrosion Rates From Linear Polarization Data," *Corrosion*, **14**, 440t.

Stratfull, R. F., 1957: "The Corrosion of Steel in a Reinforced Concrete Bridge," *Corrosion*, **13**, 173t.

"Test for Indentation Hardness of Organic Coatings," 1974: ASTM Designation D1474-68.

Timblin, L. O., and T. E. Backstrom, 1969: "A Study of Depassivation of Steel in Concrete," Report No. ChE-86, Bureau of Reclamation, Denver, Col.

Tripler, A. B., E. L. White, F. H. Haynie, and W. K. Boyd, 1966: "Methods for Reducing Corrosion of Reinforcing Steel," NCHRP Report 23.

Uhlig, H. H., 1971: *Corrosion and Corrosion Control*, Wiley, New York.

Weatherby, D. E., 1982: "Tiebacks," Report FHWA/RD-82/047, U.S. Dept. Transp. Fed. Hwy. Admin., Washington, D.C.

CHAPTER 7

STRESSING, TESTING, AND ACCEPTANCE CRITERIA

Stressing was briefly discussed in Section 2-11 in conjunction with the pre-loading requirements during the final phase of the installation procedure. In this chapter, stressing is reviewed in detail together with the testing programs and evaluation standards (acceptance criteria) that normally compliment a construction project and the long-term monitoring of an anchorage installation.

Stressing is induced by the application of load, whereas testing confirms anchor load capacity and behavior, establishes the actual factor of safety with which the design is implemented, and ensures satisfactory service performance. Acceptance criteria, based on standardized principles, provide the all important indication of suitability and effectiveness of the installed anchor as supporting unit of a structure. Quantification of the serviceability of an anchorage is thus possible to a certain degree but, more important, any errors made either in the design or introduced during construction can be identified during stressing and testing so that potentially dangerous situations can be avoided.

7-1 BASIC STRESSING TECHNIQUES

Torque and Direct Pull

Stressing is commonly introduced by torque, applied with the use of torque wrench to a suitable anchoring nut threaded on to a rigid bar tendon as shown in Fig. 7-1(a) or by direct pull, applied to the tendon by a jacking device such as the system shown in Fig. 7-1(b).

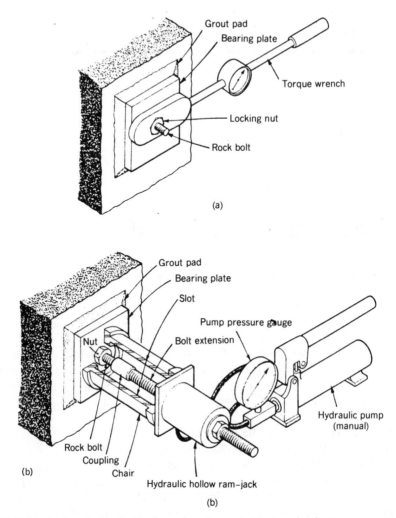

Fig. 7-1 Typical stressing methods and equipment; (a) stressing by torque wrench; (b) stressing by direct pull.

Torque application is usually restricted to relatively low-capacity anchors of bar tendons and primarily various types of rock bolts, up to 150 kN (about 40 kips). The main disadvantages are errors in the applied load (sometimes as high as 25 percent) and occasionally the introduction of torsional stresses to the tendon. The latter can be prevented by placing a friction reducing material such a lubricant beneath the lock-nut prior to stressing.

The torque T_q required to produce a tensile load T_T may be estimated from the following empirical relationship:

$$T_T = CT_q \tag{7-1}$$

where the coefficient C is derived within reasonable limits under controlled laboratory conditions. In the foregoing relation the tensile load is expressed in kilonewtons and the torque in kilonewtons-meters. Although most codes specify that torque equipment should be furnished with a calibration certificate verifying that attainable accuracy is ±5 percent, practical experience shows that under field conditions this error is much higher. The induced load is further subject to variations related to alignment control, friction between mating parts, and size of bar tendon.

Despite these drawbacks, the torque method is popular, particularly for stressing rock bolts because the associated equipment is light, compact, easy to handle, and low-cost. Complete details on equipment and load capacities are included in ISRM draft (1976).

Direct pull is the method most commonly used by anchor contractors because it is suitable for the majority of tendon types and load capacities. Where strand is used as tendon, the direct pull can be introduced either using multistrand jacks, whereby all the strands in the unit are stressed simultaneously, or monostrand (monojacking) pulls whereby individual strands are tensioned in turn. Irrespective of the system, stressing requires a bearing plate placed on the structure in a central alignment and normal to the direction of loading. Typical jacking devices for monostrand and multistrand stressing are shown in Fig. 7-2.

General Guidelines for Stressing

From the practical point of view, stressing an anchor a few days after installation and grouting serves expediency and saves time, but a factor to be considered in this case is the time required for the cement grout to develop full operational strength as is evident from Figs. 2-16 and 2-17. In general, the 7-day period is accepted as compromise between design criteria and construction scheduling. As can be seen from Fig. 2-16, the grout strength attained at this stage may vary from 20 N/mm^2 (2900 psi) for type IV cement to 30 N/mm^2 (4350 psi) for type I cement. The usual practice is to specify stressing when the grout crushing strength has attained a minimum value 25 N/mm^2 (about 3600 psi), which is close to the 7-day strength stipulated by several codes.

Since it is essential to minimize anchor head movement and seating loss, the bearing plate should be correctly bedded and in full contact with the structure or the supported rock, and have a size sufficiently large to distribute the stressing force uniformly. A bearing plate set centrally and normally to the steel tendon will prevent eccentric loading and avoid chaffing of the perimeter tendon components for multiwire or strand tendons.

For solid bar or single unit tendons the tensioning assembly can be fitted to the anchor as soon as the latter is cleaned, but with multiple-unit tendons it is essential to verify that wires or strands are not crossed or fouled in the free length before setting the stressing device. The correct alignment of

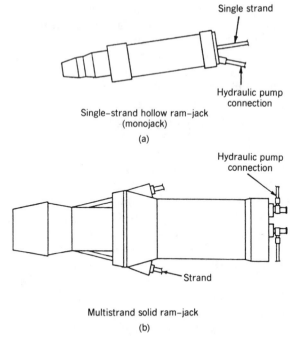

Single strand

Hydraulic pump connection

Single–strand hollow ram–jack (monojack)

(a)

Hydraulic pump connection

Strand

Multistrand solid ram–jack

(b)

Fig. 7-2 Typical jacking devices for tensioning ground anchors; (a) jack for single-strand stressing; (b) solid ram-jack for multistrand stressing; (c) view of a hollow ram multistrand stressing jack. (Ground Anchors, Ltd.)

multiple-unit tendons is maintained by suitable methods such as the comb grillage or fork shown in Fig. 7-3.

If a tendon must be initially overloaded, for example, to 150 percent the working load, the permanent grips are usually omitted from the anchor block until this stage is completed, and this requires a special arrangement. Special anchorages are also used where anchors must be detensioned and again restressed, and such requirements should be specified in the design stage and be known in advance so that the stressing equipment can be chosen and detailed accordingly.

Interestingly, tendon elongation at the top resulting from anchor stressing should be in excess of 30 mm (about 1 in) under maximum applied load, in order to allow the reusable grips or wedges to be freed on destressing. If the expected elastic extension is less than 30 mm, the jack piston should be advanced to 30 mm before placing the temporary loading head. The grips are finally homed to give a tight fit using a special ring or U-shaped hammer.

The space in front of the jack should remain accessible and free to accommodate the prestressing operations and the handling of the jacking equipment. Mechanical lifting and handling is indicated for jacks weighing in excess of 80 kg (about 175 lb). Useful data are given in Table 7-1 correlating

Fig. 7-2 (*Continued*)

Fig. 7-3 Fork used to keep correct alignment of strands. (Cementation Ground Engineering Ltd.)

**TABLE 7-1 Approximate
Weight of Hollow Ram-Jacks**

Maximum-Rated Capacity (kN)	Approximate Weight (kg)
200	20
500	40
1000	80
2000	150
3000	200
4000	300

From Littlejohn and Bruce (1977).

maximum rated capacity and approximate weight. If the anchorage installation is in a built-up area, protection of the public should be considered and provided with a small-aperture steel mesh cage enclosing the work area.

Jack systems are usually pressurized by means of a hand (manual) pump used to advance the ram, but if several tendons must be stressed better output is achieved with a motor-driven pump. If the test load must be held for an extended period, a slight drop in gauge pressure will be noted although the extension of the piston remains unchanged. This loss occurs typically, and a gendle application of pressure to the original reading will restore the extension to the initially recorded.

Lockoff or Transfer Load. When the initial stressing is completed, the double-acting ram retracts and leaves the temporary loading head ready for its removal. Thereafter, the grips are easily released, regreased, and stored for the next stressing operation.

In order to stress the tendon to the locked-off or transfer load, permanent grips must be inserted into the permanent anchor block, and this should be preferably accomplished without completely removing the jack and chair from the tendon. During stressing the chair provides a reaction head as shown in Fig. 7-4, which restricts the upward movement of the permanent gripping wedges. When the proper reading is attained, the jack ram is retracted and immediately the wedges are drawn or pulled in around the tendon as the latter tends to retract, resulting in the load being locked off. If for any reason this final load is not sufficient, the anchor may be restressed with steel spacers or shims inserted beneath the anchor block, which raise the load at lockoff by increasing the tendon extension as shown in Fig. 7-5. Whereas this stressing procedure is typical, most tendon manufacturers have standardized anchorage and stressing devices as discussed in subsequent sections.

Selection of Stressing System

Monostrand (single strand) stressing is usually preferred with tendons up to 5 and occasionally 6 strands because the operation is rapid and the jack unit

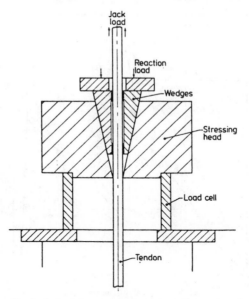

Fig. 7-4 Typical stressing arrangement at the top anchorage.

lightweight, and furthermore close control over the force induced in each strand individually can be achieved.

Despite these advantages, however, certain features inherent with monojack stressing operations remain largely unexplained and tend to inhibit the reliability of this procedure. One of these problems has been reported by Mitchell (1974) and is shown in Fig. 7-6. In this case the load fluctuation

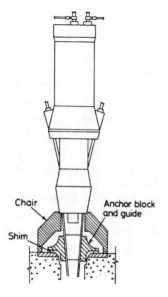

Fig. 7-5 Typical jack arrangement for shimming.

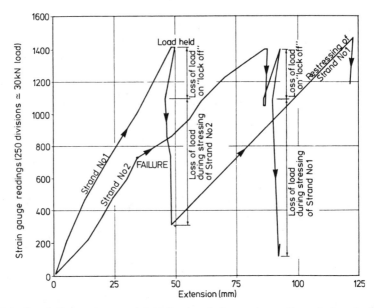

Fig. 7-6 Load fluctuations and interference between two adjacent strands during monojack stressing. (From Mitchell, 1974.)

in two adjacent strands was monitored with strain gauges and it became immediately apparent that the load in the first tensioned strand decreased steadily and considerably during the stressing of the adjacent strand. This effect was amplified in this test since the load was not incrementally applied to each strand as recommended in practice. Nonetheless, the results prompted the recommendation that after application of a seating (nominal) load to each strand the balance of the load should be applied in several increments (four to five) to each strand in alternating sequence.

The same investigator also noted that in a six-strand tendon and as each stage of incremental loading was completed, the highest and lowest load losses always occurred on the first and last strands loaded, respectively. The same observation has been made by Barley (1974), and also by Littlejohn and Bruce (1977). In practice, and following the final increment of one stressing sequence, uneven distribution of load can be minimized by a final stressing cycle to bring all strands to the required load level.

On the other hand, multistrand stressing is favored because of its simple operation. Initially, the jack must be correctly located, but thereafter it requires limited data recording and backanalysis. However, the method does not provide satisfactory control over the behavior of individual tendon units, and cannot ensure equal load in each unit at lockoff. This variation is more important if the free anchor length is less than 10 m (or 33 ft). In this case load extensions are relatively small; hence variations in the amount of wedge pull-in will represent by proportion greater load discrepancies. However, the multistrand jacking system alone can introduce the entire load in the anchor in one stressing operation. Where the anchor service involves loading and unloading in a cyclic operation, this method of stressing is easier and faster to use and particularly convenient for the destressing phase. Furthermore, a multistrand jacking system can economically provide prestressing loads in excess of 3000 kN (675 kips), and this is accomplished at a strand spacing in the anchor block smaller than the spacing required with a monojack system.

It appears from these remarks that neither system should receive preferential treatment, and a comparison should be made only when the stressing and testing requirements are established. In either case, it is essential to ascertain that the method of stressing is relevant to the application, and verify that the applied prestress is resisted along the grouted fixed zone.

7-2 EXAMPLES OF STRESSING SYSTEMS

The most frequent and common requirements of anchor performance are (a) initial tendon tensioning in increments and load lockoff; (b) lowering and again raising the tendon force (detensioning and restressing); and (c) measuring the tendon force for initial verification of anchor capacity, or periodically in conjunction with a monitoring program (surveillance anchor).

A typical stressing anchorage commercially available and used by VSL is illustrated in Fig. 7-7. As shown in part (a), this system consists basically of an anchor head proper, wedges and a bearing plate. A protective cap may also be fitted over the anchorage if this device must remain accessible for future anchor monitoring and force measurement.

In this application, all the strands of the anchor are stressed simultaneously, but locked off individually by wedges in the conical bores of the anchor head. The available range can accommodate anchors containing 1–55 strands, and the procedure is in principle the same from the smallest to the largest unit. The same anchorage can be modified and detailed to meet the installation specifications, namely, surveillance, and restressable and detensionable anchors. For example, if the specifications call for surveillance anchors, the proper arrangement can be chosen according to Fig. 7-7(c), showing diagrammatically various forms of anchorage construction, the most suitable form depending on access facilities to the anchor, the service life, and economic considerations. For solutions B and C, the stressing anchorage type E shown in part (a) is used, with a thread on its external cylindrical surface.

For a restressable anchor, alternatives A and B in Fig. 7-7(c) are suitable. To restress, the anchor head is lifted off the bearing plate and shims are inserted between them. A third solution allowing restressing involves the use of anchorage type E_R, which has a ring nut enabling the prestressing force to be adjusted, as shown in part (b) of Fig. 7-7.

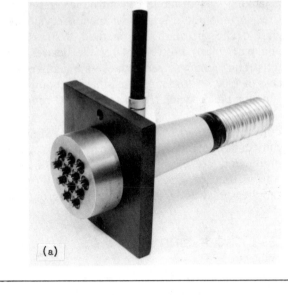

(a)

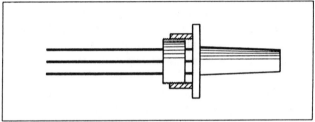

VSL anchorage type Eʀ

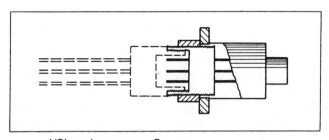

VSL anchorage type Eᴀ

(b)

Fig. 7-7 Typical stressing anchorage commercially available, designated as type E: (a) view of anchorage; (b) schematic presentation of anchorage details; (c) technical data of anchor heads for surveillance anchors. (VSL.)

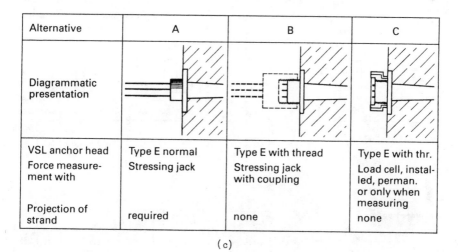

Alternative	A	B	C
Diagrammatic presentation			
VSL anchor head Force measurement with	Type E normal Stressing jack	Type E with thread Stressing jack with coupling	Type E with thr. Load cell, installed, perman. or only when measuring
Projection of strand	required	none	none

(c)

Fig. 7-7 (*Continued*)

Where an anchor must be detensioned later, VSL uses an arrangement consisting of a different type of wedge and an accessory device incorporated between the jack and the anchor head. With this arrangement, the wedges can be released and again locked at any time, thus allowing the anchor to be completely detensioned in one or more stages. If the strands must be cut off and not project beyond the anchorage, type E_A shown in part (b) of Fig. 7-7 can be used which works on the principle of an adjusting ring nut whereas a coupler is used for destressing.

Figure 7-8 shows dimensional details for the type E anchorage of Fig. 7-7. The symbols 5-1 to 5-55 and 6-1 to 6-55 indicate anchor units. The first numeral (5 or 6) is the strand diameter, 13 mm (0.5 in) or 15 mm (0.6 in), whereas the second numeral is the number of strands per unit. The same symbols designate the anchorage units; for example, the second numeral indicates the number of bores through the anchor head. The characteristic values shown in the tables of Fig. 7-8 may differ slightly depending on the applicable code and standards. The bearing plates are designed for a concrete strength according to DIN standard 1045, which specifies a 28-day strength 25 N/mm^2 (3625 psi).

7-3 LOAD AND EXTENSION MEASUREMENTS

The stressing operation has two main objectives: (a) establish a load that is relevant to the function of the anchor, such as a test load introduced to provide a real factor of safety or the prestress to be locked-off into the tendon; and (b) provide an anchor-head extension indicative of anchor per-

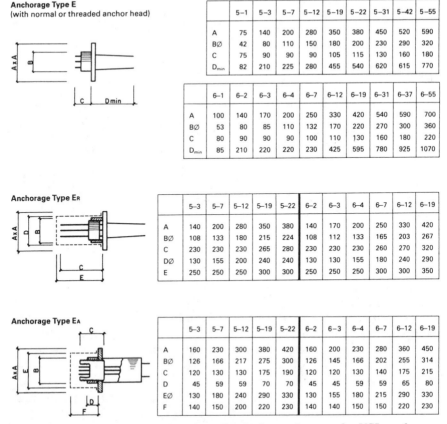

Anchorage Type E
(with normal or threaded anchor head)

	5–1	5–3	5–7	5–12	5–19	5–22	5–31	5–42	5–55
A	75	140	200	280	350	380	450	520	590
B∅	42	80	110	150	180	200	230	290	320
C	75	90	90	90	105	115	130	160	180
D_{min}	82	210	225	280	455	540	620	615	770

	6–1	6–2	6–3	6–4	6–7	6–12	6–19	6–31	6–37	6–55
A	100	140	170	200	250	330	420	540	590	700
B∅	53	80	85	110	132	170	220	270	300	360
C	80	90	90	90	100	110	130	160	180	220
D_{min}	85	210	220	220	230	425	595	780	925	1070

Anchorage Type E$_R$

	5–3	5–7	5–12	5–19	5–22	6–2	6–3	6–4	6–7	6–12	6–19
A	140	200	280	350	380	140	170	200	250	330	420
B∅	108	133	180	215	224	108	112	133	165	203	267
C	230	230	230	265	280	230	230	230	260	270	320
D∅	130	155	200	240	240	130	130	155	180	240	290
E	250	250	250	300	300	250	250	250	300	300	350

Anchorage Type E$_A$

	5–3	5–7	5–12	5–19	5–22	6–2	6–3	6–4	6–7	6–12	6–19
A	160	230	300	380	420	160	200	230	280	360	450
B∅	126	166	217	275	300	126	145	166	202	255	314
C	120	130	130	175	190	120	120	130	140	175	215
D	45	59	59	70	70	45	45	59	59	65	80
E∅	130	180	240	290	330	130	155	180	215	290	330
F	140	150	200	220	230	140	140	150	150	220	230

Fig. 7-8 Details and dimensional data of stressing anchorages for VSL anchorage type E shown in Fig. 7-7. All dimensions are given in millimeters.

formance and representing the actual deformation of the anchor–structure–ground system independent of construction effects. Extension monitoring involves measurements of relevant as well as incidental parameters lumped into a single factor; hence they are not necessarily significant and pertinent to load–extension analysis. It follows, therefore, that an extension measured after stressing and before lock-off is a "gross extension," and may include seating loss, deformations, and movements beyond the elastic behavior of the system.

As an example, the wedge grip anchorage shown in Fig. 7-9(a) may register the following extensions: (a) pullin (draw-in) of the wedge will occur at lock-off until the members become tight fit; (b) after lockoff movement may occur representing the bedding-in of the top anchor block and bearing plate (seating loss), movement of the structure and the ground, and some permanent displacement of the fixed anchor length; and (c) under load the tendon will

Type of load cell	Characteristics	Installation	Main applications	Units available*
G 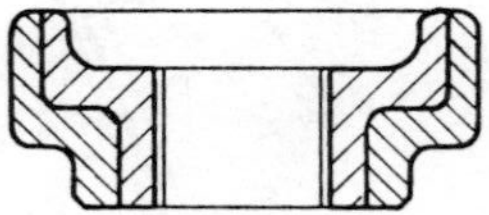	– Hydraulic cell – Remains under pressure during the measurement only	– Screwed on threaded anchor head – removable at any time	– Force control for the whole life of the structure; periodical or permanent surveillance – Permanent anchors – Test anchors	G 35, G 50, G 70 G 100, G 200, G 300, G 550, G 750, G 850
D 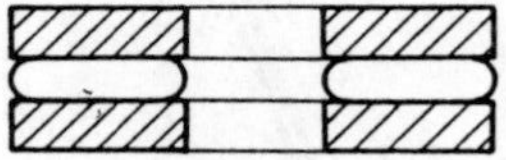	– Hydraulic capsule – Low-cost	– Between anchor plate and head	– Force control during construction – Temporary anchors	D 40, D 80, D 150
E 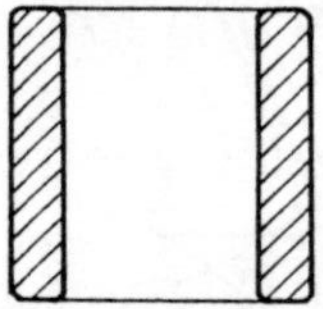	– Electric dynamometer – High precision	– Between anchor plate and head – Between pulling head and piston of stressing jack	– Tests in laboratory and on site – Test anchors	E 30, E 50, E 100 E 200, E 300, E 550 E 750, E 1000

(b)

*The numbers give the service capacity in 10··kN.

Fig. 7-10 (*Continued*)

311

extension data reflecting solely tendon elastic behavior and fixed anchor movement. In practice, this is possible only if precise observations are specified of vertical and horizontal movements of the structure and the ground. In this context, all measuring instruments should be supported independently of the supported structure to avoid interference by the prestressing operations or any ground movement associated with excavation. If an anchor is installed in competent rock and then prestressed against a bearing plate, top anchorage movement is likely to be very small and inconsequential. However, if the same anchor supports a wall to retain ground for an excavation, top anchorage movement is likely to be significant during excavation. For example, a plate movement 5 mm (about $\frac{1}{4}$ in) for an anchor with free length 4 m (13 ft) will probably be sufficient to cause prestress loss of 20–25 percent the initial prestress. In general, if the top anchorage movement is about 5 percent of the total anchor extension or less, it may be ignored in assessing the stressing operations.

Movement of Fixed Zone. It is commonly known from anchor practice that the strain developing during tensioning is not entirely a product of the stretching of the steel tendon in the free length, but also extends to a portion of the fixed anchor zone. A simple and direct method of measuring fixed anchor movement is shown in Fig. 7-11. A wire is embedded in the fixed anchor zone, and is fixed to the tendon in this section of the anchor while it is decoupled along the free length and thus unrestricted to move. The wire extends through and out of the top anchorage assembly and is subjected to some tension simply to keep it stretched. In this manner any movement of the wire registers fixed anchor movement. Likewise, a redundant tendon unit can be used in lieu of the wire, incorporated in the anchor arrangement.

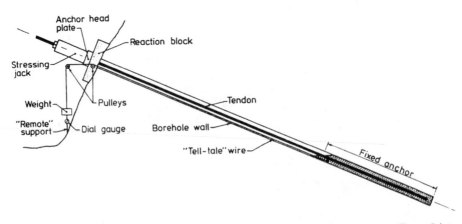

Fig. 7-11 Direct method for measuring movement of fixed anchor zone. (From Littlejohn and Bruce, 1977.)

Wedge Draw-in. Another movement mentioned in the foregoing sections involves wedge pullin of the anchorage type shown in Fig. 7-9, occurring at lockoff. In general, wedge pullin is monitored as indication of lockoff loss and the corresponding residual load at that time. This movement can usually be measured with an accuracy of ± 1 mm. With multistrand stressing system the difference between extensions recorded immediately before and just after lockoff indicates the wedge draw-in. With monostrand stressing the movement can readily be estimated by observing the strand near the jack nose during lockoff. Based on research, Littlejohn and Bruce (1977) have concluded that the amount of draw-in increases linearly with load in the tendon after an initial pullin at loads up to 30 kN per strand. At 200 kN per strand, the amount may be 6 mm. Furthermore, wedge draw-in is less in monostrand than in multistrand stressing, probably owing to the tapping home individual grip wedges prior to locking off in the monojacking operation.

Load and Extension Records. There is considerable variance as to the type, extent, and format of stressing records, which should be produced and retained for either simple or comprehensive tests. If conditions require it, prior to the start of anchorage work an all-party-approved report of the conditions of the surrounding land, including buildings, streets, ducts, springs, and any other features pertinent to the installation, should be prepared and documented as part of the permanent construction record. Likewise, after the anchorage work has been completed, an all-party-approved acceptance report on the results of the final check should be compiled and mutually documented.

For data relevant to the stressing operation, Table 7-2 gives a summary of data for inclusion in a full stressing record. The data also provide information on the type of ground anchor, jacking equipment, and operating personnel, in addition to load–movement recordings. If the stressing is introduced by torque, for a list of typical record requirements reference is made to ISRM draft document 1976.

In plotting load versus extension it is essential to identify the point of origin on the graph. In the usual cases, the point of zero extension is marked after the application of a certain seating load to the tendon rather than the point of zero load. The underlying intent is to remove the slack in the tendon and the jack and compensate for friction and other losses in the jack–pump unit in order to provide a better meaning of the load–extension process.

Interestingly, the record contains several recommendations, in principle similar but with small quantitative differences. Thus Larson et al. (1972) recommend beginning extension readings at 12 percent the working load, but also assume a zero extension 2.5 mm; Longbottom and Mallet (1973) suggest to begin at 10–20 percent that load; among the contractors, Nicholson Anchorage Company (1973) typically establishes the beginning point at 10 percent the test load; and the largest seating load published to date is 25 percent the working load (Short, 1975). Several anchor codes, including

TABLE 7-2 Recommended Items and Data to be Included in Anchor Stressing Record

General Classification Data			
Project	Contractor	Engineer	Inspector
Date	Time started	Time completed	Stressing personnel
Anchor No.	Free length	Fixed anchor length	Rock type
Tendon type	E value of steel	Working load (T_w)	Test load (T_t)
Jack type	Area of piston	Maximum rated capacity	Date of last calibration
Pump type	Pressure gauge range	Pressure gauge accuracy	Date of last calibration
Type of top anchorage assembly	Lock-off mechanism	Initial seating pressure	Strand pullin
Data Monitored during Stressing			
Permanent bearing plate movement	Tendon extension	Jack pressure	Tendon pullin at lockoff

From Littlejohn and Bruce (1977).

those of Czechoslovakia and Germany, stipulate the beginning of recording at 10 percent the test load. Whereas such guidelines may not be used in conjunction with specific acceptance criteria of anchor performance, they are simple and adequate for routine short-term tests.

7-4 FACTORS AFFECTING INTERPRETATION OF STRESSING RESULTS

Practical Aspects of Anchor Behavior

The presentation of load–extension results should ideally indicate possible errors in the measurement and assessment of the parameters involved, since the important feature of the load–extension curve to be adjusted is the elastic behavior irrespective of linear or nonlinear characteristics. Because of unavoidable limitations in the accuracy with which stressing data are obtained and recorded, it is seldom possible to obtain a truly linear graphic presentation even under the best, most controllable conditions. However, if the plotting of stressing data reveals a consistent deviation rather than erratic results, this may be explained in terms of (a) debonding in the fixed anchor zone along the grout–tendon interface and (b) fixed anchor movement. The latter is unlikely in competent rock but rather common in weak rock and in soil. Unless some provision for direct measurement of possible fixed anchor movement is made, its presence can positively be confirmed or discounted by inducing cyclic loading at least once to ensure that the load–extension characteristics of the anchor are reproducible and repeated in the same pattern.

If an adjustment has been made for top anchorage and fixed zone movement, the interpretation can then focus on the probable partial or total debonding in the fixed anchor zone. It is possible to calculate the effective free length necessary to produce the true elongation of the tendon actually monitored at different loads (Littlejohn and Bruce, 1977). In this context, special construction lines are drawn, equivalent to the extension of various free lengths, on the load–extension graph and assuming linear load–extension relationship as shown in Fig. 7-12. It can be seen that during the initial stage of the loading the curve representing the measured load–extension function tends to approximate the lines of short free length, but this trend diminishes with increasing load and the curve begins progressively to intersect lines of longer free length.

Cyclic loading is thus useful because it (a) provides a measure of fixed anchor movement, (b) facilitates backanalysis, and (c) confirms the extent

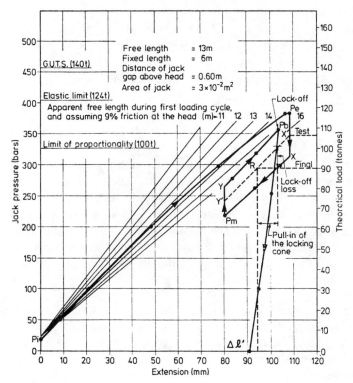

Fig. 7-12 Diagramatic presentation of the "cyclic method." (From Fenoux and Portier, 1972.)

to which the elastic load–extension characteristics are reproducible. A refined version of the method is described by Fenoux and Portier (1972), and is based on the principle that by destressing and restressing without actually changing the tendon elongation it is possible to determine the load corresponding to the measured deformations when the frictional effects are excluded and therefore estimate the actual load reaching the fixed anchor length. In this procedure a typical test includes six basic stages according to the following sequence: (a) the load is applied in increments and held constant for a brief period, (b) the load is reduced by one increment, (c) the load is reduced by two increments and the displacement recorded, (d) the load is increased by one increment, (e) the load is increased in increments, and (f) the anchor is detensioned and the residual displacement recorded.

A typical graph resulting from the test is shown in Fig. 7-12, and is subject to the following interpretation.

1. Point P_i is located on the load axis (in this case jack pressure) to represent the friction at the anchor head (in this case taken as 9 percent of the applied load).
2. After the increment specified in stage (a) the test graph is obtained, where P_e is the maximum test load.
3. Obtain point X such that P_e–X is twice the friction at the head. Point X', midpoint along $P_e X$, is therefore the maximum initial tension in the fixed anchor zone.
4. Point P_m is the minimum load (pressure) on unloading the anchor in stage (c).
5. Establish points Y and Y'. The latter is midpoint of line P_m–Y. Continue with point P_b, which is the pressure at the start of load transfer, or the lockoff load.
6. Since lines X–P_m and Y–P_b are reasonably parallel, the line X'–Y' represents the true values of loads corresponding to measured extensions since losses due to friction have been compensated.
7. Point R, constructed by intersecting $X'Y'$ with the line $\Delta l'$, gives the final load sustained by the anchor, that is, the residual tension. Note that $\Delta l'$ is moved a few mm beyond the measured value to include the effect of the stretching of the tendon between the grip of the stressing jack and the cutoff point.

Besides the foregoing test, anchor stressing in conjunction with analysis of load–extension data can indicate anchor behavior and various failure modes. A continuous cumulative permanent displacement indicated either by rapid load loss or from cyclic loading will probably mean interface failure in the fixed anchor zone. Whether this is ground–grout failure or grout–tendon failure can be verified by loading each tendon unit individually with a monojack and then comparing the load–displacement behavior.

Sources of Error. Discrepancies between the observed (actual) and the theoretical (calculated) extensions are very common. Whether they will be accepted depends on the consequences of anchor and ground movement on surroundings. Most codes stipulate a maximum amount of discrepancy between measured and theoretical anchor elongation (usually 10 percent), beyond which the difference must be rectified and the sources of error determined.

A frequent cause of this difference involves variations in the E value of steel tendon between short (laboratory) and long (field) length. Other unavoidable variations reflect differences in the testing and recording procedures in estimating the modulus of elasticity. Janische (1968) has noticed from field tests that in extension measurements on long lengths of strand (100 m or 330 ft) the associated elastic elongation yielded E values ranging from 180,000 to 220,000 N/mm², or a difference of more than 12 percent. Similar variations were mentioned in Section 2-5 for the prestressing steel used at the Wylfa nuclear reactor (Littlejohn and Bruce, 1977). A possible explanation is that stressing multistrand tendons in the field takes a longer period than laboratory testing of individual strands, and during this time plastic deformation occurs in the steel yielding a larger extension and correspondingly a lower E value. In this respect, the possible variation quoted by suppliers is closer to $\pm 7\frac{1}{2}$ percent (see also Section 2-5). Overdrilling or underdrilling of the hole is also known to affect the free length so that the accuracy and reliability of the recording procedure, distinguished from the sophistication of the instrumentation, should be checked (Littlejohn and Bruce, 1977).

A second major cause of error is friction in the free length that is likely to occur, irrespective of allowance made in the jack, especially with long sheathed tendons enveloped by a protective grout column, and around the grip assembly of the top anchorage. This friction tends to reduce the measured extension by dissipating a fraction of the applied load. This results in an extension (measured) corresponding to a free length less than the actual, since less than the total applied load actually produces tendon elongation.

Figure 7-13 shows load–extension diagrams from a stressing test reported by Hennequin and Cambefort (1966). It is evident that the measured extensions are markedly lower than those corresponding to the theoretical tendon elongation, and it can be estimated that only about 70 percent the total applied prestress was transmitted along the entire tendon length.

Types of Friction. Fenoux and Portier (1972) distinguish three characteristic types of friction in anchor systems: (a) constant value, (b) proportional to the applied load, and (c) variable and independent of pertinent parameters.

The effect of each type on the load–extension characteristics for a typical anchor stressing procedure is shown separately in Fig. 7-14 in schematic form. Friction around the top anchorage is manifested from two contributions: (a) between tendon and grout due to distortion of the tendon units

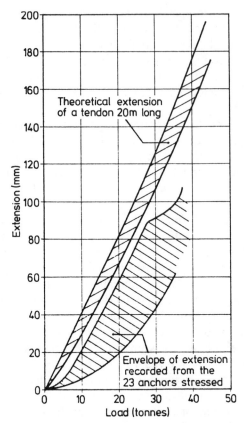

Fig. 7-13 Load–extension diagrams showing the effect of friction. (From Hennequin and Cambefort, 1966.)

under the bearing plate (usually on the order of 3–6 percent), which can be avoided by proper lubrication; and (b) between tendon and bearing plate, which may increase to 50 percent if the bearing plate and anchor block are improperly set. If friction at the top anchorage is constant, the diagram shown in part (a) of Fig. 7-14 is produced. The friction force f can be determined by carrying out a load–unload cycle as shown. More often, however, the friction force is not constant, but varies proportionally with the applied load, in which case the load–extension diagram is as shown in Fig. 7-14(b). Where there are several friction sources at random along the anchor system, the more complex diagram shown in part (c) of Fig. 7-14 is obtained.

Useful data on errors have been provided by Longbottom and Mallet (1973), and they indicate that the difference between observed and theoretical load may be as much as 15 percent. A summary is presented in Table 7-3. It is highly improbable, however, that all these errors will occur simultaneously and in the same direction; hence a statistically probable error of practical value is estimated to be $\pm 7\frac{1}{2}$ percent.

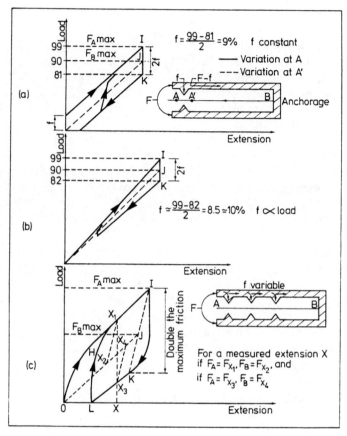

Fig. 7-14 Relationship and effects of type of friction on load–extension characteristics. (From Fenoux and Portier, 1972.)

TABLE 7-3 Estimated Possible Error between Actual and Theoretical Prestress Loads

Source	Variation (%)
Different type of manometer	±1
Typical manometer error	±2
Internal jack friction	±2
Error in reading extension	±1
Stress–strain and production tolerance of tendon	±6
Calculation error	±3

From Longbottom and Mallett (1973).

7-5   IDEAL MECHANISM OF TENDON STRESSING

Chapter 4 has demonstrated that the assumption of uniform bond resistance along the fixed anchor length is at best ideal, but in reality it may occur only in soft and loose soils. On the other hand, the stretching of the tendon steel is known to propagate beyond the free length and engage part of the fixed length. The point that appears to be the extreme point of the deforming tendon is called the "fictitious anchorage point" (FAP), and the anchor is then regarded as a flexible system with a rigid connection with the ground at the FAP. The FAP concept is illustrated in Fig. 7-15.

If the point bearing effect is ignored as being insignificant in the load-transfer mechanism, the deformations of the two types of anchorage shown in Fig. 7-15 are equal. This, in effect, means that the FAP is the center of gravity of the stresses acting on the anchorage and in this case it is located at midpoint on the fixed length.

A suggested mechanism of tendon stressing showing the stress propagation to the fixed zone is illustrated in Fig. 7-16.

The following stressing pattern is evident:

1. For relatively low load magnitude, only the front (upper) or proximal end of the fixed anchor length is under tension. The lateral friction and shear bond develops fully and no load transfer occurs further. The far (distal) end of the zone is not engaged in this interaction; hence it does not move.

2. As the load increases the resisting bond strength at the proximal end is exceeded, the engaged length increases and the fixed tendon undergoes elongation.

3. When all the bond is mobilized, the extreme (distal) end begins to move. At this stage the FAP is at midpoint. Since the bond resistance has been exceeded, the entire bond length begins to move under con-

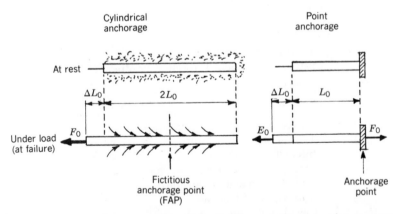

Fig. 7-15   Schematic presentation of the fictitious anchorage point concept.

Movement of fictitious anchorage point (FAP)

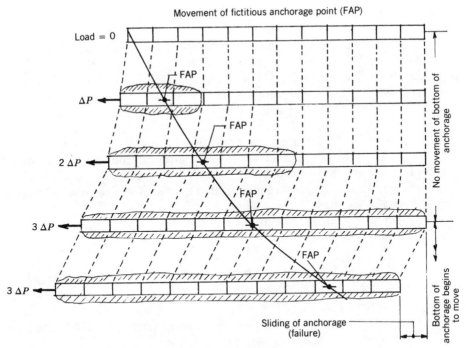

Fig. 7-16 Suggested mechanism of tendon stressing together with assumed stress propagation to the fixed anchor zone.

stant stress, and it is possible to pull the anchor out by 10 cm (4 in) before a stress decrease occurs.

7-6 PRECONTRACT TESTS

Precontract tests usually are carried out for either permanent or temporary installations and prior to use on site. There are two primary purposes: (a) to confirm that, for the particular ground type and site conditions, a particular ground anchor type can be installed and perform as expected; and (b) that manufactured components of the selected anchor, such as tendon type and top anchor assembly units, meet the design requirements and should be expected to provide satisfactory performance.

Precontract Component Testing

Component testing should be carried out at the factory or laboratory, and occasionally in the field under appropriate conditions. Ideally, it should cover all aspects of anchor behavior and performance.

Regarding the tendon steel, the engineer should request data on load–

extension characteristics, usually available from the steel manufacturer, for each batch of material delivered to the site. Test certificates and stress–strain diagrams usually must comply with applicable codes and standards.

The usual index of satisfactory stress–strain behavior is the permanent extension method with reference to a specific proof stress, already defined in Section 2-5 as the stress corresponding to 0.1 or 0.2 percent permanent elongation, and shown schematically in Fig. 4-1. This nonproportional elongation remains after the proof load has been removed. By reference to the diagram of Fig. 4-1, the 0.1 percent proof stress denoted T_G is obtained from the graph by drawing a parallel line to the straight portion of the curve (the line of proportionality) at a distance along the elongation axis equal to 0.1 percent extension. The point of intersection of this offset line with the curve defines the proof stress T_G.

In order to obtain the diagram of Fig. 4-1, the following test procedure may be followed:

1. An initial load is applied to the specimen, equal to 10 percent f_{pu} with the gauge length set at 0.6 m.
2. The extensometer is set at zero.
3. The load is increased to the specified proof stress, and held for 10 s. The total extension is read and recorded.
4. The load is reduced to just below the initial stress, and then increased to the initial stress. The permanent extension is noted.
5. The results are plotted giving the stress–strain diagram up to the maximum applied load. The modulus of elasticity can be calculated from the slope of the proportional stress–strain relationship.

Littlejohn and Bruce (1977) have provided useful data on the effect of low temperature on the ultimate strength of steel tendons. It is conceivable that a change in temperature of 1°C will produce a change in stress of about 2 N/mm² (290 psi). Lower temperature results in increased strength; hence for applications exposed to significant temperature changes, the analysis of test results may have to be adjusted to include temperature effects.

Limited data are available on fatigue resistance and impact effects on prestressing steels, and since manufacturers do not supply endurance diagrams several authors (Longbottom, 1974) suggest the investigation of a series of stress ranges each about a series of mean stresses. However, the very successful performance of prestressed concrete structures for highway and railway bridges in resisting impact and fatigue encourages optimistic expectations in anchorage performance. Useful data on this subject are provided by Lee (1973), Baus and Brenneisen (1968), and Edwards and Picard (1972).

Relevant codes and standards provide guidelines for the testing of the top anchorage combination, which includes the tendon, grips, anchor block, and distribution bearing plate. Both the grip components, which secure the bar, wire, or strand within the top anchorage, and the complete top anchorage assembly should be tested in one procedure.

Very relevant are the three procedures specified by the British Code for testing prestressing anchorages. These tests are as follows:

1. The load efficiency of steel tendon is confirmed in a short-term static tensile test. The load efficiency is the ratio of the test failure load to the average ultimate tensile strength. In general, it should not be less than 92 percent.
2. The dynamic response of an anchored tendon is tested by a fluctuating force between 0.60 and $0.65f_{pu}$ at frequency not exceeding 10 Hz applied for a minimum of 2×10^6 cycles. Loss of initial cross-sectional area of the tendon under fatigue effects should not exceed 5 percent. This test is particularly relevant where anchor function involves fluctuating stresses transmitted to the tendon.
3. The transfer of force to the load-bearing plate and block is tested by a short-term static compressive load applied to the complete top anchorage assembly. The load-bearing block must continuously support a minimum force $1.1f_{pu}$.

There are no general guidelines with reference to jacking equipment, but most jack and pump manufacturers recommend that all jacks and ancillary equipment be tested in the factory to a proof load or pressure equivalent to at least 1.25 times the rated capacity. However, overloading beyond the maximum rated capacity should never be attempted in the field, and the choice of jack should provide a rated capacity that can accommodate 85 percent f_{pu} of the largest tendon unit in the group of anchors. It should be mandatory for the manufacturer to furnish certificates about proof testing, internal losses, and load–pressure conversion factors.

Wedge Draw-in and Effect on Lockoff. This movement, discussed in Section 7-3, can cause a loss in the lockoff load. Whereas the concensus of opinion is that the amount of wedge pullin is linearly proportional to the applied load, for a given tendon type and capacity wedge draw-in is usually a fixed distance independent of the tendon unit, the nominal strand diameter and the steel grade. Assuming a linear loss of tension due to friction, the parameters w = distance along the tendon affected by wedge drawn-in, and ΔP = loss of force in the tendon, can be estimated by the following semiempirical procedure, with reference to Fig. 7-17:

$$w = \sqrt{\frac{\Delta l_c \, E_s A_s}{\Delta p}} \qquad (7\text{-}2)$$

$$\Delta P = 2 \, \Delta p \, w \qquad (7\text{-}3)$$

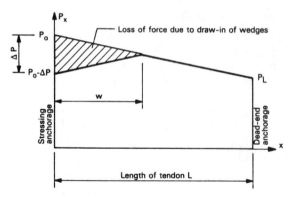

Fig. 7-17   Diagram and notation correlating wedge draw-in distance W with tendon loss of force ΔP.

where Δl_c = wedge draw-in (usually 6 mm = 0.006 m)
   E_s = elastic modulus (kN/m²)
   A_s = steel tendon cross-sectional area (m²)
   Δp = loss of force per meter (kN/m) = $(P_0 - P_L)/L$
   P_L = force in the tendon after friction loss
   P_0 = initial applied force

Example

Given a tendon composed of 12 strands, each strand 13 mm (0.5 in) nominal diameter.

   Tendon length L = 50 m
   Section of one strand A_s = 93 mm²
   Ultimate steel strength f_{pu} = 1770 N/mm²
   Elastic Modulus E_s = 1.95 × 10³ kN/m²
   Ultimate tendon force F_{pu} = 12 × 93 × 1770 × 10⁻³ = 1975 kN

The initial prestressing is selected at 70 percent F_{pu} = 1383 kN and is immediately locked off. From a separate analysis it is estimated that the friction loss is 18.6 percent so that the force remaining in the tendon is 0.814 × 1383 = 1126 kN. The effect of wedge draw-in (taken as 6 mm) is estimated as follows:

$$\text{Estimate } \Delta p = \frac{1383 - 1126}{50} = 5.14 \text{ kN/m}$$

$$\text{Estimate } w = \sqrt{\frac{0.006 \times 1.95 \times 10^8 \times 93 \times 12 \times 10^{-6}}{5.14}} = 15.94 \text{ m}$$

Estimate ΔP = 2 × 5.14 × 15.94 = 164 kN

$P_0 - \Delta P$ = 1219 kN

Precontract Anchor Testing

For all permanent anchorage installations, two or three test anchors loaded to twice the design load and then to failure will give indication of the actual factor of safety with respect to anchor pullout. The test is useful since it is carried out under known site conditions. If practical and economically feasible, on termination of the load test the anchor may be extracted for inspection and examination focusing on (a) the condition of the free length; (b) the length, shape and configuration, condition, and mode of failure of the fixed length; and (c) the condition of corrosion protection.

Since these are primarily design load tests, they should be mandatory where unusual conditions are encountered at the site for which no previous experience is available, when unusually long anchors are required, and where difficulties are expected for the drilling and grouting operations. The anchors used for the tests cannot become part of the final anchorage installation, and they are installed at an earlier stage. It may be necessary to increase the steel section, especially if the anchor is to be loaded to failure, but all other features should be the same as for the permanent anchors. This similarity is essential particularly for (a) drilling methods and anchor hole diameter; (b) length and depth of the fixed anchor zone; and (c) prefabrication, homing, and grouting methods, for which the maximum intended grouting pressure should be used.

A typical precontract anchor test may be carried out using the following stage sequence (Nicholson et al., 1982).

1. Apply a bedding–in load equivalent to 10 percent the design load in order to remove tendon slack, and check to ensure that all stress components, including pull wedges, are properly engaged.

2. Set all measuring devices to zero using the 10 percent working load initial stress as datum point for stress measurements.

3. Begin the cyclic loading using the following percentages of the design working load (loads marked * will be kept for 30 min minimum, and loads marked ** will be maintained for 24 hr).

    Cycle (a): 10, 20, 40, 50*, 25, 10.

    Cycle (b): 10, 25, 50, 75, 100*, 50, 25, 10.

    Cycle (c): 10, 25, 50, 75, 100, 125, 150**, 100, 50, 25, 10.

    Cycle (d): 10, 25, 50, 75, 100, 125, 150, 175, 200*, 150, 100, 50, 25, 10.

    Cycle (e): 10, 50, 100, 150, 200, and to anchor failure or 80 percent f_{pu} of the steel. If no failure occurs, return load to zero and record recovery.

4. Reduce loads to 10 percent the working load in the specified intervals, maintaining each load level for 5 min and recording anchor tendon recovery. Hold the same percentage at the conclusion of stages (c) and (d) for one hour before taking final reading to determine total net anchor rebound, hence elastic or plastic movement.

7-7 ACCEPTANCE TESTS OF PRODUCTION ANCHORS

Variations in soil conditions and installation procedures can cause wide differences in the actual load-carrying capacity of production anchors, and it is thus essential for each anchor to be subjected to a test. These routine acceptance tests are associated with the initial stressing operations and normally include quality control observations for a period of up to 24 h. The basis for these tests generally is derived from the precontract tests discussed in Section 7-6 and the suitability tests discussed in Section 7-8.

The first priority of the test is to establish a measured factor of safety by overloading the anchor for a short period, usually to 150 percent the working load and consistent with the governing codes. In addition, a load–extension diagram plotted for each anchor will be useful in comparing measured versus predicted performance. Finally, the test will ensure that the service load locked off after stressing is stable. Alternatively, routine acceptance tests are enhanced if the pattern and magnitude of fixed anchor movement has been established from preproduction tests, and provisions are made for monitoring loss of prestress with time.

Acceptance tests and methods of testing are covered in most codes, such as the British Code, German Code (DIN 4125, 1972), French Code (Bureau Securitas, 1972), United States (PCI, 1974), South Africa (1972), and others, including updated versions. A brief review of the tests and the testing procedures is presented in the following sections.

Acceptance Tests According to DIN 4125. According to these recommendations, each production anchor is subjected to an initial load T_0 equivalent to $0.1 T_y$ (0.1 percent proof load, approximately 83.5 percent f_{pu}), after which it is stressed to $1.2 T_w$ (working anchor load) in a single operation and held for at least 5 min in cohesionless soil and 15 min in cohesive soil, while tendon extension is monitored at the top anchorage. This procedure is identified as "type I test." Where the spacing between grouted fixed zones is less than one meter, anchor interaction should be checked by loading several adjacent anchors and observing simultaneously.

For the first 10 anchors, and thereafter one in 10 subsequent anchors, the testing involves more rigorous procedures, and the extensions are monitored from a fixed datum. Load increments are $0.4 T_w$, $0.8 T_w$, $1.0 T_w$, and $1.2 T_w$, also taking into account possible strand slippage. This test is identified as "type II." At maximum test load values observation time is as in the type I test, and on destressing to T_0, the permanent extension is observed. If the anchors are the prestressed type, the working load is reapplied and locked off. At least 5 percent of the anchors must be tested to $1.5 T_w$, noting that this cannot exceed $0.9 T_y$, and this procedure constitutes a "type III test."

Figure 7-18 shows results and load–extension graphs obtained from type II test according to DIN 4125 (1972). At $1.2 T_w$ (point X) where unloading

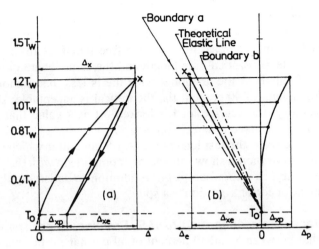

Fig. 7-18 Load–extension diagram obtained for type II stressing test. (From DIN 4125, 1972.)

is first carried out, the elastic component Δ_{xe} and permanent component Δ_{xp} of the total displacement Δ_x are clearly distinguished. The line T_0X_e in part (b) of Fig. 7-18 is taken as approximate path of the elastic displacement between the two boundary lines. The foregoing procedure applies to temporary anchors (DIN 4125, 1972).

Acceptance tests of permanent anchors are covered by DIN 4125 (1974). This document specifies that each anchor should be loaded to 1.5 T_w beginning at the T_0 initial load, with a preliminary reading at T_w. The anchor is then unloaded to T_0 and subsequently retensioned to T_w. The permanent extension is measured at T_0. For the first 10 anchors, and thereafter one in every 10, the test load is applied at increments 0.4, 0.8, 1.0, 1.2, and 1.5 T_w. The anchor is then unloaded to T_0 in the same stages, and retensioned to T_w. The displacement at 1.5 T_w from the initial loading should be measured at 1, 2, 3, 5, 10, and 15 min after lockoff. If the displacement measured between 5 and 15 min is greater than 0.5 mm, the specified period of 15 min should be extended since it may indicate possible creep.

Acceptance Tests According to Bureau Securitas. This document (1972) specifies test overloading to 1.2 T_w and 1.3 T_w for temporary and permanent production anchors, respectively. For permanent installations, it is further stipulated that 5 percent of all anchors should be tested to 1.5 T_w. Whereas the test load is not correlated to the allowable steel stress, the code recommends great caution when the elastic limit ($0.835f_{pu}$) is exceeded. The test would normally be stopped if the extension reached 150 percent the extension at 0.1 percent proof stress.

Tensioning by stages begins at 0.15–0.20 T_w; and at least five stages are recommended in order to obtain the load–extension diagram. In sand, the test load is held for 1-2 min, and during this time the displacement should not exceed 1 mm, whereas the observed free length of the tendon should fall between the theoretical free length and the same length plus 50 percent of the fixed length. If the anchor service life is less than 9 months, an observed free length of 90 percent the theoretical is acceptable. If these tests are satisfactory, the service load is locked off at a value that includes allowance for losses (see also Section 5-8).

In cohesive soils, the test load is held for 5 min, after which the displacement–time curve should show satisfactory comparison with the performance of anchors subjected to creep tests, in addition to satisfying the extension criteria mentioned for cohesionless soil.

Acceptance Tests According to U.S. Practice.
Nicholson et al. (1982) suggest that between 5 and 10 percent of all permanent production anchors should be subjected to load tests similar to the preproduction tests, but with modified loading cycles to reflect the permanent character of the work. Thus, after applying initial stress equivalent to 10 percent the working load T_w, checking all stress component parts, and properly setting all measuring devices, the following cycles should be performed:

Cycle (a): 10, 25, 50*, 75, 100**, 50, 25, 10.
Cycle (b): 10, 25, 50*, 75, 100**, 125, 150$^+$, 100, 50, 25, 10.

Likewise, these numbers express percentage of T_w. Loads marked * should be held for 15 min, loads marked ** for 30 min, and loads marked $^+$ for 4 h.

After completion of the test, the anchor is retensioned and the load locked off at the specified level. A liftoff test should be performed 24 h later, to check for load loss. If this exceeds 10 percent the lockoff load, the load should be restored and rechecked after a second 24-h period. If the load loss is still maintained above 10 percent, additional tests and investigations should be carried out to determine the cause, and whether load loss is continuous or diminishes with time, or whether the anchor should be replaced.

The remaining production anchors should be stressed to 133 percent with extension measurements taken at the following load levels (percentage of T_w): 10, 25, 50, 75, 100, 133. The highest load should be maintained for a minimum of 5 min and extensions recorded at 1, 3, and 5 min. All stress–strain data for production anchors should be plotted against control graphs obtained from other critical tests and acceptability judged accordingly. The time–displacement records will indicate creep characteristics. (See also creep tests and acceptance criteria.)

FIP Recommendations.
The FIP draft (1973) recommends limiting the tensile stress in the tendon to $0.9T_y$ or $0.75f_{pu}$, assuming that T_y is equivalent

to 0.1 percent proof stress. All production anchors should be tested to $1.2T_w$ and $1.3T_w$ for temporary and permanent works, respectively. Results and load–displacement diagrams from a typical acceptance test are shown in Fig. 7-19.

Extensions are measured at 0.4, 0.8, 1.0, and 1.3 T_w. Where the ground is not susceptible to creep, the test load is held for 2–5 min, and the anchor is accepted if (a) no noticeable displacement is measured (<1 mm) during this time, and (b) the measured total displacement at the top anchorage is in reasonable agreement with the established criteria from extended acceptance tests. The latter involve 3–10 production anchors, which at the beginning of the contract undergo special acceptance tests. The stressing program is shown in Fig. 7-20, and this test should be applied to 10 percent of all production anchors thereafter. In the extended acceptance test, the anchor is considered satisfactory if (a) the displacement of the anchor under test load has become stable within the observation period, and (b) the measured elastic extension shows reasonable correlation with the calculated value after allowance for test effects.

Acceptance Tests According to British Standards. B.S. "Draft Proposal for Ground Anchors" (1980) permits tensile testing up to 80 percent f_{pu}, and the factors of safety are in agreement with the recommendations of this author. The most common practice at present is to test load production anchors in increments up to $1.25T_w$ with minimum observation period five minutes at the maximum test load. The load is thereafter reduced to zero before retensioning in increments up to a lockoff load $1.10T_w$. Tendon extensions are monitored, but since during the first cycle the observed displacement is composite, the interpretation and analysis of data is essentially based on the load–extension diagram obtained from the second loading cycle.

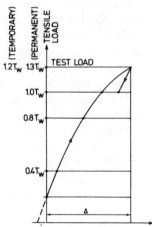

Fig. 7-19 Load–displacement diagram from acceptance load tests according to FIP draft, 1973.

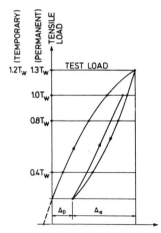

Fig. 7-20 Stressing program for extended acceptance test according to FIP draft, 1973.

If a component of a multiunit tendon fails during the stressing stage, a reduced anchor capacity may be acceptable in proportion to the components left provided all individual components have compatible stresses in service. On the other hand, if these stresses are below the allowable, it is possible to upgrade the load in each component to compensate for the loss of load in the redundant component. The same approach can be extended to other components of the system, for example, if gripping wedge failure occurs and new wedges cannot be fitted.

7-8 BASIC ON-SITE SUITABILITY TESTS

Essentially, these tests are intended to show the suitability of anchors for the particular conditions at the site. Under static loading, load–time relationships are established and used to analyze the effectiveness of the anchorage. Anchors subjected to this test can be used as production anchors in the work if necessary, or they may be additional and provided under the contract. They should be constructed in exactly the same way and located in the same ground as the production anchors. Such anchors are often termed "special" since they are more time-consuming and are costly. However, they are now widely recognized as essential to long-term installations and their usefulness may justify the cost, especially if they are incorporated in the final work. The decision to specify on site suitability tests will depend on the magnitude of the contract, the complexity of site conditions, the basic uniformity of ground characteristics, the anchor load capacity, and the experience gained in previous projects carried out under similar conditions.

In general, the basic suitability of a ground anchor system is analyzed from tests on two or more anchors in recognized types of ground. The entire work must be monitored by a specialist team, and very often this class of tests may be combined with the precontract tests discussed in Section 7-6 in one operation in the form of proving tests.

Basic and Suitability Tests According to DIN 4125. In a basic test, approximately one week after grouting the stressing is carried out as shown in Fig. 7-21, and top anchor displacements are measured from a remote datum at load levels indicated and above the initial load $T_0 \geqslant 0.1T_y$, where T_y is the elastic limit at 0.1 percent extension. Load increments are applied as shown in Fig. 7-21(a) until failure or the T_y stress of the tendon is reached. After the load level $0.3T_y$ and thereafter at each successive higher load increment, the tendon is unloaded to T_0 to obtain data on permanent displacement and also allow estimation of the effective free length.

Prior to each unloading displacements are observed under constant load in cohesionless soils until the movement diminishes but for not less than 5 min. At point $0.6T_y$ the load is held for 15 min and the corresponding displacement Δ_1 is measured [see Fig. 7-21(b)]. At $0.9T_y$ the observation time is increased to 1 h minimum, producing the corresponding displacement Δ_2. In cohesive soils the observation time at 0.6 and $0.9T_y$ is extended until the displacement during the last 2 h is less than 0.2 mm (see also Section 7-10). The maximum applied test load should be at least $1.5T_w$ but less than $0.9T_y$ for an observation time of at least 1 h, and the working load T_w should be held at least for 15 min. On completion of the stressing stage, the entire anchorage is extracted for examination.

The measured displacement of the top anchorage is likewise divided into two basic components, an elastic Δ_e and a permanent (plastic) Δ_p as shown in part (b) of Fig. 7-21. For a specified anchor load X the total displacement is Δ_x, having two components Δ_{xe} and Δ_{xp}. The elastic and permanent components of displacement are plotted in Fig. 7-21(b) for each load increment, and it is evident the failure load is at $0.94T_y$ (continuous yielding occurs). Ordinarily, however, the specified upper load limit would be $0.9T_y$, and clearly this may not always be reached in the basic test.

The stressing phase is supplemented by a technical report that assesses the anchor test characteristics and stressing results. For the observed free length, the elastic displacement curve shown in Fig. 7-21(b) should be between the boundary lines (a) and (b) as shown. Whereas this procedure is termed by the German Code a basic test, the same code stipulates that an anchor system should be subjected to suitability tests if the local ground is disimilar to that of the basic test, or if the drilling procedure and borehole diameter deviate from the basic test. In suitability tests, however, the anchors are not extracted after stressing. Interestingly, the foregoing procedures apply to temporary anchors (DIN 4125, 1972).

For permanent anchors the basic tests are essentially similar but with certain variations (DIN 4125, 1974). In this case, the load is applied in stages as shown in Table 7-4, again beginning at seating load T_0. When each loading

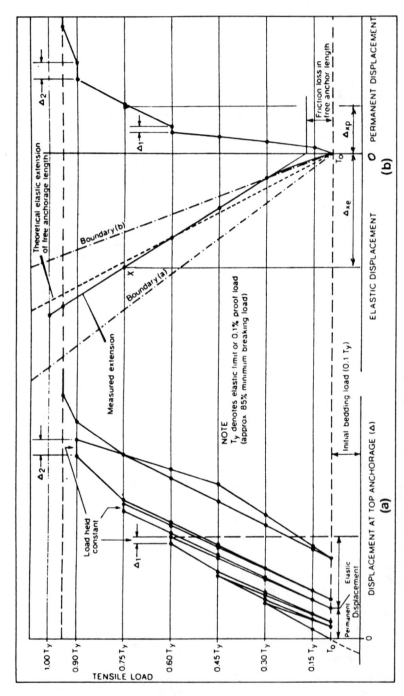

Fig. 7-21 Load–displacement data and diagrams from a basic suitability test of ground anchor according to DIN 4125 (1972).

TABLE 7-4   Load Stages and Observation Periods for Basic and Construction Site Suitability Tests for Permanent Anchors

Stage of Loading		Minimum Period of Observation	
Basic Test $T_0 \ngtr 0.1\ T_y$	Suitability Tests[a] $T_0 \ngtr 0.2\ T_w$	Coarse-Grained Soils	Fine-Grained Soils
$0.30T_y$	$0.40T_w$	15 min	30 min
$0.45T_y$	$0.80T_w$	15 min	30 min
$0.60T_y$	$1.00T_w$	1 h	2 h
$0.75T_y$	$1.20T_w$	1 h	3 h
$0.90T_y$	$1.50T_w$	2 h	24 h

[a] If the working load is not known at the time of the test or the upper limit load is uncertain, it is recommended that smaller load stages should be selected.
From DIN 4125 (1974).

stage is reached, the anchor is destressed to T_0 to obtain the elastic and permanent extensions. Typically, anchors should be stressed to $0.9T_y$ if the failure load in the fixed zone is not reached at an earlier stage.

Basic and Suitability Tests According to Bureau Securitas. This document categorizes basic tests by anchor geometry and ground type, whereas the number of test anchors is related to the number of production anchors for each category as shown in Table 7-5. If, for example, a project involves 400 anchors of which 300 are inclined and 100 vertical, two categories are introduced by geometry. If, additionally, 250 of the inclined are in gravel, and all remaining (inclined and vertical) are in clay, then the project includes three categories with corresponding test anchors as follows:

250 inclined–gravel:	3 test anchors
50 inclined–clay:	2 test anchors
100 vertical–clay:	2 test anchors

TABLE 7-5   Minimum Number of Test Anchors for a Given Number of Production Anchors According to Bureau Securitas (1972).

No. of Test Anchors	No. of Production Anchors
2	1–200
3	201–500
4	501–1000
5	1001–2000
6	2001–4000
7	4001–8000

The characteristics of test anchors must be similar to the category of anchors represented, except a tendon of greater capacity may be used to accommodate an extended test load whereby failure can be induced to the fixed anchor grouted zone.

Figure 7-22 illustrates a stressing program in ground where anchoring behavior is available and risk of creep does not exist. It is apparent that the anchor is test-loaded to the anticipated $0.75T_G$ and $0.60T_G$ for temporary and permanent work, where T_G is the symbol used by the French Code to designate the 0.1 percent elastic limit (equivalent to $0.835f_{pu}$).

In order to eliminate mechanical interference from unrelated parameters (such as tendon slack and plate bedding-in), two successive load cycles are necessary as show in Table 7-6, with pauses to record extensions. On completion of the second cycle, the stressing is carried out in stages to $0.9T_G$, with observation periods at each stage to permit detections and measurement of creep and permanent extension. After the one-hour observation period at $0.9T_G$ of the third load cycle, the load is completely removed in stages, and the anchor is then restressed to the lockoff load with pauses only for extension readings, with a lockoff load chosen to accommodate stress loss

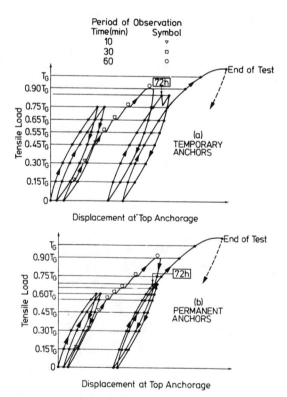

Fig. 7-22   Stressing program in soils where anchor behavior is known, according to Bureau Securitas (1972).

TABLE 7-6 Recommended Load Increments and Observation Periods for Basic Test Anchors According to Bureau Securitas (1972)

Temporary Anchors			Permanent Anchors		
Load Increment		Period of	Load Increment		Period of
Initial Two Load Cycles[a]	Third Load Cycle	Observation (min)	Initial Two Load Cycles[a]	Third Load Cycle	Observation (min)
$0.15T_g$	$0.15T_g$	10	$0.15T_g$	$0.15T_g$	10
$0.30T_g$	$0.30T_g$	10	$0.30T_g$	$0.30T_g$	10
$0.45T_g$	$0.45T_g$	10	$0.45T_g$	$0.45T_g$	10
$0.55T_g$	$0.55T_g$	30	$0.55T_g$	$0.55T_g$	30
$0.65T_g$	$0.65T_g$	30	$0.60T_g$	$0.60T_g$	30
$0.75T_g$	$0.75T_g$	30		$0.65T_g$	30
	$0.90T_g$	60		$0.75T_g$	30
				$0.90T_g$	60

[a] For these load cycles, there is no pause other than that necessary for the recording of extension data.

due to relaxation and ground creep. After 72 h the anchor is further stressed to regain its initial residual load, and then completely unloaded before final stressing, where the load is increased in load increments as before but until failure occurs by yielding, or until the extension is equal to 150 percent the extension at 0.1 percent proof stress as shown in Fig. 7-23. At this point the test is complete and the anchor is unloaded and abandoned.

Where the ground conditions are not known or prior experience of anchoring is not available (both making anchor behavior unknown), it is conceivable that anchor failure may occur at a load less than $0.9T_G$. In this case the maximum test loads for the first three cycles that are carried out without pauses are lower, as shown in Fig. 7-24(a) and (b). During these three cycles, displacement measurements are recorded whenever there is a load change of 5 percent T_G. In order to monitor creep and relaxation losses, the initial residual loads at lockoff are $0.86T_G$ and $0.7T_G$ for temporary and permanent

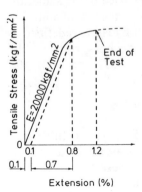

Fig. 7-23 Typical stress–strain diagram obtained from stressing procedure indicating end of test, according to Bureau Securitas (1972).

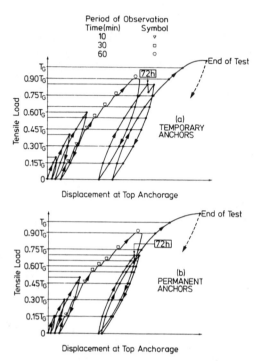

Fig. 7-24 Stressing program in soils where anchor behavior is not known, according to Bureau Securitas (1972).

anchors, respectively, held for 72 h. At this stage, if the displacement necessary to regain the loss of the initial residual load is less than 1 mm, the test continues as already described. If the same displacement exceeds 1 mm, one option is to continue the same test with a second option to initiate a second test and repeat the procedure but with a lockoff load 30 percent lower.

If the first test anchor fails at an intermediate load T_1, the stressing procedure for the second test anchor should be as shown in Fig. 7-25 (for temporary anchors). Whereas the stressing approach is basically the same, the load increments are now related to T_1 rather than T_G. If the actual minimum ultimate load for the test anchors is T_{min}, the working load is redefined as $0.67T_{min}$ and $0.50T_{min}$ for temporary and permanent anchors, respectively. If none of the test anchors fails, the working load is $0.75T_G$ and $0.60T_G$ for temporary and permanent anchors, respectively, where $T_G = 0.835f_{pu}$ approximately.

Basic and Suitability Tests According to British Standards. In Britain, the effectiveness of an anchorage installation is usually ascertained by basic tests on at least three anchors (Littlejohn, 1970). In these tests, the fixed anchor length is varied, the intent being to introduce bond failure at the

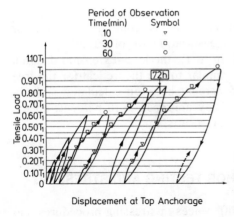

Fig. 7-25 Stressing program of temporary anchors for the second test anchor
after failure of the first at load T_1 according to Bureau Securitas (1972).

ground–grout interface in which case an estimate of the ultimate side shear
and end bearing is obtained by plotting the failure load versus fixed anchor
length. This procedure yields actual factors of safety and checks the validity
of empirical design rules.

The suitability of a proposed anchorage is usually inferred from a minimum stressing program on test anchors, as shown in Fig. 7-26. If the basis
of production anchor design must be established before the contract, it may
be sufficient to load the anchor to 80 percent f_{pu}. The anchor is first loaded
incrementally to 1.25 or 1.50 T_w (working load) for temporary and permanent
works, respectively, and this is shown as load T_t in Fig. 7-26. After an
observation period of 5 min the anchor is detensioned completely, and the
load–extension graph is obtained for the full cycle. The load is then reapplied
to T_t as shown, and the load T_x is noted at the intersection point x, this
intersection (cross-over) indicating that extra fixed anchor displacement was
necessary to mobilize T_t. If this occurs, the foregoing anchor behavior would
indicate that for the value T_t shown, T_w should have a value less than T_x in
order to minimize loss of prestress, particularly if the production anchors
are to be subjected to cyclic loading.

Subsequently, the anchor is locked off at T_t and held for at least 24 h to
record any loss of prestress. Thereafter, the anchor is progressively loaded

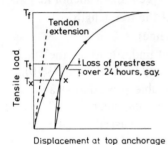

Fig. 7-26 Minimum stressing program for test anchors from current British practice. (From Littlejohn, 1970.)

for a duration as long as practicable since the intent is to detect creep and its effects on service behavior. Ultimately, the test continues to failure or $0.80f_{pu}$ in order to obtain the actual (measured) factor of safety.

During the second load cycle and up to T_t, the load–extension diagram should be expected to be essentially similar to the theoretical for the free tendon length. Discrepancies of ± 5 percent between calculated and observed results are acceptable, but if they approach ± 10 percent, an examination is appropriate to reconsile the results.

7-9 TYPICAL EXAMPLE OF ANCHOR TESTING AND STRESSING

Despite the apparent variation and differences in stressing procedures and concepts of testing among countries where anchor practice is common construction, it is interesting to note that all codes agree with the importance of overloading an anchor. The associated benefits are mainly the measured safety factor and a relevant stress history, which helps to understand anchor behavior. For the most part, differences in stressing methods relate to the staging sequence, incremental loading and observation time as well as the number of tests considered necessary. Whereas anchor practice must remain subject to local codes and standards, the broad concept of standards and procedures identified in various countries is essential to broader technical knowledge and certainly allows better engineering judgment. More important differences are evident in the interpretation of load–extension data as it relates to acceptance criteria, discussed in subsequent sections.

Typical Test Objectives

A typical anchor load testing and stressing program will include the following minimum objectives:

1. As initial construction record and prior to commencing the stressing test, prepare theoretical load–displacement diagrams with limit lines starting at the origin point for the following anchor lengths:
   (a) Theoretical free anchor length plus 50 percent fixed anchor length.
   (b) Anchor length equivalent to 80 percent the free length.
   (c) Free anchor length (theoretical elastic line).
   These lines represent the elastic movement under load for the stated length, and (a) and (b) represent the boundary lines of Fig. 7-18(b).
2. Carry out the specified tests (under applicable procedures) and use load–extension data to construct appropriate diagrams.
3. If the testing program includes creep tests (discussed in subsequent sections), draw diagrams on semilog graph paper to study creep characteristics.

4. Analyze load–extension diagrams to determine whether the load is resisted in the fixed zone and whether the anchor is fairly below the failure point. Study anchor behavior under constant load.

5. Precontract test anchors will provide design parameters, and may dictate changes in the initial theoretical criteria. Acceptance tests of production anchors will confirm the validity of precontract test results, and compare data. Basic or suitability tests will enhance this record and amplify load–extension and load–time data.

6. Where creep tests are carried out, the associated characteristics of long-term ground behavior should be established to allow predictions about potential load loss with time. The recommended procedure is discussed in the section on creep tests.

Example of Suitability Test of Permanent Anchors

Selected Accuracy of Measurements. For the measurement of axial anchor head movement, anchor plate displacement, and residual force in the anchor during the test, the selected (specified) accuracy is as follows:

For axial movement Δl of anchor head with respect to a fixed point, and anchor plate movement Δl_k in the axial direction: absolute accuracy 2 percent Δl_r (calculated theoretical elastic elongation of tendon); relative accuracy 0.5 percent Δl_r.

Movement of anchor plate Δ_s (deformation of the foundation or anchor block): absolute accuracy 2 percent Δl_r; relative accuracy 0.5 percent Δl_r.

Anchor force in the tendon (behind the anchor head): absolute accuracy 3 percent T_t (test load); relative accurracy 0.5 percent T_t.

Staging and Sequence of Test. The test anchor is tensioned in successive steps, and at each step the load–extension data are obtained and recorded. On completion of the test the anchor may be removed for inspection and examination. The operation is carried out following the process shown schematically in Figs. 7-27 and 7-28, and involves several steps.

1. An initial bedding load $T_0 = (0.1 \cdots 0.2) T_t$ is selected, and the difference or range between T_t and T_0 is divided into 6–10 equal increments ΔT, each defining a step in the loading process.

2. A fixed point is established as reference datum for measuring Δl (composite axial movement of anchor head representing tendon elongation). This movement has two components, Δl_e (elastic deformation), and Δl_p (plastic deformation).

3. The stressing and recording program is carried out to the selected maximum load (usually this test load T_t is 150 percent the working load T_w).

4. The observation time periods Δ_t are selected as follows (subject to applicable standard and code):

   (a) Rock and cohesionless ground, $\Delta_t = 5$ min minimum

   (b) Relatively cohesive materials and overconsolidated clays, $\Delta_t = 15$ min minimum

   (c) Clays and clayey silts in the normally consolidated condition, $\Delta_t =$ several hours to days.

5. After each increment, the anchor is unloaded to T_0, and the permanent extension Δl_p is recorded. The stressing, unloading, and recording process continues up to the maximum selected load T_t.

Interestingly, at each load step observations are made either of the load decrease $\Delta T'$ as shown in Fig. 7-27 with the deformation remaining constant (relaxation), or of the deformation increase $\Delta l'$ as shown in Fig. 7-28 with the load remaining constant (creep behavior).

Assessment of Results for Test Load $T_t > 200$ kN (45 kips). The anchor parameters to be evaluated from this test are the limit load T_L, the free anchor length l_f, and the plastic deformation Δl_p.

The limit load T_L is the maximum load at which the following two conditions are satisfied (taken from applicable criteria):

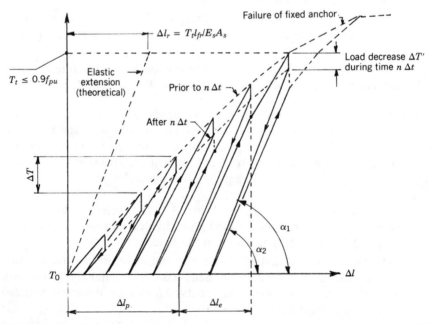

Fig. 7-27 Load–displacement diagram for a typical load test; load decreases with constant deformation (relaxation).

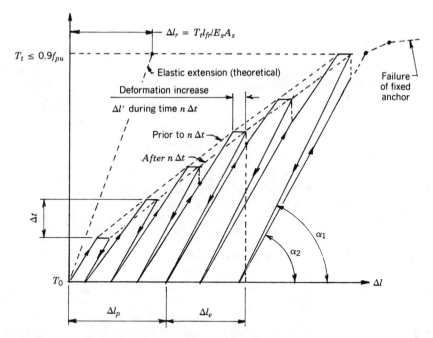

Fig. 7-28 Load–displacement diagram for a typical load test; deformation increases with constant load increments (creep).

(a) The change of deformation (creep) or load (relaxation) should not exceed the limit values given in Table 7-7. If the condition as shown in (1a) of Table 7-7 is not satisfied, the observation time is increased to $3\Delta_t$. If condition (1b) is not satisfied, the observation time is increased to $10\Delta_t$.

(b) The second condition is satisfied if the following inclination ratio is within the limit indicated:

$$\frac{\tan a_2}{\tan a_1} \geq 0.90$$

TABLE 7-7 Limit Values; Deformation Increase and Load Loss (Data to Be used for Load Test of Figs. 7-27 and 7-28).

	Observation Time (According to Step 4)	Limit Values	
Condition		Deformation Increase Δl^a	Load Loss $\Delta V'^b$
(1a)	$0-\Delta t$	Maximum 2% of Δl_r	Maximum 2% of $T_t{}^c$
(1b)	$\Delta t - 3\Delta t$	Maximum 1% of Δl_r	Maximum 1% of T_t
(1c)	$3\Delta t - 10\Delta t$	Maximum 1% of Δl_r	Maximum 1% of T_t

[a] if the load is kept constant during the observation time
[b] if the deformation is kept constant during the observation time
[c] T_t = Test load shown in Figs. 7-27, 7-28; usually 150 percent of the working load.

where (see Fig. 7-27) a_1 = angle of inclination of the unloading curve
a_2 = angle of inclination of the reloading curve

The effective free length of the anchor l_f results from the straight line $A'X$, with reference to the diagram of Fig. 7-29. This length is estimated as follows

$$l_f = \frac{\Delta l_e^x A_s}{T_x - T_0 - R} E_s \qquad (7\text{-}4)$$

where A_s = cross-sectional area of the steel tendon
E_s = modulus of elasticity of the steel
Δl_e^x = elastic deformation of the tendon under load T_x
T_0 = initial load
R = frictional force (distance AA')

Additionally, the effective free length l_f should have a value between the following limits

$$l_f \geqslant 0.9 \, l_{fr} \qquad (7\text{-}5)$$
$$l_f \leqslant l_{fr} + k l_v$$

where l_{fr} and l_v are the initial (calculated) free length and fixed lengths, respectively, and k is a numerical coefficient; $k = 0.5$ in anchor systems where the force is introduced into the anchorage body by the tendon along the anchor length.

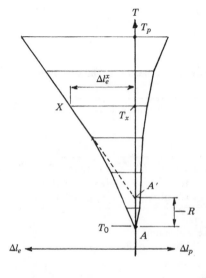

Fig. 7-29 Diagram of elastic and plastic deformations. Tests of Figs. 7-27 and 7-28.

The plastic deformation Δl_p is determined by direct reading from Fig. 7-29. The permissible value for the basic test is determined by codes, or if a criterion is not available, by the engineer jointly with the contractor.

Assessment of Results for Test Load ≤ 200 kN (45 kips). In this case the evaluation of parameters is simplified. During the test the yield stress of the tendon should not be exceeded, but it is conceivable that the test load could be large enough to cause failure of the fixed zone. The mean value of the ratio of inclination tan a_2/tan a_1 should be greater than or equal to 0.80 over at least three loading cycles.

Quantitatively, the test procedure described in the foregoing section may be modified if the criteria applied to the limit load, the free anchor length, and the plastic deformation are different, reflecting various codes and standards of practice.

7-10 CREEP TESTS

Basic and Acceptance Tests

In soils susceptible to creep, the ultimate load is assumed to have been reached when the displacement does not decrease with time but is continuous under constant load. Interestingly, in some soils such as medium to highly plastic clays of stiff consistency, creep values are known to increase rapidly at only 40 percent of the failure load (see also Fig. 4-35). In these conditions, the test load is held sufficiently long in order to allow time–displacement curves to be compared with those of anchors subjected to creep tests, in addition to complying with standard extension criteria.

One of the early codes to recognize creep effects, the Czechoslovakian draft standard (Klein, 1974) stipulates a test loading of all temporary anchors to $1.2T_w$, and a higher loading for permanent installations. The permanent displacement due to load increase from T_w to $1.2T_w$ should not exceed by more than 10 percent the permanent displacement obtained in the basic anchor test over the same load range. For creep under constant service load, the displacement should not exceed 0.135 mm/m of free tendon for every tenfold increase in time. For the specific time intervals shown in Fig. 7-30, the displacement must be less than 0.02 mm/m of free length, and the total observation period must be at least 10 min.

For the basic tests according to DIN 4125 (1972) and shown in Fig. 7-21, the limit load for minimal or acceptable creep T_k is determined by measuring the displacement under constant load prior to unloading at 1, 3, 5, 10, and 30 min, and then recording as shown in Fig. 7-31. The required minimum observation periods are as shown in Table 7-4, but can be extended if necessary until the creep coefficient K_Δ is reasonably determined. This parameter is calculated according to Eq. (4-21) and in conjunction with Fig. 4-34, where the time–displacement curve for an anchor in clay is plotted on logarithmic scale.

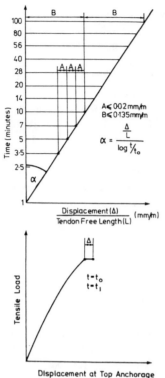

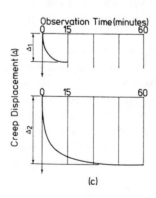

Fig. 7-30   Working diagram relevant to acceptance criteria for creep displacement according to Czech Draft Code (1974).

Creep displacement versus observation time is also shown in Fig. 7-32, where the results are recorded and plotted according to DIN 4125 (1974). All symbols correspond to the notation of Eq. (4-21), whereas the load covers the range from 0.4 to $1.5T_w$. Values of K_Δ are determined at various loading stages and recorded as shown in Fig. 7-33, and according to this code of practice the limit force T_k corresponds to a creep value $K_\Delta = 2$ mm. Following this stage, the anchor is subjected to 20 load cycles (in the range 0.3–$0.6T_y$), and the extension at the maximum and minimum loads is measured at least every five cycles. Thereafter, the anchor is detensioned to T_0, then retensioned to $0.6T_y$ with an observation period.

Fig. 7-31   Creep displacement as function of observation time, plotted according to DIN 4125 (1972).

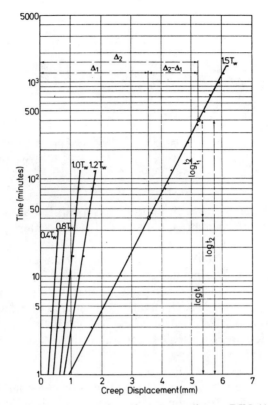

Fig. 7-32 Creep displacement versus time, according to DIN 4125 (1974).

The same aproach is likewise used for suitability tests at the site; however, these tests are carried out under the most unfavorable ground conditions. The loading stages are as shown in Table 7-4 with suitable times for observation. Twenty load cycles are carried out thereafter, and if the tests are completed in a satisfactory manner, the permanent load is locked off.

Creep tests according to Bureau Securitas are essentially an extension of the suitability tests discussed in the foregoing sections and shown in Fig. 7-24(a) and (b). The initial residual load at lockoff is held for 72 h, and if the creep displacement during this period exceeds 1 mm, it may indicate creep. In this case a second test may be carried out with a lockoff load 30 percent lower as shown in Fig. 7-34. Interestingly, the displacement value of 1 mm used as creep criterion is rather arbitrary, and the same code regards the deformation as a provisional index only.

Example of Creep Test

In this example it is assumed that an initial test has been carried out as part of a program to either check the failure strength of the ground in the fixed anchor zone or determine the limit tension of the tendon in order to eliminate

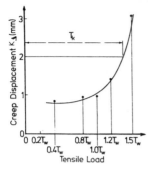

Fig. 7-33 Tensile load plotted against creep displacement. Diagram is used to determine limit force T_k. (From DIN 4125-1974).

creep along the anchorage. If, on the other hand, the purpose of the test is to check and establish a measured factor of safety, failure is not necessarily the objective or the end result.

For the second (creep) test the loads are chosen on the basis of results from the first test. The maximum test load is $0.9T_G$, where T_G is the 0.1 percent elastic limit of the French Code, approximately $0.85f_{pu}$. In this case, the maximum test load is $1.5T_w$, that is, the latter is selected with the factor of safety 2 from f_{pu}. All loads are held for one hour, but much longer for the long duration creep cycle. The general testing procedure consists of two main steps as follows:

1. Obtain creep curves from the first test for one-hour sustained loads, and determine from the creep curves the critical creep tension, or failure tension.

2. During the second test and from the critical or failure tension, determine the increments of the loading stages and the load at which long-term creep should be measured.

Load–displacement diagrams from the first test are shown in Fig. 7-35. The interpretation of test results must consider two possibilities: (a) the fixed

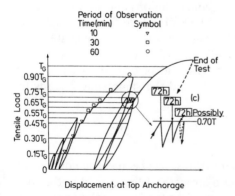

Fig. 7-34 Stressing program where creep displacement is excessive, according to Bureau Securitas (1972).

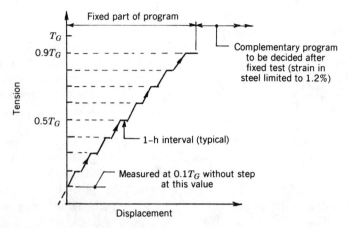

Fig. 7-35 Load–displacement diagram from first test; example of creep program.

anchor length develops its ultimate bond strength before $0.9T_G$ is reached, in which case the failure load T_f will be used as reference load for the second test; or (b) the maximum test load $0.9T_G$ is reached without failure of the fixed anchor length. If the latter is the resulting possibility, creep curves are drawn as shown in Fig. 7-36(a), where the cumulative creep displacement is plotted versus the logarithm of time. From these curves the slope tan a is measured for each load increment, and these results are plotted versus

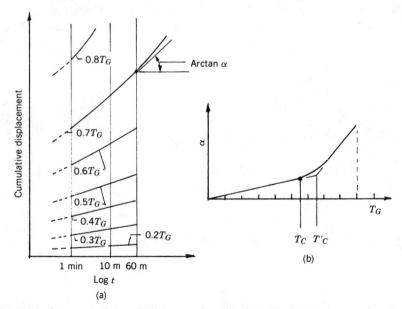

Fig. 7-36 Interpretation and diagrams of results from the first test; example of creep program: (a) creep curves, (b) critical creep tension.

the applied load resulting in the diagram shown in part (b) of Fig. 7-36. If the critical tension value T_c is not easily identified as the tangent point where the straight line converts into a curve, it can be taken as $0.9T'_c$, where T'_c is defined as the point of intersection of the two tangents to the curve as shown. Furthermore, if the plot of the a/load function is not regular and smooth as shown in Fig. 7-36(b), or if it does not consist of a straight line followed by a curve with an upward concavity, it will indicate an anomalous result which should be investigated (Pfister et al., 1982).

The second test is carried out for load increments with one-hour observation periods for creep. For a specific load value T_{ld} a partial cycle test is completed, followed by a 72-h creep observation period. The limiting tension T_{ld} is selected as follows: $T_{ld} = 0.9T_c$ or $0.8T''_c$. If, however, the first test did not reach the critical tension T_c (meaning that the diagram a/tension load is a straight line), then T_{ld} is taken as two-thirds of $0.9T_G$ or $T_{ld} = 0.6 T_G$.

During the second test, it is conceivable that a critical tension value may be reached before the limiting tension T_{ld}. This may be inferred by observing the shape of the creep curve, whereby point T_{c2} is reached where the straight line begins to curve and its slope increases with time. In this case, the long duration creep test will be necessary at the value T_{c2} rather than the initial limiting value. The end of the test may be reached and concluded as shown in Fig. 7-37.

Estimation of Working Load from Creep Test. If the creep curve for the long-duration loading stage is nearly a straight line, two criteria must be satisfied simultaneously: (a) the diagram must reasonably agree with the curve obtained for the same tension during the first test (one-hour observation period), and (b) after correction for steel creep the absolute displacement between the end of the first hour of loading and the end of the 72nd

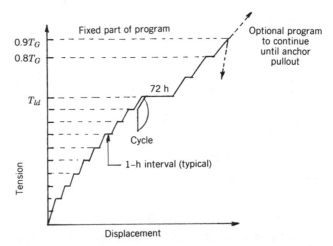

Fig. 7-37 Load–displacement diagram for second test; example of creep program.

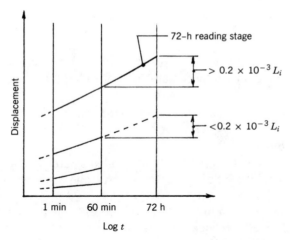

Fig. 7-38 Extrapolation of creep curves for determining limiting working load; example of creep program.

hour should be less than or equal to 0.2×10^{-3} times the free length of the anchor. In this context, the limiting (maximum) load is taken as the lowest of the following values: T_{ld}, $0.9T_{c2}$, or any failure value that may have been reached during the tests. If only one of the two criteria is satisfied, the limiting working load is taken as the nearest load for which the extrapolated displacement between 1 and 72 h is less than 0.2×10^{-3} times the theoretical free length L_i, as shown in Fig. 7-38.

If the creep curve for the long-duration loading stage is not a straight line, it is conceivable that a curved diagram with an upward concavity will appear after plotting the data. In this case, the anchor should be considered unacceptable unless a third test on a different anchor shows that the results obtained initially were not representative.

7-11   LONG-TERM MONITORING TESTS

Long-term monitoring is required to check service behavior of production anchors and to ensure anchor performance. This program includes also collection of data relating loss of prestress or creep displacement to time, type of ground, and anchor type and geometry.

Long-term losses in the anchor system are caused by a combination of steel relaxation and anchor creep. Whereas relaxation characteristics of prestressing steels are well documented and pertinent data are available from steel manufacturers, long-term creep losses can be significant and difficult to estimate. At best, they should be judged from creep tests carried out well in advance of full-scale production.

The monitoring requirements are usually determined by local codes and standards. In Great Britain, for example, periodic checks on production anchors are recommended as follows:

1. The load in all anchors should be checked 24 h after stressing to detect load loss, and this is necessary for both temporary and permanent anchors.
2. When justified by the size and scope of the contract, the first 10 anchors should be checked weekly for one month, then monthly for the following 3 months.
3. If results are satisfactory after four months, 5 percent of all production anchors should be checked at six months, and again at 12 months.

The allowable variation in anchor load is $\pm 0.1 T_w$, and restressing is not carried out unless all pertinent factors are considered. For instance, if a retaining wall is braced with several rows of tiebacks installed in weak shale, loss of prestress due to normal consolidation of the shale may be observed without accompanying movement of the wall. In this case, anchor restressing will not be required.

Bureau Securitas (1972) recommends a monitoring plan or control procedure whereby possible failures can be detected before their effects are manifested. Consequently, periodic monitoring of permanent installations for at least 10 years is mandatory under the French Code. For the first year, monitoring is scheduled at 3-month intervals, for the second year at 6-month intervals, and yearly thereafter. For each category shown in Table 7-5, the minimum number of anchors to be monitored is as follows:

10 percent of production anchors (total installed, 1–50)

7 percent of production anchors (total installed, 51–500)

5 percent of production anchors (total installed, >501)

The FIP recommendations (1973) specify that for extended acceptance tests an initial number of 3–10 anchors should be monitored, to be followed by a percentage of all others, usually 10 percent.

The most rigorous approach to monitoring probably is recommended at the present time by the South African Code (1972). According to this code, unless anchors are permanently protected by grouting, they should be tested as follows after stressing:

1. Not less than 24 h and not more than 48 h.
2. Seven days if the 24/48 h test is satisfactory.
3. One month if the 7-day test is satisfactory.
4. Monthly intervals for the first 6 months, and thereafter at 3-month intervals if the first monthly test is satisfactory.

5. After 12 months, all anchors still in service should be tested at intervals to be specified by the designer, but in no case exceeding 6 months.

Anchor monitoring in ground susceptible to creep is essentially based on the observation that creep displacement increments increase with load increase, and when the stresses at the fixed anchor–ground interface approach the ultimate strength of the ground the displacement accelerates in relation to time on a semilogarithmic scale. In this context, creep displacements may be considered stabilized when, for a constant load, the displacements are successively smaller, or they do not increase more than linearly (acceleration = 0) when plotted on a semilogarithmic scale versus time.

Useful data on the criteria of long-term performance and monitoring are provided by Buro (1972), Mitchell (1974), Australian Standard (1973), Gosschalk and Taylor (1970), Chen and McMullan (1974), Morris and Garrett (1956), and MacLeod and Hoadley (1974).

In most cases, monitoring is extended to the complete anchor–structure–ground system in the form of quality control and structure behavior. This work is particularly important in urban excavations, where it is advantageous to observe the overall performance of the project in order to control long-term effects on surroundings. Undoubtedly, overall monitoring is time consuming and expensive, but clearly essential to important civil engineering projects. The observed performance of anchored structures is discussed in the end sections of this text.

7-12 SERVICE BEHAVIOR AND ACCEPTANCE CRITERIA

Importance of Acceptance Standards

The load–extension behavior derived from basic and suitability tests has produced a variety of interpretation methods, which in substance yield acceptable limits in close agreement and in spite of differences in the basic approach. Unlike the load–extension behavior, acceptance criteria related to service behavior are widely diversified with reference to the duration of monitoring and with regard to what aspects of anchor behavior should be monitored. For example, several countries (Britain, United States, South Africa, and Australia) tend to recommend relaxation criteria (such as a prestress loss of 5 percent in 24 hs), whereas Continental Europe and Eastern Block countries tend to favor creep criteria (e.g., a creep displacement up to 4 mm in 72 h). Anchor specialists, on the other hand, recognize the arbitrary nature of these limits and suggest that inflexible controls are often irrevelant to a specific anchor and field application. As an example, the same standard may judge acceptable an anchor that shows negligible fixed zone movement but has undergone some debonding in the same zone. Likewise, an anchor which transfers only 85 percent of the applied load to the fixed zone may be considered satisfactory by several codes.

Littlejohn (1981) recommends that for economic and operational expediency the time necessary for stressing and testing should be kept to a minimum. The same investigator also suggests that a standard sequence of time intervals is ideally relevant since what should dictate the overall time of monitoring is the actual anchor behavior as opposed to other factors (e.g., a prejudgment based on the type of ground).

In order to alleviate problems of interpretation related to observed short-time behavior in basic, suitability, and routine acceptance tests, engineers must thus focus on and agree with standards and guidelines based on universally applicable criteria, and invariably these should include both load relaxation and creep displacement considerations.

Review of Load–Extension Limits from Current Practice

For type I tests according to DIN 4125 described in Section 7-7, the acceptance criteria are met when at load $1.2T_w$ the displacement stabilizes within the observation time, and when the elastic extension diagram lies between the two boundary lines a and b shown in Fig. 7-18(b). The upper boundary line a represents tendon extension for a length equal to the free length plus one-half the fixed length, or 110 percent the free length for a fully decoupled tendon with end bearing plate. The lower boundary line b is for a length equal to 80 percent the free length. The permanent displacement obtained from the approximate elastic extension line T_0X_0 in Fig. 7-18(b) should compare closely with the results of the basic test, but the permanent displacement Δp should not be greater than the observed for the basic tests in the range T_w to $1.2T_w$. For type II and III tests, DIN 4125 states that anchors meet the criteria when at maximum test load the creep displacement stabilizes within observation time, and the free length together with the permanent displacement have been proved as in type I test.

For permanent anchors, DIN 4125 considers the acceptance test satisfactory if the elastic extensions likewise fall between the two boundary lines a and b, and if the creep is less than 2 mm at a load $1.5T_w$; otherwise a limit load is established based on this creep value (see also Section 7-10).

According to current British practice, anchor testing should focus on monitoring loss of prestress with time as opposed to monitoring creep displacement favored by the French and the German codes. A check is carried out immediately after lockoff to measure the residual load in the anchor, usually $1.10T_w$, and then again after 24 h. A loss of as much as 5 percent is acceptable mainly on account of possible errors in the method of measurement. Thus, if the load at 24 hours is less than $0.95T_w$, the anchor should be replaced or the discrepancy rectified. If the load is between 0.95 and $1.05T_w$, the tendon should be restressed to $1.10T_w$ and retested after a second 24-h period. If after three such tests the anchor fails to retain a load $1.05T_w$, it should be rejected and replaced. If mutually agreed, however, the same anchor may be derated in which case the load is reduced until prestress loss is no longer observed over a period of one week. If a stable load is obtained in this manner, a safe working load is derived using the applicable factors of safety from this reduced stable load.

According to current French practice, acceptance criteria of permanent anchors relate to (a) displacement of the fixed length during proof testing, and (b) apparent free length. For the proof test, the load is held for 6 min if the fixed zone is in ground not susceptible to creep. For plastic soils the load is maintained for one hour on the first anchors, and 14 min for the remainder.

The time for measuring fixed length displacement is shown schematically in Fig. 7-39 (Pfister et al., 1982). Measurements are made 4 min apart. If Δe denotes displacement, $\Delta_{e1,2}$ is the displacement occurring between t_1 and t_2, and so forth. For soils not susceptible to creep, the proof test is satisfactory if $\Delta_{e1,2} \leq 2$ mm (0.079 in). If $\Delta_{e1,2}$ is larger, the test is still satisfactory if $\Delta_{e2,3} \leq 1$ mm (0.039 in). If the second condition is not satisfied, then the proof test at the specified load T_t is continued until time $t_n = T_0 + 58$ min, and the anchor is accepted if $\Delta_{e2,n} \leq 4$ mm (0.157 in). If the last displacement is larger than 4 mm, the anchor is not accepted in which case the options are to (a) derate the anchor to a lower load, (b) regrout in order to increase fixed anchor zone capacity, or (c) discard and replace with another anchor. Each of these solutions is accepted with discretion.

For soils susceptible to creep, the proof test for the first anchor is continued for a time $t_n = t_0 + 58$ min, and the anchor is accepted if the following two criteria are satisfied simultaneously: (a) $\Delta_{e2,n} \leq 4$ mm, and (b) the creep–

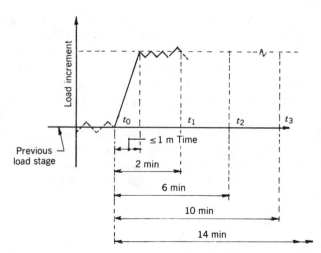

Fig. 7-39 Definition of measurement times for recording fixed-length displacement. (From Pfister et al., 1982.)

time diagram does not exhibit an upward concavity. The remaining production anchors are considered satisfactory if $\Delta_{e2,4} \leqslant 1.5$ mm. If this value is greater, the proof test is continued until time t_n, and the acceptance criteria specified for the first anchors must likewise be met.

The second acceptance criterion relates to the apparent free length. The test, in this case, must verify that the FAP (fictitious anchorage point shown in Fig. 7-16) falls between the proximal end and the mid point of the fixed anchor length.

General Recommendations

Littlejohn (1981) has put forward specific proposals relating to the service monitoring of complete anchorages as part of on-site suitability tests. Initially, the observation period should be long enough to justify a predictive method for long-term service behavior. In this context, both relaxation and creep displacement are relevant, but load is proposed as the main parameter to be monitored since anchors are essentially structural systems. If an anchor exhibits loss of load, this should therefore be a matter of concern. Additionally, anchor function is sensitive to fixed anchor displacement, hence creep should be also measured.

For practical purposes a time interval Δt equal to 5 min is reasonable, and a sequence of 3, 10, 30, 100 Δt, and so on is admissible (Huder, 1978). These intervals will accommodate short-term acceptance testing of 50 min, and for each such interval a simple relaxation or creep criterion is proposed. Stabilization can be assumed to have been reached when the data plotted on a semi-logarithmic scale give a straight line. Furthermore, whereas the test period and the intermediate time intervals should at best be determined from relevant field experience, fixed limits are arbitrary and should be adjusted provided sufficient data are accumulated to permit assessment of service performance in relation to acceptance criteria.

Proposed acceptance criteria for residual load/time behavior are shown in Table 7-8 (Littlejohn, 1981). The 6 percent load loss at day 1 is based on current practice, and for the time intervals recommended the rate of prestress loss should reduce to 1 percent initial residual load or less before the monitoring period ends.

The proposed alternative to monitoring load relaxation is presented in Table 7-9, showing creep displacement criteria. In this case, 1 percent Δ_e is the displacement equivalent to the amount of tendon shortening caused by a prestress loss 1 percent of the initial residual load.

Criteria for On-Site Suitability Tests

The general recommendations reflect a widely accepted practice, including mandatory provisions within the terms of the contract for on-site tests to confirm the suitability of a proposed anchorage for the conditions at the site.

TABLE 7-8 Acceptance Criteria for Residual Load–Time Behavior

Period of Observation (min)	Permissible Loss of Load (% Initial Residual Load)
5	1
15	2
50	3
150	4
500	5
1,500 (e.g., 1 day)	6
5,000 (e.g., 3 days)	7
15,000 (e.g., 10 days)	8

Thus, the anchors should be installed and constructed in the same ground and manner as the production anchors. At least three anchors should be subjected to suitability tests, with further tests for each category proposed for the project, namely, geometry and ground type. The proof load (maximum test load) should be $1.25T_w$ and $1.50T_w$ for temporary and permanent anchors, respectively.

Load–extension data should be plotted for each load increment for the test load range, with load increments not greater than $0.2T_w$ provided extensions are carefully monitored and satisfactorily obtained. During unloading, extensions at loads not less than two load increments in addition to datum should be measured, but preferably at the third points. The recommended loading and unloading sequence and cycles are shown in Table 7-10.

TABLE 7-9 Acceptance Criteria for Displacement–Time Behavior at Residual Load

Period of Observation (min)	Permissible Displacement (% of Elastic Extension, Δ_e, of Tendon at Initial Residual Load)
5	1
15	2
50	3
150	4
500	5
1,500 (e.g., 1 day)	6
5,000 (e.g., 3 days)	7
15,000 (e.g., 10 days)	8

From Littlejohn (1981).

TABLE 7-10 Recommended Load Increments and Observation Periods for On-Site Suitability Tests

| Temporary Anchorages | | Permanent Anchorages | | |
| Load Increment (% T_w) | | Load Increment (% T_w) | | Period of |
1st Load Cycle[a]	2nd, 3rd Load Cycles	1st Load Cycle[a]	2nd, 3rd Load Cycles	Observation (min)
20	20	20	20	5
	40		40	5
50	60	50	60	5
	80		80	5
100	100	100	100	5
	120		120	5
			140	5
125	125	150	150	15
100	100	100	100	5
50	50	50	50	5
20	20	20	20	5

[a] For this load cycle there is no pause other than that necessary for the recording of extension data.

From Littlejohn (1981).

If after 15 min the proof load does not decrease by more than 5 percent, allowing for temperature changes and movement of the anchored structure, the anchor should be considered satisfactory at this stage. If the loss is greater than 5 percent, the cause should be determined and the results rectified. An alternative is to maintain the proof load and measure anchor head displacement after 15 min. If the observed creep is less than 5 percent Δ_e, the anchor should be considered acceptable at this stage.

Residual Load–Time Data. Recording data should begin at $1.1T_w$ and continue for 10 days with observation periods according to Table 7-8. If the anchor fails to attain a stable (constant) load after allowing for temperature, structural movement and tendon relaxation, the same test should be extended for a period up to 30 days or until the load becomes stable, with monitoring at 7-day intervals.

Provided the loss of load is monitored accurately, the rate of loss from the initial load should decrease to 1 percent for each time interval for the observation periods shown in Table 7-8. Where the monitoring is relatively accurate, losses should not exceed 6, 7, and 8 percent the initial residual load at 1, 3, and 10 days, respectively.

Displacement–Time at Residual Load. As an alternative, Littlejohn (1981) proposes displacement/time criteria, beginning at $1.10T_w$ and continuing for 10 days with observation periods according to Table 7-9. If the displacement does not reach a stable value after allowance for temperature effects, structural movement, and creep of the steel, the same test should be extended for a period up to 30 days or until the displacement stabilizes, with 7-day monitoring periods. The displacement rate should decrease to 1 percent Δ_e or less per time interval for the observation periods shown in Table 7-9.

Where the monitoring accuracy is relative, the displacement should not exceed 6, 7, and 8 percent Δ_e at 1, 3, and 10 days, respectively.

Final Lockoff. If the test anchors are to become units of the production anchors and on completion of the suitability tests, they should be restressed and locked off at $1.1T_w$, provided the cumulative relaxation or creep does not exceed 5 percent initial residual load or 5 percent Δ_e, respectively. In these test Δ_e is the elongation of the free length (elastic) under the initial residual load.

Criteria for Acceptance Tests

For routine acceptance tests on production anchors, Littlejohn (1981) likewise recommends proof loads 1.25 and $1.50T_w$ for temporary and permanent anchors, respectively.

Load–extension data should be plotted continuously over the corresponding range, starting at $0.2T_w$ and for load increments not greater than 25 percent T_w provided extensions are carefully monitored. During destressing extensions at two load decrements in addition to datum should be obtained, and preferably at the third points. The recommended load increments and observation periods are shown in Table 7-11. After completion of the second cycle, the anchor should be retensioned to $1.1T_w$ and locked off, with an immediate reading after lockoff to establish the initial residual load. This reading is the zero time for monitoring load–displacement–time behavior. Proof load loss and displacement/time behavior at proof load is acceptable provided it conforms with the same criteria as in suitability tests.

Residual Load–Time Data. Suggested monitoring of residual load is at 5, 15, and 50 min. If the rate of load loss decreases to 1 percent or less per time interval for these observation periods (with allowance for temperature effects, movement of structure, and steel relaxation), the anchor is accepted. If the rate of load loss exceeds 1 percent, additional monitoring is necessary at observation periods of up to 10 days according to Table 7-8. If after 10 days the anchor fails to hold its load according to Table 7-9, it is unacceptable and deemed to have failed. In this case, there are three options: (a) the anchor is discarded and replaced, (b) its capacity is derated to lower load, and (c) anchor capacity is increased by regrouting with a subsequent restressing program.

TABLE 7-11 Recommended Load Increments and Observation Periods for Acceptance Tests of Production Anchors

Temporary Anchorages		Permanent Anchorages		
Load Increment (% T_w)		Load Increment (% T_w)		Period of Observation (min)
1st Load Cycle[a]	2nd Load Cycle	1st Load Cycle[a]	2nd Load Cycle	
20	20	20	20	5
50	50	50	50	5
	75		75	5
100	100	100	100	5
			125	5
125	125	150	150	15
100	100	100	100	5
50	50	50	50	5
20	20	20	20	5

[a] For this load cycle there is no pause other than that necessary for the recording of extension data.

From Littlejohn (1981).

Displacement–Time at Residual Load. Likewise, the alternative is to obtain displacement–time data for the same observation periods (5, 10, and 50 min). Restressing or constant load procedures can be used to monitor displacement at initial residual load. With accurate monitoring techniques, if the displacement rate decreases to 1 percent or less per time interval for the foregoing observation periods (with allowance for temperature, structural movement, and creep of the tendon), the anchor is acceptable. If the rate exceeds 1 percent Δ_e, the readings must continue up to 10 days as shown in Table 7-9.

If the monitoring techniques are less accurate and the total displacement at 1 day is less than 6 percent Δ_e, the anchor is again acceptable. Further observations are necessary if the displacement exceeds the 6 percent value, and are taken at 3 days and again at 10 days, at which times the displacement should not exceed 7 and 8 percent Δ_e, respectively.

If after 10 days the anchor continues to fail in keeping displacement in accordance with Table 7-9, it is deemed to have failed, and the remedy consists of any of the three options.

Final Lockoff. If, on completion of the test the anchor satisfies the acceptance criteria, it is restressed and lockoff at $1.1T_w$.

Relationship between Relaxation and Creep Criteria

The relationship between acceptance criteria for load–time (Table 7-8) and acceptance criteria for displacement–time (Table 7-9) is shown in Table 7-

12. Also shown are respective sensitivities to initial residual load (100 kN and 1000 kN) and free length (5, 10 and 20 m), for observation periods 5, 15, 50, and 1500 min (one day).

The tendon details for this example are as follows:

Nominal area of single strand:	100 mm^2
Elastic modulus:	200 kN/mm^2
Initial residual load (1 strand):	100 kN
Initial residual load (10 strands):	1000 kN

It appears that in the typical range of free tendon lengths, either criterion (load loss or creep) is admissible and relevant. For relatively short free lengths (<5 m or 17 ft), prestress loss becomes the significant factor hence the appropriate criterion, whereas for long free lengths (>30 m or 100 ft) clearly the displacement rate governs anchor performance. For the example illustrated in Table 7-12, a single creep criterion of 0.05 mm/m of free length is appropriate. On this basis, it is expedient to specify a limiting creep criterion for contracts with a wide variety of tendon lengths.

7-13 AVAILABLE REMEDIES FOR FAILED ANCHORS

The recommended factors of safety for failure at the ground–grout interface discussed in Section 5-10 may be reduced if a measured factor of safety is

TABLE 7-12 Relationship between Load–Time and Displacement–Time Acceptance Criteria

Period of Observation (min)	Free Tendon Length (m)	Limiting Loss of Load		Limiting Creep Displacement	
		Single Strand (kN)	Ten Strands (kN)	Single Strand (mm)	Ten Strands (mm)
5	5	1	10	0.25	0.25
	10	1	10	0.5	0.5
	20	1	10	1	1
15	5	2	20	0.5	0.5
	10	2	20	1	1
	20	2	20	2	2
50	5	3	30	0.75	0.75
	10	3	30	1.5	1.5
	20	3	30	3	3
1500	5	6	60	1.5	1.5
(e.g., 1 day)	10	6	60	3	3
	20	6	60	6	6

From Littlejohn (1981).

available, determined from suitability and basic tests. Thus, if an anchor fails at the ground–grout interface, a first reassessment of the new load can be made by dividing the observed maximum load at failure by the factor of safety. This can be taken as 1.6 and 2.0 for temporary and permanent anchors, respectively.

If anchors have satisfied the proof-loading criteria but have failed to meet relaxation and creep criteria, a provisional reduction factor of 1.2 is recommended (Littlejohn, 1981) in the absence of more field data. Service monitoring should be repeated under the same guidelines stipulated in Sec. 7-12 but under the new reduced load.

If a remedial stressing program has been found appropriate, the initial residual load $1.1T_w$ is obtained by stressing, followed by service monitoring. This procedure has been used successfully in stiff and/or hard clay where the preliminary stress history introduces a preloading effect (Littlejohn, 1970). In this manner, the ground surrounding the fixed zone is locally consolidated, and this improvement results in a better anchor performance for the subsequent service.

Where prestress gains are observed, monitoring should continue to finalize stabilization of prestress within a load increment 10 percent T_w. If the gain exceeds this increment, the cause must be determined generally by monitoring the structure–anchor–ground system. For example, if overloading continues to increase owing to insufficient anchorage capacity in design or because of unanticipated earth loads, the overall system can become stable only by additional supports.

Whereas these remarks highlight anchor performance and the available remedies in case of unsatisfactory service, it must be emphasized that routine testing and collection of data on relaxation and creep will continue to be essential to improved design procedures. In particular, the criterion of suitability and acceptance in ground where creep is unavoidable should always be reevaluated considering the conditions at hand, and the guidelines recommended by Ostermayer (1974, 1976) should be given appropriate consideration.

7-14 REQUIREMENTS OF STRESSING AND MONITORING EQUIPMENT

As part of acceptance criteria for anchor tests, certain controls should be specified and introduced with reference to the choice of appropriate equipment and its operational characteristics.

Stressing Equipment. Among the types reviewed in foregoing sections for wire, strand, and bar tendons, the preferred stressing devices should tension the entire tendon in one operation, although single- and multiunit systems are commonly used.

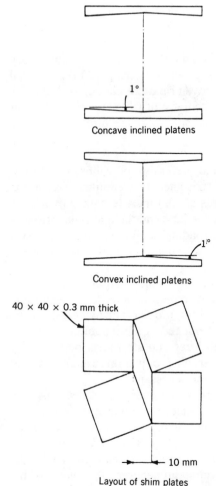

Concave inclined platens

Convex inclined platens

40 × 40 × 0.3 mm thick

10 mm

Layout of shim plates

Fig. 7-40 Typical types of platen to simulate uneven bedding in calibration of load cells. (From MacLeod and Hoadley, 1974.)

Jacks should be designed to allow tendon elongation at every stage to be measured with accuracy appropriate for the test. For example, the degree of accuracy may be ±0.2 mm for short duration (<1 h) when testing the rate of relaxation or creep.

Hydraulic pumps should be rated to operate for the pressure range of the stressing jack, and the controls of the pump should allow tendon extensions to be adjusted to the nearest millimeter. The pressure gauge should be mounted in a manner such that this apparatus is vibration free during pumping.

Load Cells. Although the basic characteristics are usually established by the manufacturer, certain tests should be introduced in order to simulate the service conditions under which the cell will operate. Among these are routine calibration tests using (a) concave inclined platens, (b) convex inclined platens, and (c) 0.3 mm sheets with irregular spacing to simulate uneven bedding (all shown in Fig. 7-40).

For routine calibration the load cell should be delivered to the laboratory in advance for the cell to attain the proper ambient temperature. The cell should be subjected to centric loading betwen rigid flat platens with the use of a testing machine with an absolute accuracy of 0.5 percent. Useful information and guidelines on the testing and calibration requirements of load cells is provided by Littlejohn (1981).

Frequency of Calibration. Jacks should preferably be calibrated at least once a year with the help of properly designed test equipment ensuring accuracy of 0.5 percent. The test records should include data showing the relationship between the load carried by the jack and the hydraulic pressure with the jack in the active mode in the loading condition, both increasing and decreasing. Prior to tensioning, it is essential to check the jack calibration and prepare a calibration curve.

It is recommended to have the calibration extended from zero over the entire working range of the jack, and cover the opening (load rising) and closing (load falling) jack operation to allow the friction hysteresis to be established when repeated cycles are introduced in the tendon.

Pressure gauges should be calibrated after 100 stressings or at 30-day intervals, whichever occurs first. However, if a group of three gauges is used in parallel, this calibration frequence is not necessary.

Calibration of load cells should be carried out every 200 stressings or at 60-day intervals, whichever occurs first, unless complementary pressure gauges used simultaneously show that no significant variation has taken place. In this case the calibration intervals can be extended for a period not to exceed one year when routine calibration should be mandatory with the help of appropriate test equipment providing an accuracy (absolute) of the order of 0.5 percent.

REFERENCES

Antill, J. M., 1965: "Relaxation Characteristics of Prestressing Tendons," *Civ. Eng. Trans. Inst. Austral.*, 7 (2), 151–159.

Barley, A. D., 1974: Private Communication.

Barley, A. D., 1978: "A Study and Investigation of Underreamed Anchors and Associated Load Transfer Mechanisms," Thesis, Univ. Aberdeen.

Baus, R., and A. Brenneisen, 1968: "The Fatigue Strength of Prestressing Steel," *Proc. FIP Symp. on Steel for Prestressing*, Madrid (June 6–7), pp. 95–103.

British Standards Institution, London:
BS 18, 1962: "Methods For Tensile Testing of Metals."

BS 3617, 1963: "Stress Relieved Seven Wire Steel Strand for Prestressed Concrete."

BS 2691, 1969: "Steel Wire for Prestressed Concrete."

BS 4545, 1970: "Methods for Mechanical Testing of Steel Wire."

BS 4447, 1973: "The Performance of Prestressing Anchorages for Post Tensioned Construction."

CP 115, 1969: "The Structural Use of Prestressed Concrete in Buildings. Part 2, "Metric Units").

CP 110, 1972: "The Structural Use of Concrete" (Part 1).

British Standards Institution, 1980: "Draft Proposal for Ground Anchors," London.

Bureau Securitas, 1972: "Recommendations Regarding the Design, Calculation, Installation, and Inspection of Ground Anchors," Editions Eyrolles, Recommendation TA72, Paris.

Buro, M., 1972: "Rock Anchoring at Libby Dam," *West. Constr.* (March), 42, 48, 66.

Chen, S. C., and J. G. McMullan, 1974: "Similkameen Pipeline Suspension Bridge," *ASCE, Transp. Eng. J.*, **100** (TEI, Feb.), 207–219.

CIRIA, 1978: "Prestressed Concrete—Friction Losses during Stressing," Report No. 74, Feb., Constr. Industry Research and Information Assoc., London, 52 pp.

Comte, C., 1965: "L'Utilisation des ancrages en Rocher et en Terrain Meuble," *Bull. Tech. de la Suisse Romande*, (Oct. 22), 325–338.

Deutsche Industrie Norm (DIN), 1972: "Verprebanker fur vorubergehende Zwecke in Lockergestein, Bemessung," *Ausfuhrung und Prufung*, DIN 4125, Vol. 1.

Deutsche Industrie Norm (DIN), 1974: "Verprebanker fur dauernde Verankerungen (Daueranker) Lockergestein," *Ausfuhrung und Prufung*, DIN 4125, Vol. 2.

Eastwood, W., 1957: "Fatigue Tests on Prestressed Concrete Beams," *Civ. Eng. Publ. Wkly. Review*, **52** (July), 786–787.

Edwards, A. D., and A. Picard, 1972: "Fatigue Characteristics of Prestessing Strand, *Proc. ICE*, **53,** Part 2 (7534), pp. 323–336.

Fenoux, G. Y., and J. L. Portier, 1972: "La mise en Precontrainte des Tirants," *Travaux*, **54** (449–450), 33–43.

FIP, 1974: "Recommendations for Approval, Supply and Acceptance of Steels for Prestressing Tendons," Cement and Concrete Assoc., London (Ref. 15.321).

Geotechnical Control Office, 1980: "Design Guide and Specification for Ground Anchors (Draft)," New Works Division, Hong Kong.

Gosschalk, E. M., and R. W. Taylor, 1970: "Strengthening of Muda Dam Foundation using Cable Anchors," *Proc. 2nd Cong. Int. Soc. Rock Mech.* Belgrade, pp. 205–210.

Hennequin, M., and H. Cambefort, 1966: "Consolidation du rembial de Malherbe," *Revue Generale des Chemins de Fer* (General Review of Railroads) (Feb.).

Huder, J., 1978: "Boden-und Felsanker: Andforderungen, Prufung und Bemessung, Die neke Norm, SIA 191," *Schweizerische Bauzaltung*, **96** (Jahrgang Heft 40), (Oct. 5), pp. 753–761.

Hutchinson, J. N., 1970: "Contribution to Discussion on Soil Anchors," *Proc. Conf. Ground Eng. Inst. Civ. Eng.*, London (June 16), pp. 85–86.

International Society for Rock Mechanics, 1976: "Suggested Method for Rock Bolt Testing."

Klein, M., 1974: "Draft Standard for Prestressed Rock Anchors," *Symp. on Rock Anchoring of Hydraulic Structures, Vir Dam,* November, 86–102.

Larson, M. L., W. R. Willette, H. C. Hall, and J. P. Gnaedinger, 1972: "A Case Study of a Soil Anchor Tie Back System," *Proc. Specialty Conf. on Performance of Earth and Earth Supported Struct., Purdue Univ.,* Vol. 1, 1341–1366.

Littlejohn, G. S., 1970: "Discussion on Paper Soil Anchors," *Proc. Conf. Ground Engineering,* Inst. Civil Engineers, London, June 16, pp. 33–44, Discussion pp. 115–120.

Littlejohn, G. S., 1981: "Acceptance Criteria for the Service Behaviour of Ground Anchorages," *Ground Eng.,* **14** (3), 26–36.

Littlejohn, G. S., and D. A. Bruce, 1977: *Rock Anchors—State of the Art,* Foundation Publ. Ltd., England.

Littlejohn, G. S., and D. A. Bruce, 1979: "Long-Term Performance of High Capacity Rock Anchors at Davenport," *Ground Eng.,* **12** (7), 25–33.

Littlejohn, G. S., D. A. Bruce, and W. Deppner, 1977: "Anchor Field Tests in Carboniferous Strata," Specialty Session No. 4, *Proc. 9th Int. Conf. on Soil Mechanics and Foundation Engineering,* Tokyo, pp. 82–86.

Littlejohn, G. S., and I. M. Macfarlane, 1974: "A Case History Study of Multi-tied Diaphragm Walls," *ICE Conf. on Diaphragm Walls & Anchors,* London, Sept., pp. 113–121.

Longbottom, K. W., 1974: Discussion on Paper "Prestressing Steels," *Struct. Eng.,* **52** (9), 357–362.

Longbottom, K. W., and G. P. Mallett, 1973: "Prestressing Steels," *Struct. Eng.* **51** (12), 455–471.

MacLeod, J., and P. J. Hoadley, 1974: "Experience with the Use of Ground Anchors," *Proc. Tech. Session on Prestr. Concrete Found. and Ground Anchors,* pp. 83–85, 7th FIP Cong., New York.

Mitchell, J. M., 1974: "Some Experiences with Ground Anchors in London," ICE Conf. on Diaphragm Walls & Anchors, Lond. Sept. Paper No. 17.

Morris, S. S., and W. S. Garrett, 1956: "The Raising and Strengthening of the Steenbras Dam (and Discussion)," *Proc. ICE,* Part 1, **5** (1), 23–55.

Nicholson Anchorage Co. Ltd., 1973: "Rock Anchor Load Tests; for Sheet Pile Bulkhead for the General Reinsurance Corp. Project," Greenwich, Conn., Unpublished Report (8 pp.)

Nicholson, P. J., D. D. Uranowski, and P. T. Wycliffe-Jones, 1982: "Permanent Ground Anchors, Nicholson Design Criteria," Dept. Transp. Fed. Hwy. Admin. Office of Research and Development, Washington, D.C.

Ostermayer, H., 1974: "Construction Carrying Behaviour and Creep Characteristics of Ground Anchors," *ICE Conf. on Diaphragm Walls and Anchorages,* London, Sept., pp. 141–151.

Ostermayer, H. 1976: "Practice in the Detail Design Application of Anchorages, A Review of Diaphragm Walls—Discussion," Institution of Civil Engineers, London, pp. 55–61.

Otta, L., M. Pantucek, and P. R. Goughnour, 1982: "Permanent Ground Anchors, Stump Design Criteria," Office of Research and Development, Fed. Hwy. Admin., U.S. Dept. Transp., Washington, D.C.

PCI Post-Tensioning Committee, 1974: "Tentative Recommendations for Prestressed Rock and Soil Anchors," PCI Chicago, 33 pp.

Pfister, P., G. Evers, M. Guillaud, and R. Davidson, 1982: "Permanent Ground Anchors, Soletanche Design Criteria," Offices of Research and Development Fed. Hwy. Admin., U.S. Dept. Transp., Washington, D.C.

Saxena, S. K., 1974: "Measured Performance of a Rigid Concrete Wall at the World Trade Center," ICE Conf. on Diaphragm Walls and Anchors, London, Sept., Paper No. 14.

Short, A., 1975: "DoE Studies Creep and Stress Behaviour at Navy Complex," *Constr. News* (Sept. 4), 28–29.

South African Code of Practice, 1972: Lateral Support in Surface Excavations. The South African Inst. of Civ. Engineers, Johannesburg.

Standards Association of Australia, 1973: "Prestressed Concrete Code CA 35-1973," Section 5—"Ground Anchorages," pp. 50–53.

Walther, R., 1959: "Vorgespannte Felsanker," *Schweizerische Bauzeitung*, **77** (47), 773–777.

Zienkiewicz, O. C., and R. W. Gerstner, 1961: "Stress Analysis of Prestressed Dams," *Proc ASCE Power Div*, **87** (POL), Part 1, 7–43.

CHAPTER 8

USES AND APPLICATIONS

8-1 ANCHOR–STRUCTURE GROUPING

Anchor-structure systems may be grouped according to the following classification:

1. Vertical Wall Systems. These are used to support open excavations and include steel-sheet pile walls, soldier piles with lagging, and cast-in-place or prefabricated–concrete slurry walls.

2. Intermittent Structures. These are discontinuous modular units used to support vertical cuts or slopes in rock and in stable ground. The prestressed anchors ensure stability, whereas the units are basically used to transfer and distribute the load. Examples are shown in Figs. 1-4 and 1-5. The progressive excavation is protected by cast-in-place or precast concrete members forming horizontal strips, or vertical beams connected with a concrete cladding.

3. Anchor/Shotcrete Supports. These have broad uses in the stabilization of large underground caverns. Examples are the hydropower projects introduced in the late 1960s.

4. Massive Structures. In this case, anchorages are necessary to balance unusual forces, or as a result of strengthening or raising a structure. The anchor system induces a vertical or subvertical force through the structure to increase resistance to uplift, overturning, or sliding. This category includes gravity dams, massive abutments of cable bridges, ski-jumps, and massive foundations.

5. Free-Standing Anchorages. Innovative schemes have been developed for marine structures such as cofferdams and dry docks. The anchors can

be installed through the dock floor, or tensioned through a body of water. In offshore installations anchors must accommodate cyclic loading if the anchored structure is free to oscillate at sea. Examples are anchorage mooring schemes proposed for submerged tunnels.

8-2 ANCHOR-WALL CHARACTERISTICS AND APPLICABILITY

Anchored Sheet Pile Walls. These are suitable in soft clays, organic materials, and dilatant soils of low plasticity. Steel sheeting forms a seal at the base of the excavation if it is driven to interlock. The system provides resistance to ground movement, particularly below excavation level, but its inherent flexibility makes sheet piling more suitable for relatively shallow excavations or where some ground movement can be tolerated.

In hard ground or where boulders and other obstructions are encountered, driving sheet piling can be difficult and even impossible. In congested sites, depth limitations may be imposed by available headroom, whereas noise and vibrations are objectionable and may impose the use of silent pile drivers (Hunt, 1974). Sheet-pile walls are relatively expensive, but some of the cost is recovered if the piles can be pulled out for reuse.

Anchored sheet-pile walls have, however, limited load–bearing capacity, a problem that can be remedied either by extending the sheet piles to full resistance in which case a deep wall will result, by placing intermittent sections on stilts or other suitable foundation elements, or by choosing a relatively flat anchor inclination to reduce the vertical load component. Since sheet-pile walls usually serve temporarily, until the permanent underground structure is in place, the use of detensionable or extractable anchors is a normal requirement.

Anchored Soldier Pile Walls. These offer flexibility in a variety of ground types except soft clays and loose sands that have a tendency to run. The system is economically attractive, and represents a time-tested ground support, adaptable where ground movement can be tolerated and the groundwater level is controlled by dewatering. Structurally the support is flexible, and below excavation level it provides limited resistance to ground movement. Like sheet piling, the installation is more economical if the piles can be withdrawn for reuse. If they are left in place, they may be incorporated in the permanent structure. Soldier piles are suitable at sites where the presence of underground utilities does not favor other methods.

Problems may arise if it is necessary to underpin existing foundations or where the excavation is carried out in water-bearing ground. A usual problem is ground loss in granular soils associated with preexcavation to install the piles, open lagging or overcut behind lagging, and surface or groundwater migration. In these conditions, predraining of saturated soils is essential, particularly if materials have a tendency to run. Difficulties will also arise if these soils are underlain by rock or by impervious layers within the proposed excavation depth, since this sequence almost precludes dewatering to the lowest extent of the water bearing formation. A useful review of soldier pile systems is provided by Wosser and Darragh (1970), and by Donolo (1971). Concrete soldier piles with concrete lagging are reportedly popular in Sweden (Broms and Bjerke, 1973). These are fairly watertight; hence, they are economical if they can become part of the permanent structure.

Concrete Slurry Walls. These systems (also called diaphragm walls) have become a standard construction method for supporting open excavations since the 1960s. A complete treatment of the subject is provided by Xanthakos (1979). Maximum economy will result if the wall can transplant the temporary sheeting and the permanent ground support, or where the excavation requires underpinning and groundwater control, and the problem can be remedied by the diaphragm wall.

When they are used in conjunction with anchorages, diaphragm walls have broad uses and applications in retention systems, underpinning, waterfront installations, underground perimeter walls, traffic underpasses and depressed roadways, and underground circular enclosures. Wall types that are suitable for anchoring are (a) the continuous cast-in-place diaphragm wall; (b) prefabricated panels installed in slurry trenches, and with a wide variety of size and structural configurations; (c) certain categories of bored pile walls; (d) composite walls consisting of vertical steel members (I beams) and concrete panels between them; (e) diaphragm wall panels interlocked with bored piles; (f) circular and polygonal enclosures; (g) posttensioned wall panels that are braced at the top by anchors; and (h) special combinations consisting of buttressed walls, cells, and arched structures. A complete review of these systems, including details and construction considerations, is given by Xanthakos (1979).

8-3   INTERMITTENT STRUCTURES

Applicability. Deep slope cuts in soil or in highly fractured rock usually require considerable removal of material on the high side to stabilize the excavation and prevent earth slides and slips. The excess removal necessary to maintain a stable slope can be markedly reduced by building an anchored retaining wall, and this solution is typically favored where space is at a premium. If a continuous wall is to be built, a temporary support will probably be required and will add to the total construction cost. Alternatively, the support can be broken into slabs and intermittent sections that are installed progressively and individually in a downward process, provided local instability is prevented.

These units can be constructed inclined or in a near vertical position. The procedure usually selected is to underpin by lateral support horizontal strips 1.5–3 m (5–10 ft) high depending on ground conditions. In this scheme, the stability of the cut is ensured by the prestressing action of the anchors whereas the wall units provide a cover and distribute the forces. These members have, therefore, the minimum practical thickness, and can be cast-in-place or prefabricated. An example of this method is shown in Fig. 1-4.

If the intent is to stabilize and simultaneously underpin, the procedure deviates from the concept of stable horizontal strips. In this case, the construction proceeds with progressively concreting and simultaneously anchoring vertical ribs or columns, and thereafter protecting the ground between them by gunned concrete or filter layers. A cladding or facing is then applied. The usual height of these sections is 3 m (10 ft), and this is the normal bracing interval for excavations supported by anchorages. Thus, each section requires two anchors, and each step ensures the stability of the next by stressing the anchors before the next step is carried out. Likewise, prefabricated members can be used as columns.

An anchored retaining wall of intermittent sections will also provide a rational solution for a generally stable rock face, since it ensures continuous protection against loosening and crumbling of rock materials caused by weathering because of rain, snow, or frost.

Optimum Anchor Inclinication. In most instances the structural element is selected to provide a stressing bed, or a pad foundation on which the stressing anchorage is mounted. A second design aspect is the optimum anchor inclination, for which the specified stabilizing action is provided with minimum installation cost. A solution is usually derived on the basis of certain assumptions, and is illustrated in the following example.

Figure 8-1 shows a geologic profile and data for a rock stabilization project at Hauetli-Alpnachstad, Switzerland (VSL-Losinger, 1978). Portions of this slope (3:2 in soil and 1:1 in rock) slipped and cracked further about 18 months after construction, indicating the possibility of extensive ground movement and deep seated failure. Remedial steps included installation of rock anchors on concrete mats. In this case, optimum angle of inclination was determined from the following considerations.

1. The slip surface forms a plane in the region of the anchorage installation.

2. The shear strength conditions along the slip surface are represented by the relation $\tau = \sigma \tan \phi$ (c = 0)

3. In any given profile, the anchors are parallel for simplicity in the drilling operations.

For an average anchor with its head located as shown in Fig. 8-2(a) the free anchor length L_f is

$$L_f = t/\sin\delta \tag{8-1}$$

where   t = distance from entry point to the plane through the upper end of the fixed length

   δ = angle between axis of the anchor and the slip surface

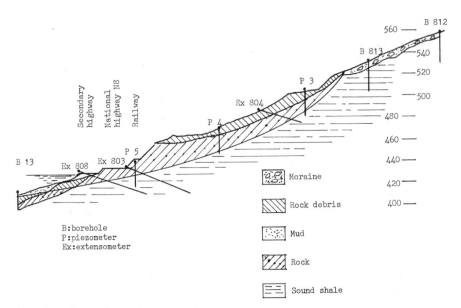

Fig. 8-1   Geologic profile of slope, stabilization project at Hauetli, Alpnachstad, Switzerland. (VSL–Losinger, 1978.)

For cost comparison, the resisting force B exerted by the anchor along the slip surface is assumed constant. From Fig. 8-2 part (b), the anchor force A is

$$A = B/(\cos \delta + \sin \delta \tan \phi) \qquad (8\text{-}2)$$

The condition $dA/dB = 0$ yields $\delta = \phi$, which implies a minimum value for the anchor force A. This solution takes effect if the objective is to minimize the anchor force rather than unit cost.

The construction cost K_1 for the free length is now obtained from

$$K_1 = EAL_f \qquad (8\text{-}3)$$

where E is the price of the free length per unit length and unit force (including drilling, steel tendon, and corrosion protection). Entering the values of L_f and A from Eqs. (8-1) and (8-2) into Eq. (8-3) we obtain

$$K_1 = EBt/(\sin \delta \cos \delta + \sin^2\delta \tan \phi) \qquad (8\text{-}4)$$

Again, setting $dK_1/d\delta = 0$ yields $\delta_1 = \pi/4 + \phi/2$ for which K_1 is a minimum.

The total anchor cost, however, must include the cost K_2 for the anchor head, bearing plate, bond length, and anchor stressing. These parameters are proportional to the number of anchors so that the anchor force A can

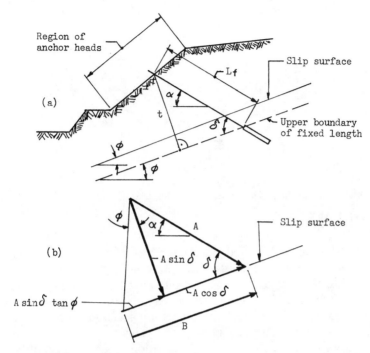

Fig. 8-2 (a) Profile geometry depicted diagrammatically: (b) resisting force B along slip surface. (VSL–Losinger, 1978.)

be obtained from the simple relation $K_2 = DA$ where D is the price for the above operations and per unit force. The factor K_2 becomes a minimum if $dA/d\delta = 0$, or $\delta_2 = \phi$.

The total construction cost-per-anchor is now

$$K = K_1 + K_2 \qquad (8-5)$$

and evidently K is a minimum if the summation $K_1 + K_2$ becomes a minimum, which yields the solution δ_{opt} for the K function. The values δ_1 and δ_2 represent the upper and lower boundaries for δ_{opt}, and they are independent of anchor geometry and unit cost. On the other hand, the solution δ_{opt} is dependent on geometry, anchor type, and unit cost. These variables can be plotted versus δ as shown in Fig. 8-3, showing results of cost comparison for one of the profiles. For the instance, $\delta_{opt} = 53°$.

Construction Considerations. For a slope in unstable equilibrium, a continuous rigid wall will not provide the best solution particularly with considerable scatter in soil and rock properties. A better remedy is a discontinuous, elastic anchored system that confines interaction to the area

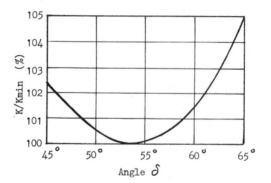

Fig. 8-3 Construction costs plotted versus angle δ. (From VSL–Losinger, 1978.)

prestressed by the anchor, and gradual removal of rock mass is followed by immediate protection. Thus, differential deformations and ground movement are limited to the area acted upon by the individual units, and this imparts to the system an acceptable ground response. In this context, local variations in rock pressures and nonuniform foundation conditions are considered in optimum engineering and economic terms. This form of construction will require, however, a monitoring and observation program to quantify ground response. If anchor load adjustments and restressing become necessary, they will have only a small effect on neighboring units, and thus an *in situ* measured factor of safety can be obtained for each panel and structural section.

An example of rock stabilization is shown in Fig. 1-4. For the uppermost excavation strip (step 1), the material is removed in a single run along the entire length of the slope. Next, the filter concrete is placed, the anchor is installed, the concrete slab is cast, and the anchors are stressed 7 days later. For stages 2 through 5 the procedure is somewhat different. For these stages the excavation is carried out for the entire slope length at an inclination that approaches the natural angle of repose up to the anchored slabs of the row above. Along the same strip, the slope is further cut to the specified inclination in alternate panels that form a checkerboard pattern. The intervening material is removed only after each alternate slab is in place and anchored. A general view of this construction is shown in Fig. 8-4.

For rock that is relatively stable at the surface, the supporting wall elements are vertical concrete tie beams as shown in Fig. 1-5, anchored at one or two points, a process that is also underpinning. For unusually high cuts the construction is limited to two staggered layers at a time, and if stability conditions require each beam layer is limited to 1.5 m (5 ft). Where the slope has relatively small inclination, the strips can be wide and thus suitable for cast-in-place concrete, which is economical and also allows the bearing face to be adapted to the excavated profile for full contact. If the slope is steep, it is better to choose small-height tie beams utilizing prefabricated elements

Fig. 8-4 General view of an intermittent slab wall during construction. (VSL.)

that allow immediate stressing after installation. A staggered anchored tie-beam wall under construction is shown in Fig. 8-5.

8-4 SHOTCRETE USED WITH PRESTRESSING

The introduction of prestressing in conjunction with a thin membrane (usually a layer of shotcrete) enables construction of massive openings as high as 150 ft (about 45 m). This method is widely used to stabilize large underground chambers for galleries, powerhouses, tunnels, and caverns. The rock/anchor/shotcrete system is adapted under *in situ* conditions but corrected during excavations to respond to local variations. Stresses are mon-

Fig. 8-5 A staggered anchored tie-beam wall under construction. (VSL.)

itored by special anchors to enable a continuous check on the stabilization process.

Stabilization of Caverns. An application of the anchor-shotcrete method is shown in Fig. 8-6, and involves an underground powerhouse for the Hongrin underground pumped storage station (Buro, 1970). The excavation was completed in eight stages as shown. The three small sections shown in Fig. 8-6(a) provided access for drill rigs used to inject grout below the cavern area. The three tunnels were located above the groundwater table to ensure their stability without supports. After the initial stage, the top section was excavated as shown in (b), and three sections were enlarged as shown in (c) to allow installation and stressing of anchors. Thereafter, the three outer tunnels were connected as shown in (d), and more anchors were installed and prestressed. The remaining excavation sequence is shown in (e) through (h) and is self-explanatory. Shotcrete and wire mesh were applied to the rock surface in successive layers 1½-in thick until a total thickness of 6 in was obtained.

The tension in the tendons was checked as the work progressed, and stabilization of the rock mass was assumed to have been reached when changes in lift-off stress values could no longer be registered. The final cumulative rock strains relative to the fixed anchorages are shown in Fig. 8-7 for a representative transverse section. This deformation pattern does not

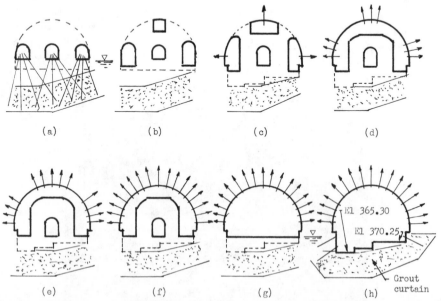

Fig. 8-6 Construction stages and cross sections, Hongrin project; underground powerhouse 100 ft wide and 90 ft high. (From Buro, 1970.)

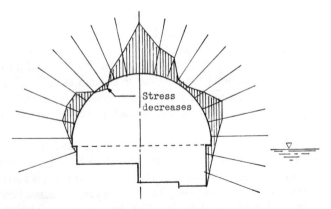

Fig. 8-7 Variations in anchor strain, longitudinally cumulated; cavern of Fig. 8-6. (From Buro, 1970.)

reflect the total movement of the cavern since the fixed anchor zone is in rock with relative mobility and freedom to move. Anchor strain increases by 8 in at the zenith of the section, and this implies an average cumulative stress increase in the rock anchors of about 10 percent.

Interestingly, this example demonstrates the value of instrumentation. Visual checking is equally valuable, however, in identifying critical zones although it is not a quantitative check. Hairline cracks, in particular, may indicate the occurrence of progressive movement.

Stabilization of Tunnels. A derivative of the anchor-shotcrete technique is the New Austrian Tunneling Method (NATM). The combination of these two systems is, however, an independent process, and does not imply interaction between the two supporting elements.

The need for immediate sealing of rock surface in tunnel work is satisfied with the use of shotcrete. Additionally, the construction is enhanced by two remarkable supplements: the introduction of the reciprocal relationship of required lining resistance and deformations (Fenner, 1938), and the theory proposed by Rabcewicz (1969) that the time-dependent behavior of rock mass is fundamental for predicting tunnel behavior. Further progress reflects (a) the formulation of the shear failure theory for tunnels under high overburden; (b) improvement in the installation of semi-rigid linings; (c) the introduction of a semi-empirical design approach using *in situ* measurements to supplement the main work; and (d) the incorporation of rock and soil in the carrying structural support system.

Whether NATM is a construction method or a tunneling concept is irrelevant and academic. On the other hand, the basic characteristic is not necessarily unique to the technique, since observed ground reaction is a tunneling experience on a worlwide basis. A unique constructual procedure inherent in NATM makes it, however, distinct from other tunneling methods.

Thus, a tunnel contract provides for support systems (rock anchors, shot-crete, and steel ribs) that are adaptable to varying ground conditions. The support method is changed if conditions at any time are not the same as those that provided the basis for support design. For general purposes, the anticipated ground conditions are covered in certain classifications, usually six or less (Daly and Abramson, 1985). Each ground type provides a characteristic response and behavior during and after excavation (See also Chapter 9).

The goal during construction is to activate the ground (mainly rock) to a load bearing (compression) ring that enables it to become a support component of the total structural system. This is achieved if (a) provisions are included to assess the anticipated geomechanical response and behavior during excavation and after prestressing; (b) the engineer chooses the most suitable and favorable tunnel profile yielding maximum composite advantages; (c) the design provides for control of unfavorable stresses and deformations by means of suitable support combinations installed in a proper sequence; (d) support resistance, offered by the ground as function of allowable deformations, is fully optimized; and (e) field measurements are used to monitor the entire process.

In this context, a tunnel becomes a composite structure consisting of the immediate rock ring around the excavation, and secondary supports or strengthening elements such as rock anchors, shotcrete linings, and often steel ribs. During excavation the opening remains active, and the initial stage of equilibrium is merely changed to a new condition of stability or secondary equilibrium. The construction sequence is planned to influence this transformation in a manner that is technically sound and economically attractive. The required control of rock behavior is achieved if deformations remain relatively small since this is necessary to prevent rock loosening and a corresponding decrease in strength, and at the same time they manifest a pattern that is compatible with the active load-bearing ring around the opening.

8-5   TALL AND MASSIVE STRUCTURES

In these instances, anchors serve to counteract the overturning commonly manifested in structures subjected to eccentric loading. Examples are systems acted upon by wind, water pressure, earth pressure, waves, ice pressure, earthquake forces, uplift, and incidental forces of unusual severity and intensity. Stability against overturning may not be ensured unless special remedies and compensating effects are introduced. A solution is to increase the size and weight of the structure, which is often difficult or costly, or produce a tie-down effect by prestressing.

Dams. Gravity-type dams are solid concrete or masonry structures with a cross section roughly triangular. A dam depends primarily on its own weight and cohesion with the foundation substratum for stability against overturning and sliding. Despite the impressive bulk density and gravity, extensive structural measures are often necessary to compensate for the overturning effects. If erosion and cavitation damage are allowed to occur, sometimes unavoidable with old deteriorated structures, unusually high static and dynamic uplift forces will result and require costly remedial measures.

Stability against overturning is ensured if anchors are installed vertically and close to the upstream face as shown in Fig. 8-8. This gravity dam, built in Leila Takerkoust, Morocco, was raised because of heavy silting in the reservoir that deprived the lake of storage capacity. The new crest elevation is 9 m (30 ft) higher. A total of 54 vertical anchors were installed through the crest and along the 250 m long spillway. The anchors have an average working load of 6.2 MN, and provide a total vertical anchoring force of 340 MN, or 1.35 MN per meter of dam length (93 kips per foot).

Resistance against sliding is increased as shown in Fig. 8-9. This example is from the Muda dam in Malaysia. The dam is founded on strata underlain by weak layers. Anchors improve stability in two ways, first by increasing the shear resistance at the base of the dam and along the underlying layers, and then by providing a net horizontal force opposing downstream movement.

Tall Structures. Typical examples where the overturning effect is intensified are ski jumps and lighthouses. The latter must provide stability against tilting and sliding, as illustrated in the lighthouse of Fig. 8-10. The critical horizontal force produced by wave heights of 9–10 m (30–33 ft) gives rise

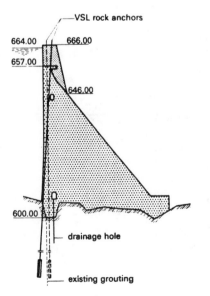

Fig. 8-8 Section through the raised structure; Leila Takerkoust Dam, Morroco. (VSL–Losinger, 1978.)

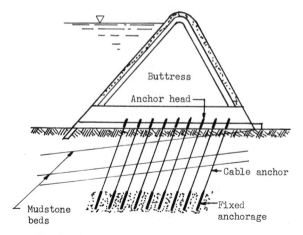

Fig. 8-9 Section through Muda Dam, Malaysia.

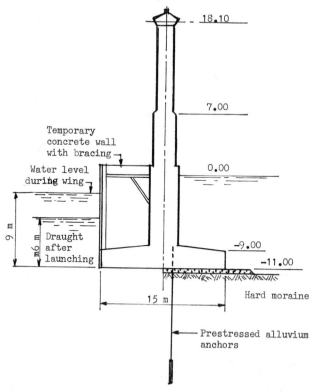

Fig. 8-10 Section through structure; Kullagrund Tower, Sweden. (VSL–Losinger, 1978.)

to a lateral thrust of 3600 kN and a lifting force of 800 kN (800 and 180 kips, respectively). The weight of the structure is 8300 kN.

The tower is anchored with 6 permanent soil anchors with ultimate strength of 1463 kN per unit, or a total ultimate anchoring force of 8800 kN. After losses, the available prestressing working force is 4000 kN. The factor of safety F_s against sliding is estimated assuming a coefficient of friction μ = 0.45 between the concrete base and the ballast.

$$\text{Without anchors } F_s = (8300 - 800)0.45/3600 = 0.94$$

$$\text{With anchors } F_s = (8300 - 800 + 4000)0.45/3600 = 1.45$$

For the same example, the ultimate factor of safety against sliding, based on ultimate anchor capacity after losses, is

$$F_s = (8300 - 800 + 8000)0.45/3600 = 1.94$$

8-6 USE OF ANCHORS TO IMPROVE SLOPE STABILITY

Slopes, rock faces, and embankments can reach the stage of instability as a result of certain natural phenomena such as water penetration, icing and thawing, and ground erosion. In other instances, the cause of instability is associated with new construction that changes the ground form or the loading conditions. The use of prestressed anchors in conjunction with a proper drainage system will offer a good solution to the stability problem and will allow deeper slopes to be excavated. Slope stabilization may be necessary in the construction of new roads, abutments of dams, open-cast mining, portals in new tunnel work, and miscellaneous works such as scour prevention.

In fractured rock, the prestressing force is applied to the unstable layers at the surface, and increases friction in the fracture planes. In this manner the upper layers are secured to the deeper sound rock mass whereas the ground load bearing capacity remains intact. In loose rock and in soil, the application of prestress will require supporting elements at the surface reviewed in the foregoing sections. These distribute the forces while they are anchored back. Their form, size and dimensions depend largely on the type and nature of the ground, anchor spacing and distribution, and prestress level.

Example of Consolidation of Rock Projection. Figure 8-11 shows a section through rock projection adjacent to Simplon Pass, Switzerland (VSL-Losinger, 1978). The rock face rises nearly vertically for 40–50 m (150 ft), and then continues its ascent at about 45°. In thic configuration, large cuts into rock mass would be required to accommodate the widening of the adjacent roadway.

The rock consists of very hard granitic gneisses, divided into blocks by various fracture systems. The position and orientation of fracture geometry would most likely affect rock stability if a cut was to be made. From Fig. 8-11(a) this geometry includes (a) a first system I near the top, sloping down toward the valley as shown, that could conceivably be the origin of slips; (b) an almost vertical system II that separates the lower half of the rock projection from the remaining mass; and (c) a system III with a flat downward inclination toward the mountain side, that did not present an immediate risk although it might lead to some overbreak.

The inclination of the first fracture system in relation to the rock cut is shown in Fig. 8-11(b), and for a specific fracture plane 38–45°. For the entire rock mass, the 38° slope becomes critical for stability, although the lower part of the fracture sloping at 45° would tend to open if movement propagated toward the valley. A further unstable condition could result if isolated blocks showed tendency to move away from the rock mass as shown in Fig. 8-11(c).

The final stabilization scheme consisted of two operations, first anchoring the entire rock system with prestressed anchors, and then removing the rock face and consolidated materials in stages. A typical section of the anchoring plan and rock removal work is shown in Fig. 8-12. After anchoring, the rock

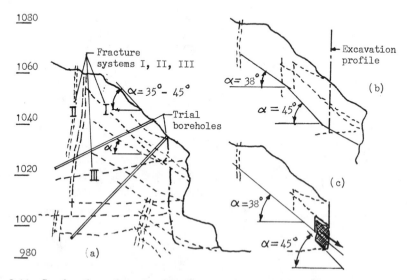

Fig. 8-11 Section through rock projection; southern slope of Simplon Pass, Switzerland. (VSL–Losinger, 1978.)

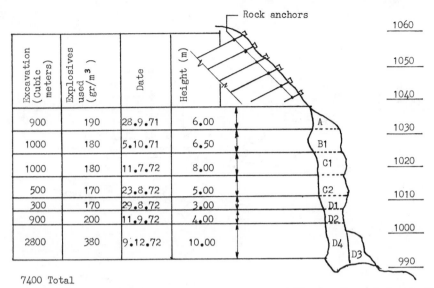

Excavation (Cubic meters)	Explosives used (gr/m³)	Date	Height (m)		
900	190	28.9.71	6.00		A
1000	180	5.10.71	6.50		B1
1000	180	11.7.72	8.00		C1
500	170	23.8.72	5.00		C2
300	170	29.8.72	3.00		D1
900	200	11.9.72	4.00		D2
2800	380	9.12.72	10.00		D4

7400 Total

Fig. 8-12 Stages of rock removed; southern slope of Simplon Pass, Switzerland. (VSL–Losinger, 1978.)

cut was completed in several blasting stages, whereas unstable exposed surfaces were protected by Gunite.

Example of Scour and Landslide Prevention. Figure 8-13 shows a stabilization scheme that is part of the construction program of the Tarbela dam in Pakistan. This project includes two spillways over which large volumes of water are discharged particularly during extended rainy seasons. Without appropriate protective works, water turbulance would cause scour and ground erosion.

As shown in Fig. 8-13(a), protection is provided by a concrete wall descending from the flip bucket at 45° and extended into underlying rock. The slope wall is secured by rock anchors that have 740 kN ultimate strength and are arranged in checkerboard patterns. The anchors are connected to the drainage galleries as shown so that the 4.3 m bond length is located partly in the slope wall and partly in rock. A total of 2000 anchors were used, 17 m long.

Anchorages are also suitable in stabilizing hillside slopes adjacent to roadways. These slopes continue to move plastically under the effect of creeping soils, resulting in distress and damage to the roadway. Prestressed anchors increase resistance along the sliding surface by applying a normal force. Applications are reported by Henke (1974) and involve the use of counterforts supported by prestressed anchors. The same investigator has presented examples of landslide stabilization in weathered clay shales.

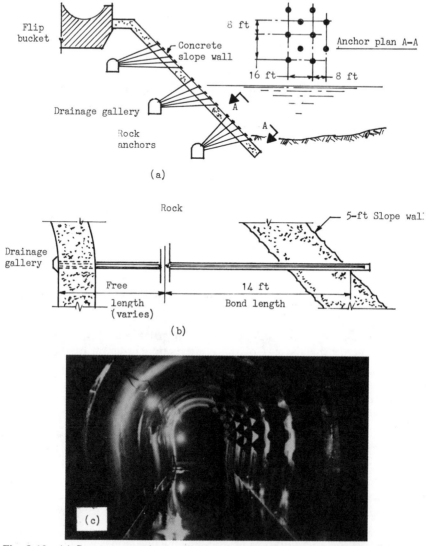

Fig. 8-13 (a) Scour prevention, spillway of the Tarbela dam; (b) rock anchors and connection to drainage gallery; (c) drainage tunnel with anchorage blocks. (VSL–Losinger, 1978.)

8-7 APPLICATIONS FOR DAM STRENGTHENING AND RESTORATION

Prestressed anchors generally replace dead concrete weight; hence, their adaptation to dam engineering becomes a matter of practicality and economics for new and old dams. Certain topographic features and foundation conditions can, however, make their choice even more attractive. For example, a relatively flat cannon floor can eliminate twisting between adjacent blocks, and thus prevent changes in the uplift characteristics that can affect changes in anchor design values.

Most foundation materials will be acceptable if the proposed dam height does not exceed 50 ft, and the difference between headwater and tailwater is less than 20 ft. For taller structures, sound rock foundation is typically required, and tests should be made to establish the consolidation characteristics. Rock deformability is not necessarily a deterent to design, but if anchorages are contemplated settlement and deformability should be investigated because of the effect on anchor-load loss. Likewise, where anchors are used to tie down a dam, bearing capacity or shearing resistance of the foundation materials are not considered in the same order of importance. A dam may tend to slide but this tendency can be resisted as shown in Fig. 8-9. Anchors add, however, to the weight effects of the structure, so ample bearing capacity should be available at the base.

Example of Dam Restoration. The Pacoima dam, showsn in Fig. 8-14, was at the time of its completion in 1929 the highest concrete arch dam in the United States. Situated above the San Fernando Valley in southern California, the dam was severely shaken during the earthquake of February 1971. The structure, 370 ft high, remained structurally sound but the rock mass supporting the left abutment was disturbed and moved upward during the earthquake. The stability was investigated by borehole photographic methods, seismic surveys, and three-dimensional stress analysis techniques. The extent of the problem was confirmed by studies on the dynamic response of the dam, abutment stability analyses, and stereographic projection techniques in conjunction with a system of rock anchors.

Remedial work consisted of an anchorage installation that ties the rock mass on the left abutment securely to underlying formations. The anchors are spaced 20–30 ft apart in rows across the top of the abutment, and are inclined at 60° with the horizontal. Anchor length varies from 130 to 200 ft. A view of the left abutment while work is under way is shown in Fig. 8-14.

Example of Dam Strengthening. The concept of dam strengthening using anchorages was pioneered with the Cheurfas dam in Algeria, mentioned in Section 1-2. An important step was the design and construction of the Allt-Na-Lairige dam in Scotland in the early 1950s that utilized bar anchors. This arrangement ties down the structure to competent granite bedrock.

Fig. 8-14   View of left abutment, Pacoima dam, while anchorage work is under way. (VSL, 1976).

The Laing dam, located west of East London, showed signs of instability after a protracted rain in 1970 raised the water flow 5.2 m (17 ft) above the spillway. This dam lies in a fairly steep gorge, and rises 38 m (130 ft) above its foundation. At the most potentially unstable position, however, the structure is 45 m (148 ft) high. The dam was analyzed for the new hydrological criteria and with the following objectives: (a) improve the dam structurally with the use of prestressed anchors installed as shown in Fig. 8-15; (b) raise the nonoverflow sections to increase spillway discharge capacity; and (c) maintain actual storage capacity at the same level. Anchor design criteria specified a maximum initial working load of 6000 kN (1350 kips), a multipull single jacking system, and provisions for postgrouting.

The Muda dam shown in Fig. 8-9 demonstrates the use of anchorages in new dam construction. This structure is founded on quartzite intercepted by layers of mudstone. Although the original design was based on a shear angle of 30° for the foundation materials, during excavation the mudstone was found to be more extensive, weaker, and interbedded with clay layers, prompting a redesign for a residual shear angle of 18°. In the final solution, the original concrete buttress design was kept, but dam stability against sliding was improved with a unique arrangement of prestressed steel tendons installed as shown in Fig. 8-9. All 205 units are the restressable type, and are fixed 100 ft (30 m) below the base of the dam. Anchor working load is 600 kips, and anchor inclination is 15°. More details for this construction are given by Cox (1978).

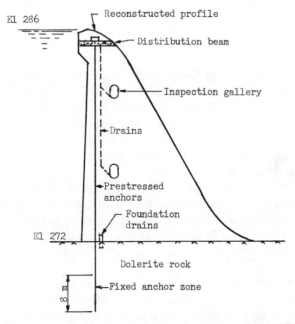

Fig. 8-15 Laing dam; typical cross section. (From Cox, 1978.)

8-8 SOIL PRECONSOLIDATION AND SOIL HEAVE CONTROL

Figures 1-6 and 1-7 show two typical examples of soil preconsolidation where the action of anchors is essentially similar to the prestressing effect that holds down a structure. Unlike, however, the prestressing effect intended to resist uplift, prestressing in soil preconsolidation is used to produce the same effect as preloading.

In general, settlement of structures in compressible ground is prevented by extending the foundation to a solid layer, and, where this is not practical, by using deep foundations such as piles, caissons, and slurry wall panels. For exceptionally large areas such as airport pavements, new highways on embankments, and construction on new landfills, consolidation by preloading is practical and less expensive although time consuming. The usual solution is to apply a surcharge fill, generally soil, and leave it in place until the consolidation settlement is completed.

In certain instances, however, and where time is of essence, preconsolidation settlement can be induced by prestressing the foundation layers with rock and soil anchors, but these effects must be manifested largely in the stratum that is most susceptible to settlement. Soil compaction under load will result in a corresponding anchor prestress loss; hence several restressing stages will be necessary until the final settlement is achieved.

An example of preconsolidation is for the extension or duplication of an existing structure, where the long-term settlement of the old and the new section will be out of phase, and consolidation settlement of the old section may be complete when the settlement of the new section is just beginning.

Prefabricated Stressing Sections. Precast, prestressed concrete slabs are suitable as stressing elements at ground surface. They are reusable, at both the same and different locations, are easily transported, and can accommodate scheduling and prestressing sequence because of their direct mobility. Used with anchors, they are intermittent structures. They are available in squares requiring one anchor per unit, or in rectangular configurations with two or three anchors per unit.

The application is more economical if corrosion protection is not needed, and the testing program involves essentially load monitoring and anchor prestressing. An example of soil preconsolidation is the Alexandra Bridge in Canada, where 24, 300-ton anchors were used to induce settlement before the bridge loads were in place.

Soil Heave Control. Swelling or heave occurring at the base of an excavation is essentially the reverse problem, and can be resisted in the same manner. Swelling is equally common in soft deposits of clay shales, and occurs on unloading in a process that is accompanied by absorption of water. Since the permeability of these materials is low, swelling occurs over a long period. Most observations indicate that the bulk of swelling is manifested in the upper layers and near the surface. Although several test methods have been proposed to quantify the amount of swell as a function of confining pressure (Richards, 1977; Aitchison, 1972; Stapeldon, 1970), their relevance to anchorages as means of heave and swelling control should be judged with caution. Examples of swelling prevention in tunnel floors using rock reinforcement are presented in subsequent sections.

8-9 ANCHORAGES FOR CONCENTRATED FORCES

Concentrated forces are manifested in suspension bridges, cableways, Olympic tents, penstocks, gantry cranes, beam brackets, galleries, and pile test loads. In these instances, anchorages can transfer the loads to firm ground or rock. The use of prestress can also reduce deformation under loading, and where dynamic conditions are expected the prestress can be increased to provide adequate safety against fatigue failure.

Rock Fall Galleries. Areas and sites exposed to rock falls are usually protected by a gallery that is either sloped to divert falling rocks beyond the protected area, or it has a structural roof to withstand the impact from falling objects.

Figure 8-16 shows a rock fall gallery for the Axenstrasse along Lake

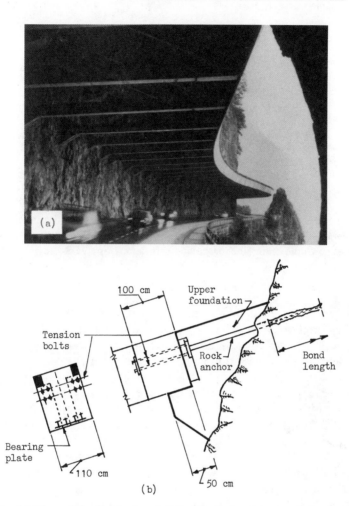

Fig. 8-16 (a) View of the finished rock fall gallery, Axenstrasse, Lake Lucerne; (b) upper anchorage for gallery, detail, and section. (From VSL–Losinger, 1978.)

Lucerne in Switzerland. The roof is cantilevered out from the rock face, and is built with main girders and lower struts as shown in part (a). The cover consists of precast concrete panels. In the upper part the concrete is pressed directly against the face by prestressed rock anchors as shown in (b). Each anchor has a stressing force of 1500 kN (340 kips).

Cable Crane Anchorages. During construction of the Jiroft dam in Iran, two cable cranes, each 520 m long, with 20-ton capacity, were used for materials handling and incidental operations. On one side of the valley the suspension cables were secured at a fixed point anchored into limestone. A recessed concrete block was built as shown in Fig. 8-17, and attached to

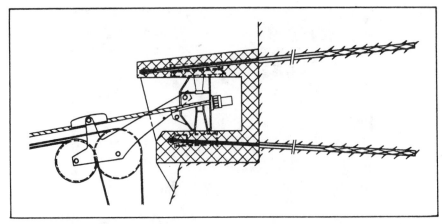

Fig. 8-17 Cable crane anchorage at Jiroft dam, Iran (VLS–Losinger, 1978).

adjoining rock with eight anchors carrying a maximum tension 640 tons (1440 kips). The anchors were 22 m long, with a downward inclination of 10°. Because of the long construction period, all units were classified as permanent and provided with double corrosion protection.

Soil Anchorages for Concentrated Forces. Heavy tensile forces from hanger installations, Olympic tent-type roofs, and heavy lifting assemblies can be anchored into firm soil by suitable anchoring schemes. Among these are (a) anchor weight blocks that resist vertical loads with minimum uplifting of the foundation; (b) prestressed concrete trench walls of T section; and (c) groups of soil anchors. The choice depends on the use, magnitude, and inclination of the force to be resisted (See also Section 9-17).

A vertical heavy tensile force can be resisted by (a) a massive concrete block filled with earth materials; (b) a concrete column on a large footing built deep enough so that resistance is provided by dead weight and mobilization of shear along an appropriate surface; (c) a base slab on tension piles; and (d) a base slab tied down with anchors. Details are shown in Fig. 9-69.

Weight blocks have been used to hold peripheral ropes of stadium tent-type roofs (Soos, 1972) that involve large forces (5000 tons) with nearly flat inclination (7°). The action provided by these blocks is discussed in Section 9-17. A modified block used as foundation base for heavy tensile forces is shown in Fig. 8-18. Its face is supported by the stem and base plate, whereas the weight that mobilizes friction along the base is increased by the weight of the fill.

Inclined concentrated forces can also be resisted by a trench wall, that has a transverse T section. Trench walls are reviewed in Section 9-17, and

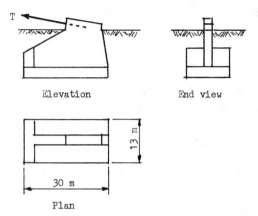

Fig. 8-18 Block bases for peripheral ropes; tent-type roof of the Olympic stadium in Munich. (From Soos, 1972.)

a typical configuration is presented in Fig. 9-71. A single unit can resist an inclined force of 900 tons (2000 kips). If large loads must be transferred, several T units are built alongside. Trench walls offer structural advantages for relatively flat inclinations, but these advantages tend to diminish when the angle with the horizontal exceeds 60°.

A group of anchors converging into a concrete block can be used as shown in Fig. 8-19(a) and (b) to resist concentrated tensile forces with inclined and

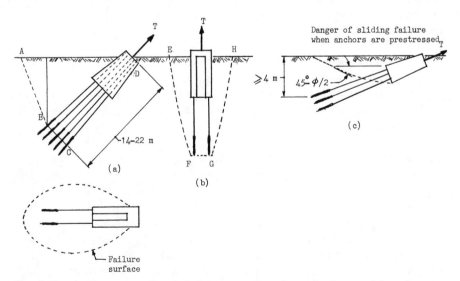

Fig. 8-19 Anchor group for resisting concentrated tensile forces: (a) anchor arrangement for inclined force; (b) anchor arrangement for vertical force; (c) probable failure of earth mass under load. (From Soos, 1972.)

vertical direction, respectively. The total holding-down capacity in this case may not depend on individual anchor capacity, but may be determined from group action. Increased tension is feasible if the group as a unit mobilizes a larger earth mass defined by lines ABCD and EFGH. Groups of anchors offer advantages when the inclination of the tensile force with the horizontal exceeds 30°. With flatter inclination, the 4–5 m overburden specified as minimum cannot be provided, and the risk of slide under the base increases as shown in Fig. 8-19(c).

Anchorages of this type were used in the Olympic tent-type roof in Munich, and resisted forces of 700 tons (1575 kips). Total anchor-group capacity was determined by shear resistance in the soil mass, and ground failure thus took precedence over anchor failure.

8-10 ANCHORAGES TO SECURE CAVERNS

A principal feature in using underground caverns for power stations and related uses is the large size and complex configuration that is needed to accommodate the mechanical layout, equipment, and services. The stabilization of these openings is also complicated if the nature, orientation, and extent of joints, faults, and shear joints in the rock mass are unfavorable for structural support. If the problem is confined to spalling and individual block stability, grouted or mechanical rock bolts are a good solution. On a large scale, however, high-capacity prestressed anchors may be the only way to improve overall stability, and this method has been used in many underground power stations in Europe, South Africa, and North and South America.

Cavern Waldeck II. This project was constructed as a pumped storage scheme between 1968 and 1975. The cavern for the underground powerhouse, shown in Fig. 8-20, is 106 m (348 ft) long, 54 m (177 ft) high, and 33.5 m (110 ft) wide. The rock consists of sandstone and slate sedimentary formations.

The structural support of the cavern is provided with 716 rocks, with working load of 1300 kN (290 kips), and average length 20 to 28 m. The free anchor length is unbonded, and hence free to move elastically. Anchor hole diameter is 116 mm (4.5 in). Each borehole was water-tested, and if necessary, grouted and redrilled. Production anchors were tested to 130 percent the working load, and one week later the stressing force was measured, followed by secondary grouting. All anchors have a double plastic sheathing with grout-free gap that provides telescopic action between rock and anchor to ensure that the latter does not respond to small rock movement within the stressing zone.

Ventilation Station for Seelisberg Tunnel. This station is an enlargement of two regular parallel tubes of the tunnel near Lucerne, Switzerland. The

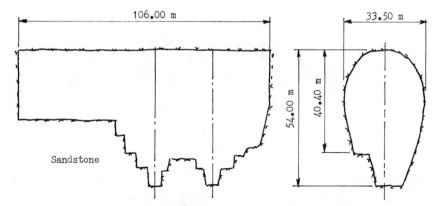

Fig. 8-20 Waldeck II; configuration and dimensions of cavern. (VSL–Losinger, 1978.)

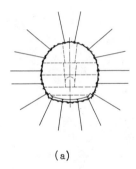

(a)

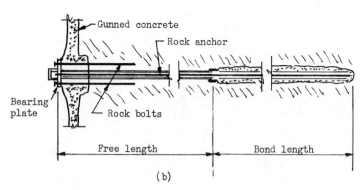

(b)

Fig. 8-21 (a) Cross section of ventilation station, showing arrangement of rock anchors; (b) diagrammatic presentation of the support system showing the gunite, rock bolts, and rock anchors. (VSL–Losinger, 1978.)

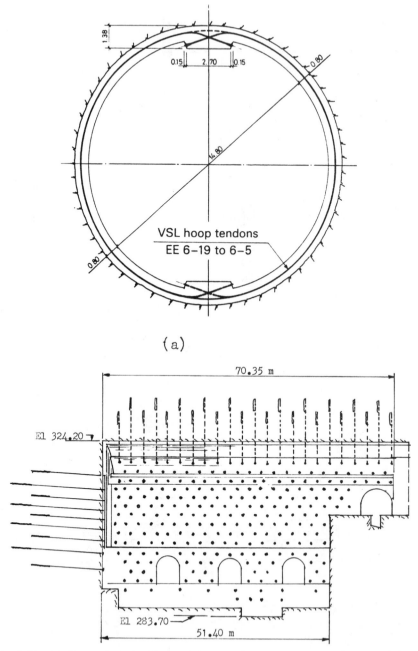

Fig. 8-22 (a) Cross section of headrace tunnel, Taloro, Sardinia, showing the post-tensioned concrete lining; (b) underground powerhouse cavern showing the arrangement of rock anchors. (VSL, 1977).

caverns are 53 m long and 18 m wide, and house mechanical and electrical equipment. Structural support is provided by rock anchors arranged along a 4.5-m lattice, or 16 units per section as shown in Fig. 8-21(a). Anchor ultimate strength is 2365 kN (530 kips), and fixed length is 4.5 m. Fig. 8-21(b) shows the composite support system consisting of gunned concrete, rock bolts, and the supporting anchors. Primary and secondary grouting was carried out in a single operation. The bearing plate was designed and sized to keep the pressure exerted on rock during test loading to less than 13 N/mm^2 (1900 psi). Seven days after grouting, production units were stressed to 1650 kN, or 70 percent f_{pu}, and locked off at 800 kN, or 34 percent f_{pu}. Interestingly, the relatively low design lock-off load has the two-fold function of support and confining pressure. The latter depends on the deformation that must occur within the rock mass to mobilize self-supporting action.

Pumped Storage Scheme, Taloro, Sardinia. The main components of this project are the headrace tunnel, approximately 1890 m long, and the underground powerhouse, a cavern 41 m high. Fig. 8-22(a) shows a cross section of the headrace tunnel, lined with posttensioned concrete using circular tendons. The cavern for the underground powerhouse is shown in Fig. 8-22(b). The support consists of system rock anchors and bolts, that also attach a concrete arch lining in the vault section firmly to rock. Where the powerhouse is adjacent to a second cavern, the anchors are not bonded to rock but pass through it and are fixed on the rock face of the adjacent cavern. These anchors have, therefore, stressing anchorages at both ends. Anchor capacity at service load is 1000 and 1500 kN, and the units are installed in 120 mm-dia. boreholes.

8-11 ANCHORAGES FOR TUNNELS

Unlike past techniques that emphasized heavy reinforced linings, horseshoe shapes, and random blasting, the current state-of-the-art relies on control of deformations so that a stable state is attained with minimum artificial support resistance. Simultaneously, the goal is to avoid and prevent excessive loosening so that the strength of surrounding ground mass can be activated and maintained as a directly usable support component.

Mt. Lebanon Tunnel, Pittsburgh. This tunnel lies within geologic strata consisting of repetitions of claystone, siltstone, shale, sandstone, limestone, dolomite, and coal with intermittent interbedding. Tunnel alignment falls in weathered and unweathered rock zones. This geology produced two engineering options A and B, shown in Fig. 8-23(a) and (b), respectively. For both options, rock behavior during excavation, primary support requirements, and ultimate rock loads on final lining were determined by rock joint characteristics, partings, and weak planes.

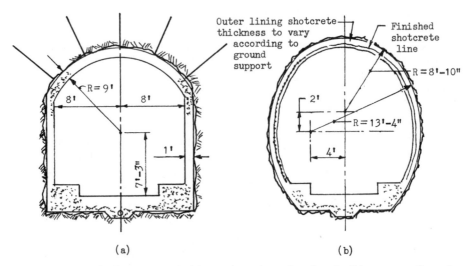

Fig. 8-23 Mt. Lebanon tunnel; (a) cross section of option A; (b) cross section of option B. (From Daly and Abramson, 1985.)

Option A is for zones of rock weathering, particularly in the portal areas, where primary support by rock reinforcement was not considered effective. For these locations rock support with structural steel ribs was specified, and special portal treatment included fully grouted, low-angle spiles to improve stability ahead of the excavation. Elsewhere in the tunnel, the optimum primary support system included pattern rock reinforcement to counteract structural defects in the immediate roof. This system consists of five, 9-ft long, size 8 rebar dowels on 4.5-ft centers as shown in Fig. 8-23(a). The final lining for this option is cast-in-place concrete designed to support a vertical load of 4.5 kips/ft^2.

Option B reflects the principles of NATM. This option specified contractual procedures whereby support systems could be adapted to varying geologic and ground conditions, and established a format for correlating ground type classification with support method (Daly and Abramson, 1985). Under this arrangement, a particular support system is specified for each ground type listed in Table 8-1. The final inner lining, including rock bolts, rebar straps, and invert concrete, represent a separate item. This lining consists of 4-in-thick reinforced shotcrete resisting the loads compositely with the outer shotcrete lining. Interestingly, options A and B intend to reflect the contrast of design and construction principles represented in each method; hence they are not technically optimal or economically equatable. Structurally, however, they are similar since their design is governed by shear and axial thrust force considerations. For the most part, tunnel construction was implemented under option B, which had a lower bid than option A.

TABLE 8-1 Mt. Lebanon Tunnel: Specifications for Option B Ground Types

	Ground Type		
Specification	I	II	III
Excavation method	Full face	Full face and top heading and bench (where necessary)	Top heading and bench
Allowable round lengths	Unrestricted (based on test blasts)	8 ft	Top heading, 5 ft bench, 9 ft
Minimum shotcrete thickness[a]	2 in	4 in	5 in, reinforced
Other support[a]	Engineer-ordered rock bolts (occasional)	Engineer-ordered rock bolts and rebar straps (occasional)	Steel ribs at 4 ft-0 in O.C. and engineer-ordered rock bolts and spiles (occasional)
Estimated length	2876 ft (linear)	1686 ft	190 ft

[a] Initial support only; reinforced shotcrete lining was required for final support.
From Daly and Abramson (1985).

Soil Reinforcement in Soft Ground Tunneling. The use of grouted spiles to reinforce tunnels in weak rock and tie together loose zones is standard practice in rock excavations, but can also be applied to reinforce tunnels in soft ground although on a limited basis because of lack of a proven design methodology.

Essentially, the reinforcement is installed as shown in Fig. 8-24. A thin shotcrete layer reinforced with wire mesh is sprayed after excavation, and spiles are added to reinforce the soil. A final concrete lining may also be added, but the initial shotcrete remains an integral part of the overall support system (Bang and Shen, 1982). Applications include the Nuremberg, Bochum, and Washington, D.C. subways.

The underlying principle is to stabilize a weak formation by installing reinforcing elements into *in situ* soil masses as excavation proceeds. Unlike, however, rock anchoring, the support mechanism does not relate to the formation of an active ring around the opening but the intent is to mobilize shear resistance between the spile and the enveloping soil. This interaction serves to prevent ground loosening (immediate stabilization), and contributes to the permanent stabilization by limiting deformations (Korbin and Brekke, 1975, 1978).

In pure cohesionless soils, the method is more effective if used in conjunction with grouting, since the excavation must preceed the installation

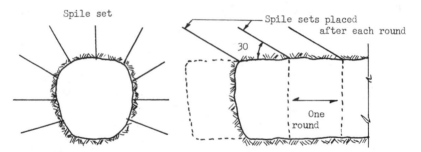

Fig. 8-24 Spiling reinforcement ahead of the face in soft-ground tunneling. (From Bang and Shen, 1982.)

of the support. In silty and clayey deposits, however, spiling is adequate without supplementary action provided the soil has sufficient standup time.

Anchors to Affect Swelling in Rock Tunnels. Swelling, discussed briefly in the foregoing sections, is defined as the time-dependent volume increase of natural ground caused by stress changes, increase in water content, or combinations therefrom. Einstein and Bischoff (1977) summarize the following processes, according to the interaction and sequence of causes:

1. A change in the state of stress, such as erosion, overburden, valley cutting, or excavation of an underground opening, can lead to volume increase with time.

2. Absorption of water because of difference in concentration, unsaturated or partially saturated bonds, and differences in potential can lead to time-dependent volumetric changes.

3. Stress changes can be the cause of water absorption with further volume increases. Volume increases due to stress changes or water absorption can occur simultaneously or sequentially.

4. This is the inverse of process 3. Time-dependent volume increase due to water absorption leads to stress changes with further volume increases. Likewise, simultaneous or sequential occurrence of the two types of volume increase are possible.

5. The associated weakening of bonds and reduction in effective stresses can lead to time-dependent reduction of shear strength, which in turn causes displacements showing similar characteristics as swelling, particularly in underground openings, although they are basically creep phenomena. The transition from this process to squeezing is gradual, but the latter is slower and accompanied by smaller displacements. This process can occur in conjunction with processes 3 and 4.

The mechanism of swelling in tunnels is documented from case histories, most of them in Swiss tunnels, and leads to an important observation: most of the displacement appears to take place in the form of invert heave,

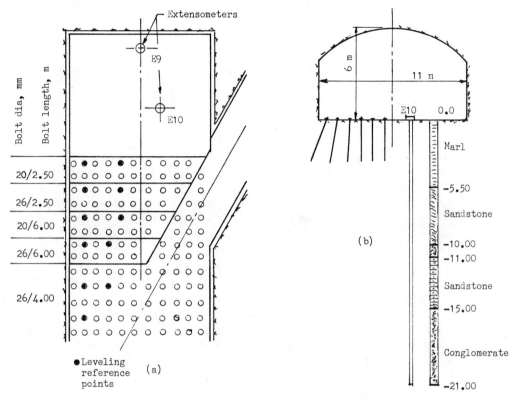

Fig. 8-25 Tunnel in marl; (a) plan view; (b) cross section. (From Einstein and Bischoff, 1977.)

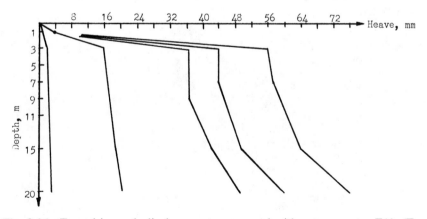

Fig. 8-26 Tunnel in marl; displacements measured with extensometer E10. (From Einstein and Bischoff, 1977.)

TABLE 8-2 Measured and Predicted Invert Heave for Unbolted Invert.

Location	Center of Invert	Near Sidewall
Predicted heave	93 mm	37 mm
Measured heave, 28 Aug., 1970– 21 March, 1974	75 mm	20 mm

From Einstein and Bischoff (1977).

whereas inward movement of the lower sidewall is not in reality lateral swell but the indirect effect of invert heave. Tunnel excavation induces stress changes leading to direct time-dependent volume increase and attraction of water, both causing swelling.

Proper construction procedures can reduce swelling by inducing artificial counterstresses, and by invert arches, grouting, and trimming. Among these, anchored or bolted invert slabs fixed below the main swelling zone produce counterstresses and thus limit the heave process. In addition, the use of anchors results in a decrease of the structural span.

An example of swelling control for a tunnel in marl is shown in Fig. 8-25. Perfobolts were used to induce counterstresses in a system pattern as shown in part (a), and displacements were measured by means of two extensometers marked E9 and E10. Computations indicated that swelling would extend about 3.5 m below invert at the center, and 0.9 m below invert near the sidewalls. These boundaries are confirmed by direct measurements with the use of extensometers as shown in Fig. 8-26. Predicted and measured displacements are shown in Table 8-2, and compare closely.

The effect of bolts is shown in Table 8-3 for various fixed lengths. Predicted displacements are based on an assumed linear stress reduction along

TABLE 8-3 Measured and Predicted Invert Heave for Bolted and Unbolted Invert Slabs in Tunnels

Area	Bolt Length (m) Bolt Diameter (mm)			2.5 20	26	4.0 26	6.0 20	26	Unbolted Invert
Center	Predicted heave	Absolute	(mm)	85	81	77	80	73	93
		Relative[a]	%	91	88	83	86	79	100
	Measured heave	Absolute	(mm)	32	26	12	18	13	37
	1/25/71–11/17/75	Relative[a]	%	86	70	32	49	35	100
Near sidewall	Predicted heave	Absolute	(mm)	33	30	29	31	27	37
		Relative[a]	%	90	83	79	86	73	100
	Measured heave	Absolute	(mm)	15	12	7	9	6	12
	1/15/71–11/17/73	Relative[a]	%	125	100	58	75	50	100

[a] Relative Heave = $\dfrac{\text{Heave of Bolted Invert}}{\text{Heave of Unbolted Invert}}$

From Einstein and Bischoff (1977).

the fixed length. The difference between observed and predicted behavior is, therefore, due to time lapses in the measurement period and the assumed linear stress transfer, which may deviate considerably from the actual.

8-12 ANCHORAGES FOR UNDERPINNING

An example of lateral protection in lieu of the conventional underpinning is shown in Fig. 8-27. This work became necessary because of excavation below the subway line for the foundation and the basement of a 60-story tower rising above the subway. The lateral protection was provided by intermittent diaphragm wall panels tied together with four levels of anchors, and was supplemented by pregrouting of the zone beneath the subway. In the final scheme the wall panels were used as foundation elements for the tower.

A different example of underpinning by lateral protection is shown in Fig. 8-28, and involves an excavation 20 m (65 ft) deep. On one side, underpinning

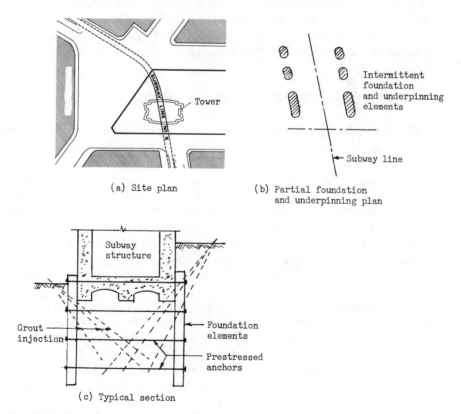

(a) Site plan

(b) Partial foundation
and underpinning plan

(c) Typical section

Fig. 8-27 Use of anchorages and strip panel walls to underpin an existing subway structure.

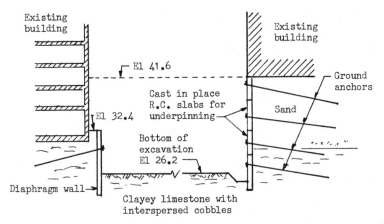

Fig. 8-28 Underpinning of existing buildings, Murat III office deelopment, Paris. (From Fenoux, 1971.)

was carried out with horizontal concrete strips braced laterally with four rows of prestressed anchors. A diaphragm wall supported the left side of the excavation, braced at the top with prestressed anchors and at the bottom by sufficient embedment into firm soil.

8-13 ANCHORAGES FOR DEEP EXCAVATIONS

Basic Principles. Vertical excavation walls are common in urban sites or other locations where demand for space is at a premium. Irrespective of any other considerations, an excavation is likely to involve underpinning and control of ground movement. As mentioned in the foregoing sections, there are various wall types for lateral support, and within each type there is a variety of systems and configurations for the particular requirements of the project. Interestingly, anchor capacity necessary to secure the sides of an excavation is nominal as long as there are no unusual lateral stresses acting against the support. Depending on anchor spacing and vertical distance between rows, a typical anchor capacity range is between 200 and 1000 kN (45–225 kips).

In many instances anchors for deep-basement excavations are temporary, and remain until the permanent framing is in place. Whereas this may require only temporary corrosion protection, it also suggests that extractable or restressable anchors should be preferred. A typical problem is the vertical load component induced by the anchor force, and for 45° inclination this component is essentially the horizontal pressure. We know, however, from experience that loads as high as 150 tons/m have been transferred to dense fine sand from 80-cm-thick diaphragm walls. Walls 30 in (75 cm) thick have been used as load bearing elements to carry a load of 25 kips/ft yielding a bearing pressure of 10 kips/ft^2, and at working loads this transfer occurs as side resistance along the back of the wall. In this context, anchors can safely be designed to induce vertical loads, and the walls can easily be adapted to carry loads from the superstructure after the anchors are destressed.

Construction Requirements. Since wall movement will begin almost simultaneously with excavation, the uppermost anchor bracing level should be determined accordingly. The sequence of excavation stages and bracing levels should, likewise, be determined from similar considerations. If the below-ground structure includes sections and members that are to be used to brace the walls and eventually replace the anchors, the final interaction should be considered and determined before construction. In most cases, the permanent bracing is introduced from the bottom up. Such permanent members are usually provided with a jacking system at the boundary-wall face to adjust for strains and deformations that originate from construction tolerance or elastic shortening. If the introduction of load and transition from anchors to permanent members is not coordinated, it may interfere with the upper anchorages and cause uneven redistribution of loads.

Examples of Anchored Walls. An inclined bored pile wall is shown in Fig. 8-29, supporting the excavation for a cut-and-cover extension of the Munich subway. The wall inclination in this case was dictated by tight alignment

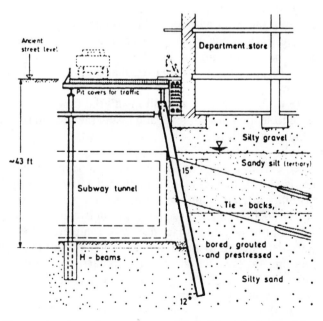

Fig. 8-29 Excavation for subway construction in Munich; inclined bored pile wall strutted at the top and anchored in the lower levels. (From Littlejohn, 1982.)

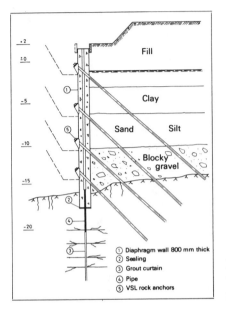

Fig. 8-30 Typical section, deep excavation for building in Stockholm. (VSL–Losinger, 1978.)

and minimum clearance, which precluded the use of other methods for lateral support and underpinning. This construction was carried out in the following stages:

1. Install bored pile wall with an inclination as shown.
2. Install steel H columns using the prefounded column method.
3. Install temporary decking at street level.
4. Excavate to just above existing foundation and install struts as uppermost wall bracing.
5. Excavate to first anchor level and install the first row of anchors.
6. Excavate to second anchor level and install the second row of anchors. Prestress anchors at both rows.
7. Excavate to final level.

An anchored cast-in-place diaphragm wall for a deep building excavation in Stockholm is shown in Fig. 8-30. This design satisfies the following criteria: (a) feasibility of combining the temporary support with the permanent structure; (b) protection of the base from groundwater effects, uplift pressures, and bottom swelling; and (c) feasibility of completing the work without effects that are detrimental to surroundings. The excavation accommodates a five-story basement 22 m (72 ft) deep, and was carried out without pumping. The wall surrounds the entire site along its perimeter, and is sealed with rock sockets. A grout curtain formed below the base seals the excavation and relieves the bottom slab from uplift pressures. After the permanent interior framing was in place, the four rows of anchors were destressed. Anchor working load varied from 1000–24000 kN (225–540 kips).

8-14 ANCHORAGES FOR LONG EXCAVATIONS

Long excavations are carried out to depress motorways and railways, or to provide grade separation at intersections, and they usually have a normal configuration and standard section. Permanent anchorages are commonly required to tie back retaining walls supporting the cuts, or to protect existing buildings. Subways and subway stations are more complex, however, and require a combination of structural supports often supplemented by ground strengthening.

Depressed Roadways. The typical traffic underpass and depressed roadway shown in Fig. 8-31 accommodates four lanes with safety walks. For normal vertical clearance, the excavation depth is about 20 ft (6 m). Although this is structurally within the range of vertical cantilevers, excessive lateral movement is likely to result if the walls are not braced. Top bracing reduces this movement and also the wall embedment below excavation level. In the covered portion top bracing can be provided by the permanent deck, usually constructed prior to excavation. The use of anchors in the open section is practical provided a permanent anchorage zone can be ensured beyond the right of way.

For the usual height, configuration, and function of the exposed walls, prefabricated panels offer distinct advantages. Walls laterally supported at the top and bottom are also ideally suitable for posttensioning.

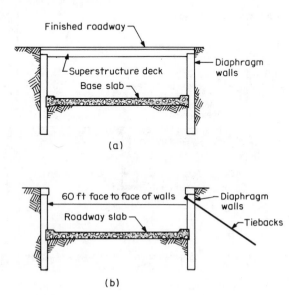

Fig. 8-31 Anchored diaphragm walls for a typical traffic underpass; (a) section at crossing; (b) section in open roadway.

Underground Railways. For subways and underground roadways built in cut-and-cover, the temporary decking and the permanent structure become part of the bracing system, especially where the construction is confined within a narrow cut. Anchorages are used, however, where a depressed railway is in open cut and its width or height restricts interior bracing, where the retaining system is temporary and separate from the permanent structure, and in special complex subway stations that have a configuration adaptable to anchorages. Representative examples of anchored walls for underground transportation systems are given by Xanthakos (1979).

8-15 ANCHORING OF FOUNDATION STRUCTURES

Unlike the structural schemes discussed in Section 8-9 where the intent is to transfer a tensile concentrated force to a resisting medium, foundation slabs and mats must be anchored if they are subjected to an upward loading originating from uplift or from overturning effects of eccentric forces.

An example where the condition of uplift is remedied without tie-down schemes is shown in Fig. 8-32. In this instance, the anchored perimeter enclosure walls are extended to an existing impervious layer. This isolation is combined with pumping inside the excavation to provide permanent

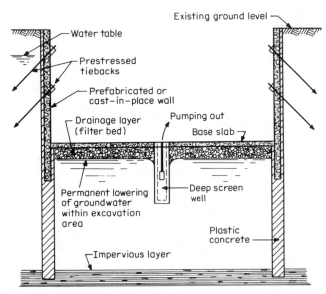

Fig. 8-32 Protection of excavation from groundwater and uplift by lowering the water table permanently within the excavation area.

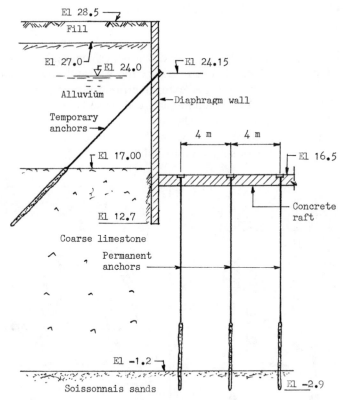

Fig. 8-33 Roche Building; typical cross section for the basement excavation. (From Fenoux, 1971.)

groundwater lowering within the protected area. If a natural impervious layer does not exist close to the base, such a layer can be created by grouting.

An example of anchored foundation slab is shown in Fig. 8-33 (Fenoux, 1971), subjected to a hydrostatic head of 8.4 m (almost 28 ft) for a corresponding uplift pressure of 1.4 kg/cm² (1700 lb/ft²). The permanent prestressed anchors have working loads 240 tons (540 kips), and a fourfold protection in the free length.

The effect of the prestress application and the resulting ground response are fully confirmed in practice. Prestress causes consolidation, leading to settlement with a corresponding loss of prestress equivalent to the reduction of elastic extension of the tendon. However, this process converges rapidly, and equilibrium between the two phenomena is soon reached. Since the elastic extension of the tendon generally is of an order of magnitude greater than that of settlement, the state of equilibrium corresponds to a small loss of prestress.

8-16   WATERFRONT INSTALLATIONS AND OFFSHORE STRUCTURES

Anchorages are used to tie back or tie down waterfront structures such as rivers and dock walls. Notable examples are the Alexandra Dock extension in Bombay, and more recently the extensive flood protection program on the River Thames in London. Anchorages are also used to provide resistance to uplift in the construction of cofferdams and dry docks. Likewise, solutions can be developed to accommodate basin-shaped structures such as service reservoirs and water tanks subjected to uplift.

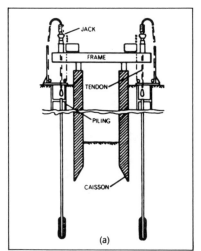

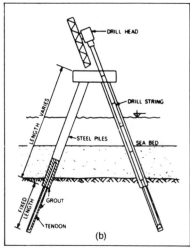

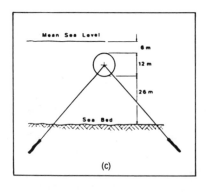

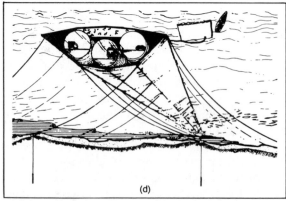

Fig. 8-34   (a) Offshore caisson sinking for highway bridge, Yokohama Bay, Japan; (b) mooring dolphin stabilized by anchorages; (c) anchorage design proposed for Bristol Cylinder Wave Energy Device; (d) anchorage for proposed submerged tunnel design, Messina Strait.

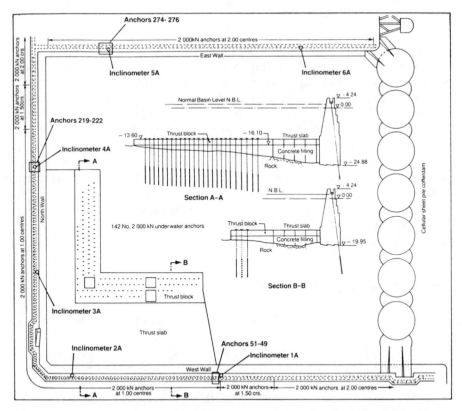

Fig. 8-35 Anchorage layout for the Devonport Submarine Refit Complex. (From Littlejohn and Bruce, 1979.)

Figure 8-34(a) shows an offshore caisson sinking with the use of hydraulic jacks. In this case, anchorages provide the jack reaction as the thrust load is applied. Fig. 8-34(b) shows schematically a mooring dolphin constructed to accommodate laden tankers of 150,000 tons or larger. These ships induce unusually large forces compounded by impact, and require relatively flexible moorings made of groups of vertical and raking piles as shown. Where it is not practical to increase pile embedment for shear resistance, the stability is improved by anchorages that tie down the structure to underlying rock.

In other offshore applications anchorages have evolved to innovative designs such as the stabilization of the wave energy system shown in Fig. 8-34(c). The sketch illustrates the prestressing of the NEL Oscillating Water Column Bottom Standing Device, which requires anchors that can accommodate a cyclic loading of 1600 tons in tension and 500 tons in compression with a period of 8 to 10 s while the cylinders (72 m long and 12 m dia.) oscillate at sea. Anchorage moorings have also been proposed for the submerged tunnel designed to link the Messina strait in the Mediterranean, a distance of 3 km, shown in Fig. 8-34(d).

A maritime project of unique design and size is the Devonport Submarine Complex in England, shown in Fig. 8-35, consisting of twin dry docks constructed in an existing basin 140 m square (460 ft) and surrounded on three sides by mass concrete retaining walls founded on bedrock. The construction required dredging and dewatering, and a south perimeter enclosure with cellular cofferdam. Existing basin walls were stabilized against overturning with vertical anchors, whereas part of the dock floor was tied down by prestressed anchors. The horizontal thrust slab shown in Fig. 8-35 is designed to give additional lateral support to the old basin walls at the northwest corner.

Approximately 330 anchors were used for the walls, fixed in underlying rock, and with inclination 7 to 15° with the vertical. This project is discussed also in Chapter 10 and 11.

REFERENCES AND ADDITIONAL BIBLIOGRAPHY

Abraham, K. H., 1973: "Construction Progress at Waldeck II Plant," *Water Power* (Dec.), 464–466.

Abraham, K. H., and A. Pahl, 1976: "Bauwerksbeobachtung der grossen Untertageraume des Pumpspeicherwerks Waldeck II," *Die Bautechnik* **5**, Seiten 145–155.

Agar, M., and F. Irwin-Childs, 1973: "Seaforth Docu, Liverpool; Planning and Design," Proc. Inst. Civ. Eng., Lond., 1, p. 54

Aitchison, G. D., 1972: "The Qualitative Definitions of the Physical Behaviour of Expansive Soils—An Engineering Viewpoint," *Symp. on Physical Aspects of Swelling of Clay Soil*, Univ. New England, Adelaide, N.S.W., pp. 63–70.

Bang, S., and C. Shen, 1982: "Soil Reinforcement in Soft Ground Tunneling," U.S. Dept. Transp., Office of Univ. Research, Washington, D.C.

Bassett, R. H., 1970: "Discussion," *Conf. on Ground Engineering*, Institution of Civ. Engineers, London, pp. 89–94.

Braun, W. M., 1972: "Post-tensioning Diaphragm Walls in Milan," *Ground Eng.* (March), London.

Broms, B. B., and H. Bjerke, 1973: "Extrusion of Soft Clay through a Retaining Wall," *Can. Geotech. J.*, **10** (1), 103–109.

Bureau Securitas, 1972: "Recommandations concernant la conception le calcul l'execution et le controle des tirants d'ancrage," Editions Eyrolles, Paris.

Buro, M., 1968: "Realisation de soutenement de la voute de la Centrale au moyen de tirants precontraints en rocher," *Annexe Bulletin Losinger*, No. 35.

Buro, M., 1970: "Prestressed Rock Anchors and Shotcrete for Large Underground Powerhouse," *Civil Eng.* (May).

Buro, M., 1972: "Rock Anchoring at Libby Dam," *West. Constr.* (March), 42, 48, 66.

Corbett, B. O., and M. A. Stroud, 1975: "Temporary Retaining Wall Constructed by Berlinoise System at Centre Beaubourg," Paris Institute of Civil Engineers, *Proc. Conf. on Diaphragm Walls and Anchorages*, London, Paper No. 13.

Cox, B. D., 1978: "Strengthening of Laing Dam," VSL Intra-Company Symposium.

Daly, W. F., and L. W. Abramson, 1985: "Mt. Lebanon Tunnel, NATM Comes to America, Tunneling Techn.," U.S. National Comm., March, pp. 1–11, Washington, D.C.

Donolo, L., 1971: "Mur de Soutenement Special," *L'Ingenieur*, Montreal (Nov.).

Druss, D. L., 1984: "Development and Applications of the New Austrian Tunneling Method," M.S. Thesis, Pittsburgh, Pa., Univ. of Pittsburgh, 113 pp.

Einstein, H. H., 1979: "Tunneling in Swelling Rock," *Underground Space*, **4** (1), 51–61.

Einstein, H. H., and N. Bischoff, 1977: "Design of Tunnels in Swelling Rock," *Proc. 16th Symp. Rock Mech.*, Univ. Minnesota, ASCE, pp. 185–195.

Federation of Piling Specialists, 1973: "Specification for Cast-in-Place Concrete Diaphragm Walling," *Ground Eng.*, **6** (4, July).

Fenner, R., 1938: "Untersuchungen zur Erkenntnis des Gebirgsdruckes," *Gluckauf, Ann.* **74** 32, 33.

Fenoux, Y., 1971: "Deep Excavations in Built-up Areas," Soletanche Enterprises, Paris.

Fenoux, G. Y., 1974: "La Paroi Prefabriquee, Ses Applications, The Precast Wall, Its Application," *Annales de L'Institut Tech. du Batiment et des Travaux Publics* (313, Jan.), pp. 198–219.

Fisher, F. A., 1974: "Diaphragm Wall Projects of Seaforth, Redcar, Bristol and Harrow," *Proc. Diaphragm Walls Anchorages*, Inst. Civ. Eng., London.

Goldberg, D. T., W. E. Jaworski, and M. D. Gordon, 1976: "Lateral Support Systems and Underpinning," Fed. Hwy. Admin., U.S. Dept. Transp., Washington, D.C.

Golser, J., 1981: "The New Austrian Tunneling Method," the Atlanta Research Chamber, U.S. Dept. Transp., UMTA, Washington, D.C.

Golser, J., E. Hackl, and J. Jostl, 1977: "Tunneling in Soft Ground with NATM," *Int. Constr.* (Dec.)

Henke, K. F., 1974: "Stabilization of Landslides in Weathered Clay-Shale Using Pretensioned Grouted Anchors," *Proc. 2nd Cong. Int. Assoc. Eng. Geology*, Sao Paulo, Vol. 2, Theme V-9, 8 pp.

Hodgson, T., 1974: "Design and Construction of a Diaphragm Wall on Victoria Street, London," *Proc. Diaphragm Walls Anchorages*, Inst. Civ. Eng., London.

Hunt, H. W., 1974: "Design and Installation of Pile Foundations," Associated Pile & Fitting Corp., Clifton, N.J.

Irshad, M., and L. H. Heflin, 1987: "WMATA'S Soft-Ground NATM Tunnel Designs, Tunneling Technology," U.S. National Comm., Sept., pp. 1–9.

Jobling, D. J., 1975: "Diaphragm Walls and Secant Piles in Subway Construction," U.S. Dept. Transp., Urban Mass Transp. Adm. Proc. Semin. Underground Constr. Probl., Techniques Solutions, Chicago.

Khaoua, M., B. Montel, A. Civard, and R. Lange, 1969: "Cheurfas Dam Anchorages. 30 years of Controls and Recent Reinforcement,'' 7th Int. Conf. Soil Mech. Found. Eng., Specialty Sessions 14, 15, Mexico City.

Korbin, G. E., and T. L. Brekke, 1975: "A Model Study of Spiling Reinforcement in Underground Openings," Tech. Report MRD-2-75, Missouri River Div., Corps of Engineers, April.

Korbin, G. E., and T. L. Brekke, 1978: *Field Study of Tunnel Prereinforcement, ASCE*, Vol. 104.

Leonard, M. S. M., 1974: "Precast Diaphragm Walls Used for the A13 Motorway, Paris," *Proc. Diaphragm Walls Anchorages*, Inst. Civ. Eng., London.

Littlejohn, G. S., 1982: "The Practical Applications of Ground Anchorages," *Proc. 9th FIP Congress*, June, Stockholm.

Littlejohn, G. S., and D. A. Bruce, 1979: "Long Term Performance of High Capacity Rock Anchors at Devonport," *Ground Eng.* (Oct.), London.

Lombardi, G., 1968: "The Influence of Rock Characteristics on the Stability of Rock Cavities," Soil Mech. Convention, Lugano.

Lombardi, G., 1969: "Der Einfluss der Felseigenschaften auf die Stabilitat von Hohl-raumen," *Schweizerische Bauzeitung*, 3 (Jan.).

McKay, A., 1970: "Ground Anchors at Orfordness," *Civ. Eng.* (Feb.), 3–6.

Mohamed, K., B. Montel, A. Civard, and R. Luga, 1969: "Cheurfas Dam Anchor-ages: 30 Years of Control and Recent Reinforcement," *Proc. 7th Intern. Conf. on Soil Mech. and Found. Eng.*, Mexico City, Specialty Session No. 15, pp. 167–171.

Newmark, N. M., 1942: "Influence Charts for Computation of Stresses in Elastic Foundation," *Univ. Illinois Bull.* **40** (12).

Ostermayer, H., 1970: "Erdanker-Tragverhalten und konstruktive Durchbildung," *Vortrage der Baugrundtagung in Dusseldorf*, Essen.

Ostermayer, H., 1975: "Construction Carrying Behavior and Creep Characteristics of Ground Anchors," Inst. of Civ. Eng., Conf. on Diaphragm Walls and An-chorages, Paper No. 18, London.

Paul, S. L., A. J. Hendron, E. J. Cording, G. E., Sgouros, P. K. Saha, 1983: *Design Recommendations for Concrete Tunnel Linings* (Report No. UMTA-MA-06-0100-83-1, U.S. Urban Mass Transp. Admin.). National Technical Information Service, Springfield, Va., pp. 436–440.

Peck, R. B., 1969: "Deep Excavations and Tunneling in Soft Ground, State of the Art Report," 7th ICSMFE, Mexico City, *State of the Art Volume*, pp. 225–290.

Peck, R. B., 1970: "Observation and Instrumentation, Some Elementary Consid-erations," Lecture Notes, Met. Section ASCE, Seminar on Field Observations in Found. Design and Construction, New York, April [reprinted in *Hwy. Focus, U.S. Dept. Transp., F.H.W.A.*, **4** (2), 1–5 (June 1972)].

Port Authority of Allegheny County, 1983: "Contract Forms and Specification, Mt. Lebanon Tunnel," Contract No. CA260, Stage 1, Light Rail Transit System.

Prestressed Concrete Institute (PCI), 1974: "Tentative Recommendations for Pre-stressed Rock and Soil Anchors," Chicago.

Rabcewicz, L. V., 1969: "Stability of Tunnels under Rock Load," *Water Power* (June, July, Aug.)

Rabcewicz, L. V., and J. Golser, 1973: "Principles of Dimensioning the Supporting System for the New Austrian Tunneling Method," *Water Power* (March), 88.

Richards, B. G., 1977: "Pressure on a Retaining Wall by an Expansive Clay," *Proc. 9th Int. Conf. on Soil Mech. and Found. Eng.*, Tokyo, Paper 2/72, pp. 705–710.

Sandhu, B. S., 1974: "Earth Pressure on Walls Due to Surcharge," *Civ. Eng., ASCE*, **44** (12 Dec.), 68–70.

Schultze, E., and A. Horn, 1966: "Der Zugwiderstand von Hangebrucken-Widerlagern," *Vortrage der Baugrundtagung in Munchen*, Essen, pp. 125–186.

Soletanche, 1970: "Strengthening of Subway in Paris, Job Report," Paris.

Sommer, P., and F. Graber, 1978: "Felsanker zur Sicherung des Tosbeckens Nr. 3 in Tarbela Dam (Pakistan)," *Schweizerische Bauzeitung, Vorabdruck*.

Soos, P. von, 1972: "Anchors for Carrying Heavy Tensile Loads into the Soil," *Proc. 5th Eur. Conf. on Soil Mech. and Found. Eng.*, **1**, 555–563, Madrid.

Spangler, M. G., 1940: "Horizontal Pressures on Retaining Walls Due to Concentrated Surface Loads," Bulletin No. 140, Iowa Eng. Exp. Sta.

Stapeldon, D. H., 1970: "Changes and Structural Defects Developed in Some South Australian Clays and their Engineering Consequences," *Proc. Symp. of Soils and Earth Structures in Arid Climates*, Adelaide, pp. 62–71.

Terzaghi, K., 1945: "Stability and Stiffness of Cellular Cofferdams," *Trans. ASCE*, **110**, 1083–1202.

Terzaghi, K., 1954a: *Theoretical Soil Mechanics*, Wiley, New York (7th printing), 510 pages.

Terzaghi, K., 1954b: "Anchored Bulkheads," *Trans. ASCE*, **119**, Paper 2720, pp. 1243–1324.

Terzaghi, K., 1955: "Evaluation of Coefficient of Subgrade Reaction," *Geotechnique*, **4**, 279, London.

Terzaghi, K., and R. B. Peck, 1968: *Soil Mechanics in Engineering Practice*, Wiley, New York.

Thompson, C. J., 1978: "Laing Dam, East London," *J. Concrete Soc. South. Africa* (9 March), 20–22.

University of Osaka, 1971: "Tests on the Efficiency of Construction Joints in Diaphragm Walls," Spec Bull.

VSL–Losinger, 1970: "Unique Post-Tensioned Cables Anchor Muda Dam to Foundation," *Engineering News Record*, Aug. 6.

VSL–Losinger, 1971: "Record Tendons Anchor Old Dam to Foundation," *Engineering News Record*, Feb. 11.

VSL–Losinger, 1976: "Post-Tensioned Cables Reinforce Dam Abutment," *Calif. Builder & Engineer*, April 23rd.

VSL–Losinger, 1977: "The Prestressing Anchors of the Waldeck II Pumped Storage Scheme," Bern.

VSL–Losinger, 1978: "Soil and Rock Anchors. Examples from Practice," Bern.

Ward, W., 1978: "Ground Support for Tunnels in Weak Rock," *Geotechnique*, 28 (2).

White, R. E., 1974: "Anchored Walls Adjacent to Vertical Rock Cuts," *Proc. Diaphragm Walls and Anchorages Conf.*, Inst. of Civ. Eng., London, pp. 181–188.

Wickham, G. E., H. R. Tiedemann, and E. H. Skinner, 1972: "Support Determination Based on Geologic Predictions," *Proc. Rapid Excavation and Tunneling Conf.*, Vol. 1, pp. 43–64, Society of Mining Eng., AIME, Littleton, Col.

Wosser, T. D., and R. D. Darragh, 1970: "Tiebacks for Bank of America Bldg. Excavation Wall," *Civ. Eng.*, **40** (3), 65–67.

Xanthakos, P. P., 1974a: "Underground Construction in Fluid Trenches," Colleges of Engineering, Univ. Illinois, Chicago.

Xanthakos, P. P., 1974b: "Diaphragm Wall Construction and Slurry Trench Applications in U.S.A.," *Ground Eng.* **7** (5): 31–33.

Xanthakos, P. P., 1979: *Slurry Walls*, McGraw-Hill, New York.

CHAPTER 9

DESIGN PRINCIPLES OF ANCHORED STRUCTURES

9-1 DAM STABILIZATION BY PRESTRESSING

A dam structure, together with its foundation and the medium supporting it, whether soil or rock, is treated as a continuous system. Functionally, its performance is ascertained by the ability to transfer stresses through the system without restraint. In this respect the interface between foundation and supporting ground is of particular interest since failure to develop net tensile forces across this joint can lead to a partially open base on the upstream side. Alternatively, lack of sufficient weight may compromise stability against sliding under extreme loading conditions. The useful contribution of prestressed anchors to dam stability was discussed in Sections 8-5 and 8-7 for rehabilitation work as well as in new construction. In particular, the Muda dam shown in Fig. 8-9 was prominently mentioned because of its unique anchorage design incorporated in a shell with interior buttresses.

Method of Analysis

Essentially, the dam types to which the technique of prestressing is applied form gravity structures, either solid or of hollow configuration, which depend primarily on their own weight for stability. The analysis is usually carried out by the gravity method, but this is entirely applicable where adjoining blocks are not made monolithic by keying and grouting the joints between them. Thus the following assumptions are made: (a) each block acts independently, (b) the load is transmitted to the foundation by cantilever action and is resisted by the weight of the cantilever, and (c) no load is transferred to the abutment by beam action. For simplicity, a gravity section is assumed to have vertical or nearly vertical upstream face and a constant downstream slope. A typical layout for a gravity dam is shown in Fig. 9-1.

Design Criteria and Loads. Frequently, the question arises as to whether dams should be designed for loading combinations whose simultaneous occurrence is highly improbable, since this may result in overly conservative and uneconomical structures. A sound criterion is therefore to avoid unintentional allowances for safety beyond specified safety factors. In general, loads commonly considered are (a) dead weight, (b) temperature stresses, (c) internal hydrostatic pressure, (d) silt and ice loads, (e) earthquakes, and (f) forces from the prestressed anchorages.

Volumetric increases due to temperature rise are considered unrestrained if the contraction joints between blocks are left ungrouted. With grouted joints horizontal thrusts caused by volumetric increases due to rising temperatures will cause load transfer across the joints with twisting effects gradually moving toward the abutments, and this will result in redistribution of anchor loads. Internal hydrostatic pressure has two main components: the

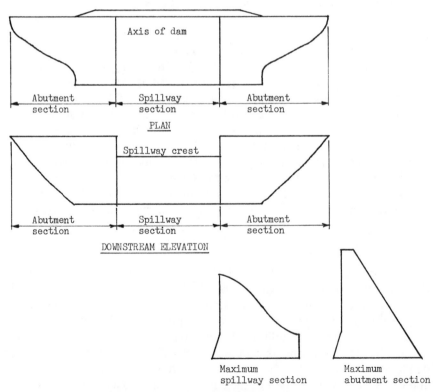

Fig. 9-1 Typical layout for a gravity dam.

lateral thrust acting behind the dam due to the water level in the reservoir and the uplift along the base of the structure, which is assumed to have a distribution varying linearly from full reservoir pressure at the upstream face to zero or tailwater pressure at the downstream face. Silt and ice loads are considered where conditions warrant and estimated according to local codes.

Since gravity dams are elastic structures they may be excited to resonance by seismic ground accelerations. The zone of ground undergoing vibrations is influenced by the relevant ground properties and mainly the ability to transmit seismic or vibratory waves. The seismic loads acting on a dam are therefore determined from the input ground motion at the site and the dynamic response of the dam to vibratory motions, discussed in the following sections. The last category of loads is the vertical force induced by the prestressing of the anchors, usually located close to the upstream face, and has a magnitude determined from stability requirements.

A detailed analysis and design is beyond the scope of this text. A basic review is, however, necessary in conjunction with the estimation and optimum utilization of the prestress force. Load combinations are thus categorized as usual, unusual and extreme (Golze, 1977). For example, normal design reservoir water level plus the usual temperature combined with appropriate dead loads, ice, and silt should be considered usual load combinations. However, an unusual load combination will include the foregoing loads but with the maximum reservoir water level. The usual load combinations acting together with a maximum credible earthquake should be considered extreme load combinations.

The two stability cases involving the effect of prestress are (a) overturning of the dam about its toe and (b) sliding along the concrete–rock interface or at any weak plane within the foundation zone. Sliding stability can be determined by applying a shear-friction factor. For the three main load categories the recommended factor of safety should be 3.0, 2.0, and 1.3 for the usual, unusual, and extreme groups, respectively. Whether a tensile strength of lift surfaces should be included in the analysis is questionable and academic in most instances, since the amount of prestress necessary for stability against sliding and overturning normally would be sufficient to keep the entire contact area in compression.

The design criteria are based on the following assumptions: (a) the concrete in the dam is homogeneous, isotropic, and uniform elastic material; (b) there are no differential movements occurring at the dam site; (c) all loads are resisted by gravity action of vertical, parallel-side cantilevers that receive no other support from adjacent elements on either side; (d) unit vertical pressures, or normal stresses on horizontal planes, vary linearly from the upstream to the downstream face; and (e) horizontal shear stresses vary as a parabolic function. These assumptions are substantially correct, except for horizontal planes near the base of the dam, where foundation yielding affects stress distribution. Stress changes that may occur as a result of foundation yielding are usually small for low- and medium-height dams, but large and critical for high dams.

Earthquake Hazard. During an earthquake a dam becomes a vibrating structure excited into motion by the ground shaking. The procedure usually employed for earthquake design is to include a constant 5 percent *g* or 10 percent *g* acceleration force, and then proceed with the static analysis. However, this procedure is difficult to correlate with the actual earthquake forces that a dam may experience. The difficulty of expressing earthquake effects by an equivalent static horizontal and uplift force means that it is not possible to know the true factor of safety under the equivalent static method; hence other procedures are suggested based on dynamic analysis. However, this approach also has inherent difficulties since all the physical parameters are not accurately known, but unlike the static method, a parametric solution can be obtained until the results approach recorded earthquake motions.

When an earthquake occurs at a dam site, the additional uplift effects are manifested as vibrations of short frequency. The ground shaking may also subject the anchorages to a special motion from which considerable loss of prestress may result although the system has not necessarily suffered structural damage. Thus when establishing earthquake design criteria it is important to consider both allowable design stresses and prestress load in conjunction with the design earthquake.

Design Procedure

The forces acting on a cantilever element (single block) are shown in Fig. 9-2, including uplift at the base and the artificially induced prestress force. For simplicity, the upstream face is assumed vertical.

The notation used is as follows:

O = origin of coordinates at downstream edge of section

δ = angle between face of element and the vertical

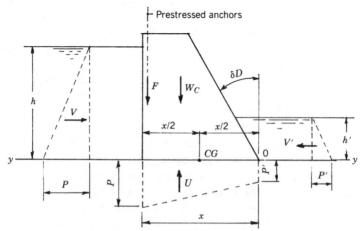

Fig. 9-2 Section through a gravity dam showing loads acting on structure.

x = horizontal distance (width of dam) at the base

CG = center of gravity of base (midpoint)

w_c = unit weight of concrete

w = unit weight of water

h or h' = reservoir or tailwater height

W_c = dead load weight above base

V = horizontal water pressure from reservoir = $wh^2/2$

V' = horizontal tailwater pressure = $wh'^2/2$

U = total uplift force on horizontal base

F = prestressed force from anchorage (assumed to act at the up-stream face)

In addition to the loads shown in Fig. 9-2, the stability analysis should consider silt and ice loads where applicable. Temperature effects are determined from the temperature distributions expected to exist within the dam during the construction operations and later through the operating life of the project. Pertinent data include weather and climatic conditions, projected reservoir water temperature, thermal properties of concrete, and the effects of solar radiation.

Where hydrodynamic effects of earthquake shock must be considered, the first step is to determine the dynamic response of the anchored structure, assuming that the system acts elastically during an earthquake. The method of analysis utilizes lumped masses, generalized coordinates, and mode superposition. For a more detailed discussion of dynamic analysis reference is made to Golze (1977), Clough (1970), Clough and Bureau of Reclamation (1963), and Myklestad (1944). In general, there should be a horizontal and a vertical earthquake component, designated as P_E and V_E, respectively. Each block in the dam system can be assumed to have certain motions inhibited, in which case the degrees of freedom are two and possibly three, the block moving vertically, laterally, and in the direction of the dam.

Because of the uncertainties associated with earthquake vibrations, the corresponding effects can be understood and better estimated by an empirical record, but currently this record is very limited. Published material includes the analysis of seismic damage of anchored bulkheads reported by Kitajima and Uwabe (1979), showing that most anchor failures occurred as a result of overloading.

For a dynamic response analysis of an anchored dam, the following guidelines are useful:

1. Although little data are available on the damping in concrete gravity dams, a reasonable assumption is to take this factor as 0.05 (Chopra and Chakrabarti, 1970).
2. In many instances anchor overloading may be caused by the opening of joints in the rock mass, causing the tendon to stretch, especially

along the free length. In this context, relatively long anchors should be preferred to short ones since for the same stretch the corresponding load change will be smaller.

3. If the ground motion is toward the structure, the result may be elastic shortening in the tendon with a corresponding loss of prestress.

4. During earthquake dam stability will also depend on the shear force manifested along its base. This force should be determined for a model with 3 degrees of freedom. The analysis may attain a higher degree of accuracy if the profile of the dam is divided in the vertical direction.

5. If the design appears to be controlled by the vertical component of the seismic load (overturning effects), the profile of the dam should be divided in the horizontal direction since the frequencies of vertical vibrations are now more important.

6. The seismic waves traveling under the dam and across its base generate the vertical force which will tend to lift the structure as does the hydrostatic uplift.

7. An analysis can be carried out for the vertical and horizontal vibrations connected and their frequencies estimated jointly, but for the initial approximation the vertical frequencies may be estimated independently (Sinitsyn and Medvedev, 1972).

8. The amplitude of the vertical vibration component of the dam is much smaller than the vertical ground vibration. The amplitude of the vertical ground vibration should be expected to decrease with depth so that the fixed anchor may be located within a zone reasonably unaffected by the seismic event.

Stability against Overturning. For the loads and forces shown in Fig. 9-2 stability against overturning is maintained if the resisting moments about point 0 exceed the overturning moments by the designated factor of safety. This is expressed as

$$\frac{MW_c + MF + MV'}{MV + MU} \geq F_s \tag{9-1}$$

where MW_c denotes the moment of force W_c about point 0 and F_s is the appropriate factor of safety for the particular category of loads. Where other loads and forces act, such as P_E and V_E, the corresponding overturning moments are added accordingly under the modified factor of safety.

Stability against Sliding. The shear-friction factor is expressed as

$$SF = \frac{Ac + (W_T - U_T)f}{V_T} \tag{9-2}$$

where SF = shear-friction factor
       A = contact area at the base of the block
       c = unit shear resistance at the base
       W_T = total vertical load (summation of all vertical forces)
       U_T = total uplift
       f = coefficient of internal friction at the base
       V_T = total horizontal force causing sliding

With reference to both overturning and sliding, all possible conditions of loading should be investigated. Interestingly, a large margin of safety against sliding is indicated by high shear-friction factors; hence the allowable minimum values of this factor for use in design should be not less than the values specified in the foregoing section.

Ordinarily, the analysis should include estimation of normal stresses on horizontal planes, shear stresses on horizontal and vertical planes, normal stresses on vertical planes, and direction and magnitude of principal stresses for any point within the boundaries of the cantilever element including stress concentration at the origin of the prestress application.

It should be noted that before bodily overturning of a gravity dam can take place, other types of failure may take precedence such as crushing of the toe material and cracking of the upstream material resulting in increased uplift pressure and decreased shear resistance. Earthquake forces may not always be considered as contributing to the overturning tendency if it can be established that they only have oscillatory nature.

Downstream Stabilization

Where the rock formation is nearly horizontal, an alternate design is to provide a rock embedment as shown in Fig. 9-3 and increase resistance to sliding by utilizing the passive resistance against the downstream face mobilized along the embedment depth h_e. If the location, orientation and nature of the joints and bedding planes in the rock mass have a favorable configu-

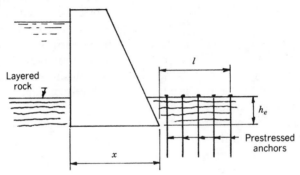

Fig. 9-3 Downstream stabilization of a gravity dam to increase resistance to sliding.

ration, the downstream section can be strengthened by prestressed anchors as shown. The tying together of individual rock layers will prevent laminar failure and individual buckling. The prestressing action will produce a solid block of rock mass acting as a unit and responding to the tendency of the dam to slide by developing passive resistance along the contact with the downstream face of the structure.

Under these conditions resistance to sliding is provided by two elements, the shear-friction factor at the base and the direct bearing at the downstream dam–rock interface. However, the interaction of these two factors and their simultaneous consideration under a single factor of safety will depend on a compatibility of displacements. As a first approximation the passive resistance developed as direct bearing at the downstream face of the dam can be estimated from the following (Underwood and Dixon, 1976).

$$P_b = \frac{\pi^2 EA}{[l/(h_e/2)]^2} \tag{9-3}$$

where P_b = passive resistance
E = modulus of deformation of the rock
A = contact bearing area downstream

and l and h_e are as shown in Fig. 9-3. For an exact estimation, the analysis should consider also the crushing strength of rock together with anisotropy, inhomogeneity, and rock mass discontinuities.

9-2   ANCHORAGE REINFORCEMENT OF SOIL IN SOFT GROUND TUNNELING

The basic principle of this application was discussed in Section 8-11, and an example is shown in Fig. 8-24. In general, it involves the stabilization of a weak soil mass by installing reinforcing elements as excavation proceeds. A reinforced zone adjacent to the opening is thus formed and minimizes the instantaneous instability as well as the permanent deformation.

The factors influencing the effectiveness of the spiling technique have been investigated by Bang and Shen (1982) in a parametric study followed by a centrifuge model testing. In this program, the incremented excavation and the soil-reinforcement composite behavior were closely incorporated by developing constitutive models for the spiling reinforcement, the shotcrete, the soil, and their geometric arrangement.

The spiling reinforcement system is modeled according to the unit cell approach. In principle, the system is considered an isolated small unit completely exhibiting the composite characteristics. The arrangement is shown schematically in Fig. 9-4.

The shotcrete lining reinforced with steel bars is modeled in the analysis

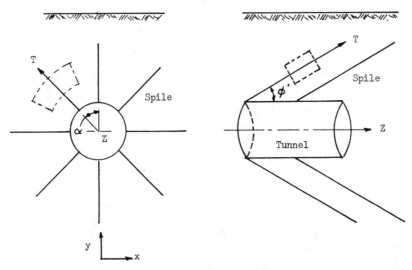

Fig. 9-4 Unit cell for spiling reinforcement system, schematic presentation; anchorages in soft ground tunneling. (From Bang and Shen, 1982.)

of membrane elements, consistent with the fact that this lining is relatively thin and flexible. Its effect in supporting the excavation can be directly superimposed to the element stiffness matrix of the composite spiling element adjacent to the shotcrete membrane.

For this study field conditions are simulated for in situ stresses, the installation of the support, and the sequence of tunnel excavation. A nonlinear inelastic model is used to describe the constitutive behavior of the soil provided it does not experience stress or strain-induced anisotropy.

Comparison between Reinforced and Unreinforced Tunnel. The spiling anchorage system in soft ground is better understood if a comparison is made between the reinforced and the unreinforced tunnel. In this case the comparison is limited to a single horizontal circular excavation, and involves two soil types: silty clay and low-plasticity clay. The geometric parameters of the tunnel are as follows:

Depth to tunnel center $H = 50$ ft
Diameter of tunnel $D = 20$ ft
Length of spiles $L = 20$ ft
Spacing of spiles $S = 5$ ft
Diameter of spiles (grouted) $d = 4$ in

Interestingly, these parameters were chosen to reflect actual case histories. For example, the depth to the tunnel center in most shallow tunnel projects is approximately 50 ft, and this includes the Washington D.C. (1975), Frankfurt (1975), and Brussels metros (1970). The tunnel diameter commonly ranges from 16 to 20 ft. Furthermore, the zone most susceptible to deterioration is within one-half radius from the tunnel opening, where approximately 60 percent of the total deformation occurs (Korbin and Brekke, 1975). Beyond the zone defined by one radius from the opening, the strains are relatively low whereas the confinement is high so that an increase of strength related to the presence of anchorage reinforcement is small. In this respect, there are no benefits in extending the spiling reinforcement beyond this zone.

Table 9-1 shows structural parameters for the support system. The spile inclination angle is 30°, giving an equivalent earth-reinforced zone one radius from the opening. Tunnel deformation for the reinforced and the unreinforced opening is shown in Fig. 9-5 for silty clay with cohesion 2.8 psi and friction angle 33°. In order to highlight the effectiveness of spiling anchorages, the percentage reduction in total displacement is calculated for the crown and the bottom of the tunnel and is shown in Table 9-2 for silty clay and low-plasticity clay, respectively. In this case

$$\% \text{ reduction} = \frac{\delta_{\text{unreinf.}} - \delta_{\text{reinf.}}}{\delta_{\text{unreinf.}}} \times 100$$

where δ denotes displacement. From these data it follows that the spiling anchorages reduce the crown settlement and bottom heave by approximately 90 and 50 percent, respectively.

Both the vertical and the horizontal surface displacement is also markedly reduced where spiling reinforcement is present, as shown by the graphs of Fig. 9-6(a) and (b) and in Table 9-3.

From Table 9-3 it is evident that the magnitude of ground surface displacement both vertically and horizontally for the reinforced system is only one-tenth the corresponding displacement of the unreinforced tunnel. Interestingly, ground surface displacement can be interpreted in terms of vertical and horizontal distortion, defined as the differential movement between two points divided by the actual distance between them. Surface distortions have always been the index for assessing structural damage of buildings and

TABLE 9-1 Additional Parameters, Tunnel of Fig. 9-4

Thickness of lining	4 in
Spacing of bars in the lining	4 in
Diameter of bars in the lining	$\frac{1}{6}$ in
Modulus of lining material (shotcrete)	2,000 ksi
Modulus of bars in the lining	30,000 ksi
Composite modulus of spiling anchorage	1,875 ksi
Composite yield stress of the spiles	3,125 ksi

From Bang and Shen (1982).

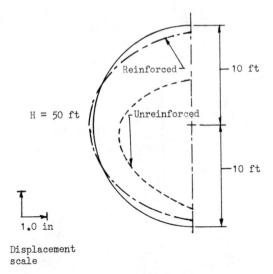

Fig. 9-5 Tunnel deformation of reinforced and unreinforced excavation. (From Bang and Shen, 1982.)

facilities near an excavation (Grant et al., 1972; O'Rourke et al., 1976; Polshin and Toker, 1957; Skempton and MacDonald, 1956).

The vertical distortion of the ground surface is shown in Fig. 9-7 for the low-plasticity clay as function of the distance from the tunnel center line. The maximum distortion, occurring at some distance from the center line, is again markedly reduced for the reinforced tunnel.

The effect of spile inclination is negligible and diminishes as the soil becomes stronger. In this case spile inclinations of 20°, 30°, and 40° with the tunnel axis produced virtually no difference. However, for the weaker soil,

TABLE 9-2   Comparison of Tunnel Deformation; Reinforced and Unreinforced Opening

	Unreinforced Tunnel (in)	Reinforced Tunnel (in)	Reduction (%)
Silty Clay			
Crown settlement	−4.44	−0.57	87
Bottom heave	1.40	0.72	48
Low-Plasticity Clay			
Crown settlement	−8.85	−0.79	91
Bottom heave	2.53	1.25	51

From Bang and Shen (1982).

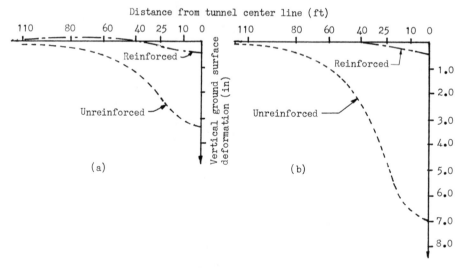

Fig. 9-6 Vertical ground surface deformation for reinforced and unreinforced tunnel; (a) tunnel in silty clay; (b) tunnel in low-plasticity clay. (From Bang and Shen, 1982.)

the angle of 30° produces least tunnel deformation, ground surface displacement, and overall distortion.

The effect of the depth of the tunnel was investigated in the range 1.5–3 times the tunnel diameter, which is the range within which full arching effects should not be manifested, and for spile spacing of 4 ft. The results show that the deformation increases with increasing tunnel depth, but the rate of deformation increase decreases with depth because of increasing arching action.

The most important single factor determining the effectiveness of soil

TABLE 9-3 Comparison of Maximum Displacement at Ground Surface, Reinforced and Unreinforced Tunnel

	Unreinforced Tunnel (in)	Reinforced Tunnel (in)	Reduction (%)
Tunnel in Silty Clay			
Maximum vertical displacement	−3.25	−0.32	90
Maximum horizontal displacement	0.89	0.09	90
Tunnel in Low-Plasticity Clay			
Maximum vertical displacement	−6.92	−0.46	93
Maximum horizontal displacement	1.78	0.12	93

From Bang and Shen (1982).

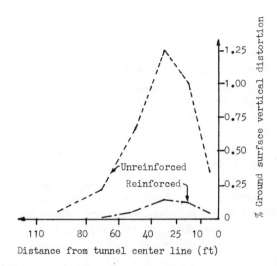

Fig. 9-7 Vertical ground surface distortion, reinforced and unreinforced tunnel. (From Bang and Shen, 1982.)

reinforcement is probably the spile spacing, since it measures directly the level of strengthening. In general, this spacing has a marked influence on the deformation of the system, including the crown settlement, the bottom heave, the lateral movement of the tunnel, and the ground surface movement. However, this is much greater in weaker soil and tends to diminish in stronger soil, although it is still significant when compared with the behavior of the unreinforced system.

9-3 SOIL PRECONSOLIDATION BY PRESTRESSING

General Principles

Preloading as a technique to improve soils for the support of structures and loads should be considered in conjunction with other feasible solutions, and compared for cost, benefits, and construction time requirements. Proper attention must be focused on potential consequences and unexpected behavior with regard to amount and rate of settlement both during and after the precompression period. The main risks that must be recognized in designing preconsolidation by preloading are (a) the predicted preload duration may be too short to achieve the desired precompression or pore pressure dissipation; (b) shear failure may occur in applying the preload; (c) after the preload is removed and the structure is completed, shear failure may not be prevented; and (d) postconstruction settlement may be larger than predicted.

Prestressing can be selected as a form of precompression where subsoils are weak and highly compressible. Soil types considered suitable for the technique include organic and inorganic silts and clays, peaty soils, and miscellaneous fills. Materials that have been successfully precompressed have water content of 20–100 percent, and include both sensitive and insensitive soils. However, the nature of these soils raises questions about the expected foundation stability, and suggests the importance of predicting the magnitude and time rate of consolidation.

Usually, the preload is intended to increase the shear strength in order to obtain a higher factor of safety against shear failure of the completed structure. Occasionally, however, a shear strength increase may be required for the stability of the preload. A further aim of preloading is to limit post-construction settlement to tolerable amounts, although settlement can occur without consolidation due to plastic deformations. Precluding this settlement requires extra care particularly for soils where the measured shear strength is strain-rate-dependent. Interestingly, the lack of an accepted design procedure for preloading soils that exhibit secondary compression is the principal deficiency in the current state of the art for preloading. The design for secondary compression must thus be empirical by necessity, and suggests the value of continuous postconstruction observations.

The economic advantages of precompression usually depend on the direct availability of low-cost and effective methods for applying preloads. In this context, the use of anchors to apply prestress must be compared with the cost and effectiveness of (a) earth fills, (b) water loading in tanks, (c) groundwater lowering, and (d) vacuum mats (Kjellman, 1952). Earth fills and water tanks are common methods, whereas groundwater lowering is indicated in sand with high groundwater table. Prestressing, however, may be the preferred and most practical form of precompression in special instances—for example, where subsoil conditions are poor whereas loads are light to moderate and relatively uniform, or where the overall function of the project is the controlling element and relegates construction economy to a secondary consideration.

A disadvantage of precompression is the relatively long time necessary to achieve the desired results, which usually ranges from months to more than a year. Considerable time and cost must also be allocated to subsoil investigations and testing. Professional attitude is also a factor in choosing prestressing for soil preconsolidation in lieu of a different foundation scheme.

Theoretical Considerations

In this discussion precompression or preconsolidation means the compressing of a soil under an externally applied pressure prior to placing or completing the structure load, final static or moving. Prestressing is the application of precompression using rock or soil anchors as discussed in Sections 1-5 and 8-8, prestressed after installation and restressed as needed. Where the term "primary consolidation" is mentioned, it refers to the Terzaghi concept. Secondary compression is volume change, which continues after completion of primary consolidation, and is characterized by a straight-line relation between volume change and logarithm of time. It may represent a plastic readjustment of stress between soil grains, and appears to be related to the presence of shear stresses (Johnson, 1968).

Unlike the more conventional means of precompression where the applied load is maintained constant and unchanged, prestressing is the application of force or unit pressure which is likely to be subjected to variations and changes with time (normally decrease) in response to the associated ground behavior. Where the prestressing is applied to consolidate weak compressible soils, the resulting settlement may be exceptionally large (of the order of several feet), thus exceeding many times the initial elastic extension of the cables. Hence, in most instances the prestress loss should be expected to be significant and even complete so that several restressing operations will be required until the process finally converges and equilibrium is reached between the two phenomena. Since no advance theoretical concept is available to describe the system behavior under a varying residual prestress force caused by restressing and prestress loss cycles, it follows that it is more practical to use restressable anchors, monitor prestress loss, and maintain the prestress to a nearly constant level. Under these conditions predictive techniques can be based on the same principles as the process of preloading.

Effect of Internal Drainage Layers. The presence of a thin sand or silt layer within the compressible zone becomes significant where a small structure is proposed or if a relatively narrow area is to be prestressed, as shown in Fig. 9-8(a), since this will permit rapid consolidation. On the other hand, if the prestressing area covers a long and wide site as shown in Fig. 9-8(b), the presence of sand or silt layer may have minor effect on accelerating the time rate of consolidation beneath the central section of the prestress since in this location consolidation may occur almost exclusively as a result of vertical drainage. Accordingly, sand drains or other means may be specified beneath the central part of a large prestressed area.

Precompression Design

Primary Consolidation. Since the initial presentation of consolidation theory for vertical settlement and vertical drainage proposed by Terzaghi, the

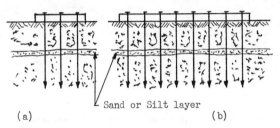

Sand or Silt layer
(a)   (b)

Fig. 9-8 Effect of structure size relative to thickness of compressible layer within precompression zone; (a) small or narrow structure; (b) long or wide structure.

fundamental concepts have been expanded to include two- and three-dimensional behavior, pore pressure coefficients (Skempton and Bjerrum, 1957), and stress path methods (Lambe, 1964). Pertinent work on the consolidation process is reported by Gray (1945), Harr (1966), Schiffman (1958), Abbott (1960), Bjerrum (1967), Davis and Raymond (1965), Barden (1965), and deJong (1968). Procedures for achieving precompression where sand drains are used to accelerate consolidation are discussed by Johnson (1970). A case expected to arise frequently is precompression by prestressing without sand drains intended to eliminate primary consolidation under permanent loading. This case is reviewed in this section.

Figure 9-9(a) shows a temporary prestress of unit intensity P_s applied in excess of the permanent load of unit intensity P_f, causing more settlement to occur in addition to that caused only by the permanent load. When sufficient settlement has occurred, the surcharge prestress is removed, and evidently this can be done when the compressible stratum is only partially consolidated under the effect of the permanent plus surcharge loading. Johnson (1968) suggests using the degree of consolidation at the center of the compressible stratum as the criterion for removing the surcharge load, although this is somewhat conservative. Under this condition part (b) of Fig. 9-9 refers to settlement of the 1-ft-thick layer at the center of the stratum.

Furthermore, the excess pore water pressure and degree of consolidation at the time the prestress load is removed likewise apply to the center of the compressible stratum. Based on these concepts and referring to Fig. 9-9, the required degree of consolidation U_{f+s} at the center of the layer when the surcharge (prestress) load is removed is

$$U_{f+s} \, \Delta H_{f+s} \; = \; \Delta H_f \tag{9-4}$$

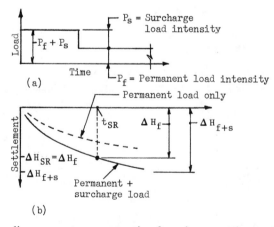

(a)

(b)

Fig. 9-9   Preloading concept; compensation for primary settlement by applying temporary surcharge load, prestressed intensity P_s

or

$$U_{f+s} = \frac{\Delta H_f}{\Delta H_{f+s}} \tag{9-4a}$$

where ΔH_f and ΔH_{f+s} refer to settlement of the 1-ft-thick layer at the center of the compressible stratum. For normally consolidated soils ΔH_f and ΔH_{f+s} can be expressed in terms of relevant parameters, namely, the layer thickness H, initial void ratio e_0, compression index C_c, initial in situ overburden stress P_0, permanent load intensity P_f, and surcharge (prestress) load intensity P_s (Johnson, 1968), from which a relation is obtained for U_{f+s} as function of these parameters.

The required degree of consolidation U_{f+s} can be determined more directly with the help of the diagrams of Fig. 9-10 for normally consolidated soils. The procedure for preconsolidated soils is similar. When U_{f+s} has been determined for assumed values P_s/P_f and known values of permanent load ratio P_f/P_0, the corresponding time factor T_u can be obtained from the graphs of Fig. 9-11 for $Z/H = 1.0$. The time for surcharge removal t_{SR} can be computed from the relation

$$t_{SR} = \frac{T_u H^2}{c_u} \tag{9-5}$$

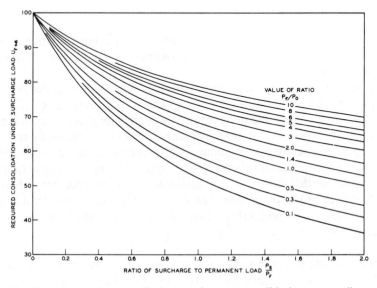

Fig. 9-10 Precompression to eliminate primary consolidation, normally consolidated soils. (From Johnson, 1968.)

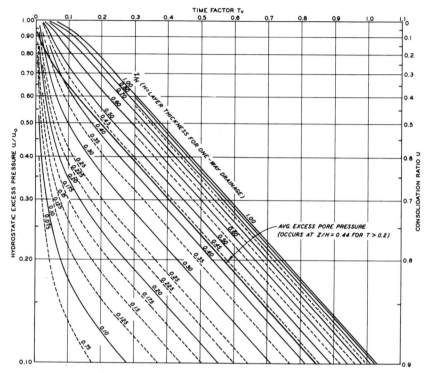

Fig. 9-11 Consolidation by vertical drainage; hydrostatic excess pressure as function of time factor T_u and position Z/H. (From Johnson, 1968.)

where c_u is the coefficient of consolidation. With the time for surcharge load removal predetermined, the above procedure is reversed to find the required surcharge loading ratio P_s/P_f.

The simplified loading conditions assumed in the foregoing analysis apply to highways, airfields, and similar work, where P_f corresponds to final permanent loading. Where the precompression is in connection with structures, it may be necessary to compute stresses from foundation loads at several depth points vertically in the compressive layer since the center conditions may not govern precompression design. Furthermore, with structures, recompression settlement caused by addition of structure loadings should be considered. In order to minimize rebound following surcharge load removal the magnitude of P_f preferably should be not less than about one third P_s (Johnson, 1968).

Secondary Compression. The intent of the precompression technique is to eliminate settlement expected to result from primary consolidation prior to construction of a structure or pavement. If primary consolidation is eliminated, construction and postconstruction settlement may involve a slight recompression and postconstruction settlement as structure loads are applied, and subsequent settlement resulting from secondary compression. The latter may be significant and amount to more than 1 ft during the economic life of the structure; hence it should be considered in the design. While these settlements cannot be fully prevented, they can be reduced through surcharge loading to inconsequential amounts. For suggested methods of design of secondary compression reference is made to Johnson (1968), Taylor (1942), Gould (1949), Bjerrum (1967), Parsons (1968), Lowe et al. (1964), Casagrande (1964), Jonas (1964), Aldrich (1964), and more recently Stamatopoulos and Kotzias (1985).

9.4   CONTROL OF SWELLING IN ROCK TUNNELS

Swelling Mechanism in Rock

The swelling mechanism in tunnels is associated primarily with displacement taking place in the form of invert heave (see also Section 8-11). The process has been documented in anhydrite, shale, and marl.

Anhydrite swells following hydration, which transforms the material into gypsum. The associated volume increase can reach 60 percent. The state of stress is known to affect swelling, whereas combinations of water content increase and stress change affect the time-dependent volume increase. Interestingly, most of the Swiss and southern German tunnels in anhydrite rock where significant swelling displacement has been observed have a dry appearance on the rock surface. In this case, the water must be supplied from the interior of the rock mass (Einstein and Bischoff, 1977).

Swelling of shale and marl is to large extent swelling of the clay minerals. Two basic types of swelling can be distinguished: (a) hydration swelling where the polar water molecules are absorbed at the exterior of the mineral and (b) osmotic swelling due to ion concentration difference between the double layer water and the free water. In addition, there is swelling due to the elastic rebound of particles. Hydration swelling in constant volume conditions produces pressure of several thousand kilograms per square centimeter, whereas osmotic swelling several tens of kilograms per square centimeter.

Shales can exist in two conditions: stable where only reduction of in situ stresses leads to swelling, or meta-stable where swelling is not completely counteracted by the present in situ stresses but where sufficient water is not available to permit swelling. An incidental mechanism is cracking due to stress changes which facilitates access of water and thus promotes swelling. The process can also be affected by a combination of water content increase and stress changes as for anhydrite.

Swelling Mechanism in Tunnels

Several procedures are available and can be used to identify factors affecting swelling, measure swelling displacements and pressures directly, and correlate other rock properties to the swelling potential. Brekke and Howard (1973) have expanded these methods to tunnel design in fault gouge by correlating index properties with observed tunneling problems and well pressures. Einstein and Bischoff (1977) suggest testing procedures developed by Huder and Amberg (1970). These involve swell and swell pressure measurements, in situ tests, and a special oedometer swell test on undisturbed cylindrical specimen. During the latter test a reduction of the axial stress leads to a particular curve and a point that is the point of no swelling since it corresponds to the natural stress state at which swelling does not occur.

These laboratory testing procedures provide the necessary parameters for the analysis and design of tunnels in swelling rock and also demonstrate the factors and the extent to which they affect swelling. Thus, swelling in the invert of a tunnel (and by analogy at other locations) can be compared to the aforementioned oedometer test since lateral displacements are largely prevented and the axial stress reduction corresponds roughly to the effect of excavation.

Figure 9-12 shows a swell curve obtained from the oedometer test. The specimen is subjected to a load–unload–reload cycle (curves a, b, c, d) in the axial direction and in a dry state. Water is added at point D' leading to a swell displacement d to point D. If the axial stress is kept at σ_D, no further swell displacement will occur. A stepwide reduction of the axial stress produces swell curve s, and point A is the point of no swelling. Interestingly,

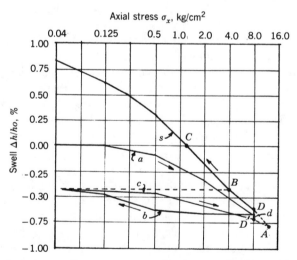

Fig. 9-12 Oedometer swell test curve, swelling in tunnels. (From Einstein and Bischoff, 1977.)

several states in the swelling process can be observed. For example, point *B* is the point where the displacements due to swelling are equal to the displacements caused by complete unloading without swelling.

Eistein et al. (1972) have shown that the state of stress at *B* (axial and lateral stresses) is equal to the primary state of stress if both are expressed by the first stress invariant I_1 (primary state of stress = original state of stress before excavation of the tunnel), or

$$I_{1,\text{swell}} \quad \text{at} \quad B = I_{1,\text{primary}} \tag{9-6}$$

Swelling in the range σ_{ax} swell $> \sigma_{ax}$ swell at *B* continues until the stress state due to swelling reaches the primary state of stress, and it is called "initial swelling." Swelling in the range σ_{ax} swell $< \sigma_{ax}$ swell at *B* occurs under distinctly different conditions: the state of stress due to swelling expressed by $I_{1,\text{swell}}$ is always less than the primary state of stress, and correspondingly large swell displacements can occur called "main swelling."

These considerations can be applied to an idealized but typical ($\sigma_v > \sigma_H$) underground opening shown in Fig. 9-13 where stress changes due to excavation produce zones of different swell effects. Thus zone 1 is the zone of initial swelling corresponding to the range $\sigma_{ax} > \sigma_{ax}$ at *B* in the oedometer test and where $I_{1,\text{swell}}$ becomes equal to $I_{1,\text{primary}}$. Zone 2 is the zone of main swelling corresponding to the range $\sigma_{ax} < \sigma_{ax}$ at *B* in the oedometer test and where $I_{1,\text{swell}} < I_{1,\text{primary}}$. The boundary between zones 1 and 2 defines a virtual opening which can deform only as much as an opening of the same shape and size without any swelling.

For the tunnel of Fig. 9-13 there are only unloading displacements and no swell displacements at the sidewalls for the idealized conditions ($\sigma_v > \sigma_H$) and the particular shape. Additional displacement due to swelling oc-

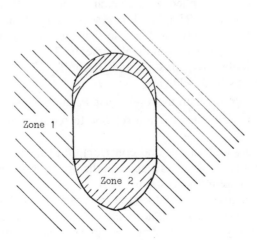

Fig. 9-13 Swell zone in tunnels. (From Einstein and Bischoff, 1977.)

curs, however, at the crown and at the invert. The stress state under a horizontal invert is unfavorable, whereas the conditions at the crown are more favorable since the first stress invariant is not as small because of the greater tangential stresses. These observations would indicate that by varying the configuration of the tunnel it is possible to reduce swell displacements. Appropriate measures include the following: (a) select a tunnel shape that produces a small zone 2, (b) introduce artificial counterstresses transforming zone 2 into zone 1, and (c) adopt other approaches appropriate for the type of rock.

Design Procedure

Einstein and Bischoff (1977) recommend the following steps for swelling control in rock tunnels:

1. Determine the primary state of stress.
2. Determine the swell zone around the opening based on the primary state of stress and the stress changes caused by the excavation stages.
3. Perform laboratory tests in the oedometer on specimens taken from the swell zones determined in step 2. Note that case histories confirm that the oedometer test conditions reasonably represent actual behavior.
4. Determine time-swelling properties in the same oedometer swell test by measuring the time displacement relations for several stress steps.
5. Determine swell displacement as shown in Fig. 9-14, assuming that the stress conditions in the oedometer are equal to the in situ. The stress difference between the primary state of stress and the stress state after excavation yields a swell strain in the swell diagram of Fig. 9-14 which is multiplied by the height h_1 of the relevant ground layer. This is essentially a simplified inverse settlement analysis, carried out for each pertinent ground layer so that the total heave is obtained as the sum of all partial swell displacements. This procedure has several inherent limitations (Einstein and Bischoff, 1977), since it represents a one-dimensional behavior.
6. Obtain in situ measurements for displacements and swell pressures.
7. Select a suitable construction scheme to reduce swelling.

Among the design features and construction procedures that are known to reduce swelling, emphasis is placed on the invert arch, anchored or bolted invert slabs, and counterstress slots.

An arch-shape invert will reduce the zone of main swelling. Furthermore, the natural counterstresses will act more effectively than under a horizontal invert. Structurally, the arch shape is appropriate, whereas the liner will provide resistance and cause the formation of artificial counterstresses.

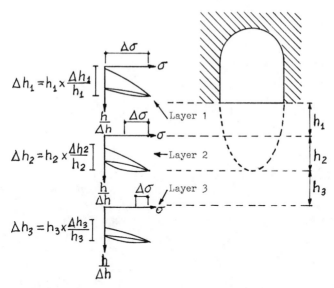

Fig. 9-14 Determination of swell displacement in a tunnel. (From Einstein and Bischoff, 1977.)

As mentioned in Section 8-11, anchoring or bolting of invert slabs should extend below the zone of swelling to produce counterstresses and control the process. Pressure grouting counterstress slots can reduce swelling by providing constrainment in any direction. Although the method is not widely used, it provides the feasibility of introducing such constrainment by cutting and pressure grouting of slots normal to the invert shape extending beyond the main swell zone. Interestingly, on many occasions most of the water causing swelling has its source in the interior of the rock mass. However, exposure of the free rock surface to water should be avoided, and an effective drainage facility is an essential feature as countermeasure to control swelling.

From the design standpoint swelling is only one of many time-dependent phenomena encountered in tunneling. The emphasis on the stress release and water increase aspect is because the most common swelling rocks (i.e., anhydrite, shales, and marls) are subjected to this form of swelling. Where creep and squeezing are present, they could lead to further volume increase and must also be considered.

Whether anchors, bolts or rebars are the most appropriate and economical measure will depend on their depth, which must be beyond the main swelling zone; otherwise they will serve minimal purpose. In some instances it may be cheaper to use an invert arch rather than longer anchors. The anchors may be installed immediately after the opening is made or some time later. If they are installed immediately, the swell deformation is reduced but the swell pressure is increased causing higher stresses on the anchors or on the invert arch. If swell deformations are not totally detrimental, the process can be allowed to occur to some degree before the supports are installed in which case the stresses are reduced.

9-5 ANCHORAGES FOR ROCK CAVERNS AND TUNNELS BY SEMIEMPIRICAL DESIGN

Rock openings cover a wide spectrum of configurations and uses, and examples are transportation tunnels, hydropower production, pressure headrace tunnels, machine hall chambers, pumped-storage projects, energy storage schemes, and the vast field of mining operations. Invariably, rock mechanics and support design have direct relationship to the engineering aspects of the project, since all underground work is typically done in rock that is initially stressed so that openings will cause changes in the initial state of stress.

There are several methods of analysis and design, all tested and with proven importance in practice. Methods chosen to be reviewed in this text include the semiempirical approach based on observed cavern performance (Cording et al., 1971), the Q system (rock mass quality) introduced by Barton et al. (1977), rock tunnel reinforcement by equivalent support method (Bischoff and Smart, 1977), spiling reinforcement techniques (Korbin and Brekke, 1976, 1978); rock tunnel support by convergence–confinement method; elastic theory methods for openings in competent rock based on the Kirsch solution; rock anchoring using the exponential formulations by Lang and Bischoff (1981); and the development of block theory for hard, jointed rock based on a three-dimensional approach (Goodman, 1989).

Observed Cavern Performance

From the analysis of a large number of case records it becomes evident that a useful correlation exists between the type and amount of support and the rock mass quality Q with respect to rock stability. Thus Q appears to be function of six rock parameters, each with a specific rating of importance. These are the RQD index, the number of joint sets, the roughness of weakest joint, the degree of alteration or filling along the weakest joint, and two parameters accounting for rock loads and water inflow. Combined, these factors represent the rock block size, the interblock shear strength, and the active rock stress.

Anchorages used as support develop their action according to the following mechanisms: (a) suspending the dead weight of a slab from the rock above or fastening loose blocks to stable layers, (b) providing a normal stress on the rock surface to clamp bedding planes together and develop composite beam action, (c) introducing a confining pressure to increase shear resistance with arch response whereby the rock becomes selfsupported, and (d) preventing key blocks from becoming loose. These mechanisms form the basis for developing suitable support methods, but give no indication of support capacity because of the difficulty of assessing the rock strength and the stresses acting in the rock mass. Both these factors depend on displacement along discontinuities. As displacement occurs the rock strength reaches peak values but then it drops to residual values. Simultaneously, stresses are reduced in the zone adjacement to the opening and rebuilt at some distance from the excavated face. On the walls of the opening the radial and shear stresses are zero since this is a free surface. The tangential stress varies within a range defined in terms of the principal stresses.

Tables 9-4 and 9-5 compile a summary of design and performance data from large rock caverns (Cording et al., 1971). Displacement measurements at the cavern surface provide criteria of cavern behavior and indicate the degree of stability. Potentially unstable conditions are identified by (a) large displacement as compared to predicted, (b) high displacement rates unrelated to construction and persisting after excavation, and (c) displacement exceeding support capacity.

Figures 9-15 and 9-16 show three types of displacement, in this case observed is cavities I and II, Nevada Test Site, and typically occurring in large caverns (Cording et al., 1971). The cavities are supported by rock bolts 20 ft long at the crown and 16 ft long on the sidewalls. The three types of displacement are elastic, shallow slabbing, and deep-seated movement. Measured elastic displacement (Fig. 9-15) agrees fairly with the predictions of elastic theory, although near the cavern surface actual displacement is greater than predicted because of the normally lower deformation modulus as the cavern is excavated and the rock disturbed. Elastic displacement indicates stable conditions.

Where shallow slabs within 5–7 ft of the cavern surface continue to become loose, a condition commonly encountered when irregular sections and rock slabs protrude into the opening, fractures formed in the slabs are most prone to open. The displacement is now within the range of shallow slabbing, but is likely to continue at decreasing rate with times and does not indicate instability as long as the anchorage can hold the slabs together. Where this condition develops, the crown is usually gunited at the end of construction to stop further drying and cracking of the loose slabs and possible breakup of rock beyond the anchorage point.

Deep-seated movement is shown in Fig. 9-17 for cavity II. Displacement as much as 1–2 in (see also Fig. 9-15) appears concentrated at depths of 10–30 ft. This movement probably involves the entire plane face before stability is reached. As the process occurs, joints and bedding plane weaknesses disrupt the continuity of the face and form wedges likely to move into the opening.

TABLE 9-4 Observed Displacement in Large Caverns

Location, Size and Dimensions	Rock Properties	Measured Field and Laboratory Moduli E, $\times 10^6$ psi	$\dfrac{E_{\text{field}}}{E_{\text{lab}}}$
Nevada Test Site, Cavities I and II: hemisphere on end; height $H = 110$ ft, width $B = 80$ ft, length $L = 120$ ft, depth $D = 1300$ ft, vertical stress $\sigma_v = 1000$ psi, horizontal stress $\sigma_h = 500$ psi	Massive, bedded tuff, RQD: Excellent, unconfined compressive strength $q_u = 1500$ psi, water content $= 21\%$	$E_{\text{lab}} = 0.5$, $E_{\text{lab, sonic}} = 1.5$, $E_{\text{seismic}} = 1.0$	1.0 est.
Nevada Test Site, Cavity III: hemisphere on end; $H = 80$ ft, $B = 50$ ft, $L = 75$ ft, $D = 350$ ft, $\sigma_v = \sigma_h = 400$ psi	Granite (quartz monzonite), iron stained joints. blocky, RQD: fair to good (75% average); major joint set parallel to wall, occasional shear zones	$E_{\text{lab}} = 10$	0.4 est.
Tumut I: $H = 110$ ft, $B = 77$ ft, $L = 300$ ft, $D = 1100$ ft, $\sigma_v = 1500$ psi, $\sigma_h = 1800$ psi	Granite, granite gneiss, $q_u = 20{,}000$ psi, RQD: fair-good. high angle fault intersects one wall. joint spacing 1–5′.	$E_{\text{lab}} = 8$, $E_{\text{lab, sonic}} = 8$, $E_{\text{flat jack}} = 5$, $E_{\text{pressure chamber}} = 2$	0.25–0.6
Tumut II: $H = 110$ ft, $B = 60$ ft, $L = 300$ ft, $D = 1000$ ft	RQD: good to excellent	$E_{\text{lab}} = 8$, $E_{\text{lab, sonic}} = 8$, $E_{\text{flat jack}} = 6$, $E_{\text{pres. chamber}} = 3$, $E_{\text{plate beam}} = 1$	0.13–0.75
Morrow Point Powerplant, Colorado: $H = 100$–138 ft, $B = 57$, $L = 207$ ft, $D = 400$ ft, $\sigma_v = 400$–2000 psi	Micaceous quartzite, mica schist, $q_u = 6000$–16000 psi, RQD: good to excellent	$E_{\text{mica schist lab.}} = 1.3$, $E_{\text{mlc. quartzite lab.}} = 4$, $E_{\text{plate bearing}} = 1.3$	0.33–1.00
Oroville Powerplant, California: $H = 120$ ft, $B = 69$ ft, $L = 550$ ft, $\sigma_v = \sigma_h = 500$ psi	Amphibolite, RQD: fair to good	$E_{\text{lab}} = 13$, $E_{\text{lab, sonic}} = 16$, $E_{\text{seismic}} = 5$–15, $E_{\text{flat jack}} = 1.5$–$16$, $E_{\text{plate bearing}} = 1.2$–$1.8$	0.12–1.0 0.12
Oroville Tunnel: $H = 35$ ft		$E_{\text{lab}} = 13$	—
Poatma Power Station, Tasmania: $H = 85$ ft, $B = 45$ ft, $L = 300$ ft, $D = 500$ ft, $\sigma_v = 1200$ psi, $\sigma_h = 1800$–2400 psi	Thin to massive bedded mudstone; $q_u = 5000$ psi, water content $= 1.5\%$	Loaded perpendicular to bedding: $E_{\text{lab}} = 4.5$, $E_{\text{flat jack}} = 2.4$	0.53
		Loaded parallel to bedding: $E_{\text{lab}} = 6.3$, $E_{\text{flat jack}} = 3.2$	0.51
Kariba Powerplant, Rhodesia: $H = 132$ ft, $B = 75$ ft, $L = 468$ ft, $D = 200$ ft	Biotite gneiss, RQD: fair to good	$E_{\text{lab}} = 9$, $E_{\text{seismic}} = 11$, $E_{\text{plate bearing}} = 0.9$	0.1
Lago Deho: $H = 195$ ft, $B = 68$ ft, $L = 620$ ft, $D = 520$ ft, $\sigma = 800$–1600 psi	Fine grained peleozone gneiss, near vertical foliation normal to axis of cavern, local weathered zones	$E_{\text{seismic}} = 7.6$–8.6, $E_{\text{pressure chamber}} = 1.4$–$3.8$, $E_{\text{plate bearing}} = 1$–$3$	0.3 0.25
Kisenyama, Japan: $H = 165$ ft, $B = 83$ ft, $L = 200$ ft, $D = 810$ ft	Chert, sandy slate	$E_{\text{seismic}} = 8.5$, $E_{\text{plate bearing}} = 0.7$	0.08
Nagata Tunnel: $H = 50$ ft, $D = 300$ ft	Bedded limestone, shale, roof in hard limestone, wall in shale, sandstone, and dolomite	$E_{\text{lab}} = 1$–10	—

From Cording et al. (1971)

Observed Cavern Displacement δ, (in)	Modulus Determined from δ, (in)	$\dfrac{E_\delta}{E_{\text{lab}}}$	$\dfrac{E_\delta}{E_{\text{field}}}$	Remarks
Elastic, crown and sidewalls; 0.15–0.4	0.5	1.0	1.0	
Shallow, crown: 1 to 2	(0.05–0.15)	(0.1–0.3)	(0.1–0.3)	Displacement occurred at 3′ to 5′ depth behind shallow slabs
Deep-seated Cavity II wall: 1 to 2	(0.05–0.15)	(0.1–0.3)	(0.1–0.3)	Displacement occurred at 10′ to 30′ depth over area 80′ × 100′ of wall. Stabilized with additonal bolts
Crown 0.015–0.035	4	0.4	1.0	
wall, upper: 0.05	3	0.3	0.75	Larger displacements due to opening of joints 3′ to 5′ behind and parallel to wall
wall, middle: 0.125	2	0.2	0.50	
Crown	2*	0.25	0.50	*Determined from strains in machine hall roof
Wall at springline: 0.15	2	0.25	0.50	
Wall near fault at springline: 0.4	(0.7)	(0.09)	(0.18)	
Crown	5*	0.6	1.2	*From strains in machine hall roof
walls: 0.2–0.6	2	0.25	0.50	By exact survey
Walls	0.8–7.0	0.1–0.9	0.2–1.7	From extensometers
Crown: 0.1–0.4	0.5–2.1	0.17–0.7	0.5–1.0	*Shear zone in a-line wall formed wedge which moved into cavity
b-line wall: 0.3	3.0	1.0	1.0	
a-line wall: 2.3*	(0.4)	(0.13)	(0.2)	
Crown*: 0.002–0.059	8.2 ave.	0.63	1.0	*Measured with 20′ extensometers
Wall: 0.054–0.247**	2	0.15	0.2–1.0	**Measured with 40′ extensometers
Left Wall: 0.250**			0.2	**Vertical shear intersected wall at shallow angle.
Crown extensometers	0.6–7.5	0.05–0.58	—	
Crown: 0.10	1.6	0.36	0.67	*After horizontal shear developed at intersection of crown and haunch
*0.18	(0.9)	(0.20)	(0.37)	
Wall: 0.4	2.5	0.4	0.78	
Wall: less than 0.5	0.5–1.0	0.1	1.0	Arch concreted. rock bolts used as required.
*Wall: 1.5	(0.2)	(0.02)	(0.2)	*Fault zone at NW corner of power house reinforced with anchor tendons
*Crown: 0.5	0.7	0.09	0.36	* Occurred within 25′ of surface
Wall: 0.7	1.2	0.15	0.60	
**End wall: 1.0	0.5	0.06	0.24	**End wall is parallel to foliation
Crown: Excellent rock: 0.08″	6	0.70		*Chert layer separated from slate causing large displacement.
Poor rock: 0.28	0.8	0.10		
*0.60	(0.4)	(0.05)		
Walls, left: 0.4	5	0.59		**Struts installed across cavern because of risk of collapse when displacement occurred. Struts were removed later and wall was concreted.
right: 1.5**	(1)	0.12		
Crown: 0.05	2	0.5	—	Steel rib supports, heading and bench
Wall: (1.8)	(0.1)	(0.05)		Horizontal offset on bedding planes observed

TABLE 9-5 Support Types Used in Large Caverns

Cavern Size and Dimensions	Rock Properties	Support Type Used
Nevada Test Site, Cavities I and II: width $B = 80$ ft, height $H = 140$ ft, length $L = 120$ ft, depth $D = 1300$ ft, $\sigma_v = 1000$ psi, $\sigma_h = 500$ psi; hemisphere on end; crown is dome-shaped	Tuff RQD 95–100%, $\gamma = 125$ pcf; water content = 21%, $q_u/\sigma = 1.5$	Tensioned bolts, 8 ft anchored in grout, remainder free Crown $1\frac{1}{8}$ in dia., 3×3 ft spacing, 32 ft long Walls $1\frac{1}{8}$ in dia., 6×6 ft spacing, 24 ft long After stabilization, Cavity II, wall $1\frac{1}{8}$ in dia., 3×3 ft spacing, 48 ft long
Tumut I, Australia: $B = 77$ ft, $H = 110$ ft, $L = 300$ ft, $D = 1100$ ft, $\sigma_v = 1500$ psi	Granite, granite gneiss; RQD fair to good, $q_u = 20,000$ psi, $q_u/\sigma = 13$	Crown 1 in dia., 4×4 ft spacing, 15 ft long, 23 kips yield ungrouted Wall 1 in dia., 5×5 ft spacing, 12 ft long
Morrow Point Powerplant, Colorado: $B = 57$ ft, $H = 100$–138 ft, $L = 207$ ft, $D = 400$ ft, $\sigma_v = 400$–2000 psi	Micaceous quartzite, mica schist, $q_u = 6000$–16,000 psi, RQD good to excellent, $q_u/\sigma = 6$–16	Crown 1 in dia., 4×4 ft spacing, 20 ft long Walls 1 in dia., 4×4 ft spacing, 12 ft long Stabilization of A-line wall grouted rebars 60–110 ft long, 25–250 kip tendons to drainage gallery, 27–$1\frac{3}{8}$ in dia. bolts 26–78 ft long to hold corner at wedge (60 kips, tensioned); total support pressed added
Oroville Powerplant, California: $B = 69$ ft, $H = 120$ ft, $L = 550$ ft, $\sigma_v = \sigma_h = 500$ psi	Amphibolite, RQD fair to good (est.)	Crown 1 in dia., 4×4 ft spacing, 20 ft long Wall 1 in dia., 6×6 ft spacing
Poatma Power Station, Tasmania: $B = 45$ ft, $H = 85$ ft, $L = 300$ ft; $D = 500$ ft, $\sigma_v = 1200$ psi, $\sigma_h = 1800$–2400 psi	Thin to massive bedded mudstone, $q_u = 5000$ psi, water content 1.5%, $q_u/\sigma = 2.5$	Crown 3×3 ft spacing, 14 ft long, bolts doubled using two 3×3 ft patterns to hold slabbing rock Haunches and top of walls, 3×3 ft spacing, 12 ft long Middle of wall bolts, 14 ft long
Morad, Cheyenne Mountain Colorado: $B = 45$ ft, $H = 60$ ft; intersecting chambers; $L = 600$ ft, $\sigma_v = 1200$ psi	Biotite granite, $q_u = 10,000$–20,000 psi; RQD fair to good; $q_u/\sigma = 8$–16	Crown 1 in dia., 4×4 ft spacing, 10 ft long Wall 1 in dia., 4×4 ft spacing, 10 ft long Intersections 2×2 ft to 6×6 ft spacing, 24–30 ft long
Boundary Dam Powerplant, Washington: $B = 76$ ft, $H = 175$ ft, $L = 476$ ft, $D = 500$ ft	Bedded limestone, $q_u = 10,000$ psi; RQD good to excellent	Crown 1 in dia., 6×6 ft spacing, 15 ft long Haunches alternate 15 ft and 20 ft bolts, 5×5 ft spacing Walls bolted as required Draft tube pillars 8 anchors per pillar, 636 kips capacity
Ronco Val Grande, Lake Maggiore: $B = 67$ ft, $H = 195$ ft, $L = 624$ ft	Gneiss, partially clefted	Crown 2.5-ft-thick concrete arch Wall anchor tendons 6 to 12 ft in spray concrete Upper 50 ft of walls 200 kips, 10×10 ft spacing, 54 ft long Lower 140 ft of walls 80 kips, 20×20 ft spacing, 51 ft long
El Toro, Chile: $B = 80$ ft, $H = 126$ ft, $L = 335$ ft	Granodiorite orthogonal points	Crown anchors, 400 kips, 20×20 ft spacing, 49–55 ft long Plus bolts, 40 kips, 8×8 ft spacing, 13 ft long Total Walls 400 kips, 20×20 ft spacing, 50 ft long

From Cording et al., (1871)

Anchor length span L/B or L/H	Pi at yield psi	n/m	Remarks
			Gunite added to prevent drying and cracking in crown, after excavation was completed; Cavity I wall stable
0.40	40 at yield	0.5	under design bolting; Cavity II had joints and bedding
	20 tensioned	0.25	planes intersecting wall that formed an unstable
0.17	10 at yield	0.09	wedge
	5 tensioned	0.045	
0.33	20	0.18	
0.20	10	0.11	Arch concreted, supplementary steel sets placed at contractor's request
0.11	6	0.05	
0.35	13	0.20	1–5-in-thick shear zone (A) dips 32°, exposed on a line
0.12	13	0.11	wall; 100-ft-wide wedge moved 2 in into cavern, along
1.0	1.2	0.01	17° component of dip of shear zone A; rebars installed
1.0	6.0		during excavation stopped the movement; anchor
0.5	1.0	0.07	tendons and long bolts were added after excavation
	8.0		was completed; support pressure computed for 7500 ft^2 area of wall moving into cavern
0.29	13	0.16	4 in gunite and wire mesh
4	6	0.04	
0.31	10	0.20	Compressive strength parallel to bedding, 3-ft-deep stress-relief slots used in crown to reduce
0.3	20	0.4	compressive stresses and slabbing of beds in crown;
0.14	10	0.1	shear failure developed on horizontal bedding planes at intersection of haunch and crown, resulting in $\frac{1}{8}$-in
0.16			displacement of haunch into cavern; 4-in gunite and wire mesh in crown.
0.22	11	0.21	Remedial measures to hold wedges moving into
0.17	8	0.11	intersections; wedges were formed by major high-
0.45	6 to 50	0.10–0.80	angle joint sets angling across intersections
0.20	6	0.07	Additional bolts in crown 30 ft long, installed where
0.20–0.26	8	0.10	joints appeared in form of wedges; wire mesh used; some gunite; attention was given to reinforcement of rock around reentrants; anchors used in draft pillars to support jointed and slickensided rock
			Initial wall support 25 kips, 10 × 10-ft spacing; 16-ft-long bolts, resin anchors, $L/H = 0.08$, $p_i = 1.8$ psi
—	—	—	
0.28	14	0.06	
0.26	2	0.006	
0.60	7	0.08	
0.16	6	0.14	
—	13	—	
0.4	5	0.03	

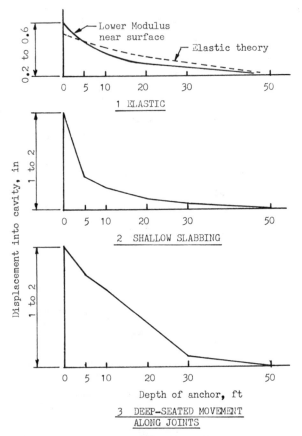

Fig. 9-15 Typical observed displacement in cavities I and II, Nevada Test Site. (From Cording et al., 1971.)

The deformation modulus shown in Table 9-4 is estimated from observed displacement. Modulus values shown in parenthesis have questionable validity since the associated displacement is related to opening and shearing of discontinuities, and thereore it does not express elastic behavior. Columns 7 and 8 compare modulus values determined from observed displacement E_δ to values obtained in laboratory (E_{lab}) and in situ (E_{field}). For elastic rock movement the moduli from in situ tests and observed displacement should be nearly equal, hence the ratio in column 8 close to unity. Low values of E_δ/E_{field} indicate large inelastic movement likely to be associated with specific geologic conditions. Large displacement occurs where joints, shears, or foliation trended parallel to the walls of the cavern, but can also be caused by construction procedures. Joints and shear planes intersecting the surface form unstable wedges prone to movement and deformation.

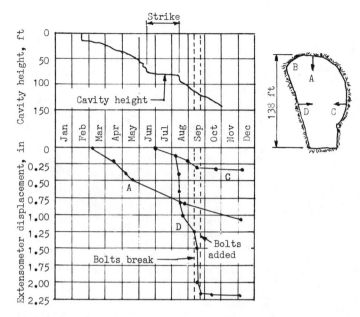

Fig. 9-16 Displacement as function of the excavation process, cavity II, Nevada Test Site. (From Cording et al., 1971.)

Factors Affecting Support Requirements

For anchorage design, the first guideline is constractibility: the support method must ensure a construction process compatible with the ground conditions. Where reinforcement systems are considered suitable, they may consist of anchors, bolts, and rebar dowels. Shotcrete, if added, provides a membrane preventing surface degradation. If the design also calls for interior concrete lining, it is likely to be unrelated to the initial rock reinforcement.

Stresses acting on the rock are divided into two groups: (a) stresses concentrated around the opening originated from the natural state of stress and (b) stresses developed from body forces as gravity loads in the rock blocks. For rock sensitive to creep time-dependent movement, if restrained, will result in additional pressures. Around the opening natural stresses are assumed to be relieved as blocks displace and move toward the opening. After the installation of the support, therefore, the only cause of movement is the combination of gravity loads from the blocks. However, under the action of gravity, the interlayer slip regions may become loose, progressively destroying interlocks, until the support is in place. It is not always feasible to construct supports on time or with sufficient pressure to stop formation of slip regions, in which case progressive enlargement of these regions by loosening of the rock will result in excessive overbreak and in extreme conditions

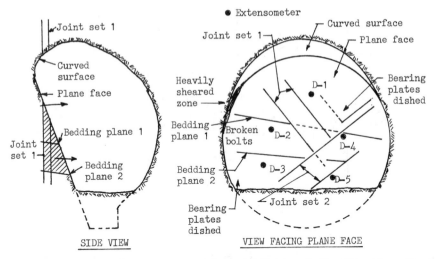

Fig. 9-17 Deep-seated movement zone in cavity II, Nevada Test Site. (From Cording et al., 1971.)

cavern collapse. On the other hand, the initial state of stress can affect initial movement and fracture patterns, and under favorable boundary conditions this movement may continue to act on the support.

From the foregoing remarks it follows that flexible supports will have to carry the weight of extensive slip regions corresponding to low friction angles for loosened rock (Goodman, 1989). At the other end of the spectrum, a prestressed support can be designed to maintain tangential stresses for at least part of the opening by preventing interlayer slip.

Estimation of Gravity Loads. Assuming that the forces causing block movement are related to gravity loads only, the internal pressure P_i necessary to maintain stability in rock with frictional resistance and cohesion is

$$P_i = nB\gamma - (\text{rock cohesion}) \tag{9-7}$$

where n = parameter depending on rock friction
        B = width of opening
        γ = unit weight of rock

and all symbols are as shown in Fig. 9-18.

Rock mass cohesion depends on the ratio of the joint spacing to opening width B. For a given joint spacing increasing B decreases cohesion and vice versa. Thus, a small opening may be stable without support, whereas if the opening is large rock mass cohesion is almost zero, in which case

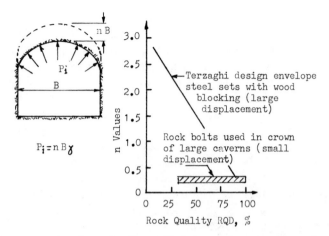

Fig. 9-18 Relation of rock quality Q and displacement to support requirements; (a) underground opening showing geometric configuration and internal pressure diagram; (b) correlation between rock quality RQD and load parameter n. (From Cording et al., 1971.)

$$P_i = nB\gamma \qquad (9\text{-}7a)$$

The factor n is function of four rock properties (a) rock quality, expressed by the RQD; (b) displacement along discontinuities; (c) strength and orientation of discontinuities; and (d) the ratio of intact unconfined compressive strength to the maximum natural stress q_u/σ_{max}. The variation of n is shown in Fig. 9-18, and evidently large displacement or low rock quality yield higher n values. The RQD corresponds to the initial Terzaghi rock classification (Proctor and White, 1946; Monsees, 1969). The graph shows the zones of two main support types: steel ribs with blocking, and rock bolts or anchors. Interestingly, with steel sets the support requirements are higher because large rock displacement is allowed during and subsequent to support installation, causing more rock loosening and increasing pressures accordingly. With bolts, the anchorage zone capacity shown in Table 9-5 for the crown is based on smaller n values compared to those of the steel sets for the same cavern, reflecting a trend to carry out the excavation incrementally and install the anchors close to the face, a procedure allowing the rock to be supported while excavated to prevent its loosening.

Effect of Discontinuities. Factors affecting support requirements are the orientation of primary joint sets and the strength of shear zones relating to the cavern surface. If clay fills a shear zone, its residual strength becomes critical if no irregularities exist on the surface to improve resistance to deformation. Residual shear strength for soils with high plasticity and clay content may have a corresponding friction angle as low as 6°–15°.

Horizontal layers tend to force the roof of the opening to open up, although the rock remains tightly compressed in the walls. When the strata are dipping, the zone of interbed separation and potential buckling moves off center and the walls can be undermined by sliding. Thinner strata near the roof tend to detach from the main rock mass and form independent beams. Their stability will depend on the stress pattern and their stiffness. Thus, diagonal tensile cracks develop as shown in Fig. 9-19(d) resulting from fixed-end moments exceeding the strength of the beam, and their growth releases a complete block on the roof. In this case, the end anchors in the system reinforcement should be installed diagonally at an angle as shown to defend the thin block against diagonal tensile and shear failure. The behavior of openings with relatively flat roof is discussed in more detail in the following sections.

Figure 9-19(a) shows a two-dimensional wedge assumed to act above the crown of a cavern. The effect of joint orientation on stability is considered by letting the wedge be displaced. The following relationship exists

$$P_i = P_N \left(1 - \frac{\tan \phi}{\tan \theta} \right) + \frac{\gamma B}{4 \tan \theta} \tag{9-8}$$

where θ = one-half the wedge angle
    P_i = internal pressure necessary to balance the wedge load
    P_N = normal pressure acting on wedge per unit length along axis of cavern

As the wedge is displaced P_N tends to decrease and reach a minimum value depending on the weight of the additional rock loosened around the wedge, or proportional to the width B.

In assessing wedge effects the following comments are useful.

1. If the water level stands above the cavern by a distance more than $3B$, P_N is considerable. In this case the wedge cannot be forced into the opening if θ is smaller than ϕ.
2. If the cavern is shallow, the absence of large horizontal stresses in the arch may cause rock block fallout regardless of the value of ϕ.
3. Tight, irregular joint surfaces have friction angles greater than 45°. In this case a wedge that could invade the opening should have an angle $2\theta \geq 90°$ as shown in Fig. 9-19(b).
4. A tight joint system intersecting the cavern surface at relatively flat angle or one that is parallel to the crown will impose additional support requirements.
5. If the discontinuity consists of a thick shear zone at residual strength, ϕ will be close to 15°, requiring a wedge equilibrium angle close to 30° and a steep wedge ready to move into the opening as shown in Fig.

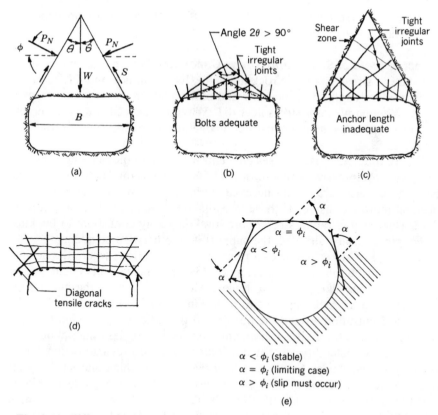

Fig. 9-19 Effect of joint orientation on the stability of underground openings.

9-19(c). If anchors are used to support the cavern, they should be exceptionally long.

6. When rock layers slide with respect to each other, the interlayer forces are inclined at an angle ϕ_i with the normal to the layers as shown in Fig. 9-19(e).

For static equilibrium the resultant interlayer force cannot have inclination greater than ϕ_i with the normal to beds, except a direction parallel to them, in which case shear resistance needs not be mobilized. For the circular opening shown, the normal and shear stresses are zero at the surface and therefore the tangential stresses constitute the resultant force across the layer. At the periphery of the opening the tangential stresses must be inclined less than ϕ_i from the normal to the layers or be parallel to them. In any other case the layers must mobilize cohesion or slip. The latter redistributes the stress or changes the tunnel shape, or both. This interlayer slip enhances sliding where the layers dip downward toward the opening, and flexure in the opposite case. Otherwise stated, in regularly layered rock, the stress flows around the opening as if its shape is different from that assumed initially (Goodman, 1989).

Effect of State of Stress and Stress Concentration. In a general context, the state of stress is a basic rock attribute. Its magnitude and direction can affect rock strength, deformability and other relevant rock mass characteristics. Where any major stress near the excavation exceeds $\frac{1}{4} q_u$, new cracking can occur as a result of construction operations irrespective of protective measures.

Where the tangential stresses concentrated around the opening approach the intact compressive strength of rock, new extension fractures can form near the cavern surface during excavation, and blast damage is likely to become more extensive. If these stresses are developed along a relatively flat surface where radial stresses are low, they may contribute to buckling or popping of slabs although the support may otherwise be adequate.

The state of stress becomes a prime factor in sections where the desired stress redistribution cannot develop until large displacements occur. Major rock wedges are likely to move toward the cavern if q_u in the intact zone or in the altered rock adjacent to a discontinuity is low compared to the stress concentration around the opening. If the ratio q_u/σ_{max} (where σ_{max} is the maximum natural stress, horizontal or vertical) is less than about 5, it may promote instability so that new extension fractures can occur during excavation and wedge failure can take place by crushing and shearing off of irregularities. Ratios of q_u/σ_{max} are shown in Table 9-5 for several large caverns. For cavities I and II, Nevada Test Site, this ratio is very small (close to 1.5) and is considered the reason of large extension fractures in the cavern walls during excavation (Cording et al., 1971).

In noncircular openings stresses appear to concentrate at corners and concave bends of small radius but decrease at convex bends. Stress concentrations are usually high in the sidewalls and low where the line of action of the largest initial stress intersects the opening. Stress concentrations are inconsequential in shapes that are smooth, without corners and reentrants. They are also minimized in openings where the major axis is aligned to the major principal stress and if the ratio of width to height is proportional to $K = \sigma_h/\sigma_v$. Stress concentration values are given in Table 9-6 for various shapes and under vertical stress only ($K = 0$). Stress concentrations for other values of K can be derived by superposition.

Design Considerations

Crown Support Pressure. For relatively small openings (width $\leq$ 40 ft or 13 m), rock anchors and bolts are determined with regard to the spacing necessary to support and hold the rock slabs. Pattern anchoring with close spacing is indicated in poor quality rock, whereas in sound rock anchorages

TABLE 9-6   Stress Concentration around Opening under Vertical Stress Only (K = 0)

Shape	Height/Width	Stress Concentration (σ_θ/σ_v)	
		Roof	Side
Ellipse	$\frac{1}{2}$	−1.0	5.0
Oval	$\frac{1}{2}$	−0.9	3.4
Rectangle (round corners)	$\frac{1}{2}$	−0.9	2.5
Circle	1	−1.0	3.0
Ellipse	2	−1.0	2.0
Oval	2	−0.9	1.6
Rectangle	2	−1.0	1.7

From Goodman (1989) σ_θ = tangential stress; see Fig. 9-39(a).

are needed occasionally. Anchor tendons equivalent to 1-in-dia. bolts are usually adequate, and when tensioned they will exert enough pressure to keep the rock mass stable.

With larger openings the anchorage pressure also increases, and larger-size anchors or bolts in special patterns are required. Thus, the first design consideration is to determine the support pressure necessary for stability before selecting anchor size and spacing. There is an infinite number of combinations regarding anchor size and capacity, spacing, and length. The optimum selection may be based on the guidelines and criteria outlined in Chapter 5. Anchor spacing and the type and amount of intermediate support should be correlated to the length of the advance and the quality of rock. A feasible, and often economical, combination is to use long, high-capacity anchors on a wide pattern and install shorter low-capacity units or shotcrete to support the cavern between the primary anchor. A summary of caverns supported in this manner is included in Table 9-5.

Figure 9-20 correlates cavern width with crown average support pressure used for large caverns. Values of n vary in the range 0.10 to 0.25, which is equivalent to a cavern 100-ft-wide supporting 10–25 ft of rock. A large number of caverns have arched crowns with span/rise ratios (B/R) in the range 2.5–5.0. Flatter roofs will require higher support pressures. Furthermore, unless the crowns are formed predominantly in rock of fair to excellent quality, major shear zones should be expected above the crown requiring support pressures higher than the range $n = 0.25$.

Sidewall Support Pressure. Support conditions along the sidewalls are often more complex than the crown. The sidewalls have larger planar surfaces, facilitating slabbing and buckling of rock wedges, and there is higher probability of discontinuities adversely oriented intersecting the sidewall

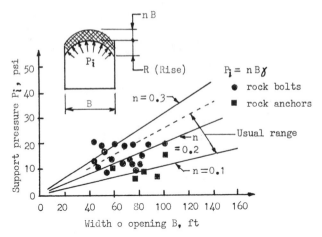

Fig. 9-20   Correlation of cavern width with average support pressure used in the crown of large caverns. (From Cording et al., 1971.)

surface to form unstable wedges. Thus, unlike crown support the sidewall support requirements depend more on joint geometry. A further problem is related to excavation procedures where economy dictates large and deep bench cuts exposing substantial sidewall sections before a support is installed.

Figure 9-21 shows support capacities selected for the sidewalls of caverns in conjunction with the height of the cavern. Ignoring rock cohesion, the sidewall pressure P_i is determined from the expression

$$P_i = mH\gamma \qquad (9\text{-}9)$$

where m = design parameter and H = height of cavern. By comparison, support pressures for the sidewalls are lower than those for the crown, although this will depend also on the relative H/B ratio. The values of m are generally in the range 0.05–0.13. In most cases the anchorages are concentrated in the upper portions of the sidewall, and lighter supports are provided in the lower sections.

Anchorage Length. Factors determining anchor length are discussed in Section 5-6. In addition, the supporting units should extend through any wedges liable to be displaced toward the opening. Anchorages should also extend through the zones where tensile stresses act normal to the cavern walls. Where large planar surfaces are present, the stresses normal to the surface may be in low compression away from the face, and any planes of weakness parallel to the cavern surface will in this case promote buckling of slabs into the cavern. In general, anchorage length should be related to the opening size and should be greater for flat than for curved surfaces.

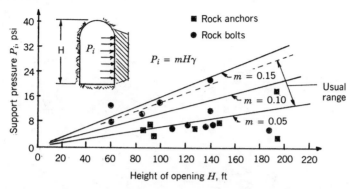

Fig. 9-21 Correlation of cavern opening height with average support pressure used in the sidewalls of large caverns. (From Cording et al., 1971.)

Figure 9-22 shows anchor lengths used for the crown of large caverns ($B > 40$ ft). This length appears to be within the relatively narrow range $0.15B$ to $0.35B$. Since most caverns have B/R ratios between 2.5 and 5.0, selecting anchor lengths about $0.35B$ is reasonable and conservative for supporting all wedges except very deep ones formed by low-friction shear zones.

Anchorage lengths used to support the sidewalls of large caverns are shown in Fig. 9-23, and generally they vary from $0.1H$ to $0.5H$. For these examples anchorages consist mainly of bolts except in the upper region of the graph, which represents anchor tendons extended beyond the minimum required length. Shorter bolts are considered adequate if deep wedges formed by major discontinuities do not exist around the opening.

Installation Requirements

In general, it is expedient to install the anchorage close to the face as excavation proceeds. In competent rock the time requirements may be somewhat relaxed provided the overall construction sequence is favorably affected. For example, for the tunnels at the Oroville dam the bolt installation time was extended from 3 to 4 hr after blasting (Kruse, 1970). Anchorages should be monitored and checked for possible damage if they are installed and tensioned near blasting operations (see also subsequent sections for effects of blasting).

Construction procedures including the sequence and method of excavation and support should be carefully controlled near benches, portals, and intersections. Preanchoring of buttresses may often be desirable to preserve the rock integrity. In certain cases, when benching down around a reentrant it may be difficult to install horizontal anchors because of limited access. In such cases vertical units or grouted dowels can be used in a line just behind the final excavation surface.

Portals and intersections are zones where difficulties are commonly en-

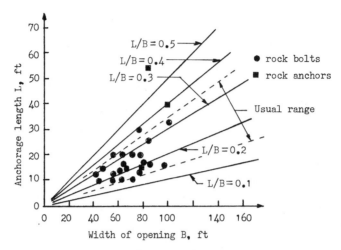

Fig. 9-22 Anchorage length used in crown of caverns. (From Cording et al., 1971.)

countered. When portalling a tunnel from a cavern it is necessary to install a ring of long anchors about 2 ft outside and parallel to the smooth-wall perimeter holes to support the intersection and hold the tunnel crown in place as the first rounds are advanced. Whereas flexibility is often the key to successful construction, in intersection and reentrant areas it is often necessary to specify detailed and restrictive provisions including controlled blasting methods.

The *Q* System and Numerical Rating of Rock Parameters

Beginning with the analysis of about 200 case records and by grouping relevant statistical data Barton et al. (1977) have developed an approach based on a weighting process. Accordingly, all factors affecting estimation of support requirements, including the positive and negative aspects of a rock mass, are assessed and assigned an index value. The procedure can be used in conjunction with the analysis presented in the foregoing section or as a supplement to it. The method is implemented according to the following steps:

1. Classify the rock mass quality by numerically rating the following rock parameters: *RQD*, number of joint sets, roughness of most unfavorable joint set, degree of alteration or filling of most unfavorable joint set, rock load resistance, and water inflow.
2. Select optimum excavation scheme.
3. Estimate an appropriate permanent support (shotcrete, anchors, concrete lining) or combinations therefrom.

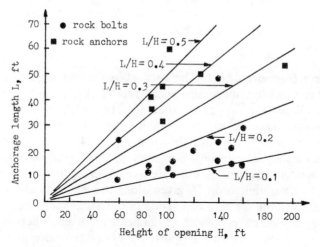

Fig. 9-23 Anchorage length used in cavern walls. (From Cording et al., 1971.)

The rock mass quality Q, already defined at the beginning of this section, is correlated to the six index parameters as follows:

$$Q = \left(\frac{RQD}{J_n}\right)\left(\frac{J_r}{J_a}\right)\left(\frac{J_w}{SRF}\right) \qquad (9\text{-}10)$$

where RQD = rock quality designation (Deere, 1963)
$\quad\quad\quad J_n$ = joint set number
$\quad\quad\quad J_r$ = joint roughness number
$\quad\quad\quad J_a$ = joint alteration number
$\quad\quad\quad J_w$ = joint water reduction number
$\quad\quad SRF$ = stress reduction factor

The three factors on the right side of Eq. (9-10) are crude measures of the following rock characteristics:

$$\frac{RQD}{J_n} \qquad \text{rock block size}$$

$$\frac{J_r}{J_a} = \tan\phi \qquad \text{interblock shear strength}$$

$$\frac{J_w}{SRF} \qquad \text{active rock stress}$$

Barton et al. (1977) give detailed summaries for the rock mass descriptions and ratings for each of the six parameters, which are numerically entered in Eq. (9-10) to provide a possible value for Q encompassing the entire spectrum of rock mass quality from heavy squeezing ground to sound unjointed rock.

Excavation Support Chart. The rock classification used to estimate its quality Q is based on successive reanalysis of case records from which a consistent relationship has been obtained between Q, the relevant excavation dimension, and the support actually used. These three variables are interelated in the support chart shown in Fig. 9-24. The box numbering 1–38 indicates a support category. Barton et al. (1977) provide a detailed summary of support measures appropriate to each category.

The equivalent dimension D_e given along the left-hand axis of the chart is a function of both the size and scope of the excavation. For the analysis of roof support the cavern width or diameter are used, whereas the height or diameter are used for wall support. An excavation/support ratio ESR is used to modify these dimensions reflecting construction practice and factors relevant to safety.

Excavation/Support Estimates. Barton et al. (1977) provide detailed summaries of support estimates for each of the 38 block categories of Fig. 9-24. These are prepared to fit the largest number of case records available that plot within the same support category. However, small variations in support methods are expected to occur because of rock mass differences and due to the combination of different variables. Interestingly, the support recommendations emerging from these summaries are not confined to one particular support system but give optimum combinations of anchors, bolts, shotcrete, mesh reinforcement, cast concrete arch, and steel reinforcement.

This approach, also known as the "classification–support method," can be used independently yielding a design based on precedent. Where the method has been tested the results were satisfactory, hence for experienced designers it represents a significant design tool.

Support Pressure Estimates. Fig. 9-25 presents a chart for estimating permanent radial support pressures required to stabilize the crown or the walls of a cavern. For a given value of Q, the pressure also depends on the dilational properties of the weakest joint set, described by the J_r value. Barton et al. (1977) suggest a probable pressure range confined within the shaded envelope. The pressure (expressed in kg/cm²) may also be estimated from the following relationships:

$$P_{i(\text{roof})} = \frac{2}{3} \cdot \frac{1}{J_r} \cdot J_n^{1/2} \cdot Q^{1/3} \tag{9-11a}$$

$$P_{i(\text{wall})} = \frac{2}{3} \cdot \frac{1}{J_r} \cdot J_n^{1/2} \cdot Q_w^{1/3} \tag{9-11b}$$

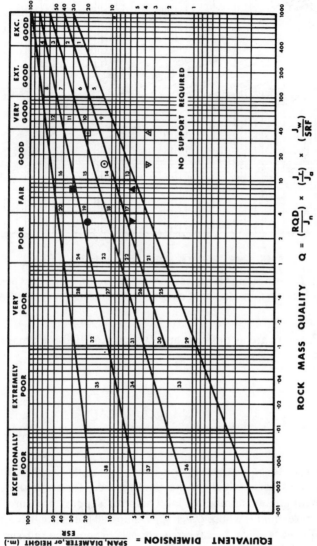

ROCK MASS QUALITY $Q = \left(\frac{RQD}{J_n}\right) \times \left(\frac{J_r}{J_a}\right) \times \left(\frac{J_w}{SRF}\right)$

Fig. 9-24 Excavation support chart correlating rock quality Q, relevant excavation dimensions, and actual support used from case records. (From Barton et al., 1977.)

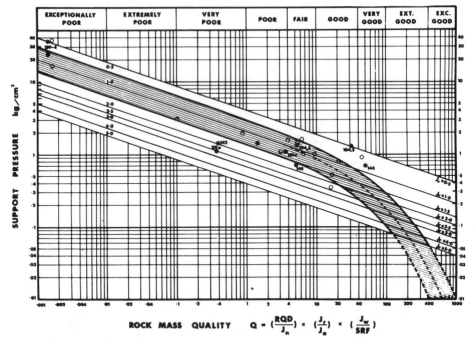

Fig. 9-25   Chart for estimating permanent support pressures. Numbered points refer to case records described by Barton et al., 1974a. (From Barton et al., 1977.)

where all parameters are as defined, and Q_w is a modified rock mass quality as follows for

$$Q > 10 \qquad Q_w = 5.0Q$$

$$10 > Q > 0.1 \qquad Q_w = 2.5Q$$

$$Q < 0.1 \qquad Q_w = 1.0Q$$

Equations (9-11) are equivalent to the chart of Fig. 9-25 when there are exactly three joint sets, which is the limiting case for three-dimensional movement. With greater number of joint sets the support pressure is likely to increase. Interestingly, where Q is higher than 100 the estimate of pressure becomes meaningless since the cavern is almost certain to be self-supported.

Barton et al. (1977) propose these relationships between support pressure and rock mass quality for static as well as dynamic loading in rock caverns. A suggested increase resulting from the passage of seismic waves is 20 percent for the case of lined excavations based on case records reported by Glass (1973). An increase in support pressure to respond to dynamic loading can be made in the rock mass classification. Reducing, for example, Q by 50 percent will yield a dynamic/static stress ratio of about 1.25.

9-6 ROCK TUNNEL REINFORCEMENT BY EQUIVALENT SUPPORT METHODS

This method allows estimation of a system of pattern rock reinforcement (anchors or bolts), which is structurally equivalent to an internal tunnel support. It is based on the assumption that the internal support is acted upon primarily by thrust. In order for the two systems to be structurally equivalent, the rock reinforcement must carry an equal amount of thrust. The thrust capacity of a zone of reinforced rock is function of its effective thickness and the increase in strength, resulting from the confining pressure exerted by the anchorage. The structural equivalence is therefore defined as a reinforcement pattern that will provide sufficient confinement to the rock mass to increase its own capacity by an amount equivalent to the support capacity provided by the internal support. The procedure has been proposed by Bischoff and Smart (1977).

Thrust Capacity of an Internal Support

Figure 9-26 shows a circular tunnel excavation together with an internal support equally spaced along the tunnel alignment. If the supports are sufficiently blocked so that rock pressures are uniform around the opening, flexural deformation of the support is small so that the assumption can be made that the only stresses induced in the system are axial thrusts.

The unit thrust capacity T_s of the support can be expressed as

$$T_s = \frac{\sigma_s A_s}{M} \tag{9-12}$$

where σ_s = yield stress of the internal support
       A_s = cross-sectional area of the support
       M = distance (spacing) between supports

For structural equivalence the anchorage must provide additional strength to the rock mass equal to T_s.

Thrust Capacity of an Anchored Rock Arch

The stability of the opening essentially depends on the shear strength characteristics of the jointed rock mass, its unconfined compressive strength, and the state of stress as excavation proceeds. Failure is imminent when the rock strength is exceeded. Radially installed reinforcement (anchors, bolts, or rebars) ties together the rock mass by increasing the confining pressure in the rock adjacent to the opening, making the rock mass self-supported. The increase in the confining pressure occurs in two ways. Initially, by tensioning of the anchors the rock mass is compressed between

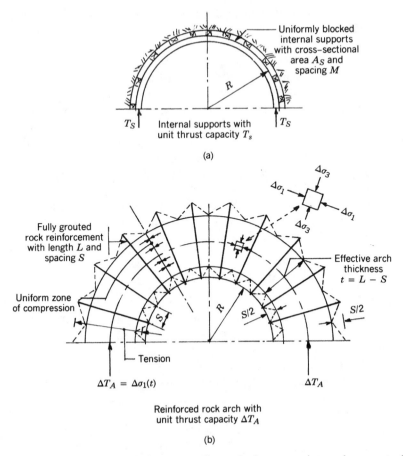

(a)

(b)

Fig. 9-26   Reinforced rock arch structurally equivalent to an internal support when $\Delta T_A = T_s$. (From Bischoff and Smart, 1977.)

the anchor ends. Thereafter, tensioning of any unstressed reinforcement occurring with dilation of the rock mass after excavation produces the same effects (see also Section 9-7).

Figure 9-26(b) shows the anchored zone of rock around a tunnel opening. Geometrically the two cross sections in parts (a) and (b) are similar. The rock arch is reinforced with grouted anchors of a given length and spacing. The additional unit thrust capacity of the reinforced zone depends on its thickness and the effective increase in axial stress which the rock can sustain as result of increasing the rock confinement. This function can be expressed as

$$\Delta T_A = \Delta \sigma_1 t \tag{9-13}$$

where ΔT_A = increase in rock unit thrust capacity per unit length
$\Delta \sigma_1$ = effective increase in allowable axial rock stress
t = effective thickness of reinforced rock zone

The relationship between $\Delta \sigma_1$ and the rock reinforcement confining pressure is function of the friction angle ϕ, and is shown in Fig. 9-27. If $\Delta \sigma_3$ is the increase in confining pressure induced by the anchorage at orientations normal to the boundary of the tunnel opening, the increase $\Delta \sigma_1$ can be estimated from the following

$$\Delta \sigma_1 = \tan^2 \left(45° + \frac{\phi}{2} \right) \Delta \sigma_3 \qquad (9\text{-}14)$$

Equation (9-14) shows the dependence of $\Delta \sigma_1$ on the rock mass friction angle ϕ. For low values of normal stress ϕ can be quite large (Barton, 1973; Goodman, 1974), and this is illustrated by the slope of the shear strength envelope in Fig. 9-27. A large angle of internal friction results in large effective increase in allowable arch stress even with small increase in confining pressure.

The parameter $\Delta \sigma_3$ can be expressed in terms of the geometry and the strength properties of the rock reinforcement as

$$\Delta \sigma_3 = \frac{\sigma_b A_b}{S^2} \qquad (9\text{-}15)$$

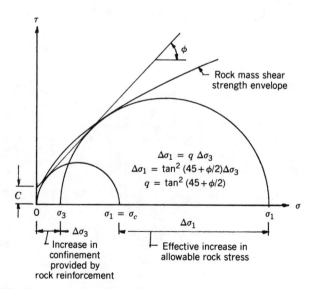

Fig. 9-27 Effective increase in allowable rock stress with increase in confinement. (From Bischoff and Smart, 1977.)

where σ_b = yield stress of the anchorage (per unit area)

A_b = cross sectional area of tendon or bolt

S = spacing of reinforcement pattern, both transverse and longitudinally to the tunnel axis

If the anchorage length is large relative to the tunnel radius, S^2 should be replaced by the product of the transverse spacing at the center of the reinforced zone and the spacing of the anchors longitudinally along the tunnel. Combining Eqs. (9-13), (9-14), and (9-15) yields

$$\Delta T_A = \left[\tan^2\left(45° + \frac{\phi}{2}\right)\right]\left[\frac{\sigma_b A_b}{S^2}\right] t \tag{9-16}$$

If L is the anchorage length, the effective thickness t of the arch can be taken as $t = L - S$. Setting also $\tan^2(45° + \phi/2) = q$ yields

$$\Delta T_A = q \frac{\sigma_b A_b}{S^2} (L - S) \tag{9-17}$$

Equation (9-17) can be expressed graphically in terms of anchor length and spacing and for rock masses with different friction angles as shown in Fig. 9-28. In these diagrams the thrust capacity of a reinforced rock arch ΔT_A is plotted versus the spacing S and the anchor length L, and for a given friction angle and anchor–bolt size.

The foregoing method of estimating anchor/bolt capacity takes into consideration the stress increase in the radial direction, which is most effective

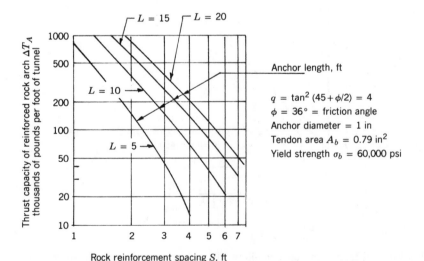

Anchor length, ft

$q = \tan^2(45 + \phi/2) = 4$

$\phi = 36°$ = friction angle

Anchor diameter = 1 in

Tendon area $A_b = 0.79$ in^2

Yield strength $\sigma_b = 60,000$ psi

Thrust capacity of reinforced rock arch ΔT_A thousands of pounds per foot of tunnel

Rock reinforcement spacing S, ft

Fig. 9-28 Thrust capacity of reinforced rock arch for various rock reinforcement patterns. (From Bischoff and Smart, 1977.)

in rock with high-friction angle. The analysis, however, does not include the technical cohesion and the improvement that can be achieved by reinforcement, particularly in small-size openings.

9-7 EFFECT OF CONFINING PRESSURE IN ROCK CAVERNS AND TUNNELS

Introducing a confining pressure to increase shear resistance is a mechanism available by the action of anchorages. Most rocks are indeed significantly strengthened by confinement, and this is more pronounced in a highly fissured rock. Sliding along a fissure is possible if the rock can displace normal to the average surface of rupture as shown in Fig. 9-29, but under confinement the normal displacement requires additional stress effort.

It is not uncommon for fissured rocks to undergo strength increases as much as 10 times a small increment in mean stress, and this characteristic makes anchorages ideally effective in supporting tunnels in weathered rock. However, as the mean pressure is increased and reaches a value known as the "brittle-to-ductile transition pressure," the rock behaves plastically; in other words, after this point continued deformation of rock can occur without any increase in stress (Goodman, 1989). This behavior, also referred to as "stress-hardening," can sometimes be observed at even higher pressures suggesting that the rock is actually strengthened as it deforms without a real peak stress.

Interestingly, the brittle-to-ductile transition occurs at pressures beyond the confining effects introduced by anchors in tunnels, with the exception of evaporite rocks and soft clay shales where plastic behavior can be exhibited at service loads. Transition pressures are shown in Table 9-7 for common rock types. Without confining pressures, a rock specimen can form more than one fracture, and where the ends are not smooth the specimen can split in two. As the confining pressure is increased the fractured specimen develops faults, but in soft rock this can occur even without confinement. If the specimen is short, continued deformation beyond the faulting zone will produce complex fracturing with associated strain-hardening behavior. At pressures beyond the brittle-to-ductile transition, failure is not explicitly manifested, but the rock exhibits parallel inclined lines that represent the intersection of inclined rupture surfaces and the surfaces of the specimen. The effect of confining pressure is also evident by the changing volumetric strain response that is observed in triaxial compression tests.

9-8 THE EXPONENTIAL FORMULATION THEORY FOR ROCK OPENINGS

The methods presented in Section 9-5 are semiempirical and to some extent based on precedent. If a cavern is contemplated in conditions not previously

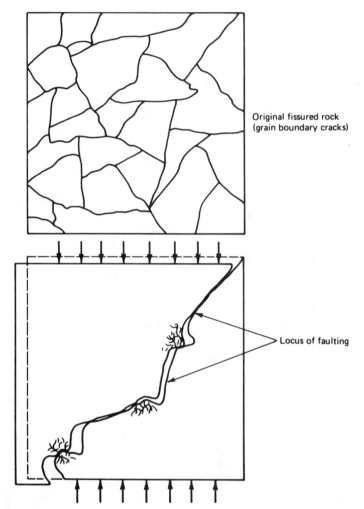

Original fissured rock
(grain boundary cracks)

Locus of faulting

Fig. 9-29 Dilatancy caused by roughness of the rupture surface. (From Goodman, 1989.)

encountered, the past experience must be extended by resorting also to observations to confirm the design assumptions. It appears that the support of underground openings is sometimes chosen with greater confidence knowing that many others in the past have been supported in the same way and performed satisfactorily.

Several theories have been proposed to explain the behavior and failure mechanism of the cavern roof, all beginning with the fact that excavation of the opening allows the virgin strains in the rock to relax so that the rock deforms inwards toward the opening and the stresses in the rock mass around

TABLE 9-7   Brittle-to-Ductile Transition Pressures for Various Rocks

	Gauge Pressure	
Rock Type	(MPa)	(psi)
Rock salt	0	0
Chalk	<10	<1500
Compaction shale	0–20	0–3000
Limestone	20–100	3000–15,000
Sandstone	>100	<15,000
Granite	⩾100	⩾15,000

From Goodman (1989).

the excavation are modified. These include the theories for elastic and plastic behavior in a wide spectrum of structural response such as flexural, beam, plate, buckling, and arching action. There is a certain inherent validity in all these theories—noting, however, that the phenomenon of failure in rock is not explicitly clear. Indeed, total loss of cohesion and strength may or may not occur according to the way the rock is loaded and is not necessarily a real rock property.

Behavior of Openings with Relatively Flat Roof

Figure 9-30 shows an opening in rock with relatively flat roof, and with a height less than the width. In part (a) the thinner, and flexurally less rigid, layer sags and tends to separate from the bed, whereas in (b) the thicker layer supports the thin one above. If the horizontal stress is high, buckling will accompany the flexural deformation. Under bending each layer slips with respect to the adjoining beds along the mutual contact surface, restrained only by the shear strength along these planes. Tensioned anchors or bolts installed normal to the bedding planes normally would be expected to create a composite action where the rock strata act as one unit.

However, tests by Panek (1963) show that with all the beds bolted the largest bending strain is observed in the bed with the highest flexural rigidity, and failure of the rock is initiated with the failure of the stiffest (usually the thickest) bed. The tensioned reinforcement, however, compresses the beds and increases the shear resistance between them with a reduction in the magnitude of the flexural stresses. If bedding plane slip could be prevented, a composite action would take effect and all layers would respond together as a single beam. Tests show, however, that this result is not totally achieved, hence a series of anchored beds is not equivalent to a single beam of the same composite thickness and flexural rigidity (Panek, 1955).

Based on these observations and confirmed behavioral pattern, a rock reinforcement factor has been proposed as the ratio of the minimum bending

Fig. 9-30 Two-member beam in a relatively rectangular opening in rock, height smaller than width. (From Lang and Bischoff, 1979.)

stress for an unsupported and supported roof. This is a function of the thickness of the bolted strata, the friction between adjoining layers, the weight of rock materials, bolt tension, pattern spacing, and the opening width (Panek, 1962b, 1964; Panek and McCormick, 1973). Further studies of the shear theory have been carried out by Fairhurst and Singh (1974), but they are directed to untensioned bolts acting mainly as shear dowels or keys between layers.

Modified Stress Field. When a rock mass is acted upon by an anchor or bolt in tension, it develops a compression zone normal to the line of tension somewhat between the bulk of the anchorage zone and the bearing plate. Studies of photoelastic and mathematical models (Lang, 1958, 1959, 1962) show that the extent to which the units interact is dependent on the length/spacing ratio. Typical stress trajectories are shown in Fig. 9-31. When the anchorage interaction is complete, it produces a relatively uniform compression zone near the central area formed as shown in Fig. 9-32.

These diagrams are for length/spacing ratios between 1.2 and 2.0. The boundary zone can be approximated by 45° lines originating at the anchoring ends as shown. It is apparent that the thickness of the compressed area increases with higher length/spacing ratio, but it is less than the anchorage length. The compression zone in turn generates compressive stresses normal to the anchorage direction as long as adequate restraint is available at the ends. This stress pattern is superimposed on the construction stresses induced from the excavation process, yielding a modified stress field. Thus, near the surface the latter are principal stresses, with the major and intermediate principal stresses being tangential to the new surface.

Reinforced Rock Structure, Stability, and Modes of Failure

Arch Effects. Ideally, the effectiveness of rock reinforcement is maximized if it can be installed in advance of the opening, and in practice this condition is approached by spiling anchors installed diagonally or from small drifts (see also the following sections).

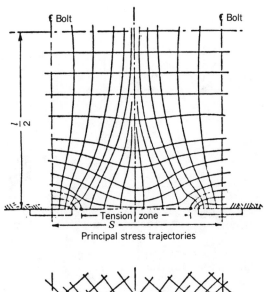

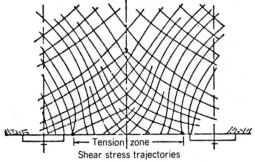

Fig. 9-31 Typical stress trajectories caused by tensioned bolts in rock. (From Lang et al., 1979.)

Usually, however, the installation of rock reinforcement follows the advance of the face. In this case the behavior of the roof strata follows a sequential pattern initiated by downward deflection already described as beam and shear action. Rock is inherently weak in tension, hence the beam effects dominate rock behavior until the tensile strength of rock is reached and exceeded or the process is disrupted by the presence of joints and discontinuities. This becomes evident when tension cracks are observed near the rock surface or close to joints. At this stage the reinforced roof is no longer an active beam, and the bending process merges into arching action expressed by zone *A–B–C* in Fig. 9-33.

The shape of the arch must be consistent with the loads acting on it, which are defined assuming the reinforced rock zone to act as separate structure from the surrounding rock mass. From Fig. 9-33 these loads are identified as follows:

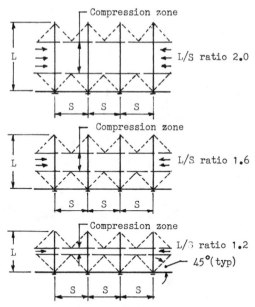

Fig. 9-32 Effect of *L/S* ratio on the thickness of compression zone, tensioned bolts in rock. (From Lang et al., 1979.)

1. Weight of arch *A–B–C*.
2. Weight of zone *D* below the arch suspended by the anchorage.
3. Weight of zone *E* just above the upper arch boundary but within the anchored zone.
4. Surcharge G acting on top of the anchored zone, from fractured or loose material.

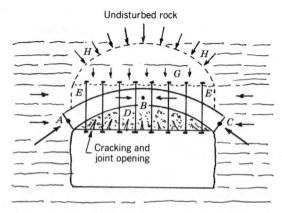

Fig. 9-33 The arch effect in a reinforced roof of a rock opening. (From Lang et al., 1979.)

5. Initial pressure along boundary *H–H* above which the rock remains undisturbed.

6. Abutment horizontal reactions at the ends *A* and *C* from horizontal stresses.

In conditions nearly uniform and where the effect of wedges, joints, and fractures is not magnified, the arch shape will approach a parabolic or circular configuration. With joints and fractures present, the arch has a modified shape similar to a segmented or voussoir unit, usually approximated by a three-hinged arch with hinges at points *A*, *B*, and *C*. On the other hand, the restraining and confining effect provided by the surrounding rock enhances the carrying capacity of the reinforced arch and should be considered.

Modes of Failure. A basic reinforced rock unit (RRU) is shown in Fig. 9-34. Its essential parameters, together with the tension, are the length *l* and spacing *S*. These parameters must be chosen so that a unit cannot behave independently, and near the abutments the unit is stable for the imposed loads and conditions.

The most common types of failure have their origin in joints, faults, and

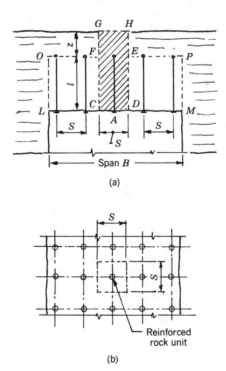

(a)

(b)

Fig. 9-34 Reinforced rock unit; (a) sectional elevation; (b) partial plan. (From Lang et al., 1979.)

structural fracture zones, and occur more frequently in massive shales or laminated sandy shales. These defects can weaken the roof structurally, or create a system with low shear strength and deformation modulus (Cox, 1974). Thus, failure can occur as follows:

1. The roof of the opening may fall as one unit by completely detaching from the rock mass. In this case the thickness of the fall will be roughly the area covered by the anchorage length.
2. Failure can occur not as complete collapse but in the form of fallout of a coherent block or anchored stratum. Such failure is generally sudden, and occurs with little or no warning.
3. Failure may be exhibited in the development of excessive and continuing deflection of the roof until the corresponding reduction in the rise of the arch affects its carrying ability.
4. Deformation or deflection of the roof can cause localized raveling between anchored units, with progressive fallout of rock fragments.

Anchorage Tension

Length and Spacing of Anchors or Bolts. Unlike conventional anchorages where anchor length is determined in conjunction with the presence of a competent zone for the fixed or bonded section, the required length for roof support initially depends on the number of falls and the thickness of rock strata. Useful guidelines are given by Panek and McCormick (1973). A common rock unit is 4 ft × 4 ft or 4 ft × 5 ft, but these suggested patterns are not necessarily ideal for every opening. Extra reinforcement is usually provided near the ends, and some units are installed diagonally as shown in Fig. 9-19(d). The spacing is furthermore related to the minimum tension since the required compression in the rock is inversely proportional to the RRU.

Installation Time. The consensus of opinion is that the reinforcement should be installed as soon as possible in order to arrest or stop unstable behavior (Lang, 1971). The significance of the time factor is demonstrated in field observations showing that about 60 percent of bedding plane slips occur before the reinforcement is in place (Fairhurst and Singh, 1974). Interestingly, considerable dilation of the roof occurs not immediately after blasting but within the second hour (Tincelin and Sinou, 1957).

Basic Anchorage Tension. Lang et al. (1979) and Lang and Bischoff (1981) have carried out analytical studies and tests to determine the optimum installation tension for bolts in mine roofs. Concepts and equations have been derived taking into account the parameters involved. This program has been supplemented by sensitivity analysis and physical model investigations. Because of its expedience and relevance to the general problem of rock caverns, the associated theory is reviewed in this section.

Figure 9-35(a) shows a typical reinforced rock unit acted upon by pertinent stresses. Let

σ_v = vertical stress in the unit at distance y from FE

σ_h = horizontal stress = $K\sigma_v$ if K is constant

q = stress superimposed on top surface FE by overlying strata $EFGH$, or other loading

τ = shear strength in the unit = $\sigma_h \tan \phi + c = K\sigma_v\mu + c$

μ = friction coefficient of rock

c = apparent cohesion of rock mass

γ = specific weight of rock

S = bolt spacing

σ_0 = vertical stress at CD

Noting that CD is a free surface, the value of σ_0 is given by

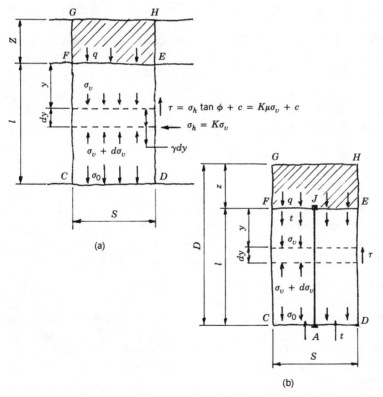

Fig. 9-35 (a) Stability of a reinforced rock unit; (b) stability of reinforced rock unit with bolt support. Suggested value of $K = (1 - \sin \phi)/(1 + \sin \phi)$. (From Lang et al., 1979.)

$$\sigma_0 = \left(\frac{\gamma S}{4} - c\right) \frac{1}{K\mu} (1 - e^{-4K\mu(z + l)/S}) \tag{9-18}$$

From Eq. (9-18) it follows that $\sigma_0 S^2$ is the basic direct support that must be provided by a prop under the rock unit to prevent its fallout from the surrounding rock. We note, however that if the term $[(\gamma S/4) - c]$ is zero or negative, theoretically support is not required. The value of σ_0 is thus dependent on the cohesion term c in intact rock, which is known to decrease rapidly along joints and discontinuities as deformation occurs. In deriving Eq. (9-18) the assumption is made that, provided fallout of the rock material is prevented, the rock unit $CDEF$ has undergone a deformation leading to failure according to Coulomb's criteria.

RRU—Passive Reinforcement. If a stabilizing pressure t is selected equal to σ_0, the total force T necessary to support the rock unit is

$$T = \sigma_0 S^2 \tag{9-19}$$

where σ_0 is computed from Eq. (9-18). The bolt tension introduced in A as shown in fig. 9-35(b) will, however, be transfered through the bolt unit so that an equal reaction will appear at the anchorage J; thus, for a passive support pressure t of CD an equal stress is added at FE from the bolt anchorage J. Under these conditions σ_0 at CD is now given by

$$\sigma_0 = R + te^{-4K\mu l/S} \tag{9-20}$$

where the term R is the second part of Eq. (9-18). From $t = \sigma_0$, it follows that

$$t(1 - e^{-4K\mu l/S}) = R \tag{9-21}$$

where R is likewise the second term of Eq. (9-18). The rock bolt load is now $T = tS^2$, where t is estimated from Eq. (9-21), and is just sufficient for a passive action after the deformation of the rock unit $CDGH$ has developed full shear resistance along its periphery.

RRU—Active Reinforcement. If the bolts are installed and tensioned immediately after excavation exposes surface CD in Fig. 9-35(b) so that significant deformation is prevented, they will apply an active stress t to the rock unit at the surface CD and at the anchorage plane FE. In this case the average compression in the rock between the bolt ends will be the average stress intensity t contributing actively to equilibrium. Setting $z + l = D$, the equation determining t is now

$$t(1 - e^{-4K\mu l/S}) = \alpha \left(\frac{\gamma S}{4} - c - h\mu\right) \frac{1}{K\mu} (1 - e^{-4K\mu D/S}) \tag{9-22}$$

and the tensile load in the rock bolt is $T = tS^2$. The parameter α in Eq. (9-22) varies from 0.5 to 1.0, and depends on the time lapse between exposure of the roof and anchorage installation. The factor h is an assumed horizontal field stress acting normal to the bolt line.

Structural Member. Equations (9-18) through (9-22) express the protection against fallout of an individual rock unit. The entire opening, however, is initially a beam merging to an arch; hence the bolt tension is affected by several factors in addition to the requirements of the relative RRU stability. Thus the bolt tension may include a suspension component induced by the tendency of the beds to separate; a confinement component necessary to enhance the compressive and shear strength of the confined rock layers; a dilation component induced as tension in the bolts resulting from volumetric strain of the rock; time effects related to the time lapse between roof exposure and support construction; and a system component associated with the length of bolts and pattern spacing.

For a circular or parabolic arch the bolt length l is approximately proportional to the square of the opening width B and inversely proportional to the rise h in Fig. 9-36. This is expressed as

$$l = C \left(\frac{B^2}{h} \right)^{1/2} \tag{9-23}$$

in which C is a constant usually approximated as 1.0.

Parametric and Sensitivity Studies

The general exponential formulation theory reviewed in the foregoing sections has been further investigated by Lang and Bischoff (1981) in parametric and sensitivity studies involving the bolt tension. These studies indicate the following.

1. The rate of increase of bolt tension T as the spacing S increases is relatively low if S is less than $l/2$, that is, when the ratio l/S is greater than 2. With greater S values the rate of increase of bolt tension accelerates and approaches very high values if S exceeds l.
2. With increasing bolt length l, the bolt tension approaches a minimum constant value according to the friction angle ϕ. As this angle increases

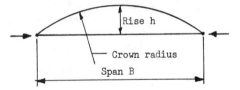

Fig. 9-36 Arch rise and span.

the bolt length l decreases, indicating very long bolts are required in special conditions.

3. As the height of distressed rock above the roof increases, the bolt tension T for a given bolt length and spacing approaches a limiting value that decreases as the friction angle increases.

4. The bolt tension increases with the time of installation after the roof is exposed.

5. The bolt tension decreases with increasing ratio $K = \sigma_h/\sigma_v$, but the rate of decrease becomes smaller with increasing friction angle.

6. The bolt tension T is quite sensitive to the apparent cohesion and the in situ horizontal stress σ_h. The consensus of opinion is, therefore, that for the initial design of the reinforcement system these parameters should be ignored.

7. Apart from the physical characteristics of rock affecting bolt tension, the essential parameters determining T are the ratios of bolt length and depth of distressed rock to bolt spacing.

9-9 SPILING REINFORCEMENT IN ROCK TUNNELS

Prereinforcement, installed in advance of the opening, extends the stand-up time by preventing loosening and contributes to the permanent stabilization of the excavation by restricting deformations. There are two forms of pre-reinforcement, both involving untensioned fully grouted steel bars that are placed ahead of the excavation, shown in Fig. 9-37. In part (a) the reinforcement is installed from small drifts, whereas the reinforcement shown in (b) consists of spiles diagonally placed. Both systems have similar characteristics and effects, but to a different degree.

The behavior of an unreinforced or unsupported opening is idealized in the diagram of Fig. 9-38, showing rock mass deterioration as time-dependent process. The mass responds to creep effects produced partly or wholly by compaction, consolidation, rebound, shearing, fractures, and crushing of

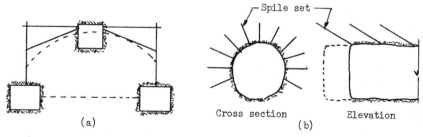

Fig. 9-37 Prereinforcement in tunnels; (a) small drifts; (b) spiling reinforcement ahead of face. (From Korbin and Brekke, 1976.)

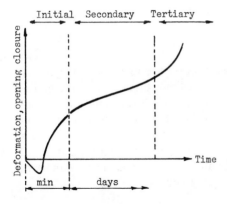

Fig. 9-38 Idealized deformation history of unstable unreinforced or unsupported opening. (From Korbin and Brekke, 1976.)

intact rock or material between intact rock blocks. If the ground is of poor quality and the construction loads are high, the accelerating deformation in the tertiary region will culminate in the collapse or failure of the opening.

Results from model studies carried out by Korbin and Brekke (1976) show that prereinforcement changes the behavior of the opening markedly. Comparison of radial and axial deformations between the unreinforced or unsupported and the prereinforced model shows that the latter exhibits significantly reduced accelerating rates of deformation. However, the prereinforcement system shown in Fig. 9-37(a) is superior to the spiral arrangement shown in (b) with reference to the immediate stabilization of the opening.

Physically the influence of prereinforcement can be explained as a strength increase contributed to the reinforced region. The major component of this increase is generated by deformation-induced tension in the reinforcement proportional to the radial strain distribution confirmed by field investigations (Korbin and Brekke, 1976). With a distribution of strength increase rather than a mere increase in cohesion, the unconfined material strength increases considerably. By placing grouted spiles ahead of the tunnel face, an effective reinforcement is installed before any significant deformation can occur. Furthermore, the spiles are contained within ground that has undergone relatively minor disturbance. With reference to time, this explains why prereinforcement is more effective than other forms of support, and why placed ahead of the face is more effective than spiles at the face.

Interestingly, reinforcement will not work if the intent is to pin and tie together loose rock masses, but only if it is to prevent loosening. Since the region most susceptible to deterioration is within one-half radius from the opening, the reinforcement should extend well beyond this region. In the region which is more than one radius from the opening the strains are low whereas the confinement is high so that an increase in strength associated with the presence of spiles will be small. Field studies of tunnel prereinforcements are presented in Chapter 11.

9-10 ELASTIC AND PLASTIC BEHAVIOR OF ROCK TUNNELS

Elastic Solution

A tunnel excavation can be idealized as a circular opening in a homogeneous mass. If the rock is stressed below the elastic limit, taken about one-half its compressive strength, designers usually consider elastic solutions. These involve a plane strain equivalent of a hole in a plate of isotropic and continuous material. The stress derivation is known as the Kirsch solution, but another version using real stress functions is presented by Obert and Duvall (1967). A point located at polar coordinates r and θ near an opening with radius a as shown in Fig. 9-39 has stresses given by

$$\sigma_r = \frac{p_1 + p_2}{2}\left(1 - \frac{a^2}{r^2}\right) + \frac{p_1 - p_2}{2}\left(1 - \frac{4a^2}{r^2} + \frac{3a^4}{r^4}\right)\cos 2\theta \quad (9\text{-}24a)$$

$$\sigma_\theta = \frac{p_1 + p_2}{2}\left(1 + \frac{a^2}{r^2}\right) - \frac{p_1 - p_2}{2}\left(1 + \frac{3a^4}{r^4}\right)\cos 2\theta \quad (9\text{-}24b)$$

$$\tau_{r\theta} = -\frac{p_1 - p_2}{2}\left(1 + \frac{2a^2}{r^2} - \frac{3a^4}{r^4}\right)\sin 2\theta \quad (9\text{-}24c)$$

where σ_r = radial stress
σ_θ = tangential stress
$\tau_{r\theta}$ = shear stress
p_1 = applied horizontal stress
p_2 = applied vertical stress

Setting $r = a$ in Eq. (9-24) yields $\sigma_r = \tau_{r\theta} = 0$; thus, the radial and shear stresses are zero since this is a free surface. The tangential stress σ_θ varies from a maximum value $3p_1 - p_2$ at $\theta = 90°$ to a minimum $3p_2 - p_1$ at $\theta = 0°$. From Fig. 9-39(b) it can be shown that stress concentrations fall off quickly away from the opening.

Radial and tangential stresses plotted in Fig. 9-39 are expressed as ratios of the appropriate stress (radial or tangential) to the in situ vertical stress for various values of K (coefficient of the at-rest pressure). It can be seen that σ_r and σ_θ converge to the at-rest values. As the r/a ratio approaches 4, the tangential stress concentrations can be as much as three times the at-rest value at the boundary. Since the tangential stresses are carried by the rock itself and the net radial stress at the boundary is zero, the opening theoretically can remain stable without internal support.

The solution presented in Eq. (9-24) allows estimation of potential effects of joints around the tunnel. If in a given position and orientation a joint results in no changes in the stress field, the shear and normal stresses along its surface are compared with the limiting values of shear stresses consistent with the criteria of peak shear strength (Goodman, 1989). In this manner a

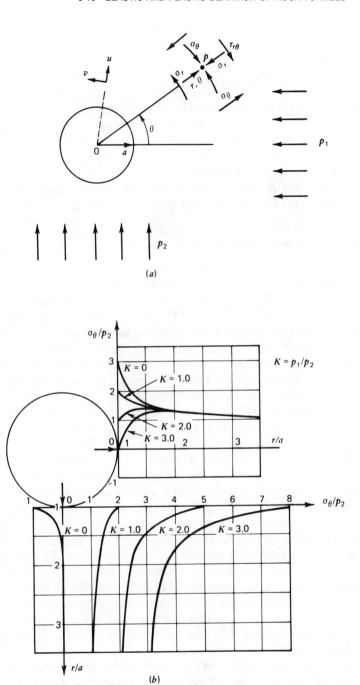

Fig. 9-39 Stresses distributing around a circular hole in isotropic. linearly elastic. homogeneous continuum. (From Goodman, 1989.)

region of joint influence is defined and helps isolate potential problem areas. However, Eq. (9-24) does not fully represent effects of gravity, which, besides creating a vertical stress in the form of p_1 or p_2, also exerts a redundant force on loose rock near the roof (see also Section 9-5). Clearly, this effect cannot be modeled by any set of loads on the boundary. Gravity loads could be the weight of rock in the region of joint slip calculated according to the Kirsch solution, and assigned to the system of supports, in this case anchors, bolts, and shotcrete. By omitting gravity loads, the Kirsch solution cannot isolate tunnel size effects, hence it is not consistent with the fact that a small tunnel is more stable than a large one. Consideration of size effects in tunnel design is necessary not only in connection with gravity loads acting on the rock near the opening, but also because of size effects on material strength.

Plastic Response around Tunnels

In general, the plastic region is assumed to have been reached when a developing stress approaches a certain fraction of the rock strength. For example, Rabcewicz (1969) suggests that the rock reaches the elastic limit when the tangential stress concentration at the border of the opening exceeds the compressive strength. Deere et al. (1969) define this limit as the point at which the same strength is exceeded by the vertical in situ stress. Ward (1978) proposes an elastic limit that is reached when the boundary deviator stress equals the compressive strength. Irrespective of these definitions, when the tangential stress becomes greater than about one-half the unconfined compressive strength, cracks will begin to form. These cracks should not be related to the usual rock breakage due to construction and an accompanying zone of relaxation around the opening. The new cracks indicate rather the formation of separate slabs parallel to the periphery. For a partially yielded rock mass, assumed for simplicity to be annular in shape and with a uniform radius, elastic–plastic analysis is indicated and therefore the circumstances under which the rock yields become irrelevant.

Plasticity is commonly experienced in squeezing ground when the gradual depletion of strength drives the zone of broken rock deeper into the tunnel until the supports undergo a gradual buildup in pressure. In this case two distinct forms of behavior may be exhibited as shown in Fig. 9-40. If the rock tends to arch and the supports can provide sufficient resistance to stop progressive deterioration, the inward displacement of the walls will decrease with time approaching an asymptote (stable condition). If the erection of the support is delayed or the rock induces very high loads, the inward movement will accelerate creating unstable conditions.

Theoretical expressions for the radial and tangential stresses in the plastic region and also for the radius of the plastic–elastic zone have been derived by Bray (1967) assuming the theoretical model shown in Fig. 9-41. If the construction of the tunnel creates intolerable conditions of stress and deformation, the rock is assumed to fail according to the Mohr–Coulomb the-

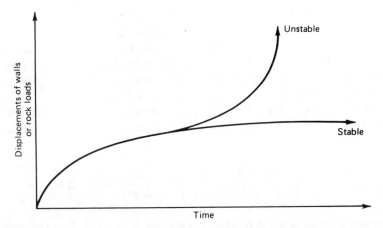

Fig. 9-40 Convergence between the walls of tunnel corresponding to stability and instability. (From Goodman, 1989.)

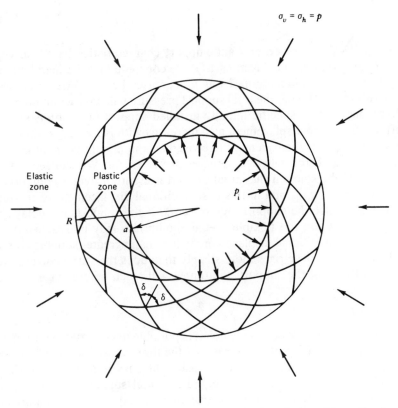

Fig. 9-41 Conditions assumed for Bray's elastic–plastic solution. (From Goodman, 1989.)

ory. In this model the fracture lines are log spirals inclined at δ degrees with the radial direction, a solution considered appropriate and acceptable for clays and shales. All three parameters (R, σ_r, and σ_θ) are expressed in terms of the initial rock stress ($\sigma_v = \sigma_h = p$), the unconfined compressive strength q_u of intact rock, and the internal pressure p_i provided by the support (see also Goodman, 1989).

If the broken rock inside the plastic zone contains log spiral surfaces with shear strength $\tau_p = s_i + \sigma \tan \phi_i$, the radius R of the plastic–elastic boundary is

$$R = a \left(\frac{2p - q_u + [1 + \tan^2(45 + \phi/2)]s_i \cot \phi_i}{[1 + \tan^2(45 + \phi/2)](p_i + s_i \cot \phi_i)} \right)^{1/Q} \tag{9-25}$$

where the quantity Q is given by $Q = \tan \delta / \tan(\sigma - \phi_i) - 1$. For a complete solution an additional support pressure is included for the portion of the weight of the overlying material, which is equal to $m\gamma(R - a)$ where m is a coefficient ≤ 1 so that the total pressure to be provided by the support is

$$p_{i,\text{total}} = p_i + \gamma (R - a)m \tag{9-26}$$

Figure 9-42 shows criteria and support characteristics based on Bray's elastic–plastic solution. From part (a) we conclude that the load increment due to rock loosening decreases as p_i increases because R/a increases inversely with p_i. Hence, the total support pressure will have a minimum value corresponding to a specific initially installed support pressure as shown in part (b). Since the displacement also increases with increasing ratio R/a, the curve of p_i versus displacement likewise has a minimum value at $u_r = u_{\text{crit}}$ as shown in part (c), where u_r denotes displacement. We can see now that a tunnel will be stable if the initial support pressure provides equilibrium at a value of $u_r < u_{\text{crit}}$, and in this case additional displacement of the tunnel walls will reduce the support requirements. If the supports are delayed too long or if they are too flexible, initial equilibrium may be reached at $u_r > u_{\text{crit}}$, and any additional displacement will demand greater support pressure. If the $p_{i,\text{total}}$–u_r curve rises more steeply than does the curve corresponding to the stiffness of the support, the tunnel will reach failure stage.

Example

Table 9-8 shows data from which it is possible to construct a design curve as in part (c) of Fig. 9-42. For $m = 1$ the three values of initially designed support pressure of 40, 5, and 1 psi yield final support pressures 64.6, 58.7, and 104.5 psi, respectively. For initial support pressure >40 psi supplied quickly with a reserve load-carrying capacity, the support would begin to acquire additional load after installation and if it posseses sufficient stiffness the tunnel would reach equilibrium. For a support pressure

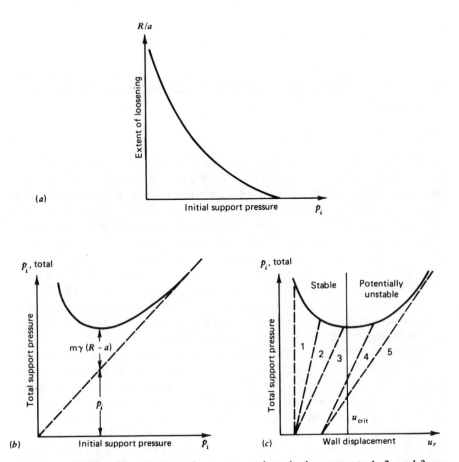

Fig. 9-42 Criteria for selection of supports and methods; supports 1, 2, and 3 are safe; support 4 is potentially unsafe; support 5 is unsafe. (From Goodman, 1989.)

TABLE 9-8 Data Used to Construct a Design Curve. Plastic Solution for a Tunnel

p_i (psi)	u_r (in)	R/a	$R - a$ (ft)	$m\gamma(R - a)/144$ (psi)	p_i(total) (psi)
40	0.62	3.47	19.8	20.6	64.6
5	1.55	7.44	51.5	53.7	58.7
1	4.26	13.42	99.4	103.5	104.5

$a = 8$ ft, $\gamma = 150$ lb/ft^3, $m = 1$.

From Goodman et al. (1989).

supplied too slowly or in conjunction with a very flexible support, the rock may become loose and reach the potentially unstable condition of Fig. 9-42(c) or even collapse.

9-11 ROCK TUNNEL REINFORCEMENT BY THE CONVERGENCE–CONFINEMENT METHOD

Fundamental Tunnel Behavior

The methods reviewed in Sections 9-5 through 9-10 reflect specific aims as well as engineering precedent. The American practice, for example, emphasizes the prevention of movement in order to preserve the strength of intact rock. The main justification is that tunnels in this country are constructed largely in hard elastic rock in which maximum strength is preserved if loosening can be avoided. Unlike this trend, rock reinforcement development in Europe has been stimulated by the construction of formidable size hydropower caverns and the parallel need for large-diameter transportation tunnels in the Alps and in Scandinavia. This work is primarily in hard ground, including plastic and squeezing rock. In these conditions, controlled movement is encouraged so as to mobilize the rock strength and achieve a corresponding reduction in support requirements.

From the foregoing it follows that the internal support of an opening does not act independently under a given set of loads, but its behavior is also governed by the characteristics of the surrounding ground. On the other hand, we know from experience that grouting (i.e., filling the voids and the annular space) consolidates the surrounding rock so that asymmetrical loads are not generated. As long as a good contact exists the support tends to redistribute the loads back to the ground. Axial stiffness, on the other hand, appears to be more important than flexural strength. The latter is small compared to the shear strength of rock; hence it does not significantly change ground behavior. However, increasing support flexibility enables the support to adapt successfully to ground deformation.

If the rock around an opening is acted upon by prestressed reinforcement, the ground will respond by a nonlinear resistance to compression as a fully confined medium. A confining pressure induced in the rock mass at some appropriate time can inhibit ground deformation. Consequently, an internal support cannot be loaded by ground deformations that occurred before the support was installed. A support erected after the ground has deformed and become stable will resist only subsequent loads such as groundwater pressure, long term creep and affects of continuing construction.

This summary sets forth the basic principles of tunnel support design and introduces movement as a prime factor in addition to stress and loads (Kuesel, 1986). By delaying the installation of the support until an optimum condition is reached, ground movement is allowed to continue until the rock strength is mobilized (for the elastic–plastic analysis of squeezing rock this would be the u_{crit} point in part (c) of Fig. 9-42). The real problem is, therefore, manifested as a complex ground–support interaction in which time becomes an important element.

Time-Dependent Behavior. Although there is no rational method for predicting the time to failure, commonly known as "standup time," engineers and contractors estimate this time through the use of the geomechanics classification. Essentially, this takes into account the rock mass rating and the active span, defined as the minimum of the tunnel width and the unsupported length of the face.

Noting that no theories can express the loosening of layered rocks or plastic behavior around tunnels in terms of an explicit time factor, this dependence is commonly understood by observing the full effect of excavation for a few days and sometimes a few weeks. Typically, crack propagation and stress redistribution are manifested within such a time frame, but other phenomena can also cause a tunnel to squeeze (e.g., rock load changes, water effects such as swelling, and rock behavior as viscous or viscoelastic).

Although complete theoretical proof is not available, a linear viscoelastic response is often assumed when we make predictions of tunnel displacement rates. Such a model, however, should be used with caution noting that with time dependency involving changes in geometry of the rock mass due to crack growth, the response is nonlinear.

Principles of NATM

Although similar "yielding ground" systems have been mentioned in Britain, Switzerland, and Scandinavia (Kuesel, 1981), the NATM (New Austrian tunneling method) has probably attracted more attention because of the documented applications. The method relies initially on the strength of rock for support and tunnel stability. Shotcrete, anchors, and bolts provide additional support and surface protection. A further objective is to avoid damage to the surrounding cavity caused by construction operations, especially blasting with intent of fast advance. Allowing some yielding and rock displacement to occur is part of the concept so that by certain limited deformations a new secondary stage of equilibrium can be reached and then maintained by controlled pressure release. Excavation and support installations are coordinated with the time-dependent stress rearrangement through appropriate measurements of stress and deformation.

Shotcrete as Temporary Support. Shotcrete is the thin-membrane system mentioned in Section 8-4. It is a semirigid lining usually closed by an invert. By adhering directly to rock it prevents loosening, whereas its early high strength helps attain good bearing capacity rapidly. The shotcrete is applied by pneumatically spraying a thin concrete mix in layers thicker than 25 mm

(1 in) containing coarse-grained aggregate [>10 mm ($\frac{3}{8}$ in) in size]. Strength development is accelerated by the use of special additives which also enhances adherence to rock (Nussbaum, 1973).

The application of shotcrete to exposed rock surfaces seals open joints and fissures, and helps inhibit movement. Stress measurements at the lining surface show that the shotcrete absorbs the tangential stresses close to the face.

System Anchoring and Bolting. A thin semiflexible lining of shotcrete is sufficient as protective or strengthening element for openings with sections 14–20 m² (150–220 ft²) in competent rock. For larger openings and relatively weak rock the support should be expanded with systems anchoring or bolting. Unlike rock bolts used locally, system anchoring provides rock stabilization through pattern reinforcement. Radially oriented anchors or bolts are used to hold together individual rock blocks along the tunnel perimeter and form a support arch. With prestressed anchors and following some movement of the reinforced ring toward the opening, normal stresses are developed and a support is formed having a specific carrying capacity and radial resistance. A stable state is maintained as long as the rock betewen anchorage points is firm enough to be self-supported and the lining prevents the detaching of rock sections.

If the anchors intersect unfavorably oriented shear planes, shear zones are not likely to develop. This behavior has been observed in tunnel work but is not fully understood as yet. On the other hand, the prestressed anchorage is compatible with the concept of the yielding ground. The considerable deformation and changes in shape occurring with conventional linings under lateral pressures are in this case absent, and the anchored circular section simply becomes smaller through a reduction of its periphery.

Rock-Yielding Theories

Support Reaction Curve. Figure 9-43 shows a ground–support reaction curve correlating radial displacement with support pressure. The support is installed at point D relatively close to the face where an immediate elastic displacement u_p has already occurred. When the support is installed (in this case a liner) the pressure acting on it is zero, but as the face advances a load is manifested on the support that deflects according to line DB. This is the support reaction curve, and evidently the support pressure reaches a maximum value P_e at point B corresponding to displacement u_s. This condition is reached when the support is in the plane strain zone relative to the face. The effect of support stiffness is reflected by the direction of the reaction curve DB. For stiff, unyielding support the reaction curve is almost vertical meaning that the liner pressure reaches the maximum potential value without further displacement. With increasing support flexibility the pres-

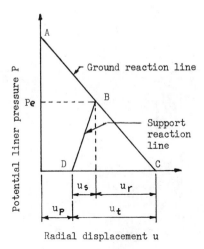

Fig. 9-43 Ground-Support Reaction diagram; elastic conditions. (From Ranken et al., 1978.)

sure on the support decreases following a corresponding increase in displacement.

The generalized solution shown in Fig. 9-43 is not entirely valid because of the three-dimensional nature of the problem. Fairly accurate approaches are, however, possible using finite element models and examples are given by Einstein and Schwartz (1980).

Zone of Yielded Rock. The rock yielding theory assumes a zone of yielded rock with an annular shape and uniform radius R as shown in Fig. 9-44 (Kastner, 1962). For the elastic–plastic model a net radial stress exists at the excavation boundary so that equilibrium is attained only when this stress is balanced by an internal pressure p_i. Likewise, the magnitude of the radial stress is dependent on p_i (Druss, 1984). The radius R of the plastic zone is given by Lombardi (1970) as

$$R = r \left[(1 - \sin \phi) \frac{p_2 + c \cot \phi}{p_i + c \cot \phi} \right]^{(1 - \sin \phi)/2 \sin \phi} \tag{9-27}$$

where r = tunnel radius
p_2 = applied vertical stress
c = apparent cohesion of the rock mass

From Eq. (9-27) we note that R is inversely proportional to p_i meaning a tunnel behavior similar to the Bray solution. An increase in R is accompanied by an inward radial displacement (convergence) at the tunnel opening. The R values from Eq. (9-27) are in fairly close agreement with values obtained from Eq. (9-25).

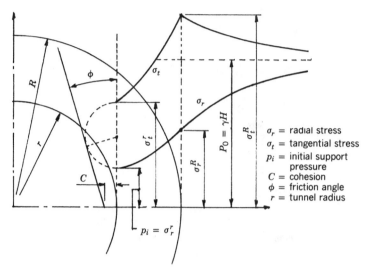

Fig. 9-44 Schematic presentation of stress distribution around a circular opening in rock; elastic—plastic solution. (From Kastner, 1962.)

Ground–Support Interaction. Since support resistance and internal pressure p_i required to provide stable conditions decreases with increasing convergence of the opening, we can conclude that flexible supports allowing some yielding of the rock will eventually carry less load than will stiff rigid linings that prevent yielding (Deere et al., 1969). Reduced support resistance with increasing convergence means that rock strength is mobilized in a process similar to the change of the at-rest pressure to an active value by causing the shear strength to be mobilized through appropriate movement.

The mechanism of rock yielding can be represented in a ground reaction curve in which p_i is plotted versus inward radial displacement as shown in Fig. 9-45. Three zones are distinguished: elastic, yielding and loosening. Interestingly, beyond a certain point in the convergence process the net value of p_i necessary for equilibrium begins to increase signaling the beginning of loosening with a breakdown of the arch effect. At this stage the load to be carried by the support is a gravity load (see also Section 9-5) representing the volume of loosened rock above the crown. Theoretically, therefore, it is desirable to install the support just before detrimental loosening occurs.

The dashed portion of the ground reaction curve represents an opening which is stable without support, and this condition is possible if the net radial stress can be balanced by the apparent cohesion of the rock (Ranken et al., 1978).

A support reaction curve, which plots the pressure–displacement function of the support, can be superimposed on the ground reaction curve to produce the composite ground–support curve shown in Fig. 9-46. The support is

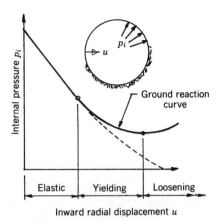

Fig. 9-45 Ground reaction curve. (From Ranken et al., 1978.)

installed after immediate displacement u^* has occurred, and thereafter the pressure begins to increase around the opening because of time-dependent effects and advancing of the tunnel face. Simultaneously, the net radial stress in the rock p^* at radial displacement u^* also decreases, until equilibrium is reached when the support reaction curve intersects the ground reaction curve. The interaction begins with the time of support installation, and when equilibrium is reached both the ground and the liner have undergone additional displacement u_1 for a corresponding pressure p_1. With infinitely stiff support, additional deflection would not occur and the support pressure would remain p^*.

Theoretical Background

Assumptions. Invariably, ground–support interaction theories assume the following: (a) plane strain conditions exist, (b) the rock is homogeneous and isotropic, (c) the shear strength parameters represent the discontinuities of the rock (joints, bedding planes, etc.), (d) the opening is sufficiently deep so that the vertical stress can be taken uniform, (e) the rock behavior is elastic–plastic, and (f) Coulomb's law is applicable.

Stress Distribution and Rearrangement. Figure 9-47 shows the assumed distribution of stress around an opening (note that this is an expanded version of Fig. 9-44), including the effects of yielding. The radial and tangential stresses represented by the solid lines are before convergence occurs, whereas the dashed line shows stresses that exist after the plastic stage is reached and the convergence has occurred. Both stresses decrease considerably at the boundary of the opening after rock yielding, and also with increasing R. The radial stress in particular is of main interest since it must be balanced by the tunnel support. The relationship between internal support

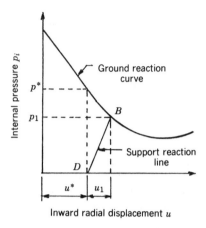

Fig. 9-46 Composite ground–support reaction curve. (From Ranken et al., 1978.)

pressure p_i is given as follows (Rabcewicz, 1964; Fenner-Talobre, 1957; Kastner, 1962):

$$p_i = -c \cot \phi + [c \cot \phi + p_0(1 - \sin \phi)] \left(\frac{r}{R} \right)^{2 \sin \phi/(1 - \sin \phi)} \quad (9\text{-}28a)$$

where $p_0 = \gamma H$ = vertical overburden stress. If cohesion is omitted, the internal support pressure is

$$p_i = p_0(1 - \sin \phi) \left(\frac{r}{R} \right)^{2 \sin \phi/(1 - \sin \phi)} \quad (9\text{-}28b)$$

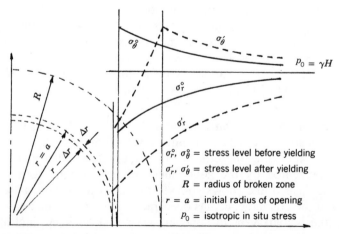

Fig. 9-47 Effect of yielding on stress distribution around an opening. (From Rabcewicz, 1973.)

The interaction between convergence, radius of plastic zone and internal support pressure is plotted in Fig. 9-48, known as the Fenner–Pacher diagram (Rabcewicz, 1973). The relations expressing the elastic–plastic behavior yield the dashed portions of the curve at the higher values of Δr, but the loosening effects are not directly quantified. Rabcewicz (1964, 1965, 1969) attributes the deviation from the theoretical behavior to decreased shear strength of the rock as progressive loosening continues. With advanced yielding, residual shear strength governs rock behavior and a decrease in ϕ and E combined with loosening effects causes the radial pressure to increase with further yielding. The net result of this process is higher load on the support as shear and compressive stresses are no longer effectively transmitted to the rock mass around the opening. Using pertinent parameters, the plot of the ground reactions curve can be obtained in finite element analyses (Druss, 1984).

Other yielding theories consider the interaction between ground and internal support a statically indeterminate system (Lombardi, 1970), whereas Ward (1978) emphasizes the shear strength properties of rock to interpret the mechanism of yielding.

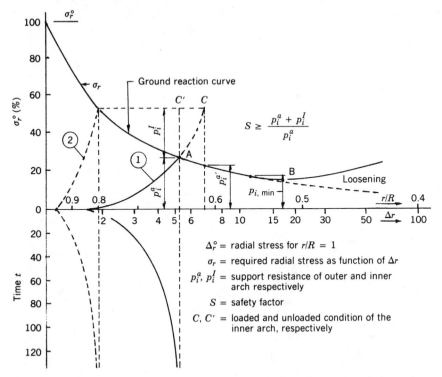

Fig. 9-48 The Fenner–Pacher diagram. (From Rabcewicz, 1973.) Schematic presentation of the $\sigma_r/\Delta r$ curve showing the reciprocal relations between σ_r, Δr, r/R, and t for support of different yields 1 and 2 and time of application.

Shear Failure Theory. When a cavity is made in rock, the accompanying stress rearrangement disturbs the original equilibrium according to the extent to which the shear strength of the rock is exceeded. The process is shown in Fig. 9-49 (Rabcewicz, 1964), and occurs in three progressive stages. Initially, wedge-shaped bodies on either side are sheared off along the Mohr surfaces and move towards the cavity and in a direction normal to the main pressure (stage I). The increased effective span thus produced causes the roof and the floor to begin a converging process as shown in stage II. The final stage (III) is characterized by increased movement where the rock buckles under continuous lateral pressure and protrudes into the cavity. Pressures arising from this action are, likewise, termed "squeezing pressures" (see also Section 9-10).

Design Requirements

The Fenner–Pacher Diagram. This diagram, shown in Fig. 9-48, provides the basis for support design according to NATM (Rabcewicz, 1973). Equilibrium is reached when the support reaction curve intersects the ground reaction curve, and this should be achieved in the descending portion of the curve. If the liner or other support should yield or fail, the effect theoretically would be a decreased load acting on the support. If the support were to be installed after detrimental loosening is manifested, any yielding or failure of the support means increased support resistance necessary for equilibrium. Hence, the intersection of the two curves must occur left of point B in the diagram. On the other hand, optimum design is achieved if the intersection is near point $p_{i,\min}$, but in practice establishing an accurate value of $p_{i,\min}$

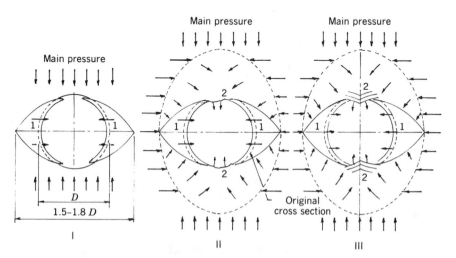

Fig. 9-49 Mechanical process and sequence of shear failure around a circular cavity by stress rearrangement pressure. (From Rabcewicz, 1964.)

will require extensive instrumentation and field measurements, the cost of which may well offset savings in the cost of support. Thus curve, marked ① in Fig. 9-48 represents a typical design case, and the support reaction to achieve equilibrium is p_i^a, commonly referred to as "initial support" or "outer lining." A final or inner lining is placed after time-dependent deformations have occurred, and provides a resistance p_i^1 as shown in Fig. 9-48 so that the total support resistance is $p_i^a + p_i^1$.

Shotcrete Design. As mentioned in the foregoing sections, the shotcrete lining is not intended to support gravity loads of loosened rock, but rather to stabilize the surrounding rock and preserve its shearing resistance so that the rock itself can become a support component through arching. With the two media well bonded, a close interaction is formed where the rock and shotcrete tend to act as one unit. The shearing resistance provided by the shotcrete lining is (Rabcewicz, 1969, 1973)

$$p_i^s = \frac{\tau_s d}{(b/2) \sin \alpha} \tag{9-29}$$

where p_i^s = shear resistance of shotcrete lining
d = thickness of shotcrete
b = height of shear zone (see also Fig. 9-50)
α = angle of shear resistance between rock and shotcrete
τ_s = shear strength of shotcrete

Additional resistance can be included if the shotcrete is reinforced by wire mesh or steel bars.

Design of System Anchoring

Figure 9-50 shows the shearing resistance of support elements, derived from the shear failure mechanism shown in Fig. 9-49. The anchors or bolts supplement the shotcrete lining if the latter alone cannot provide the required support resistance. System anchoring consists of radial units placed at regular intervals, and may be formed by expansion bolts or fully grouted rock anchors. When placed radially, the anchors create normal forces within the rock mass and induce radial stress resistance.

It must be noted that for the conditions shown in Fig. 9-50 part of the load is carried by the rock itself along the planes of shear failure, and this load is

$$p_i^r = \frac{2s}{b} (\tau_r \cos \psi - \sigma_n^r \sin \psi) \tag{9-30}$$

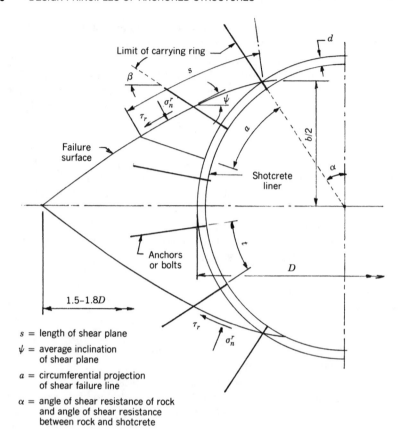

s = length of shear plane

ψ = average inclination
of shear plane

a = circumferential projection
of shear failure line

α = angle of shear resistance of rock
and angle of shear resistance
between rock and shotcrete

Fig. 9-50   Shearing resistance of support elements. (From Rabcewicz, 1973.)

where p_i^r = load carried by the rock
τ_r = shear strength of rock
σ_n^r = normal stress on the shear plane
ψ = average inclination of the shear plane
s = length of shear plane

Note, however, that this is not a net support resistance to be included in the total support capacity, since it is mobilized before the minimum pressure p_i necessary for equilibrium is estimated.

The resistance provided by the anchors is (see also Section 4-2)

$$p_i^a = \frac{a f_{pu} A_s \cos \beta}{et(b/2)} \tag{9-31}$$

where p_i^a = resistance provided by the anchors
a = circumferential projection of shear failure line

A_s = cross sectional area of anchor
f_{pu} = guaranteed limit of tendon
e, t = distance (spacing) of anchors
β = anchor inclination

The value of p_i^a is added to p_i^s to obtain the total resistance of the support.

For the general case where a shear zone is not developed but a net radial pressure still must be resisted, anchors or bolts can provide a good solution. By reference to Fig. 9-51, the load to be resisted by a single anchor is

$$P_a = et\sigma_r \qquad (9\text{-}32)$$

where P_a = load acting on anchor
σ_r = radial stress to be resisted at the boundary
e, t = distance (spacing) of anchors

The load transfer (fixed anchor zone) should be located outside the plastic zone and in a rock mass that is competent and free of movement. Anchor capacity should be checked according to Chapter 4 for yielding or rupture of the tendon steel and shear resistance at the tendon–grout and at the grout–rock interface. The effect of anchoring is to provide equilibrium with the radial stresses around the opening and thereby strengthen the rock surrounding the cavity so as to stabilize further inward deformation.

It should be noted that the Fenner–Pacher diagram provides only an initial indication of support requirements; hence it is compatible with preliminary support selection and dimensioning. After construction begins, the initial design must be verified and modified if necessary by an appropriate field instrumentation program subject to applicable contractual arrangements. A complete description of the NATM philosophy encompassing the design and contractual and construction phases of a project is given by Druss (1984).

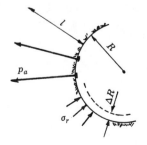

Fig. 9-51 Radial stresses and anchor load capacity; circular tunnel. (From Rabcewicz, 1973.)

9-12 UNDERGROUND OPENINGS IN BLOCKY ROCK—THE "BLOCK THEORY"

Where excavations are carried out in rock masses with several sets of discontinuities, the release of key blocks in critical locations can undermine neighboring blocks and threaten the integrity and stability of the entire scheme. Initially a block can move by falling, sliding on one or two faces, or by combined sliding and rotation. These motions become evident when certain faces begin to open, and the first warning of block movement is the widening of particular joints. Potentially dangerous blocks should be identified prior to movement in order to ensure their stability. This objective constitutes the principle of "block theory" (Goodman and Shi, 1985), which establishes appropriate procedures so that we can locate key blocks and determine support requirements (Goodman, 1989). These include optimum reinforcement schemes and excavation orientations and shapes inhibiting block movement.

Block theory divides blocks into groupings with distinct characteristics. For example, nonremovable blocks are identified by Shi's theorem (Goodman, 1989). A mode analysis, on the other hand, distinguishes other types of blocks, taking into account the direction of sliding and falling tendencies. Using limit equilibrium analysis and considering friction on the block faces, it is possible to establish the key blocks, whereas extensive use of stereographic projection enables the block theory to explain rock movement in three-dimensional models.

Applications to Underground Openings. For tunnels, the surface of the excavation can be considered a group of a family of planes parallel to the tunnel axis, hence almost every joint pyramid can produce a removable block at some location around the tunnel interior. These removable blocks are restricted, however, to particular areas of the tunnel surface. Consider, for example, the joint planes in the tunnel cross section of Fig. 9-52. If the blocks should lie simultaneously in the lower half space of each joint plane, then no block could be larger than the region *ABD*. If information about the extent and spacing of joints is not available, it will be prudent to select a support for the maximum removable block *ABD*. However, given the tunnel cross section, each joint pyramid has a maximum removable block, which can be determined from a theoretical basis using graphical procedures. These enable the designer to optimize tunnel design with respect to support requirements, particularly since microcomputer programs are available (Goodman, 1989).

9-13 PRINCIPLES OF ROCK SLOPE ENGINEERING

Rock slope excavations and cuts are required for transportation routes in mountainous terrain, in open pits, and for installations such as buildings,

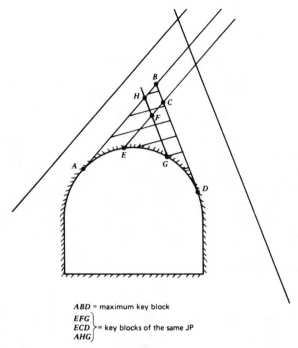

ABD = maximum key block

$\left.\begin{array}{l}\textit{EFG} \\ \textit{ECD} \\ \textit{AHG}\end{array}\right\}$ = key blocks of the same JP

Fig. 9-52 The maximum key block of a tunnel corresponding to a given joint pyramid. (From Goodman, 1989.)

powerhouses, or portals to underground projects. In most cases, the cost of failure will be considerably higher than the cost of excavation and protection. Sound engineering is therefore, necessary when planning the drainage, supports, instrumentation, and construction procedures. In soft rock, design of safe slopes is essentially direct application of rock mechanics theory since failure in this case may be by slumping or sliding along the body of main rock. In hard rock, existing discontinuities are likely to determine the direction and path of rock movement so that failure modes are not subject to conventional theories.

In designing rock slopes the main objective is to decide among two feasible solutions: select a safe angle for the rock slope, or design an unnatural slope and stabilize it with artificial supports such as anchors. The choice of safe angle requires elaborate evaluation of shear strength parameters, whereas the design of structural supports must recognize the extremely high forces generated when a rock face begins to move.

Failure Patterns of Slopes in Hard Rock. The very high strength of hard rock usually precludes failure under gravity alone unless discontinuities facilitate movement of discrete blocks. However, the conditions of failure are not manifested until excavation or rock movement removes the barrier to block translation. The movement of a block in a plane slide is shown in Fig. 9-53(a), and is possible when the resistance to sliding has been exceeded not only along the sliding plane but also at the end boundaries. In hard rock plane sliding occurs only if discontinuities or valleys transverse to the crest of the slope release the sides of the block.

The wedge slide shown in Fig. 9-53(b) is formed by two planes of weakness intersecting to define a tetrahedral block. Slip is possible without topographic or structural release of features if two discontinuities intersect within the excavation. The toppling failure shown in (c) involves overturning of rock layers resembling a series of cantilever beams. The hard rock can form columnar structure separated by steeply dipping discontinuities. Each layer tends to bend downhill under its own weight, and transfers a force downslope. If the toe of the slope is free to slide or overturn, flexural cracks formed in the layers above will release a substantial mass of rock, leading to destructive rock movement. The circular failure shown in Fig. 9-53(d) can occur in overburden soil, waste rock or heavily fractured rock without any identifiable structural pattern.

Alternatively, higher modes of failure can occur in bedded rock or in systems with complex joints where plane sliding, wedge sliding, and toppling can occur simultaneously by flexure, shear, or splitting (Goodman and Bray, 1977).

Design Methods. Failure that occurs entirely within intact rock, similar to slumping in clay, is observed in rock slopes in very weak continuous rocks and in highly weathered rocks. Pervasively fractured systems may exhibit pseudocontinuous behavior. Slopes in such rocks are usually analyzed using soil mechanics methods such as limit equilibrium. On the other hand, in rocks weakened by well-defined and regular sets of discontinuities, kinematic rigid block analysis is more appropriate.

Reinforced Rock Block. Figure 9-54 shows a rock block held in place by means of an anchor without which the block will tend to slip along discontinuity x–x; U_1 denotes the hydrostatic uplift normal to the base of the block, whereas U_2 is the hydrostatic pressure along the tension crack. The prestressed load in the anchor is F acting at an angle α with the direction of the slope.

Equilibrium is maintained if

$$W \sin \theta + U_2 - F \cos \alpha = cl + (W \cos \theta - U_1 + F \sin \alpha) \tan \phi \quad (9\text{-}33)$$

where c and ϕ denote the rock strength parameters. Interestingly, the anchor force F provides equilibrium in two ways, by resisting the sliding tendency directly and by increasing the friction along the sliding plane. Solving for F we obtain

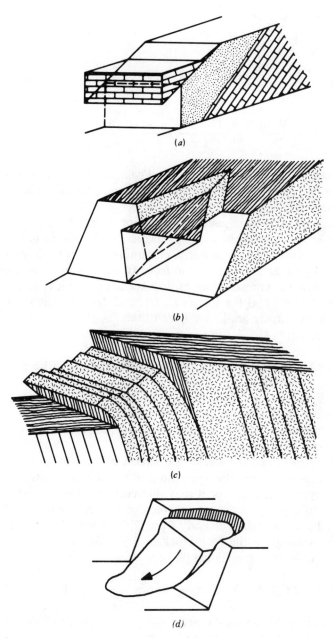

Fig. 9-53 Failure patterns of rock slopes: (a) plane slide; (b) wedge slide; (c) toppling; (d) circular failure.

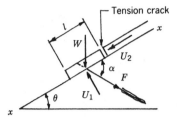

Fig. 9-54   Rock block stabilized by an anchor.

$$F = \frac{W \sin \theta + U_2 - cl - W \cos \theta \tan \phi + U_1 \tan \phi}{(\cos \alpha + \tan \phi \sin \alpha)} \qquad (9\text{-}34)$$

F becomes a minimum when $\alpha = \phi$. From Eq. (9-34) it also follows that the value of F can be further reduced if U_1 and U_2 are absent, that is, by appropriate drainage.

The above analysis is valid if the anchor is prestressed, and movement of the block is compatible with the elastic characteristics of the tendon. If no load is induced in the anchor, a certain movement (sliding) of the block is necessary to develop a reaction force in the tendon. The analysis must consider now the compatibility of displacement, and it is conceivable that the rock strength and the forces F, U_1, and U_2 will not be mobilized simultaneously (Londe et al., 1969, 1970).

9-14   ANALYSIS OF PLANE SLIDES IN ROCK SLOPES

A plane slide in a rock slope must interesect the slope surface, and the dip of the failure plane must be greater than the friction angle of the rock. These two conditions are considered in a limiting equilibrium analysis in a simple formulation of failure criteria.

Figure 9-55(a) and (b) shows the two typical cases of plane failure. A tension crack usually is the upper limit of the slide as shown. It may be at a point beyond the crest, or it may intercept the slope itself. If water fills the crack to depth Z_w, we can assume that it seeps along the sliding plane while its head diminishes linearly from the tension crack to the toe of the slope. By analogy to the rock block of Fig. 9-54, failure occurs when the shear strength along the sliding plane is exceeded, or

$$W \sin \delta + V \cos \delta = cA + (W \cos \delta - U - V \sin \delta) \tan \phi \qquad (9\text{-}35)$$

where c, ϕ = rock strength parameters (cohesion, friction)
$\qquad \quad W$ = weight of sliding wedge
$\qquad \quad A$ = area per unit width of the sliding plane
$\qquad \quad U$ = resultant water uplift along the sliding plane
$\qquad \quad V$ = resultant of water pressure along tension crack

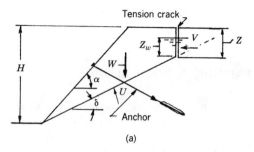

(a)

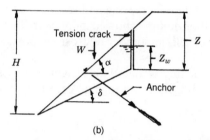

(b)

Fig. 9-55 Geometry for analysis of plane failure in rock slopes. (From Hoek and Bray, 1977.)

From the geometry of Fig. 9-55 we obtain

$$A = \frac{H - Z}{\sin \delta}, \qquad U = \frac{1}{2} \gamma_w Z_w A, \qquad V = \frac{1}{2} \gamma_w Z_w^2 \qquad (9\text{-}36)$$

If the tension crack intercepts the crest of the slope as in (a)

$$W = \frac{1}{2} \gamma H^2 \left[\left(1 - \frac{Z^2}{H^2} \right) \cot \delta - \cot \alpha \right] \qquad (9\text{-}37a)$$

but if the tension crack intercepts the face as in (b)

$$W = \frac{1}{2} \gamma H^2 \left[\left(1 - \frac{Z}{H} \right)^2 \cot \delta (\cot \delta \tan \alpha - 1) \right] \qquad (9\text{-}37b)$$

Equation (9-35) is used to generate a slope chart for design purposes, where H is plotted against $\cot \alpha$. By varying the parameters in Eq. (9-35), we can conclude that a decrease in c affects steep slopes to a greater extent than flat slopes, whereas a smaller ϕ reduces the fact of safety of high slopes more than low slopes. Drainage is effective in reducing the stabilizing requirements as is in the case of the simple block of Eq. (9-33). If an anchor is added introducing a prestressed force as shown, its effect is included as in Eq. (9-33).

9-15   WEDGE FAILURE IN ROCK SLOPES

The three-dimensional nature of the wedge failure in rock slopes usually requires lengthy solutions feasible with the aid of programmable calculators. Methods of analysis have been developed by Hoek and Bray (1977), Kalkani (1977), and Kovari and Fritz (1977). Kalkani (1977) has extended the analysis to include the ambient stress field conditions before excavation. This investigator concluded that with increasing ambient stress coefficient, the factor of safety against slope failure showed nonlinear decrease for wedges without a crack, with a crack, and under hydrostatic pressure. The required anchor force increased in all cases.

Figure 9-56 shows a sliding wedge with weight W and surface areas on the contact planes A_1 and A_2. The orthogonal coordinate system (s, n, h) is

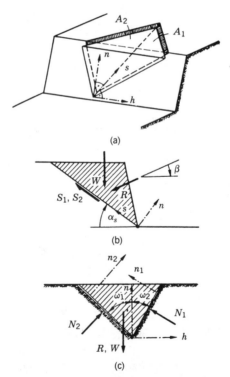

(a)

(b)

(c)

Fig. 9-56 Geometry of wedge in rock with the associated forces: (a) isometric view; (b) vertical plane through the line or intersection of the sliding surface; (c) plane normal to intersection line. (From Kovari and Fritz, 1977.)

the reference system. The axis s is the direction of the intersection of the two sliding surfaces. The axis n is normal to s and defines the vertical plane. The axis h is at right angle to s and defines the horizontal plane.

Initially we assume that the slope of the line of intersection of the two sliding planes α_s is known, as are the angles ω_1 and ω_2 between the n axis and the sliding planes shown in Fig. 9-56(b) and (c). We further assume that the resultant force R (external forces acting on the sliding body such as anchors, water, and earthquake forces) is parallel to the vertical plane (s, n) so that its direction is determined by angle β. For simplicity the friction angle ϕ is taken the same for both sliding planes. We can now write the Coulomb failure condition (Kovari and Fritz, 1977) as follows:

$$S_{max} = (N_1 + N_2) \tan \phi + c_1 A_1 + c_2 A_2 \qquad (9\text{-}38)$$

where $S_{max} = S_{1,max} + S_{2,max}$. The factor of safety is

$$F_s = \frac{S_{max}}{S} \qquad (9\text{-}39)$$

where $S = S_1 + S_2$. For equilibrium, the following three equations must be satisfied

$$\left.\begin{aligned}
S + R \cos (\alpha_s + \beta) - W \sin \alpha_s &= 0 \\
N_1 \sin \omega_1 + N_2 \sin \omega_2 - R \sin(\alpha_s + \beta) - W \cos \alpha_s &= 0 \\
N_1 \cos \omega_1 - N_2 \cos \omega_2 &= 0
\end{aligned}\right\} \quad (9\text{-}40)$$

By setting

$$\lambda = \frac{\cos \omega_1 + \cos \omega_2}{\sin(\omega_1 + \omega_2)}$$

and $\tan \phi^* = \lambda \tan \phi$, we obtain

$$N_1 + N_2 = [W \cos \alpha_s + R \sin(\alpha_s + \beta)] \tan \phi^* \qquad (9\text{-}41)$$

The resultant R (which includes the prestress force in the anchors) is obtained by combining Eqs. (9-38), (9-40), and (9-41) with Eq. (9-39) as

$$R = \frac{F_s \sin \alpha_s - \cos \alpha_s \tan \phi^*}{F_s \cos(\alpha_s + \beta) + \sin(\alpha_s + \beta) \tan \phi^*}$$
$$\times \left(1 - \frac{c_1 A_1 + c_2 A_2}{W} \cdot \frac{1}{F_s \sin \alpha_s - \cos \alpha_s \tan \phi^*}\right) W$$

$$(9\text{-}42)$$

Setting

$$K_1^* = \frac{F_s \sin \alpha_s - \cos \alpha_s \tan \phi^*}{F_s \cos(\alpha_s + \beta) + \sin(\alpha_s + \beta) \tan \phi^*}$$

and

$$K_2^* = \frac{1}{F_s \sin \alpha_s - \cos \alpha_s \tan \phi^*}$$

R is given by

$$R^* = K_1^* \left(1 - \frac{c_1 A_1 + c_2 A_2}{W} K_2^* \right) W \tag{9-43}$$

The calculations required to determine R^* involve determination of the factors K_1^* and K_2^* as functions of α_s and $\tan \phi^*$, where α_s and λ are functions of the strike azimuths and the inclinations of the failure surface. The wedge action is related to an increase in the angle of friction ϕ and to a reduction of the inclination of the slope α. The factor λ can be obtained graphically with the aid of the diagram of Fig. 9-57 as a function of the wedge angle 2ω for the case of symmetrical wedge ($\omega_1 = \omega_2 = \omega$).

The General Case of Wedge Failure. If the problem involves two different angles of friction ϕ_1 and ϕ_2 and an additional anchor force R_h in the direction of the h axis, the basic equation, Eq. (9-43), can be extended as follows:

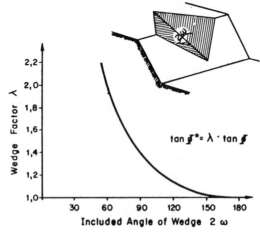

Fig. 9-57 Wedge factor λ for a symmetrical wedge as function of the wedge angle. (From Kovari and Fritz, 1977.)

$$R_{s,n} = K_1^{**} \left(1 - \frac{c_1 A_1 + c_2 A_2}{W} K_2^{**} \right) W + K_1^{**} K_2^{**} H R_h \qquad (9\text{-}44)$$

The apparent friction angle ϕ^{**} is given by

$$\phi^{**} = \text{arc tan} \frac{(\cos \omega_2 \tan \phi_1 - \cos \omega_1 \tan \phi_2)}{\sin(\omega_1 + \omega_2)} \qquad (9\text{-}45)$$

and the factor H by

$$H = \frac{\sin \omega_2 \tan \phi_1 - \sin \omega_1 \tan \phi_2}{\sin(\omega_1 + \omega_2)} \qquad (9\text{-}46)$$

The factors K_1^{**} and K_2^{**} are given by Kovari and Fritz (1977) in the same forms as the expressions derived for plane failure. From Eq. (9-46) we draw the following conclusions:

1. If the wedge is symmetrical ($\omega_1 = \omega_2$ and $\phi_1 = \phi_2$), the horizontal force component R_h normal to the direction of sliding has no effect on the failure of the wedge.

2. For an asymmetrical wedge ($\omega_1 \neq \omega_2$) but with the same friction angle on the two sliding planes, the force component R_h affects stability positively if it is directed toward the steeper of the two sliding surfaces. In this case the sum of the normal forces $N_1 + N_2$ increases.

3. A stronger anchorage in the direction of the steeper sliding surface is more effective than anchors symmetrically arranged and carrying the same loads.

9-16   OTHER METHODS OF ANALYSIS AND SUPPORT OF ROCK SLOPES

Anchoring Schemes

Figure 9-58 shows various anchoring schemes suggested by Sharp (1973) for the most common types of rocks. Possible modes of failure are also presented. The classification and choice of anchoring system is made from a consideration of the nature of the rock, the joint and discontinuity pattern, the appropriate strength values in intact rock and the zone of weakness, and the groundwater presence and effects. A sensitivity analysis is a usual part of the investigation and helps evaluate the effect of each variable on stability (Hoek and Bray, 1977).

The location and type of anchors is generally determined after this approach is completed. Possible options include a uniform distribution along

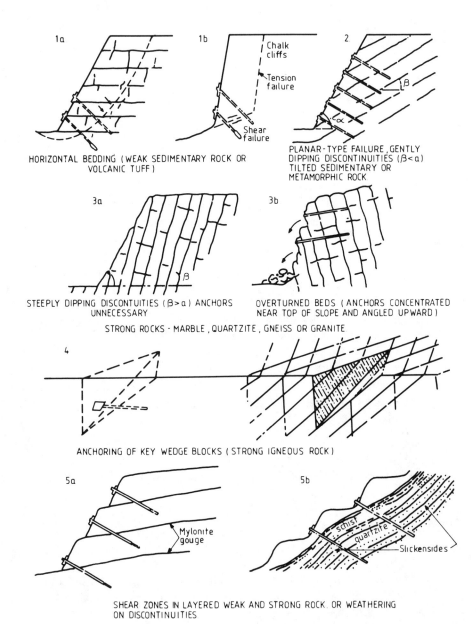

Fig. 9-58   Anchoring schemes for the support of rock slopes. (From Sharp, 1973.)

the slope or a concentration of anchors near the toe of the slope where overstressing occurs. The uniform distribution of anchors is advocated by many investigators including Barron et al. (1970). Invariably, the fixed anchor zone should be located sufficiently beyond the slip plane so that anchor resistance is mobilized in undisturbed rock. Rock support design should include provisions for effective drainage.

Among the methods of rock stability analysis and support design presently in use is the procedure suggested by Hobst and Zajic (1977) shown in Fig. 9-59(a), which is based on failure along a slip circle. The anchor is considered most effective if it is inclined at an angle $(90 - \phi')$ from a direction normal to the slide. The required force in the anchor is

$$T_A = \frac{Tlc' - (\tan \phi'/F)N}{\sin(90 - \phi') + \dfrac{\tan \phi'}{F} \cos(90 - \phi')} \tag{9-47}$$

where T_A = force in the anchor
       T = tangential component of self-weight on failure surface
       N = normal component of self-weight on failure surface

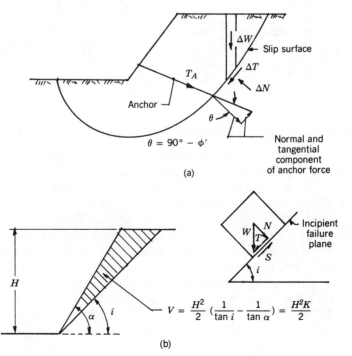

(a)

(b)

Fig. 9-59 (a) Stability analysis using a slip circle, rock slopes; (b) volume of rock requiring anchor support.

l = length of slip circle
c' = cohesion along failure surface
F = factor of safety
ϕ' = effective angle of friction

Among other methods of analysis to be mentioned is the procedure presented by Barron et al. (1970), which includes determination of support components for a deep mine cutting in rock, as shown in Fig. 9-59(b). In this case the cut slope has an overall inclination α, whereas i is the inclination of the plane along which the rock mass is stable. A stabilization scheme is based on preloading the rock surface by prestressed anchors to increase the effective stresses on planes with inclination i. In the general analysis it is assumed that the potential failure plane is inclined at i to the horizontal and that the mass of rock between the planes inclined at α and i will fail as a rigid body.

Along the failure plane the cohesion is assumed zero, whereas the anchor force is considered a point load. For a slope of height H as shown in Fig. 9-59 the angles α and i are related by the expression

$$\cot \alpha = \cot i + \left(\frac{\mu \cot i - 1}{\sin 2 i + \mu \cos 2 i} \right)$$

where μ is the coefficient of friction along the failure plane. The above relation gives the value of angle i at which excess shear stress becomes a maximum for any value of α or μ. This analysis provides the basis for developing a method to estimate anchor force, spacing and length (See Barron et al., 1970).

Inadequacy of Limiting Equilibrium Methods for Rock Slopes

Very often, as we have seen, the stability of slopes is conditional by the characteristics of joints. Essentially, these exhibit brittle behavior which is exemplified by a peak and residual strength. In such cases progressive failure becomes relevant to slope stability, and design methods based on limit equilibrium cannot provide satisfactory results. Potentially serious errors can arise in predicting the shear strength actually mobilized, since this depends not only on the normal stress but also on tangential displacement along the joints. These errors are compounded by the assumption that at near-failure conditions the maximum strength is mobilized simultaneously by all points on the slip surface, which rarely is the case (Manfredini et al., 1977).

Two possible modes of joint behavior are shown in Fig. 9-60. In part (a), as the condition of incipient instability is approached the rock mobilizes its maximum strength along the entire failure surface. The condition in (b) exemplifies brittle behavior or strain softening (Bishop, 1971) whereby under a given normal stress a peak value of strength is reached and gradually tapers

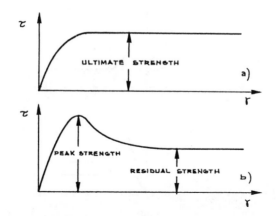

Fig. 9-60 Typical joint behavior in rock masses.

off to a residual value as the strain increases. In this condition, failure propagates along the potential failure surface starting from the most stressed area and entails a new stress distribution. If this progressive form of failure is to be considered in limit equilibrium analysis, a choice would have to be made between peak and residual strength.

For a rock slope in jointed ground, stability initially requires definition and identification of all rock wedges outlined by the joints. Then it is necessary to verify whether displacement of every wedge from the main mass is kinematically possible resorting, for example, to stereographic projections as suggested by Hoek (1974) and calculate the safety factor for each wedge.

Brittle behavior of joints has been observed on typical rocks (Goodman, 1974). In general, brittleness depends both on the amount of strength loss from peak to residual value as the transition occurs and on the rapidity of this transition as joint displacement progresses. This process can be understood by reference to the simple model of joint tangential deformability shown in Fig. 9-61. The tangential stiffness modulus K_s is a function of the normal stress σ_n and the tangential deformation δ_s previously undergone by the joint. Up to the failure point K_s is constant and independent of σ_n. As the strength decreases to residual value, the modulus is still constant but with negative value. Finally, when residual strength conditions prevail, the modulus becomes zero.

It appears that the analysis of stability of a jointed rock mass, even with simplified models, requires explicit knowledge and evaluation of the following parameters: (a) joint peak strength; (b) joint brittleness; (c) original state of stress, both vertical and horizontal; (d) joint deformability prior to failure; and (e) elastic parameters of the rock. Stress analysis using numerical techniques can in many instances eliminate the uncertain results of limiting equilibrium, particularly when the joints exhibit strong brittleness.

Interestingly, safety factors calculated from peak strength parameters de-

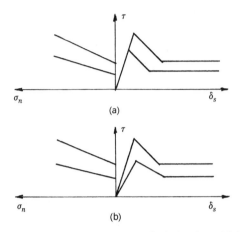

Fig. 9-61 Idealized model for stress–strain behavior of joints in rock.

viate more from actual values as slope height increases. Under other conditions safety factors decrease with increasing joint brittleness and are thus dependent on the ratio of residual and peak cohesion and on the strength drop rate.

Rock Slope Analysis by Kinematic Methods

Rock cuts are often stable on steep slopes, although they contain planes of weakness that are steeply inclined and have low strength. Stability is maintained as long as the blocks are not free to move along a weak surface because other ledges of intact rock are on the way. If the blockage is removed by excavation or by other physical causes, the slope will most likely fail immediately. Using kinematics it is possible to evaluate the directionality and orientation of discontinuous rock masses and determine whether movement of potential failure blocks can be stopped (Goodman, 1989).

The three basic elements of kinematics for a rock mass are shown in Fig. 9-62. They are the dip vector $\hat{D}_i$, pointed down the dip of a weakness plane; the normal vector $\hat{N}_i$, pointed in the direction normal to the plane of weakness; and the line of intersection $\hat{I}_{ij}$ of weakness planes i and j. The dip vector is a line bearing at right angles to the strike and plunging with vertical angle δ below horizontal. Suggested references for the principles of stereographic projection are Goodman (1976, 1989), Hoek and Bray (1977), and Priest (1985).

The line of intersection $\hat{I}_{ij}$ of the planes i and j is determined by the great circles of each plane as shown in Fig. 9-62(b). Once all three line elements $\hat{D}, \hat{N}$, and $\hat{I}$ are plotted for a critical rock mass, the kinematic requirements for potential slope failure can be examined for a rock slope of any strike and dip.

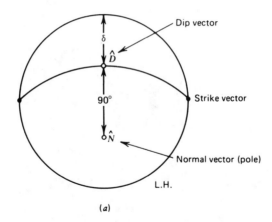

(a)

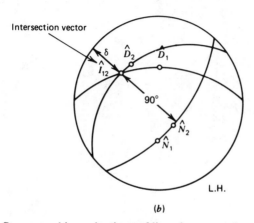

(b)

Fig. 9-62 Stereographic projections of line elements relevant to the stabilization of rock slopes. (From Goodman, 1989.)

Plane Sliding under Gravity. Figure 9-63 shows a plane sliding of rock under gravity. If a block begins to slide along a single plane surface, the movement will be parallel to the dip of the weakness plane, that is, parallel to $\hat{D}$. If the slope has an angle α with the horizontal, the conditions for a slide to occur are that $\hat{D}$ is pointed into the free space of the excavation and plunge at an angle less than α as shown in (a). Part (b) shows a cut slope plotted as a great circle in the lower hemisphere. The kinematic requirements for plane sliding are met when the dip vector of a potential sliding surface is in the ruled area above the cut slope circle (Goodman, 1989). Thus, plane 1 will accommodate sliding but plane 2 will not.

If the dip vector for a potentially weak surface is known, we can determine a limiting safe angle corresponding to a cut of assigned strike following the simple procedure shown in Fig. 9-63(c). For a cut with strike 1 the maximum

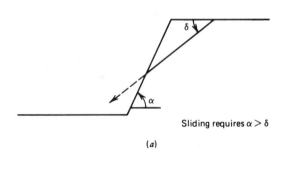

Sliding requires $\alpha > \delta$

(a)

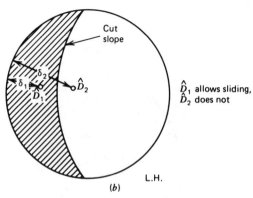

$\hat{D}_1$ allows sliding,
$\hat{D}_2$ does not

(b)

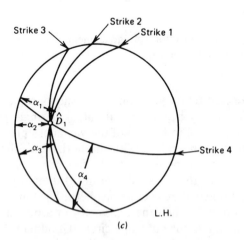

(c)

Fig. 9-63   Kinematic test for rock slope failure by plane sliding. (From Goodman, 1989.)

safe angle α_1 is the dip of the great circle through strike 1 and $\hat{D}_1$. For plane sliding on a single weakness surface, kinematic freedom for slip exists only in one-half of the set of possible cut orientations. Cut orientations almost parallel to the dip direction of the plane of weakness will remain stable even at a direction nearly vertical.

Wedge Sliding. Since this is assumed along the line of intersection of two planes, the relevant line element is $\hat{I}$. An example of kinematic analysis of wedge failure is shown in Fig. 9-64, and involves a rock mass consisting of three sets of joints. If a cut is made with strike down, potentially sliding wedges would be those formed of planes 1 and 3 or planes 1 and 2. If the cut has an inclination angle α determined by the dip of the great circle through $\hat{I}_{13}$ and having the assigned strike for the cut, only one wedge determined by planes 1 and 2 is kinematically free to slide.

Toppling Failure. As mentioned, large flexural deformations must be preceded by interlayer slip. Furthermore, the surface of the cut is the direction of major principal stresses. If the layers have a friction angle ϕ_j, slip can occur only if the direction of applied compression has an angle with the normal to the layers greater than ϕ_j. Referring to Fig. 9-65, a precondition of interlayer slip requires the normals to have inclinations less steep than a line inclined ϕ_j degrees above the plane of the slope. If the layers have a slip δ, toppling failure with a slope intersecting the horizontal at an angle α can occur only if $(90 - \delta) + \phi_j < \alpha$ (Goodman, 1989). On the stereographic projection, this implies that toppling is possible only if the normal vector $\hat{N}$ lies more than ϕ_j degrees below the cut slope. Furthermore, toppling can only occur if the layers strike nearly parallel to the strike of the slope. From these requirements, toppling appears possible on a regular, closely spaced discontinuity set if its normal plots inside the ruled zone in Fig. 9-65(b).

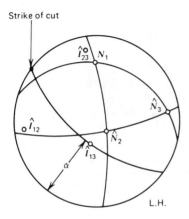

Strike of cut

L.H.

Fig. 9-64 Kinematic test for rock slope failure by wedge sliding. (From Goodman, 1989.)

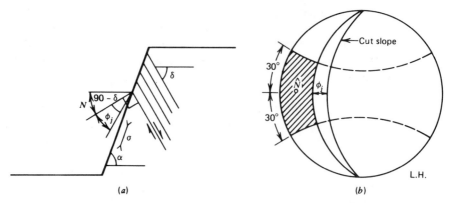

Fig. 9-65   Kinematic test for toppling: (a) $(90 - \delta) + \phi_j < \alpha$; (b) $\hat{N}$ must plot in the shaded zone. (From Goodman, 1989.)

9-17   STRUCTURES RESISTING CONCENTRATED FORCES

Horizontal Tensile Forces

Concentrated forces from anchors with nearly horizontal direction can be transferred to concrete blocks or heavy beams (deadmen) embedded in the ground, pile combinations where both rows are battered to develop better resistance to pull, or existing underground structures with reserve capacity to resist such loads. Methods for transferring concentrated horizontal forces are shown in Fig. 9-66.

For a relatively long continuous concrete block or heavy beam (length substantially greater than depth) and with embedment d_1 less than $0.5d_2$ to $0.7d_2$ as shown in Fig. 9-67, the maximum load that can be resisted is

$$F_A = (P_p - P_a)L \qquad (9\text{-}48)$$

where P_p, P_a = passive and active forces based on
    d_2 = embedment from ground surface to base of block
    L = length of block

Equation (9-48) describes a failure mechanism similar to the stability of nearly horizontal shallow anchors described in Section 5-3, shown in Fig. 5-10(a). For short anchorage blocks in granular soils the maximum force is

$$F_A = L(P_p - P_a) + \tfrac{1}{3} K_0 \gamma (\sqrt{K_p} + \sqrt{K_a}) d_2^3 \tan \phi \qquad (9\text{-}49)$$

where K_0 = earth pressure coefficient at rest (0.4 recommended). Likewise, for cohesive soils the maximum force is

$$F_A = L(P_p - P_a) + q_u d_2 \qquad (9\text{-}50)$$

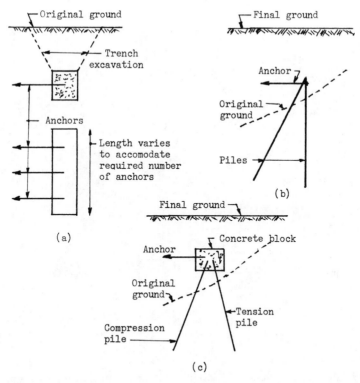

Fig. 9-66 Methods of transferring concentrated horizontal forces from anchorages: (a) cast-in-place concrete blocks and beams; (b,c) pile groups used to resist anchorage loads.

where q_u = unconfined compressive strength. From Eqs. (9-48), (9-49), and (9-50) the working anchorage load is obtained by dividing F_A by the appropriate factor of safety.

If H is much smaller than the embedment d_2, the structure should be analyzed according to bearing capacity theories as continuous footing. In this case the footing width B is taken as H and the effective footing depth D_f as $d_2 - H/2$ below the ground surface. Using the Terzaghi bearing capacity equations, the capacity of the block is obtained from

$$q_{ult} = cN_c + qN_q + 1/\gamma BN_\gamma \qquad (9\text{-}51)$$

where q_{ult} = ultimate soil bearing pressure
      c = cohesion
      ϕ = angle of friction
      $q = \gamma D_f K_0$
      $B = H$
      γ = unit weight of soil

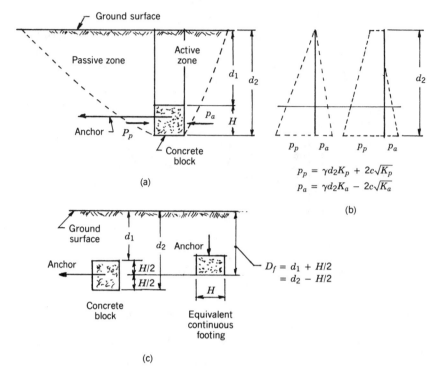

Fig. 9-67 Resisting capacity of concrete blocks and beams; (a) member located near surface with $d_1 < 0.5$–$0.7d_2$; (b) earth pressure diagram for the arrangement shown in (a) for granular and cohesive soil; (c) block or beam located with H much smaller than d_2 (in this case the member can be treated as a continuous footing).

and N_c, N_q, and N_γ are appropriate bearing capacity factors (Terzaghi, 1945) estimated from the following (based on general shear conditions):

$$N_c = \cot \phi \left[\frac{a^2}{2 \cos^2[(\pi/4) + (\phi/2)]} - 1.0 \right]$$

$$N_q = \frac{a^2}{2 \cos^2[45 + (\phi/2)]}$$

$$N_\gamma = \frac{\tan \phi}{2} \left(\frac{K_{p\gamma}}{\cos^2 \phi} - 1.0 \right)$$

$$a = e^{[\pi(3/4) - (\phi/2)] \tan \phi}$$

(9-52)

The factor $K_{p\gamma}$ is a general shear term. Values of N_c, N_q, N_γ, and $K_{p\gamma}$ can be taken directly from Table 9-9 for various values of ϕ.

If the concrete block resisting the concentrated anchor force is located near an excavation or cut as shown in Fig. 9-68, it is essential that its failure

TABLE 9-9   Bearing Capacity Factors for Use in Eq. (9-52)

ϕ	N_c	N_q	N_γ	ϕ	$K_{p\gamma}$
0	5.7	1.0	0.0	0	10.8
5	7.3	1.6	0.5	5	12.2
10	9.6	2.7	1.2	10	14.7
15	12.9	4.4	2.5	15	18.6
20	17.7	7.4	5.0	20	25.0
25	25.1	12.7	9.7	25	35.0
30	37.2	22.5	19.7	30	52.0
34	52.6	36.5	35.0	35	82.0
35	57.8	41.4	42.4	40	141.0
40	95.7	81.3	100.4	45	298.0
45	172.3	173.3	297.5	50	800.0
48	258.3	287.9	780.1		
50	347.5	415.1	1,153.2		

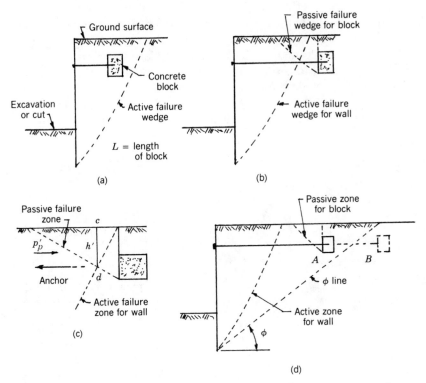

Fig. 9-68   Anchorage (block) location and effects; blocks near an excavation or a cut.

wedge should not intercept the failure wedge of the cut. For example, the block shown in part (a) is clearly within the failure wedge of the cut and thus ineffective. In (b) the two failure zones overlap and the resulting interference renders the anchorage partially effective. The enlarged diagrams of pressure wedges shown in (c) can be used to estimate the effect of this interference. The reduced block capacity is

$$F_A = [(P_p - P_a) - (P'_p - P'_a)]L \qquad (9\text{-}53)$$

where P'_p and P'_a are estimated in conjunction with depth h'. For the arrangement shown in (d) where the block is located inside the ϕ line, no reduction is necessary in the effective resistance provided by the soil–structure interaction. This location, however, tends to increase the lateral pressure since the block is essentially a surcharge load.

Vertical Concentrated Tensile Forces

Figure 9-69 shows four ways of resisting a concentrated vertical anchorage force T (see also Section 8-9). The typical dead-weight block shown in part (a) allows maximum efficiency especially if its configuration reduces possible uplift. It also results in the least soil heave δ_1, which is essentially the elastic rebound of the soil under the base only. For the methods shown in (b), (c), and (d) the shear strength of the ground is utilized in resisting the force T. Hence, in addition to the rebound of the soil under the base the vertical movement will include the elongation δ_τ due to the shear stress in (b), the elongation of the tension piles δ_p in (c), and the elongation of the anchors δ_a in (d), which can be counterbalanced by prestressing.

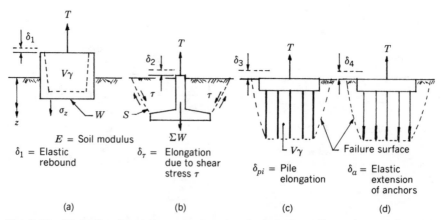

Fig. 9-69 Methods of resisting a concentrated vertical tensile anchorage force T: (a) dead-weight block; (b) deep footing with earth weight; (c) footing transferring tension to piles; (d) footing transferring tension to ground anchors.

Dead-Weight Block. From Fig. 9-69(a) the maximum force that can be resisted is

$$T = W + V\gamma \tag{9-54a}$$

where W is the weight of the concrete block, V is the inside volume of earth materials, and γ is the unit weight of soil. The amount of heave at the base is

$$\delta_1 = \int_{T_0}^{T_1} \frac{d\sigma_2}{E} \, dz \tag{9-55a}$$

Deep Footing. Likewise the deep footing shown in (b) derives its capacity from its own weight plus the weight of earth above it, but in addition it mobilizes shear resistance along the perimeter S. Thus the total capacity is

$$T = \sum W + S\tau \tag{9-54b}$$

where $\sum W$ is the sum of total dead weights, and τ is the unit shear strength along the perimeter S. The total heave is now

$$\delta_2 = \delta_1' + \delta_\tau \tag{9-55b}$$

Footing on Piles. The arrangement shown in Fig 9-69(c) has a total capacity

$$T = V\gamma \tag{9-54c}$$

where V is the volume of ground mass within the failure zone. In this case this weight is assumed to be less than the capacity of the pile group. The total heave is the sum of three components or

$$\delta_3 = \delta_1' + \delta_\tau + \delta_p \tag{9-55c}$$

where δ_1' and δ_τ are as before and δ_p is the pile elongation.

Anchored Footing. Likewise, the arrangement in (d) is similar to (c) and its capacity is derived in a similar manner, or from the total capacity of the anchorage. The total heave is

$$\delta_4 = \delta_1'' + \delta_\tau' + \delta_a \tag{9-55d}$$

where δ_a is the elastic extension of the free anchor length. We note, however, that this heave may be partially or totally eliminated if the anchors are pre-stressed.

Inclined Concentrated Tensile Forces

Weight Blocks. Concentrated forces with relatively flat inclinations (<10°) can be resisted by conventional weight blocks discussed in the foregoing section. Essentially, these are extension of end blocks used in suspension bridges that rely primrily on friction, except that the passive resistance in front of the blocks is also included in the total resisting capacity. However, the stability of the block is dependent on the position of application of the tensile force, and this effect is demonstrated in Fig. 9-70.

For the block shown in part (b) the force is applied along a direction and position well above the center of gravity of the block and tends to rotate it clockwise. This action increases the shear resistance in front of the block. If the force is applied as shown in (a), it tends to rotate the block counter-clockwise, reducing the shear resistance along the front face. If T_h and T_v are the horizontal and vertical components of T, respectively, for the block shown in Fig. 9-70(b), equilibrium is maintained if these components are resisted simultaneously, or

$$T_h = F_b + E_p = T \cos \delta$$
$$T_v = W + F_f = T \sin \delta \tag{9-56}$$

where W = weight of block
$\quad\quad\quad F_b$ = friction mobilized along the base
$\quad\quad\quad F_f$ = friction mobilized along the front face
$\quad\quad\quad E_p$ = passive resistance mobilized at the front face

From Eqs. (9-56) we see that the design is balanced if

$$\frac{F_b + E_p}{\cos \delta} = \frac{W + F_f}{\sin \delta} \quad \text{or} \quad \tan \delta = \frac{W + F_f}{F_b + E_p} \tag{9-56a}$$

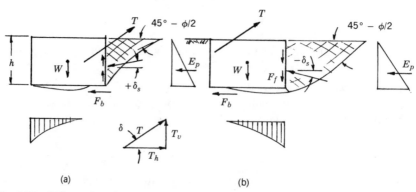

(a)   (b)

Fig. 9-70   Effect of position of concentrated tensile force on wall friction and passive pressure; weight block base.

since in this case the safe working load obtained from either T_h or T_v will be the same.

Alternatively, the solid block can be substituted by a stem and base plate as shown in Fig. 8-18. Stability is now derived from the weight of the fill material on top of the plate, which resists the vertical component and mobilizes base friction.

Trench Walls. Concrete trench walls, mentioned also in Section 8-9, for inclined tensile forces can be constructed in various configurations. As specially shaped foundation elements, they were introduced following the development of machines that perform slot excavations (Xanthakos, 1979). Hence, their dimensions and shape are limited only by the construction method and equipment available at the site, or by the required reinforcement.

Figure 9-71 shows a T trench wall that can resist as much as 900 tons of inclined tensile force (Soos, 1972). For equilibrium, the following must be satisfied:

$$\left.\begin{array}{l} T_h = T \cos \alpha = A\sigma_0 - A\sigma_u' \\ T_v = T \sin \alpha = F_w \end{array}\right\} \qquad (9\text{-}57)$$

where T_h, T_v = horizontal and vertical component of T, respectively
     $A\sigma_0, A\sigma_u'$ = passive resistance developed as shown
         F_w = friction (shear resistance) mobilized along the trench wall face, or along a modified shear failure perimeter as shown by the dashed line

Likewise, for a balanced design the following must be satisfied:

$$\frac{A\sigma_0 - A\sigma_u'}{\cos \alpha} = \frac{F_w}{\sin \alpha} \quad \text{or} \quad \tan \alpha = \frac{F_w}{A\sigma_0 - A\sigma_u'} \qquad (9\text{-}57\text{a})$$

We can reasonably assume that a trench of these dimensions is infinitely stiff and for analysis purposes it can be considered rigid. The wall rotates, therefore, about a pivot point 0 as shown. If the ground is considered an elastic medium, a modulus of subgrade reaction should be established. Methods of analysis are available in any text on laterally loaded bored piles and concrete subpiers.

For the transfer of the vertical component by shaft resistance, the following factors must be considered: (a) the development of this reaction as a function of the vertical (in this case upward) displacement; (b) the dimensions of the element, shape and configuration, relative confinement in the soil mass, and probable group action; (c) the stiffness of concrete with respect to the compressibility of the supporting soil; (d) the shear strength of

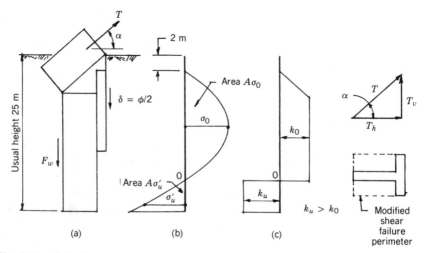

Fig. 9-71 A T-shape trench wall constructed to resist an inclined tensile force; (a) wall configuration and dimensions; (b) assumed lateral earth pressure diagram; (c) simplified trapezoidal diagram.

the ground mass; and (e) effects of the construction method, for example, whether the trench is built in the dry or in conjunction with bentonite slurry (Xanthakos, 1979).

Where the size and scope of the project justify the cost, the analysis of trench walls can be supplemented by full-scale field pullout tests. In this respect, the following comments are helpful:

1. Caution should be exercised when interpreting comparative test results. For example, certain difficulties will be encountered in simulating the sidewall configuration (obtained using the bentonite slurry process) by any form of dry construction in soils where the slurry method would normally be used.

2. In general, trench walls should be excavated, reinforced, and concreted in the minimum possible time in order to avoid soil disturbance, cavitation, sloughing, and other detrimental effects.

3. Main factors influencing the load transfer are the roughness of the concrete–soil interface, the moisture sensitivity of the supporting earth materials, the slurry displacement conditions, and soil strength changes.

4. There is no evidence to suggest that the load-transfer characteristics of slurry panels differ materially from dry members. Sidewall shear is practically reached almost at the same vertical displacement for both slurry and dry elements. However, much higher displacement will be required to mobilize full shear if the member is displaced upward.

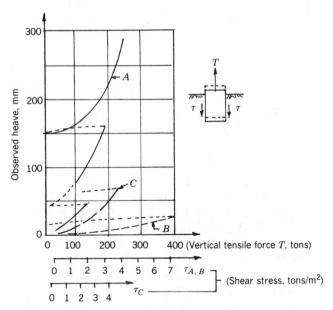

Fig. 9-72 Heave observed in trench walls as function of vertical tensile load.

Figure 9-72 shows results from tension tests (pullout) on three I-section wall panels, marked *A, B,* and *C* (Soos, 1972). Walls *A* and *B* are 9 m (29.5 ft) deep, and wall *C* is 4 m deep. The curves represent a plot of the observed heave versus vertical pullout force. Although walls *A* and *B* are similar and built in the same soil (sand and gravel), *A* has a much lower pullout resistance than *B*. Furthermore, wall *A* moved upward 25 cm in order to develop a side shear of 4 tons/m^2 (545 lb/ft^2) whereas wall *B* developed a shear resistance of 7 tons/m^2 (950 lb/ft^2) at an upward movement of 2.5 cm. This inconsistency is probably explained by a 2-day interruption in the work schedule of wall *A*, during which deep filtration and rheological blocking in the sand–gravel layers affected the development of shear resistance. Both elements reacted to horizontal forces in a similar manner.

9-18 ANALYSIS OF ANCHORED FOUNDATION MATS AND RAFTS

Types of structures subjected to permanent uplift are presented in Section 8-15. A prime concern, arising from the application of prestress, is the associated ground response and the accompanying consolidation and settlement. The result is a reduction or loss of prestress equivalent to a reduction in the elastic extension of the cables. This, however, is not as critical as in soil preconsolidation, and the process is rapidly converging until equilibrium between restressing and prestress loss is reached.

In general, construction work that can induce the condition of large uplift is associated with large volumes of excavation. If the weight of excavated materials is sufficient to balance the initial anchor prestress and the center of gravity of excavation and prestress nearly coincide, settlement due to initial prestress is substantially reduced and may even compensate for the amount that the soil rebounded from unloading as excavation is carried out. In any case, in predicting the settlement due to prestress, the weight of soil removed should be treated as negative intensity of pressure and subtracted from the prestress force.

Whether anchored mats and rafts should deliberatedly be made flexible so that they can deform, rather than crack with the settlement, is a matter of initial criterion and structural intent. Rigidity, however, ensures uniform pressure at the soil–raft interface, which is advantageous where differential settlements and losses of prestress are more difficult to control.

Criteria for Rigidity. Consider a raft of finite dimensions L and B acted on by two prestress forces F_1 and F_2 as shown in Fig. 9-73(a). The criterion for rigidity can be established on the basis of the term λL which includes the dimensions of the raft and the elastic properties of the interacting media. This term is

$$\lambda L = \sqrt{\frac{K_s L^4}{4\,EI}} \tag{9-58}$$

where $K_s = k_s B$ = modulus of subgrade reaction multiplied by raft width
   E = modulus of elasticity of raft material (concrete)
   I = moment of inertia of raft

If $\lambda L < \pi/4$ the raft can be treated as rigid with linear soil stress distribution. The condition $\pi/4 < \lambda L < \pi$ yields a flexible member producing loading at the ends as an infinite beam on elastic foundation.

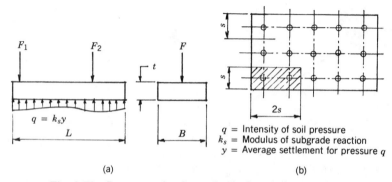

q = Intensity of soil pressure
k_s = Modulus of subgrade reaction
y = Average settlement for pressure q

(a)   (b)

Fig. 9-73 Prestressed raft on elastic foundation.

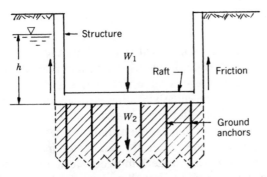

Fig. 9-74 General stability of a deep structure subjected to permanent uplift.

The rigidity of a mat acted upon by a group of prestressed anchors spaced at distance s as shown in Fig. 9-73(b) may be established if we consider an isolated portion such as the shaded corner section with dimensions $2s$ and s. This solution is approximate but acceptable considering the relatively large number of points of application of the prestressing force. Interestingly, the condition $\lambda L < \pi/4 = 0.8$ can be satisfied with small k_s values for the soil or large moments of inertia for the raft. It follows, therefore, that rigidity is attainable better in loose soil, and is difficult in firm soil.

Stability of Anchored Structures Subjected to Uplift. Figure 9-74 shows an underground structure on continuous raft under permanent uplift. Stability must be ensured in terms of failure of a mass of ground discussed in Section 5-3, and the geometry of soil mass assumed to be mobilized at failure can be considered according to Figs. 5-7 and 5-8. If the anchors have overlapping cones of influence, the geometry of ground mass engaged in resisting uplift will be as shown in the shaded area of Fig. 9-74. Equilibrium requires

$$U = W_1 + W_2 \qquad (9\text{-}59)$$

where W_1 and W_2 are total weights of structure and ground, respectively, and U is the total uplift resulting from an uplift pressure $\gamma_w h$. If conditions warrant, the friction mobilized along the sidewalls of the structure can be included in resisting uplift.

9-19 CONSIDERATION OF DYNAMIC LOADS

General Principles

Dynamic effects on anchorages and anchored structures usually result from three main sources: earthquakes, explosions, and vehicular traffic or moving loads. The last category will produce dynamic effects that are relatively small and can be considered as an impact factor of equivalent surcharge on live loads. This category is also within the definition of repetitive loads discussed in Section 4-11. Loads associated with explosions and earthquakes are more critical and can pose certain problems during and after the dynamic event.

Earthquakes and explosions have similarities and common characteristics; for example, they produce vibrating stresses of limited duration, yet they exhibit fundamental differences in terms of the type and magnitude of the associated force. Earthquakes, even shallow ones, occur at great depth, and the resulting motion and forces are manifested mainly from the upward propagation of motion from underlying rock formations. Explosions and blasting, on the other hand, are usually planned relatively close to ground surface, and the associated effects result mainly from the horizontal propagation of motion directly through the ground formation where they are generated. Because of these differences, anchorage design should be based on different approaches and treated independently.

If the ground around an anchorage is disturbed by any of these patterns, anchor performance may be influenced and changed according to one of the following ways: (a) dynamic vibrating stresses may propagate to the anchored structure and transferred to the anchors for the duration of the event thereby increasing the anchor load; (b) fault movement or dislocation of ground blocks within the anchorage may alter the initial configuration between the entry point and the fixed end of an anchor, resulting in prestress loss if the total length is shortened and load increase if the tendon is stretched; and (c) the earthquake may cause the complete liquefaction of the ground around the fixed zone (particularly in fully saturated fine sand) with a corresponding loss of soil strength and inability of the soil to sustain the load transfer.

Ground Motion

If a load applied to a soil mass changes rapidly so that inertia forces become more significant in comparison to static forces, the soil mass is set in vibration stage and waves are propagated away from the disturbance. The motion alternates with increasing distance from the origin because of geometric damping and loss of energy in deforming the soil. Figure 9-75 shows three motions of a point on the ground surface for various types of dynamic loading described by three components for displacement, velocity, and acceleration (D'Appolonia, 1968).

In part (a) the motion has constant amplitude and period. The duration is infinitely prolonged, and the number of cycles can reach millions. An example is a foundation supporting vibrating machinery, in which case satisfactory performance is dependent on the amplitude of displacement. Part (b) shows ground motion of short duration with varying amplitude and frequency, for example pile driving. Nevertheless, this type is considered pe-

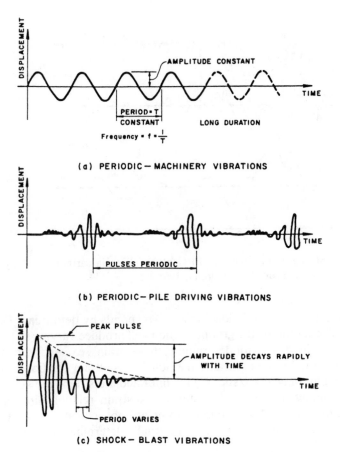

Fig. 9-75 Typical periodic and shock ground motions. (From D'Appolonia, 1968.)

riodic and treated as the motion shown in (a). Part (c) shows the time–displacement behavior at a point away from the origin of excitation, in this case a blast. The pulse rises sharply to a peak, and thereafter the motion is diminished rapidly with time. Likewise, the frequency of the motion varies with time and generally is high. The parameters to be derived from this illustration are the amplitude x_1, the time required to repeat the amplitude termed the period and designated T, and the frequency, which is the reciprocal of T, expressed as $f = 1/T$.

Figure 9-76 shows a strong motion recording from an earthquake. Similar data, with appropriate changes in maximum amplitude to reflect the estimated maximum ground acceleration at a particular site, are typically used to predict the response of a structure and assess liquefaction effects of granular soils (D'Appolonia, 1968).

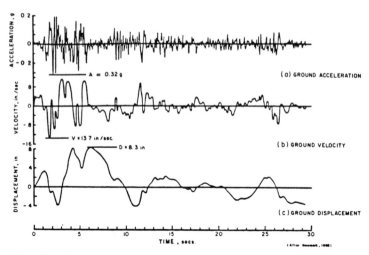

Fig. 9-76   Strong ground motion. El Centro California earthquake, May 18, 1940, N–S component. (From D'Appolonia, 1968.)

Strains. During dynamic loading a soil responds by deforming. Depending on the soil conditions, the ground motion will produce varying strains. There is no simple direct way for determining these strains, except for the following guidelines: (a) for small strains the soil stress–strain relationship is time-dependent and linear–elastic; (b) for intermediate strains the soil exhibits a time-dependent and elastic–plastic stress–strain relationship; and (c) for large strains, stress–strain relationships are not available so that stability, bearing capacity, and liquefaction are usually determined from limiting equilibrium.

Anchored Rafts and Shallow Foundations

The design of these structures for dynamic effects is based on permissible motion and involves methods that have been refined to a point where they can be used with certain confidence (D'Appolonia, 1968; Christian, 1980). It is, however, essential to understand the origin of these methods and properly judge their applicability to anchored slabs and foundation elements. Interestingly, when a structure–anchor system is subjected to vertical oscillations, the soil mass responds with the structure, increasing the damping characteristics of the soil–foundation system. On the other hand, for rocking and torsional motions the hysteric damping may be a significant portion of the total damping of the system, and its determination is open to conjecture.

   Evaluation of vibration-induced anchor overloading requires knowledge of the initial as well as the dynamic stresses and accelerations in the soil mass enveloping the system. Reliable methods, however, are not as yet available to allow simultaneous control of dynamic stresses and accelerations. Furthermore, for a complete analysis it is necessary to determine whether ground densification can occur during vibrations possibly leading to vibration-induced settlement. For controlled accelerations of less than 1 g, the process does not necessarily cause densification but produces inertia stresses that must be considered in evaluating cyclic stress changes.

Since the permanent strain due to repeated stress is highly dependent on the level of dynamic stress, predicting this excitation numerically becomes relevant. In the absence of specific data and methods of correlation, a relative quantitative estimate is possible by reference to Table 9-10. Conclusions therefrom may be valid for initial assessment of minimum relative density of granular soils subjected to dynamic loading so that an approximate correlation between dynamic settlement and density can be made.

Liquefaction. Seed (1968) has shown that liquefaction is the cause of several important failures during earthquakes. For ordinary foundation slabs and rafts, "liquefaction" means loss of bearing capacity. Although the association with anchorages is not explicit, it implies that under this process the soil around the anchorage zone can loose its strength thereby rendering the transfer of load impossible.

The general conditions for liquefaction have been qualitatively documented based on the Niigata earthquake experience (Seed and Idris, 1967). It appears that well-graded medium to fine sand having 10 percent or more particles in the silt size range is less likely to liquefy than sand having the same average grain size particles but poorly graded and containing no fine-grain materials (Lee and Seed, 1967; Peacock and Seed, 1968) Other conditions promoting liquefaction are as follows: (a) the content of silt and clay-size particles should be less than 10 percent, (b) particle diameter at 60 percent passing should be between 0.2 and 1.0 mm, (c) the uniformity coefficient should be between 2 and 5, and (d) the blow count from SPT should be less than 15.

Liquefaction is also affected by the duration of ground motion, and can occur even during earthquakes of low intensity as long as the duration of ground shaking is long. On the other hand, it is documented that with higher

TABLE 9-10 Recommended Minimum Relative Density for Granular Soils Subjected to Dynamic Loading

Soil Strain Caused by Ground Motion	Suggested Minimum Relative Density
Small	70
Intermediate	80
Large	90

From D'Appolonia (1968).

confining pressures a greater number of dynamic shear stress cycles is necessary to cause failure and this also holds for density.

From the foregoing comments two practical procedures emerge: (a) extend the anchor beyond the zone of liquefiable materials and (b) densify the soil in situ by mechanical means.

Analytical procedures have been suggested by Seed and Idriss (1967), and later by several investigators listed in the references of this chapter, for determining the depth to which soil densification should be extended in order to ensure stability during strong ground motion. The two main guidelines are (a) for the design earthquake the ratio of cyclic shear stress, necessary to produce initial liquefaction, to the dynamic shear stress should be greater than 1.5; and (b) for the maximum credible earthquake the cyclic shear stress necessary to produce 10 percent strain should be equal to a greater than the dynamic shear stress.

Anchored Walls

Walls above the Water Table. From relevant statistics it appears that complete collapse of retaining walls due to earthquake effects has been sporadic and a relatively infrequent occurrence (Seed and Whitman, 1970). In general, the performance of a wall during an earthquake event is inferred by model tests where small-scale structures are subjected to base excitation. These tests can provide visual evidence of movement and suggest failure modes.

Results of tests on model walls carried out by Seed and Whitman (1970) show that wedge failure in the soil mass is along an essentially plane surface inclined about 35° or greater with the horizontal, which is generally much flatter that the failure plane inclination under static conditions. This probable position of the failure surface suggests an extended zone of influence in the ground mass, and provides the criterion for selecting the anchorage zone for dynamic loading.

Walls Extended below the Water Table. Documented failures show extensive damage for various wall types, including gravity walls and anchored bulkheads (Kureta et al., 1965). The failure mechanism results from a combination of increased lateral pressure behind the wall, a reduction of water pressure on the front face, and a loss of strength (possibly complete liquefaction of the backfill materials). Earthquake damage of anchored sheetpile bulkheads includes extensive outward movement under increased pressure, with resulting damage to surroundings due to settlement and lateral translation of the foundation. More serious damage should be expected if the backfill is completely saturated by the pressure of an adjacent body of water, in which case wall movement with detrimental effects can be induced by dynamic earth pressures without dynamic water pressure effects. Liquefaction can contribute to wall movement and load to complete failure.

Investigation of dynamic effects through the use of model tests carried out by Seed and Whitman (1970) has established two distinct modes of wall response. Initially, the increase in lateral pressure due to dynamic effects is accompanied by outward wall movement whereas its magnitude increases with increasing base acceleration. However, after a retaining structure with granular backfill has been subjected to a base excitation there exists a residual pressure substantially greater than the initial pressure before base excitation.

Effect of Blasting and Explosions. Blasting and explosions in the vicinity of retaining structures give rise to two categories of problems illustrated in Fig. 9-77. In part (a) an air blast sweeps outward over the ground surface from the point of near surface explosion. These shock waves induce stress waves in the ground producing dynamic stresses on buried structures. This problem is minor with small ordinary explosions but is manifested into an effect of major magnitude with air blast from nuclear explosions.

Part (b) depicts stress waves originating from the point of an underground explosion and moving directly through the ground. These waves induce stresses in any wall or underground structure along their path. Dynamic stresses may be imposed on retaining walls, such as the anchored wall shown in (c), from adjacent blasting operations. The duration of stress pulses from typical small construction blasting is short, usually less than 0.10 s and often less than 0.02 s. The likelihood of severe structural damage is small in this case, although the wall may crack in extreme cases and the anchors may be overstressed or lose some of their prestress.

Tests to Study the Effect of Blasting on Anchors

The service behavior of prestressed anchors in terms of instantaneous and residual load increase resulting from major blasts in the immediate vicinity can be reasonably assessed by trial tests.

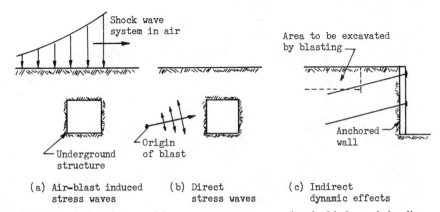

Fig. 9-77 Typical problems arising near structures, associated with dynamic loading from blasts and explosions. (From Seed and Whitman, 1970.)

One of the early arguments (Ortlepp, 1969) is that anchors or bolts in rock should be the yielding type if failure is to be prevented. This conclusion is based on field tests of a tunnel excavation supported on one side by conventional rock bolts and on the other by yielding bolts. Simultaneous detonation of explosives placed along both sides caused total damage of the conventionally supported side whereas the yielding bolts remained intact. More details of these tests are provided by Ortlepp and Reed (1970), Grob (1974), and Armando et al. (1979).

Fluctuations of anchor prestress during blasting have been investigated by Littlejohn et al. (1977) in field tests on four production anchors with capacity 1500 kN (340 kips), and in connection with a reinforced rock slope of an open coal mine where bulk blasting is a routine operation. The anchors had an inclination 45° as shown in Fig. 9-78. Tendon details are summarized in Table 9-11. Following proof load each anchor was locked off at working load plus 10 percent to allow for prestress loss. Immediately after lock off each anchor was check-lifted to establish the actual residual load, and subjected to a second series of load cycling up to 1.5 times the working load to establish the load–extension behavior.

The arrangement and layout of the test anchors and the blast areas are shown in Fig. 9-79. Typically the blast hole drilling pattern varied from 3 m

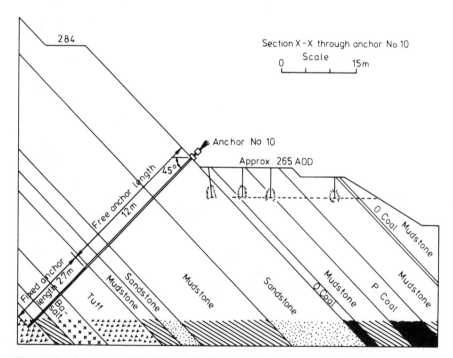

Fig. 9-78   Section *X–X* of Fig. 9-79(b) through open coal mine showing rock formations, test anchors, and blast areas. (From Littlejohn et al., 1977.)

TABLE 9-11   Details of Four Test Anchors to Study Blasting Effects

Anchor No.	Free Anchor Length (m)	Fixed Anchor Length (m)	Maximum Test Load (kN)	Initial Service Load (kN)	Remarks
9	12	4	2400	1655	Load cell installed
10	12	6	2165	1600	Two strands broke during stressing; load cell installed
11	12	4	2400	1640	—
12	12	6	2400	1605	—

From Littlejohn et al. (1977).

× 6 m to 6 m × 6 m with a depth of 6 m. Observations made during blasting are summarized and tabulated in Table 9-12, and confirm that prestressed anchors perform satisfactorily during blasting events. The most significant load increase was observed during blast C when the first line of charge holes was positioned only five meters (16.5 ft) from the anchor heads as shown in part (b) of Fig 9-79. Indeed, this blast increased the load in anchors 9 and 10 by 40 and 60 kN, respectively, representing merely 2.5 to 4 percent of the service loading. Within a few minutes, the initial prestress values were reestablished.

The instantaneous effects of blast C on anchor 10 are shown in Fig. 9-80, and were obtained with a portable tape recorder attached to the unit. The effect of the blast is greater but still moderate, representing an increase 110 kN, or 7 percent of the service load. This was recorded within one second of detonation, and following the vibratory dampening it was reduced to a stable increase of 64 kN after 10 s. This overall increase in prestress is probably caused by an opening of the joints in the rock causing tendon stretching.

Theoretical Considerations. A general rule frequently introduced to avoid damage from blasting is to limit peak particle velocity at the point of concern to less than 2 in/s. This limit is suggested by several investigators in connection with effects on nearby buildings, as shown in Table 9-13 (Wiss, 1968).

The peak dynamic stress in a wave of short duration is expressed as

$$\sigma = dc_D v \tag{9-60}$$

where σ = stress
 d = mass density

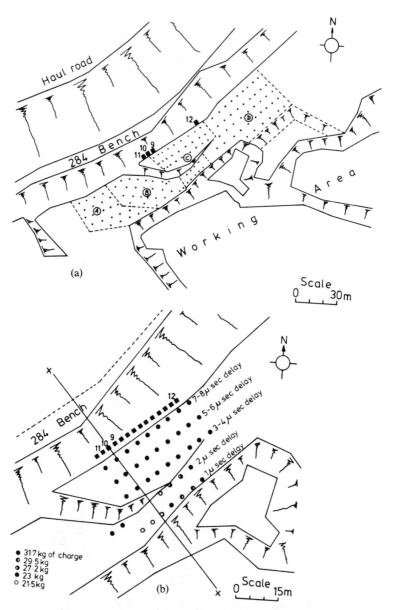

Fig. 9-79 (a) Layout and location of four test anchors and blast areas; (b) plan and layout of blast area C in relation to test anchors. (From Littlejohn et al., 1977.)

TABLE 9-12 Summary and Record of Observations Made During Blasting: Test Anchor in Open Mine

Period Observation (days)	Anchor Service Load (kN)		Remarks
	No. 9	No. 10	
0	1655	1600	Initial service load after check lift
4	1610	1580	
5	1610, 1610	1575, 1575	Blast A (730 Kg aluminum slurry). Average charge = 31.7 Kg/hole. No change in load.
6	1610	1570	
7	1610	1570	
8	1600, 1600	1575, 1575	Blast B (603 Kg aluminum slurry). Average charge = 23.2 Kg/hole. No change in load.
11	1600	1570	
12	1610	1570	
13	1610	1570	
14	1610	1570	
15	1620, 1660	1565, 1625	Blast C (1205 kg aluminum slurry); average charge = 30 kg/hole
	1650 (1 min)	1620 (1 min)	
	1645 (5 min)	1620 (5 min)	Increase in load observed = 2.5% (No. 9), 3.8% (No. 10)
	1645 (15 min)	1620 (15 min)	
18	1640	1630	
19	1625	1630	
22	1625, 1625	1600, 1630	Blast D (1518 kg AN/FO); average charge = 35.3 kg/hole; momentary load increase of 1.9% observed in No. 10 but within a few seconds reading restored
27	1625	1600	
32	1610	1585	
36	1610	1585	

From Littlejohn et al. (1977).

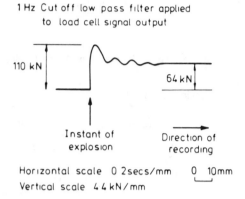

Fig. 9-80 Recorded anchor load output for anchorage 10. (From Littlejohn et al., 1977.)

TABLE 9-13 Empirical Relationship Between Ground Motions from Blasting and Damage to Nearby Structures

Peak particle velocity, in/s	Langefors (Sweden)	Edwards (Canada)	Bumines (U.S.A.)
11	Serious cracking		Major damage (fall of plaster serious cracking)
10	Serious cracking		Major damage (fall of plaster serious cracking)
9		Damage	
8	Cracking	Damage	
7	Cracking	Damage	
6	Fine cracks and fall of loose plaster		Minor damage (fine plaster cracks, opening of old cracks)
5	Fine cracks and fall of loose plaster		Minor damage (fine plaster cracks, opening of old cracks)
4	Caution	Caution	Caution
3	No noticeable damage		
2	No noticeable damage		
1		Safe Limit	Safe
0		Safe Limit	Safe

From Wiss (1968).

c_D = dilatational wave velocity

v = maximum particle velocity

Interestingly, with 5 lb of explosive per relay, the peak particle velocity is reduced to 2 in/s within 45 ft from the blast line. In most instances the fixed anchor zone would be expected to be located more than 45 ft from the nearest blast origin. Although the 2-in/s limit is certainly most conservative for buried structures and could be conceivably raised to 10–15 in/s for any part of an anchorage, it may suggest that controlled blasting near an anchored structure is unlikely to have detrimental effects.

9-20 FUNDAMENTALS OF ANCHORED WALLS

Working Principles

Anchored walls provide the support of vertical or near-vertical excavations. Limiting conditions of the system were identified in Section 5-4 in relation to fixed anchor zone location and are shown in Fig. 5-11. In general, excavation in soil mass causes unloading and local yielding of the soil. If the opening is deep enough a shear surface develops, resulting in some form of shear failure. A retaining wall is constructed against the excavation face to limit unloading of the soft ground and inhibit formation of a failure surface. The wall is acted upon by an active stress environment, and unless it is stable a resisting force must be introduced, for example, in the form of anchors, to provide the conditions of stability. On the other hand, movement (vertical or horizontal) must be restrained and confined within allowable limits.

The mechanism of an anchored wall is thus complex since the ground, wall and anchors must interact and work together in order to resist earth pressure loads and surcharges developing during and after construction, and restrict deformations to acceptable values. As the wall deflects toward the excavation under the lateral loading, the anchor stretches and initiates the load transfer in the fixed zone. The fixity imposed on the anchorage by the soil restraints further wall deflection. This movement is further controlled if anchors are prestressed.

It appears from these brief remarks that the analysis and design of anchored walls involves the following prime considerations: (a) a suitable anchorage zone must be available for the load transfer to the ground, as discussed in Chapters 4 and 5; (b) lateral earth stresses and water pressure acting against the wall must be evaluated and numerically defined; (c) the stability of the composite ground–wall system must be ensured against translation, rotation, and sliding; and (d) the anticipated horizontal and vertical movement must be predicted and, if excessive, controlled.

The current profound understanding of wall behavior suggests that the major factors affecting performance of the wall and supported ground mass include the following:

1. Vertical wall movement, under the effect of the vertical anchor component.
2. Wall stiffness, expressed in terms of the elastic modulus E, the moment of inertia I, and the wall length (depth) L between anchors (spacing).
3. Initial anchor prestress level in relation to static earth loads.
4. Internal deformation of the soil block in the zone of influence.
5. Movement of the ground resulting from the excavation.
6. Ground loss associated with the construction method and the imposed controls.
7. Volumetric strain.

Suggested Methodology

Essentially, methods of analysis and design rely on predictive techniques whereby the field situation is identified and defined, to be simplified in terms of the important parameters involved, and then the expected mechanisms are superimposed on the relevant parameters and manipulated to obtain estimates and solutions. This approach may be based on the following tasks:

1. Determine the problem and review the applicability of available knowledge on the subject.
2. Study the ground in conjunction with site characteristics and restraints. Consider the important parameters of strength, deformability, and sensitivity to damage because of construction.
3. Establish and group anticipated loads; these may include earth stresses, surcharges, and dynamic effects.
4. Consider probable mechanisms of failure and carry out investigations to determine the degree of stability (in terms of the factor of safety) provided by the design.
5. Where areas of uncertainty exist carry out sensitivity analyses (parametric studies) in which pertinent variables are introduced to assess their influence on stability.
6. Correlate the foregoing process with the design methodology of anchors discussed in Section 5-11.
7. If a field monitoring is necessary to confirm the adequacy of the design, formulate an outline and specify its objectives.

9-21   PROCEDURES FOR ESTIMATING LATERAL STRESSES AND DEFORMATIONS

General Requirements

In simple terms, the formulation of the problem of predicting lateral pressures and deformations is essentially the definition of appropriate boundary values. This requires knowledge of the initial stress conditions in the ground, the constitutive relations for the soil, and the correct or the most realistic boundary conditions for useful results.

Initial Stresses. In sedimentary soil, as the buildup of overburden continues there is vertical compression of soil because of increase in vertical stress, but there should be no significant horizontal compression. In this case the horizontal earth stress is less than the vertical, and for sand deposits formed in this manner K_0 usually ranges between 0.4 and 0.5. Thus, for initial loading the expression proposed by Jaky is confirmed by the majority of investigators (Bishop, 1958) so that

$$K_0 = 1 - \sin \phi' \tag{9-61}$$

where K_0 = coefficient of lateral pressure at rest
        ϕ' = angle of shearing resistance for effective stress

However, with the exception of certain soils such as normally consolidated clays, the initial effective stresses in a given ground are seldom known with confidence. There is also evidence that the horizontal stress can exceed the vertical if a soil deposit has been heavily preloaded, as a result of a process where the stress remained locked and did not dissipate when the preload was removed. The coefficient K_0 may now approach 3, and under certain conditions it may become close to K_p (Brooker and Ireland, 1965; Skempton, 1961).

Constitutive Equations. Although the nature of constitutive equations for sands and normally or lightly overconsolidated clays prepared in the laboratory is adequately understood, natural soils or soils placed under field conditions are not always fully represented. Obviously, natural soils may display anisotropic, nonhomogeneous, and time-dependent properties. Furthermore, discontinuities give rise to size effects in response to loading.

Boundary Conditions. These are equally essential for meaningful estimates of lateral stresses and deformations. They are more reliable if they can represent actual construction procedures and a pragmatic interaction between structure and soil, including the anchorage. In the following sections examples are presented demonstrating the difficulty in prescribing correct boundary conditions for certain categories of problems. In some instances, these conditions can only be stated in a crude idealized approximation, even where K_0 and constitutive equations are established reliably.

Where the prediction of deformations is essential, the problem is usually approached with linear elastic theory. If maximum lateral pressure or resistance is the governing factor, limiting equilibrium methods are typically used to estimate these forces. In this case little, if any, consideration is or can be given to actual deformations and associated movement. In other instances, such as braced excavations, movement is usually reduced if not entirely stopped, and this affects the distribution of lateral earth stresses. Semiempirical methods are in this case used to arrive at a reasonable solution. Likewise, anchor prestress and wall stiffness affect movement and cause changes in the magnitude and distribution of earth loads.

Elastic Methods of Analysis

This procedure involves both linear and nonlinear stress–strain relations. The former requires judgment in selecting the appropriate modulus. Nonlinear analysis on the other hand, should include studies of several stress paths so that relations can be found that are not unduly restrictive. Linear analysis can be used to calculate both small and relatively large deformations by changing the elastic modulus. Problems, however, involving large deformations and simulation of yielding are better approached with nonlinear models.

Linear Analysis. An excavation with a high factor of safety and small deformations is a good example for linear analysis. If this excavation is in clay, base failure will occur under undrained conditions when

$$\gamma H = N_c s_u \qquad (9\text{-}62)$$

where γ = bulk density of clay
   H = height (or depth) of excavation
   N_c = stability number depending on the geometry of the problem
   s_u = undrained shear strength

Terzaghi and Peck (1968) have introduced the dimensionless number $N = \gamma H/s_u$ as an index of probable base failure. If N is about 3–4, some plastic yielding can occur. According to Alberro (1969), if N is less than 4, pressures and deformations can be computed using elastic theory. If $N_c = 6$ is taken as typical for most excavations and $N = 3$–4, a criterion is manifested for the applicability (lower bound) of elastic theory (Morgenstern and Eisenstein, 1970). Until recently, however, this criterion was limited to excavations in deep soft and medium clays.

The effect of K_0 on stresses and deformations is documented by Dibiagio (1966) in finite difference schemes. In these studies the maximum heave appears to be insensitive to K_0 variations, but maximum lateral displacement is practically linearly dependent on this coefficient. If K_0 is taken as unity for soft clays with initial undrained modulus 1500 psi, the maximum lateral displacement in an excavation 25 ft deep will be 3.1 in, and for stiff clay with modulus 15000 psi it is only 0.3 in. It follows, therefore, that for typical excavations where elastic theory is used, maximum displacement should not exceed a few inches. When larger displacements are expected substantial yielding should be predicted, and either the modulus should be adjusted or nonlinear relations introduced.

The significant effect of boundary conditions is illustrated in the excavation shown in Fig. 9-81 (Morgenstern and Eisenstein, 1970). The wall supporting this excavation is rigid but smooth with respect to vertical displacement, and can simulate a slurry wall with internal bracing or anchors. Earth pressures are computed from elastic theory for differing boundary conditions. The soil behind the wall has the following parameters: $E = 65,000$ psi; Poisson's ratio $\mu = 0.3$; and bulk density $\gamma = 125$ pcf. The initial earth stresses are derived from elasticity and the relation

$$K_0 = \frac{\mu}{1 - \mu} \qquad (9\text{-}63)$$

so that for $\mu = 0.3$, $K_0 = 0.43$. Earth pressures against the wall are computed for the position of zero lateral yield when the excavation of height H is taken down to a rigid base. As expected, the calculated lateral pressures for both the rough and the smooth base are the same as the initial K_0 horizontal stresses, since neither lateral nor vertical displacement has occurred and the presence of excavation has no influence on the stress environment.

Earth pressure distribution for the condition of no lateral yield is shown in Fig 9-82 together with the pressure distribution when the rigid base is at distance $0.5H$ and H beneath the base of excavation. In the latter cases earth pressure distribution changes significantly, although the wall has not moved, because of the ability and freedom of materials to flow beneath the wall. This effect is amplified when the rigid base changes to smooth and is located

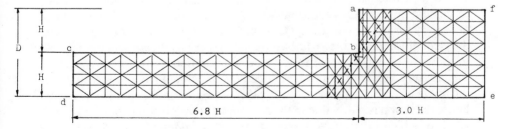

Boundary Conditions
a-b : Wall, rigid and smooth, allowed to yield
b-c : Free surface
c-d : Fixed side boundary, rigid and smooth
d-e : Base, rigid and either rough or smooth
e-f : Fixed side boundary, rigid and smooth
f-a : Free surface

Fig. 9-81 Excavation supported by a rigid, smooth wall; element idealization of a problem of earth pressure behind a yielding wall. (From Morgenstern and Eisenstein, 1970.)

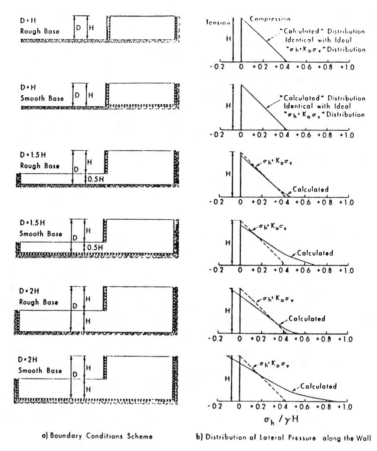

a) Boundary Conditions Scheme   b) Distribution of Lateral Pressure along the Wall

Fig. 9-82 Lateral pressure distribution for different boundary conditions, wall of Fig. 9-81; condition of no lateral yield. (From Morgenstern and Eisenstein, 1970.)

deeper below excavation level. Interestingly, the maximum horizontal pressure at the base increases while stresses at the top reverse to tension.

For the same example lateral earth stresses are computed for a wall displacement toward excavation of $0.0025H$, which is less than the displacement necessary for the active state. The results are shown in Fig. 9-83. The boundary condition along the rigid base is now most significant when it is close to the base of excavation. As excavation is carried down to the rigid base the pressure behind the wall is reduced by 50 percent from the K_0 state for the rough base, but only by about 10 percent for the smooth base. The former larger reduction is partly due to the presence of tension along the base, which is not feasible in reality. A nonlinear stress distribution is developed as the rigid base is taken below excavation level.

Likewise, lateral earth stresses are computed for a small displacement $0.0025H$ toward the ground approaching the passive state, and are shown

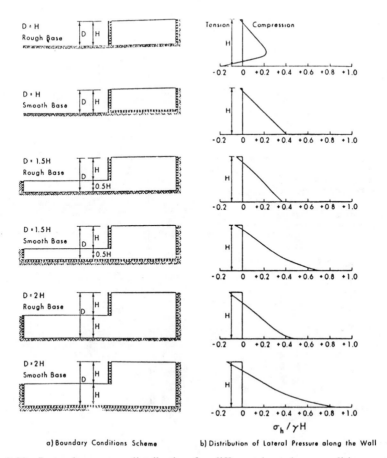

a) Boundary Conditions Scheme   b) Distribution of Lateral Pressure along the Wall

Fig. 9-83 Lateral pressure distribution for different boundary conditions, wall of Fig. 9-81: pressure diagrams for wall yielding 0.0025H towards active state. (From Morgenstern and Eisenstein, 1970.)

in Fig. 9-84. The passive resistance increases considerably owing to the presence of the rough rigid base, but the effect of conditions along the rigid base decreases as this base is moved further down below the excavation. An important conclusion is that earth pressures in the elastic range are sensitive to changes in lateral deformations when the rough rigid base is close to excavation level.

Nonlinear Analysis. For soils such as loose sands and compacted clays it usually is difficult to fit material behavior within the scope of linear analysis, and in spite of high factors of safety. Typical examples are walls associated with fill operations, in which case some form of nonlinear analysis might be considered. On the other hand, nonlinear anchor behavior and nonlinear soil

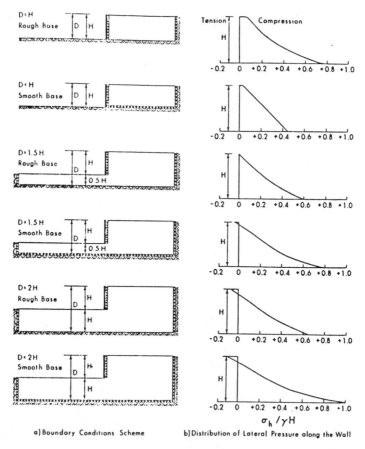

a) Boundary Conditions Scheme   b) Distribution of Lateral Pressure along the Wall

Fig. 9-84 Lateral pressure distribution for different boundary conditions. wall of Fig. 9-81; pressure diagrams for wall displacement towards the ground (passive state) 0.0025H (From Morgenstern and Eisenstein, 1970.)

response have been considered by many investigators and, if properly represented, yield satisfactory results.

Invariably, this method of analysis requires successful interpretation of results from triaxial compression tests in order to establish meaningful stress–strain relations. Conventional finite-element techniques help to solve the problem in an incremental manner.

Limiting Equilibrium Methods

A major point of interest is the very small displacement required to reduce the earth pressure in sand to values approaching the full active state. The strains required to produce active and passive conditions are found empirically in triaxial tests. Thus, for granular materials less than -0.5 percent is needed to convert pressures to active state, whereas a horizontal compression of about 0.5 percent is necessary to reach about one-half the maximum passive resistance.

For problems where it is not necessary to know the actual deformations expected to occur, limiting equilibrium methods are used. In this case movement is integrated with the choice of appropriate factors of safety, which is intended to contain deformations within empirically acceptable limits.

Invariably, all limiting equilibrium theories are formulated on the concept of soil strength expressed by the Coulomb–Mohr failure criterion. They differ, however, in the shape and location of the failure surface and in the application of statics. Coulomb's theory assumes that failure occurs along a planar sliding surface. Other theories are based on logarithmic spiral slip surfaces, slip along a circular surface, and the concept of slices. In the Sokolovski method (Sokolovski, 1965) the equations of equilibrium and the failure criterion are explicitly satisfied for each infinitesimal element responding to the failure.

The Role of Friction. When friction exists at the supporting boundary Rankine's theory must be modified. In practice, the relative wall movement with respect to the soil results in shear (frictional) stresses along the interface. For a wall in the active state friction is usually a downward force. In the passive zone, however, the soil bulges upward but either positive or negative friction can develop according to the actual motion.

Janbu (1957, 1972) considers the shear stresses that may develop along the contact wall–soil surface and their applicability to limiting equilibrium concepts. The difference between earth pressures at failure and at equilibrium is illustrated in Fig. 9-85 for an expansion zone (active) and a compression zone (passive). Symbols with the subscript f indicated failure values. The problem assumes plane strain and a smooth wall, in which case the direction of major principal stresses will be vertical in expansion and horizontal in compression for vertical walls. The corresponding critical shear surfaces are planes, and the shear stresses along these planes change sign from expansion to compression.

Since smooth walls are seldom encountered in practice, a generalized approach to earth pressure calculations should include shear stresses τ_w along the wall. If τ_e is the equilibrium shear stresses in the adjacent soil, the roughness ratio r is defined as (Janbu, 1972)

$$r = \frac{\tau_w}{\tau_e} \tag{9-64}$$

Figure 9-86 shows that the geometry of the critical shear zones varies widely with r, and analyses have demonstrated that this factor has a marked influence on the value of earth pressures (Janbu, 1957). In practice, therefore, earth pressure calculations require graphs and expressions covering the en-

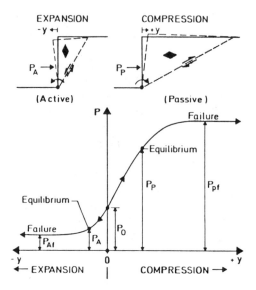

Fig. 9-85   Definition of earth pressures at equilibrium and at rest. (From Janbu, 1972.)

tire range of equilibrium conditions from design to ultimate failure values.

Two examples where the foregoing principles have direct applicability are shown in Fig 9-87. During pullout of the horizontal plate in part (a) the soil above compresses vertically and expands horizontally, thus compressing the surrounding ground. In this case, a state of passive pressure with $r = -1$ develops at failure, whereas the theoretical shape of the shear surfaces is confirmed in model tests. For the indicated movement at failure of the anchor block, $r = -1$ in the passive zone whereas $r = +1$ in the active zone. For the anchored wall shown in (b), vertical equilibrium requires that $r < 0$ in the active and $r > 0$ in the passive zone.

Semiempirical Methods

For certain classes of structures and construction methods (e.g., sheet pile walls and braced excavations) theoretical analysses often have limitations.

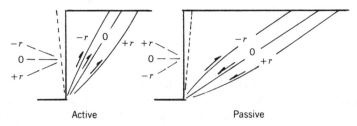

Fig. 9-86   Influence of the roughness ratio on the geometry of the zones of critical equilibrium. (From Janbu, 1972.)

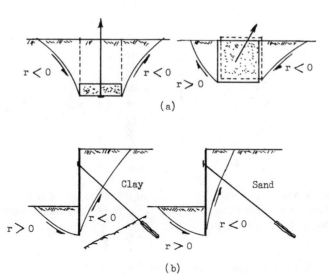

Fig. 9-87   Example of earth pressure zones with positive and negative r (From Janbu, 1972.)

Semiempirical methods have evolved in these instances and provide reasonable basis for design.

Typical examples are the apparent pressure diagrams used to estimate strut loads in braced excavations (Peck, 1969), and the "free earth support method" used in the analysis of sheet pile walls (Tschebotarioff, 1962).

9-22   EMPIRICAL GUIDELINES FOR CONTROL OF MOVEMENT

Two typical patterns of movement for anchored walls are shown in Fig. 9-88. If the top remains fixed, the deformation is as shown in (a) and is similar to an internally braced wall. Settlement, partial yielding of the anchors, gross movement of the soil mass and shear deformation produce the composite movement shown in (b). This may involve translation, rotation, and flexural bulging.

If the soil mass involved in the zone of influence deforms essentially as a unit, wall movement may be as shown in Fig. 9-88(c). This movement is not typical, but very likely with unyielding base and where the bottom of the wall is restrained against outward displacement.

In general, the deformation mode of anchored walls is complex but can be improved to the level of satisfactory performance, particularly in competent soils, if the following factors are considered:

1. In granular soils in which soil modulus increases with stress level, the prestressed soil mass engaged in the anchor interaction is made more rigid, and hence less deformable.

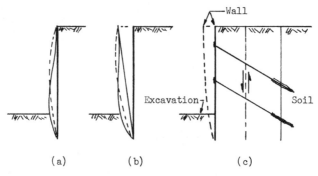

Fig. 9-88 Movement of anchored walls; (a) fixed top or slight translation; (b) rotation about bottom and flexural bulging; (c) internal shear development, horizontal shift of top relative to bottom.

2. Anchors are typically prestressed to about 150 percent the working load and then locked off at a value close to this load. This effect tends to stiffen the soil monolith. However, prestressing the anchors to compensate for movement expected to occur may have uncertain effects. In stiff clay or in dense sand anchor prestress tends to push the wall toward the ground, but if the wall is very stiff, the ground will resist this tendency more vigorously. On the other hand, in very compressible ground and with unusually high prestress levels the wall may move excessively toward the ground, and this displacement becomes permanent if the anchors have the fixed zone in rock.

3. Prestressing the anchors to draw the wall back to its initial profile and thus compensate for movement which has already occurred does not always produce the desired results. In stiff soil and with stiff walls considerable resistance is offered by the ground itself, or the wall may revert to its first profile as the anchors yield to mobilize their strength.

4. Very relevant is the position of the top row of anchors, usually dictated by the allowable cantilever moment. Usually, the first row of anchors is located 4–5 m (13–16 ft) below ground level. However, considerable wall movement has occurred during the cantilever stage and reached as much as 50 percent of the wall movement at full excavation. Thus, it is prudent to locate the first row of anchors close to the ground (1.5 m or 5 ft deep), but in this case the risk of ground failure close to the surface as the tendon is prestressed should be assessed.

5. Very important is the time-dependent movement observed for excavations in clay. Very often, this movement extends behind the wall for a distance twice the excavation depth. Long-term movement of anchored walls has reached 150 percent the movement at the end of excavation, and was recorded 18 months after construction. These observations suggest the importance of providing the permanent bracing (where contemplated), including the base slab, as soon as possible.

6. Certain improved concepts have been suggested (Goldberg et al., 1976) to reduce movement, and include the hybridization of anchors and internal bracing. In this manner, the best features of each system can be combined for best results.

7. Wall embedment below excavation level restrains movement at this location, and is most effective if it is in stiff or dense soil.

9-23 GENERAL STABILITY OF THE GROUND–ANCHORED WALL SYSTEM

The literature contains a fair amount of material on the stability of the ground–anchored wall system. Suggested references include the French Code (1972), Goldberg et al. (1976), Pfister et al. (1982), and Otta et al. (1982).

Analysis with Plane Failure Surface. A simple method based on limiting equilibrium is shown in Fig. 9-89. This method assumes that a slip failure develops of planar configuration. The force acting in this manner is compared with the resisting force developed along the slip surface, and a factor of safety is established. Deformations are not computed, but lumped into the

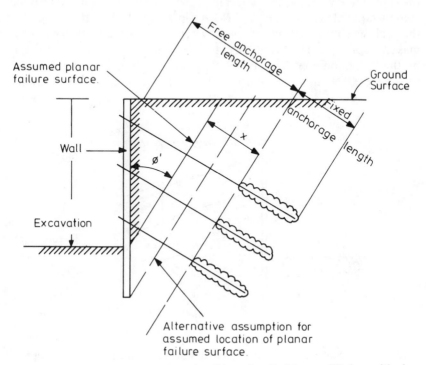

Fig. 9-89 Stability analysis of anchored wall based on limiting equilibrium with plane failure surface.

factor of safety. In the simplest form the plane failure surface begins at the base of the excavation and extends at an assumed angle. From these considerations the fixed anchor zone begins at some point beyond the assumed failure surface. Alternatively, the failure plane is assumed to begin at the base of the wall. Dimension x is determined from the foregoing considerations.

Analysis with Circular Slip Surface. Probably a more realistic assumption of limit equilibrium analysis is failure along a circular slip surface as shown in Fig. 9-90. The method most commonly used in Europe is the Fellenious slip circle, supplemented by Huder (1965), Locher (1969), and Malijain and Van Beveren (1974). Likewise, the fixed anchor zone is located outside the most critical failure surface, and additional check is necessary to ensure that the anchors are not too short for these conditions.

The Sliding Block (Kranz) Method. This procedure, initially formulated by Kranz (1953), was expanded by Ranke and Ostermayer (1968). The objective is to simplify the analysis by replacing the actual experimental failure shown in Fig. 9-91(a) by the composite surface shown in (b) and (c), which is a modified version of the method (Pfister et al., 1982). The failure prisms (*dce*) in active pressure as well as in passive (*bhg*) are replaced as a second step by equivalent forces P_a and P_p as shown in (c). With this simplification the analysis shifts from the complex system wall–ground–anchors to the soil mass M represented by the block *ecbf*. The wall and the anchors are replaced by their reactions on the mass, $-P_A$ for the wall and A for the tension in the anchor. The analysis is carried out for a unit length of excavation.

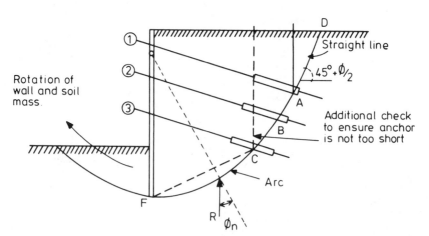

Fig. 9-90 Stability analysis of anchored wall based on limiting equilibrium with circular slip surface.

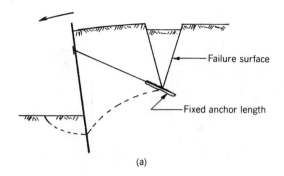

(a)

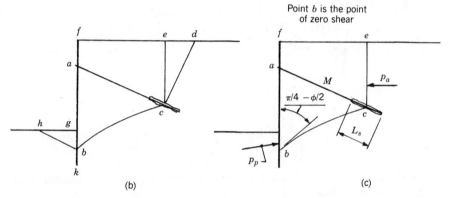

(b)   (c)

Fig. 9-91 Failure of an anchored wall–ground system; (a) actual mode of failure; (b,c) simplification of the wall anchor system for stability analysis, using the sliding block method.

Analysis with One Row of Anchors

The soil mass M is defined by planes bf and ce by a curved failure line bc, where b is the point of zero shear in the wall. Point c is located on the axis of the anchor at either of the following distances from the end:

- Half the fixed anchor length L_s if the spacing B between two adjacent anchors of the same row is less than or equal to $L_s/2$.
- Equal to B where $B > L_s/2$.

The failure curve bc represents a circle whosé tangent at point b has an angle $(\pi/4 - \phi/2)$ with the wall as shown. This assumption is probably as close to reality as possible, since a straight line between points b and c may yield conflicting results. The forces involved in the equilibrium of the block, shown in Fig. 9-92, are as follows:

1. Known Forces. These are W = weight of soil mass M, $-P_A$ = wall reaction equal to active pressure on height bf, and P_a = active pressure on

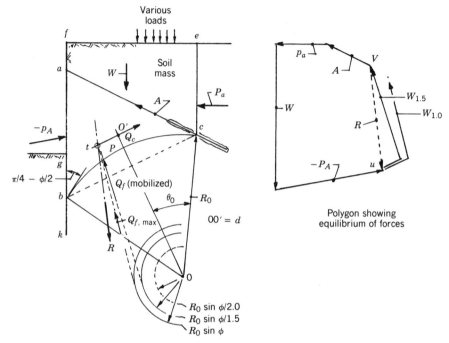

Fig. 9-92 Assumed condition, forces and equilibrium diagrams for a sliding block necessary for stability analysis.

plane *ec*. These three forces are estimated using effective stress analysis since hydrostatic pressure, which cannot stabilize the soil mass *M*, should not be taken into consideration. Hydrostatic forces, however, may have to be introduced as horizontal component in P_a and vertical component in *W*: A = the tension force in the anchor and F_e = exterior forces applied to the soil mass but having no stabilizing effects.

2. Unknown Forces. The two components due to friction and cohesion *c* of the reaction along surface *bc*. The cohesion component, designated as Q_c, is parallel to *bc* and has a magnitude $2cR_0 \sin \theta_0$ at distance $d = R_0\theta_0/\sin \theta_0$ from center *O*. The friction component, designated as Q_f, is tangent to a circle with radius equal to $R_0 \sin \phi$. The forces Q_c and Q_f as defined in this manner are limit values of the two components corresponding to the failure state (Pfister et al., 1982).

Computation Procedure. The analysis is carried out in the following steps: Determine direction and magnitude of resultant *R* of forces *W*, $-P_A$, P_a, *A*, and F_e. Establish the intersection point *t* with the Q_c axis. Initially the polygon is drawn for the forces *A*, P_a, *W*, and $-P_A$ as shown in Fig. 9-92, terminating at point *u*. A factor of safety is then selected and the polygon is completed from point *u* by adding two vectors as follows: (a) a vector

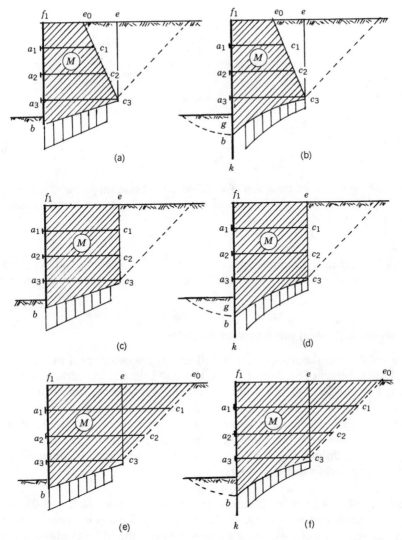

Fig. 9-93 Equilibrium condition of soil mass M for a wall supported by multiple row of anchors.

equal to Q_c/F_s, and (b) a vector parallel to the tangent drawn from point t to the circle with radius $R_0 \sin \phi/F_s$ centered at O. The actual value of the factor of safety is determined by trial-and-error procedure until the polygon of forces is closed exactly at point V. This factor is preselected by the designer.

The analysis is considered satisfactory if (a) the factor of safety F_s is at least 1.5; (b) point P where the Q_f support line intersects circular surface

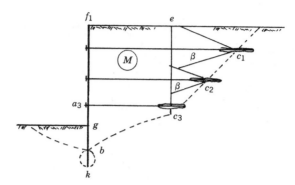

Fig. 9-94 Equilibrium condition of a multianchored wall supporting a soil mass M; unequal length anchors and location of fixed zones with respect to stability of mass M.

bc is located in the central area of bc; and (c) the average stress along bc is the allowable (working), obtained from the failure stress using a factor of safety 3.

Analysis with Multiple Rows of Anchors

The equilibrium conditions for a soil mass supported by a multianchored wall are shown in Fig. 9-93. Prior to the stability analysis certain conditions must be satisfied: (a) the wall embedment below excavation level is compatible with the passive resistance required in this area and (b) the fixed anchor length is extended beyond the unstable zone. The stability of soil mass M is checked following the same procedure as for the single-anchor wall, but modifying the rear limit of mass M, the shape of slip surface and the tension forces in the anchors.

Rear Limit of Mass M. Conceivable cases are shown in Fig. 9-93, where the actual volume of mass M is designated by the shaded areas. The rear limit of soil mass engaged in the interaction is defined by the lowest row of anchors, and is the vertical line through point c_3.

Shape of Slip Surface. If the wall embedment is small and it can be ignored in the analysis (parts a, c, and e), the mass M can be treated as a gravity wall with a surface foundation. In this case the slip surface is plane bc_3. If the embedment is considerable (parts b, d, and f), the slip surface approaches a circle with a tangent at an angle $\pi/4 - \phi/2$ at point b of the wall.

Tension in the Anchors. If the fixed anchor zone is placed beyond the boundary of mass M, it is possible that the force engaged in the stability of the soil mas will be less than the actual force in the anchor A. This case will arise if one anchor in the group is much longer than its neighbors. Under these conditions part of the cone defined by the angle $\beta = \pi/4 - \phi/2$ will pass below the rear limit of mass M. An example where this problem will not arise is shown in Fig. 9-94.

Computation Approach. Likewise, the system is acted upon as does the single-anchor row with the exception of the slip surface. With a planar surface the maximum value of Q_f corresponds to its angle $\pi/2 - \phi$ with the slip surface, and the maximum Q_f is borne by the straight line bc. Furthermore, for a planar slip surface Q_f is defined by the angle $\pi/2 - \phi/F_s$ which is made with the slip surface bc, and this is intended to represent the condition of limited wall embedment so that the procedures of surface foundations can apply (Pfister et al., 1982).

9-24 ESTIMATION OF ANCHOR LOADS

With little or no prestressing the loads developed in the anchors will correspond to the active state if sufficient movement occurs to yield this condition, or to partially at rest pressures if movement is limited. In any other case, lateral earth stresses behind the wall are manifested by the level and sequence of prestressing.

In practice prestressing is introduced to a level that is quite variable and often arbitrary. It may be related to ultimate anchor capacity in which case its magnitude is determined by the factor of safety, or it may have a theoretical basis. A review of the methods used to calculate prestress loads is presented in Table 9-14 (Clough, 1973), and evidently these methods cover limiting equilibrium and apparent pressure theories.

In general, the choice of an appropriate earth pressure diagram for estimating anchor prestress loads first depends on the tolerable wall and soil movement. If important adjacent structures are sensitive to settlement, increased prestressed levels offer a good choice in restricting excessive move-

TABLE 9-14   Summary of Methods Used to Estimate Prestress Load on Anchors

Reference	Method
Kapp	Percentage of allowable tie-rod load (20–60%)
Mansur and Alizadeh	At-rest pressures
Rizzo, et al.	Active to at-rest
Shannon and Strazer	50% anchor yield load
Clough, et al. (1974)	Terzaghi–Peck rules $(0.4\gamma H)$
Liu and Dugan	15 × height wall (in psf)
Hanna and Matallana	Pressures halfway between active and at-rest
Oosterbaan and Gifford	Active pressures
Larsen, et al.	Pressures between active and at-rest

ment. These may be the apparent pressure diagrams, the earth pressure at rest or active pressures increased by a factor of 1.5–2.0. Field monitoring of anchored walls shows that these higher pressures will remain unless a major stability problem arises. If some wall movement can be tolerated, prestress may be reduced to smaller values, for example, 75 percent the Terzaghi–Peck apparent pressure or 1.25 times the active state.

Effect of Restraints and Construction Conditions. Restraining effects are manifested by wall stiffness and wall fixity against rotation or translation. Both these mechanisms can be available in prestressed anchored diaphragm walls. Systems supporting excavations can be represented by the four basic types shown in Fig. 9-95. In (a) the wall is an anchored cantilever-type, deriving its stability from sufficient embedment. The assumed linear earth stress distribution is appropriate and consistent with the wall movement. The wall in (b) is anchored at ground surface (elastic support) and embedded at its lower end, so that its deformation may be that of a simple beam. The wall shown in (c) is likewise supported at the top, but its small embedment below excavation level allows some lateral movement so that overall deformation is as shown. Walls stabilized with multiple supports as shown in (d) are likely to undergo a nearly uniform translation. The types shown in (b), (c), and (d) have deformation characteristics producing nonlinear redistribution of lateral earth pressures as shown in the corresponding diagrams.

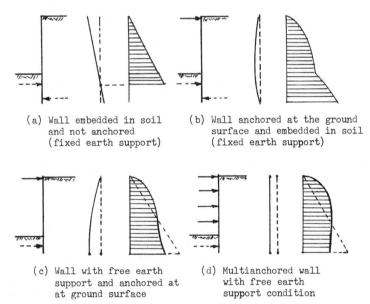

(a) Wall embedded in soil
    and not anchored
    (fixed earth support)

(b) Wall anchored at the ground
    surface and embedded in soil
    (fixed earth support)

(c) Wall with free earth
    support and anchored at
    at ground surface

(d) Multianchored wall
    with free earth
    support condition

Fig. 9-95 Basic types of wall supporting excavations; various restraints against rotation and translation.

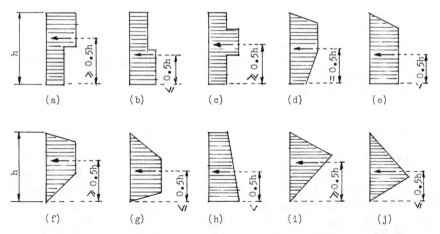

Fig. 9-96 Examples of pressure diagrams conceivable with prestressed anchored walls.

Unless anchored walls are prestressed to specific active stress levels and their movement is consistent with the requirements of the active condition at each construction stage, lateral earth pressure distribution will be essentially nonlinear, and largely determined by the interaction of local factors. These may include soil type, degree of fixity or restraint at the top and bottom, wall stiffness, special loads, and construction procedures. Clearly, in this case earth pressures cannot be calculated directly, and may resemble any of the pressure diagrams shown in Fig. 9-96. Alternatively and within certain limits (particularly in connection with rigid walls), a desired earth pressure distribution can be manifested by a particular arrangement and sequence of prestressing (Otta et al., 1982). If, for example, the distribution must be changed towards the top of the wall to produce a resultant which lies in the upper half, it will be necessary only (for multianchored walls) to make the upper anchors longer than the lower ones. The same effect can be produced if the prestress level is higher in the upper rows. Water pressures are not subject to redistribution.

Statical Analogy of Anchored Walls. For simple investigations, retaining anchored walls resemble a continuous beam with point supports at anchor locations, and bottom support beneath excavation approaching an elastic spring. Earth resistance is developed on both sides of the embedded section. The embedment is varied until the given boundary conditions are satisfied; that is, for complete restraint of the base (fixed and support) the bending moment condition governs, and for the free earth support condition the resultant of the passive resistance must equal the support forces in the beam. With this model, calculations of the anchor forces requires simply determination of the shear forces in the continuous beam. Any assumed redistribution of earth pressures induced by the prestressing of anchors is taken into consideration either by modifying the pressure diagrams accordingly or by increasing the calculated shears in the beam by an appropriate factor, usually 1.2–1.3. Among the available procedures, Blum's elastic line theory by graphical methods offers simplicity and expediency. An excellent reference is Otta et al. (1982).

The Method of Equivalent Tie Support for Multianchored Walls

This procedure has its origin in elastic theory; hence it considers load as function of deformation. Successful uses have been reported by Littlejohn and MacFarlane (1974). Although it is based on theory, the method presents essentially empirical solutions in which flexibility coefficients are used for multiple-anchor analysis (James and Jack, 1974).

For a load P that is a function of deformation the following is true:

$$P = \frac{d_y^4}{dx^4} EI = yr_s \tag{9-65}$$

where r_s is the equivalent spring stiffness of the soil. The wall may be considered a series of members connected by nodes. At these points horizontal members simulating the soil stiffness or support joints can be represented as shown in Fig. 9-97. A simple iteration routine and a simplified stress–strain relationship can be used to simulate the elastic–plastic effects of a soil as shown in Fig. 9-98. Estimates of P continue for each deflection profile until convergence to a condition of equilibrium is achieved.

Anchor forces can be predicted from a consideration of the temporary effects produced by the passive resistance at intermediate excavation stages. The position and magnitude of a resultant anchor is estimated by treating

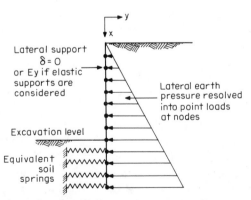

Fig. 9-97 Simulation of wall–soil interaction; anchored walls. (From James and Jack, 1974.)

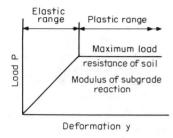

Fig. 9-98 Simplified stress–strain relation in elastic–plastic soil.

the wall as a single-anchored structure under the following assumptions: (a) mobilizing and resisting pressures correspond to the Rankine state (this is made for simplicity and is justified where appropriate movement is expected); (b) at failure there is rotation of the wall about a unique point along its plane (this enables the use of a simple procedure for calculating the additional anchor forces induced when the passive resistance is shifted to a different area during the next stage of excavation); and (c) the wall has a length sufficient to provide a factor of safety of 1 against rotation at each excavation stage (this allows estimation of the maximum resultant anchor to free earth support conditions ignoring fixity at the end).

Equilibrium conditions are shown in Fig. 9-99 for various stages. From (a) we obtain

$$\sum F_H = 0 \quad \text{is satisfied when} \quad T_1 = P_a' - P_p'$$
$$\sum M = 0 \quad \text{is satisfied when} \quad M_p' = M_a' \text{ about point } T_1 \tag{9-66}$$

Likewise from part (b) the following are derived:

$$\sum F_H = 0 \quad \text{is satisfied when} \quad T_1 + T_2 = P_a'' - P_p''$$
$$\sum M = 0 \quad \text{is satisfied when} \quad M_p'' = M_a'' \text{ about centroid of } T_1, T_2 \tag{9-66a}$$

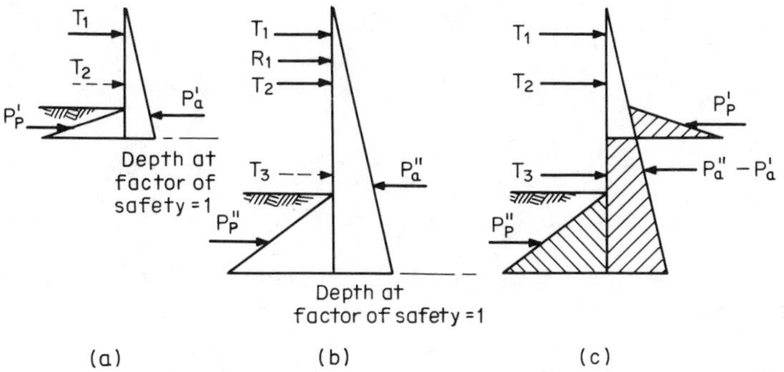

Fig. 9-99 Equilibrium conditions for multianchored wall converted to single-anchored wall.

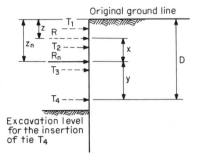

Fig. 9-100 Iteration procedure for system equilibrium, multitied wall: R = previous resultant tie force R_n = new resultant tie force, z = previous resultant tie-force level, z_n = new resultant tie-force level T_1, T_2, T_3, T_4 = individual tie forces.

If R_1 is the resultant of T_1 and T_2 (centroid), then $R_1 = P''_a - P''_p = T_1 + T_2$. Combining with Eq. (9-66) gives $P'_a - P'_p + T_2 = P''_a - P''_p$, or

$$T_2 = P''_a - P''_p - P'_a + P'_p = (P''_a - P'_a) - (P''_p - P'_p) \qquad (9\text{-}67)$$

from which we can see that the additional earth load transmitted to T_2 is the temporary support offered during the previous stage. The equivalent beam loading to T_2 is shown hatched in Fig. 9-99(c). As long as T_2 is unknown, so are the positions and magnitude of R_1. If one is known, the other can be estimated and the initial position checked.

The iteration procedure suggested to ensure convergence to the correct value of the cantilever arm is illustrated in Fig. 9-100 (James and Jack, 1974). This shows the stage with the excavation level reduced to a position for the installation of the fourth anchor. For equilibrium, the following must be satisfied:

$$\begin{aligned} \sum F_H = 0 \quad &\text{or} \quad T_4 = R_n - R \\ \sum M = 0 \quad &\text{or} \quad Rx - T_4 y = 0 \end{aligned} \qquad (9\text{-}68)$$

where $y = D - x - z$ so that

$$f(x) = Rx - T_4 D + T_4 x + T_4 z$$

Substituting in the Newton–Raphson iteration formula gives

$$x_{n+1} = x_n - \frac{Rx - T_4 D + T_4 x + T_4 z}{R + T_4} \qquad (9\text{-}68a)$$

where x_{n+1} and x_n are the new and previous estimate of x, respectively.

James and Jack (1974) have found satisfactory comparison of this procedure with results from published experimental work in estimating anchor forces. The results are also in good agreement with model and full-scale tests carried out by Tcheng and Iseux (1972). The method has also been checked in the analysis of anchored walls described by Littlejohn and MacFarlane (1974).

9-25 ANALYSIS OF ANCHORED WALLS BY FINITE-ELEMENT METHODS

Advantages and Limitations

It is evident from the foregoing that partially integrated techniques inhibit complete problem formulation since they pursue each phase independently. Thus earth stresses are determined by limiting theory, support loads are estimated empirically, and deformations are predicted by statistical data, elastic theory, and one-dimensional consolidation theory. Limiting equilibrium analysis is simple in predicting collapse loads for earth-retaining structures but does not predict deformations associated with limit loads and provides no information for conditions other than those at the limit. Finite-element analysis, on the other hand, permits solutions based on actual stress–strain relations, boundary conditions, and constitutive equations. As a predictive technique it allows consideration of structures with arbitrary shape and flexibility, complex construction sequence, and heterogeneous soil conditions. Furthermore, it is possible to analyze seepage loading and nonlinear soil–interface behavior, and also predict stress changes and deformations for both the soil and the structure for conditions other than at the limit. If instrumentation is contemplated to monitor construction, the method becomes valuable in predicting critical phases and instrumentation requirements, and provides a logical supplement to the process.

Two general mechanisms are available. In the first, predictions are made and compared with observed behavior. If a discrepancy exists and cannot be rectified, the assumptions are altered and the analysis is repeated until convergence is reached. The second approach involves parametric studies of factors influencing wall behavior, from which different designs are evaluated for different wall functions. Finite-element analyses have been carried out by Cole and Burland (1972), Ward (1972), Wong (1971), Clough (1973), Egger (1972), Tsui (1973), Barla and Mascardi (1974), Clough and Tsui (1974a,b), Clough et al. (1974), Murphy et al. (1975), Breth and Stroh (1976), Simpson et al. (1979), Stille and Fredricksson (1979), and Pfister et al. (1982).

Limitations in the use of the method are, however, imposed by inability to always prescribe appropriate constitutive behavior and determine the parameters needed for the constitutive models. Thus, accuracy is influenced by the availability of input data routinely necessary. The programs typically require soil parameters, some of them not readily available, which must be determined through extensive soil investigations and laboratory tests. It is also conceivable that application of soil–structure interaction involves certain special problems for which solutions are approximated. Other difficulties arise from the simulation of the relative movement between the soil and the structure, the special construction sequence that must be modeled, and the numerical problems that are intensified by the stress-strain pattern of the soil.

Statement of Problem. Table 9-15 shows a typical flow chart incorporated in finite-element analyses. The chart lists the steps involved in the investigation, each step representing an idealized form of the actual problem, so that the work is based on the introduction of certain assumptions.

The first two steps are typical and require merely conversion of soil and groundwater conditions into an idealized profile. Next, behavioral models are selected for the soil, the structure, and the soil–structure interface. Structural behavior is expressed in a form that can be interpreted mathematically—elastic, elastic–plastic, and so on. The soil model and the soil–structure interface model are the most difficult to define. Parameters indicating the media properties are selected next, and approximated where nonhomogeneous conditions exist. Then initial stress conditions are selected, a step particularly difficult in certain soil types such as overconsolidated clays and clay shales. Finally, the construction sequence is worked out, the finite-element mesh is drawn, and the analysis is carried out.

Simulation of Construction Sequence. The considerable influence of construction factors on wall and soil behavior is easily demonstrated in practice. In general, construction sequence simulation involves the division of the loading sequence into small increments, analysis of the effects of each increment in sequence, and superposition of the results to obtain the resultant

TABLE 9–15   Typical Flowchart and Procedure Leading to Finite-Element Analysis

Statement of problem
↓
Idealization of soil and groundwater conditions
↓
Selection of constitutive modeling techniques
↓
Selection of media properties
↓
Assumption of initial stress conditions
↓
Assumption of construction sequence
↓
Drawing of finite-element mesh to accommodate soil conditions, structural configuration, and construction sequence
↓
Analyses

stress and displacement conditions. For example, diaphragm wall modeling must consider trench excavation under slurry, concrete tremie placement, general excavation on one side (probably accompanied by dewatering and recharging), and installation of anchors.

Models for simulating excavation have been proposed by Clough and Duncan (1971), which are general and accurate in a number of situations such as excavations in soil and rock and for in-place structural elements. If the soil is assumed linear elastic, results for one and three-step excavation simulation should be comparable. On the other hand, dewatering and seepage loading constitute pressure changes on the elements in the mesh, and are merely special cases of the more complex loading produced by the excavation. Installation and prestressing of the anchors can be simulated in the form of a restraint or load change (see also case histories in the following sections).

Among the initial conditions, the simple initial at-rest stresses are difficult to conceive, whereas the construction process prompts changes in these stresses. For example, stress–strain changes are induced by certain operations such as the replacement of soil by slurry to be followed by fresh concrete in diaphragm walls, pile driving in sheeted walls, or the installation of the lagging in soldier pile walls.

Among the boundary conditions, the representation of the interface is particularly relevant to the boundary model. With diaphragm walls, the soil–slurry interaction must be inferred first, and the shear resistance mobilized at the interface must be determined next as the wall moves with respect to the soil. Invariably, these factors will influence the shear stress–deformation behavior.

Activities that are relevant to wall performance but cannot be simulated are (a) construction-induced movement in certain soft clays or loose sands; (b) overexcavation or delays in support installation; (c) vibrations caused by adjacent work, and subsidence originating in caisson construction; (d) water loss through anchor holes, pile interlocks, and wall joints; (e) remolding and undercutting of temporary berms used in lieu of other bracing; and (f) surcharge loads from moving equipment and soil.

Examples of Finite-Element Analysis

Figure 9-101 shows an anchored wall supporting an excavation 32.5 ft (10 m) deep (Tsui, 1973). The soil is homogeneous clay underlain by rock. The wall is a concrete diaphragm 2 ft (60 cm) thick, and the anchors consist of steel rods, 1 in^2 in area, with the fixed length in rock. The prestress loads are estimated from an apparent pressure diagram shown in (b). The clay has undrained shear strength increasing linearly with depth from 500 to 1400 lb/ft^2 (2.5–7.0 tons/m^2) at the bottom of the clay layer. The coefficient K_0 is taken as 0.85, and the insertion of the wall is assumed to have no effect on the initial at rest condition. The initial tangent modulus of the soil is taken

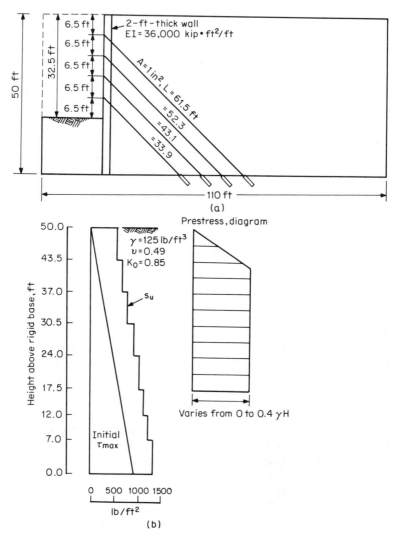

Fig. 9-101   Anchored wall in clay; (a) section through wall; (b) soil data and pre-stressed diagram. (From Tsui, 1973.)

as 400 times the undrained shear strength. The assumption of plane strain condition is considered valid for a wall 2 ft thick and anchor spacing less than 10 ft (3 m).

A nonlinear elastic model is incorporated in the analysis, and tangent modulus values are obtained for a stress–strain curve represented by a hyperbola. The interface between the wall and the soil is treated similarly on both sides using a bilinear stress–strain deformation relationship with initial shear stiffness 50,000 pcf reduced by a factor of 1000 if the yield strength of the interface is exceeded. The construction sequence is simulated by an incremented loading process based on the nine-step modeling shown in Fig. 9-102. Anchor lengths vary from 61.5 to 33.9 ft.

Figures 9-103 and 9-104 show wall and ground movement and earth pressure distribution, respectively, for the two prestress levels and with zero prestress, together with anchor loads corresponding to apparent pressure diagrams. Wall movement responds consistently to prestress level decreasing almost linearly with the amount of prestressing. Likewise, ground settlement behind the wall decreases as the prestress increases, but the effect diminishes as the next higher prestress load is introduced. Settlement is thus reduced more by the first increase than by increases that follow.

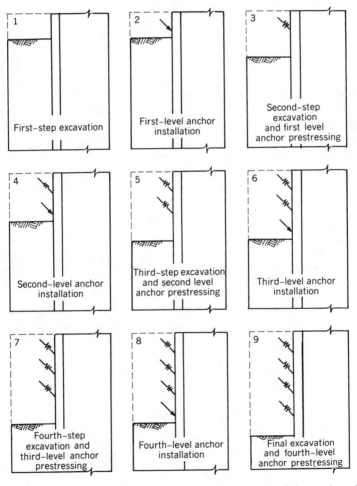

Fig. 9-102 Construction sequence; Finite element analysis of the anchored wall of Fig. 9-101. (From Tsui, 1973.)

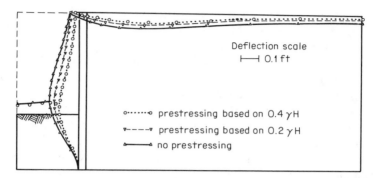

Fig. 9-103 Wall and ground movement predicted by finite-element analysis; tied-back wall of Fig. 9-101. (From Tsui, 1973.)

The predicted earth pressure diagrams shown in Fig. 9-104(a) can be compared with the apparent pressures shown in (b) obtained by distributing the anchor loads over the appropriate spans. Evidently, the predicted pressures approach the original at-rest values and exhibit a definite triangular distribution. Interestingly, there are no pressure bumps at the anchor points.

A second example of anchored wall in clay modeled by finite-element analysis is shown in Fig. 9-105 (Clough and Tsui, 1974b). Two cases are investigated, one with four rows and the other with three rows of anchors. The wall is flexible, with moderate stiffness equivalent to PZ-72 sheeting. The anchor prestress is likewise obtained from apparent pressure diagrams.

The predicted lateral pressures are more triangular than the design trap-

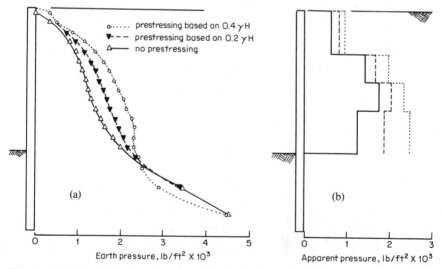

Fig. 9-104 Lateral earth pressure predicted by finite-element analysis and appearance pressure diagrams, tied-back wall of Fig. 9-101. (From Tsui, 1973..)

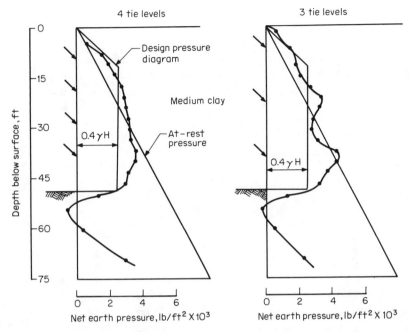

Fig. 9-105 Lateral earth pressure behind a flexible wall predicted by finite-element analysis; prestressed tied-back wall. (From Clough and Tsui, 1974b).

ezoidal diagram, and this distribution is consistent with the actual wall movement. In this example, unlike the previous case, we can notice that the earth pressures tend to concentrate slightly at each anchor level. This bulging is caused by the wall flexibility in response to the application of prestress; hence it must be distinguished from the linear stress distribution observed with the stiff wall. Its effect is to reduce the bending moments slightly.

9-26 MISCELLANEOUS TOPICS RELEVANT TO ANCHORED WALL DESIGN

Effect of Wall and Anchor Stiffness, and Anchor Prestressing

Effect of Wall Stiffness on Lateral Displacement in Clay. Wall stiffness refers not only to the structural rigidity derived from the elastic modulus and the moment of inertia but also to the vertical spacing of supports (in this case anchors). The measure of wall stiffness is defined as the inverse of Rowe's flexibility number for walls, and is thus expressed by EI/L^4, where L is the vertical distance between two rows of anchors. A plot of observed displacements correlating the stiffness EI/L^4 with the stability number $N = \gamma H/s_u$ is shown in Fig. 9-106 (Goldberg et al., 1976).

The data demonstrate what engineers know intuitively, specifically, that deformation and wall movement in excavations are functions of soil strength and wall stiffness. The contour lines of maximum lateral wall movement show this trend explicitly. These data allow qualitative examination of the relative change in expected lateral wall displacement with a change in wall stiffness or soil deformability expressed by the stability number.

By reference to Fig. 9-106 an immediate comparison can be made between a sheet pile wall and a diaphragm wall braced at 10-ft vertical intervals. The sheet pile section is PZ-38, and the concrete wall is 30 in thick. The stiffness factors are as follows:

$$\text{Steel sheeting } \frac{EI}{L^4} = \frac{(30 \times 10^6) \times (281)}{120^4} = 40.7 \text{ psi} = 5.86 \text{ ksf}$$

$$\text{Concrete wall } \frac{EI}{L^4} = \frac{(3 \times 10^6) \times (1/12)\,(12 \times 30^3)}{120^4} = 391 \text{ psi} = 56.3 \text{ ksf}$$

From the plot of Fig. 9-106 the expected maximum displacement for the sheet pile wall is 3 in, and for the stiffer diaphragm wall 1.5 in (approximately).

Theoretical Analysis. Egger (1972) has reported the results of finite-element studies in sand to investigate the effect of flexible (sheet pile) walls and stiff (diaphragm) walls. Total wall height is 10 m (33 ft) for an excavation 7.5 m (24.5 ft) high, so that wall embedment is 2.5 m (8.5 ft). The wall is supported by anchors 7.5 m long (24.5 ft). The construction sequence in-

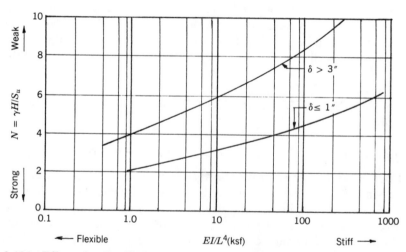

Fig. 9-106   Effect of wall stiffness and soil deformability (expressed by the stability number $N = \gamma H/s_u$ or lateral wall deflection (ksF = kilopounds per square foot). (From Goldberg et al., 1976.)

volves initial excavation to a depth of 3.5 m for the installation of the first row of anchors, followed by a second phase and third phase, each 2 m deep for the second and third row of anchors, respectively. The toe of the wall is just embedded into bedrock to prevent lateral wall movement at this level; the only possible deformation is rotation about this point. The flexible and stiff systems have a stiffness ratio of 1:100.

Predicted lateral earth pressures are shown in Figs. 9-107 and 9-108 for both the flexible and the stiff walls, and for the two prestress levels 1 and 6 tons/m, respectively. During the first excavation stage the wall acts as a free cantilever and deflects accordingly. The lateral movement, shown in Fig. 9-109, at the top is three times greater for the flexible wall, hence this wall mobilizes passive earth resistance near the surface to a much higher degree. The stiff wall, on the other hand, mobilizes passive resistance over a markedly greater zone, although the limiting equilibrium is not reached because of the smaller displacement. Likewise, the active earth pressure

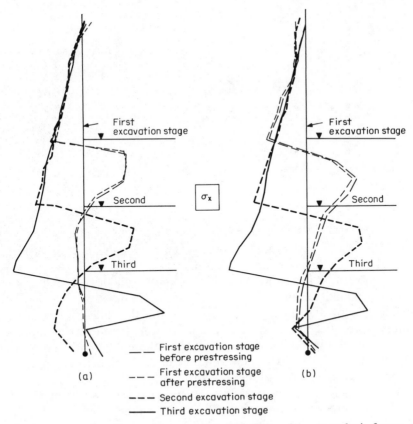

Fig. 9-107 Lateral earth pressures predicted by finite-element analysis for an excavation in sand: (a) flexible wall; (b) stiff wall. Anchor prestress 1 ton/m. (From Egger, 1972.)

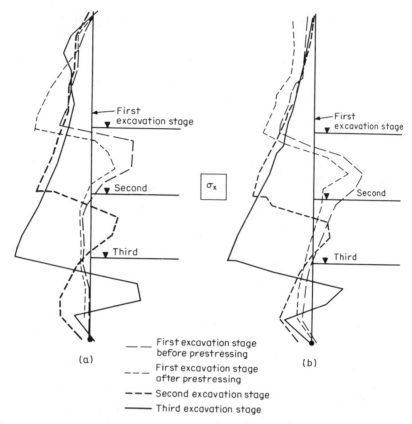

Fig. 9-108 Lateral earth pressures predicted by finite-element analysis for an excavation in sand; (a) flexible wall; (b) stiff wall. Anchor prestress 6 tons/m. (From Egger, 1972.)

approaches the at rest condition near the base of the wall much faster for the flexible wall than does for the stiff element.

As the prestress is increased from 1 to 6 tons/m, the flexible wall receives this supplementary charge as a stress concentration between 1.5 m above and 2.5 m below the anchor level. However, for the stiff wall the resulting earth pressure diagram is essentially uniform from the top down to about 6 m. Interestingly, in the last excavation stages these differences are still observable but less significant, indicating the effect of the closer anchor-row spacing on equalizing the stiffness factor EI/L^4.

The effect of anchor prestressing on wall movement is shown in Fig. 9-109. The prestress of 1 ton/m is much lower than the active level, whereas the value 6 tons/m corresponds roughly to the actual pressures existing at final excavation level. The movement of the top of the flexible wall is reduced from 2.5 to 1.7 cm (from 1 to 0.7 in) when the prestress is increased from

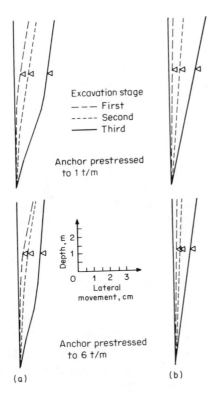

Fig. 9-109 Lateral wall movement for a wall in sand predicted by finite-element analysis: (a) flexible wall; (b) stiff wall. Anchor prestress 1 and 6 tons/m. (From Egger, 1972.)

1 to 6 tons/m. For the stiff wall, the same movement is reduced from 2.3 to 1.4 cm with the higher prestress level.

From Fig. 9-110 we can see the increase in the actual anchor force as excavation continues to final level, and for the two levels of prestress, namely, 1 and 6 tons/m. With a prestress of 1 ton/m, anchor reaction grows quickly and reaches practically the same value for stiff and flexible wall, about 7.2 tons/m. With higher prestress, the increase becomes less significant as expected, but still higher than with the lower prestress. Thus, at final excavation stage and with 6 tons/m prestress, anchor reaction is about 8.8 tons/m for both walls.

Clough and Tsui (1974b) studied the effect of prestress while maintaining wall and anchor stiffness constant. The wall is given the medium stiffness value 36,000 kip-ft^2/ft, and the anchors are assumed to be steel bars with a cross-sectional area of 1 in^2. The predicted soil and wall deformation decreases with increased prestress. This effect appears more prominent near the top of the wall where the use of prestress corresponding to a trapezoidal

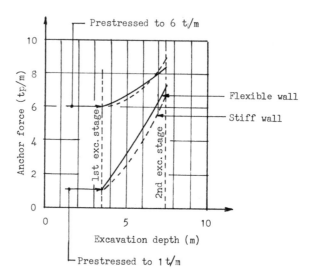

Fig. 9-110   Increase in anchor force with excavation depth for two levels of prestress; stiff and flexible wall. (From Egger, 1972.)

loading can eliminate movement almost completely. However, the prestress is only partly effective in preventing movement near the base of excavation; hence settlement may result for most conditions.

In the same investigation the effect of wall rigidity was isolated by keeping the prestress load and anchor stiffness at assigned constant values. The results confirm that wall deformation and ground movement are reduced as the wall rigidity increases, but this reduction is not linearly proportional to the increase in wall stiffness.

The effect of anchor stiffness has been studied by Clough and Tsui (1974b) in the same series of studies. The excavation is 32.5 ft deep in a homogeneous deposit of normally consolidated clay, and is supported by four rows of anchors. The strength of clay is assumed to increase from 600 lb/ft² near the surface to 1800 lb/ft² at 75-ft depth. The initial tangent modulus is taken as 400 times the soil strength. Two types of anchor tendons are considered, bars 1 in² in cross-sectional area and strand with cross-sectional area 0.1 in², so that their stiffness varies by a factor of approximately 10.

The predicted wall and soil movement is shown in Fig. 9-111 for stiff and flexible anchors. The former are seen to reduce movement on the order of 50 percent; hence reduction in wall movement is not linearly dependent on anchor stiffness change.

Combined Effect of Parameters. The combined effect of the three parameters—wall stiffness, anchor stiffness, and prestress—on anchor wall

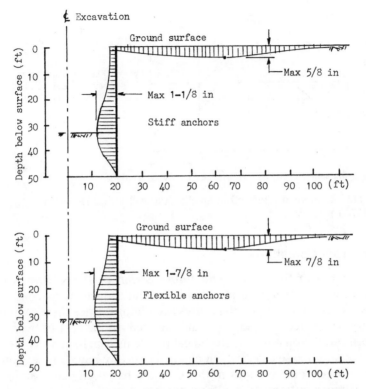

Fig. 9-111 Effect of anchor stiffness on wall and soil movement. (From Clough and Tsui, 1974b.)

behavior has been investigated by Clough and Tsui (1974b) for two cases. First, a flexible wall is analyzed but without prestress application, and with a flexible anchor system. In the second case, a relatively high prestress level is introduced based on a trapezoidal apparent diagram with ordinate 0.68 γH, combined with a very stiff wall–anchor system.

Wall movement is shown in Fig. 9-112, plotted on a dimensionless diagram as suggested by Peck (1969) for braced excavations. Movement divided by excavation depth is shown versus distance from the wall divided by excavation depth. The two solid curves in the diagram divide three zones of behavior as defined by Peck (1969). Zone I represents the best soil and construction workmanship conditions; zone II, intermediate conditions; and zone III, the worst credible conditions. Without prestress, the wall yields settlements that are large enough to bring the system into zone II. However, the effect of the combined parameter change on predicted settlement in the second case is significant, and ground settlement is not restricted entirely to zone I.

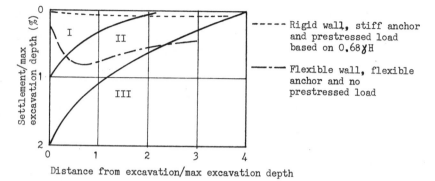

Fig. 9-112   Combined effect of parameters on wall movement. (From Clough and Tsui, 1974b.)

Plane Strain Conditions

Assumptions of Symmetry and Plane Strain or Plane Stress Conditions. From the analytical standpoint it is advantageous to consider axisymmetric, plane strain, or plane stress conditions, since these require only two degrees of freedom per node and thus reduce the cost of the analysis. Anchored walls constitute geotechnical problems which, in strict terms, are not plane strain cases, but which characteristically have a repetitious load pattern. In general, solutions in this case are based on plane strain conditions. Three questions arising from these assumptions are (a) how close anchored walls are to plane strain conditions considering the wide spectrum of system stiffness; (b) how the stiffness of discontinuous components, namely, anchor tendons, is best represented in the analysis; and (c) how continuous planar walls can simulate the behavior of discontinuous walls such as soldier beams with lagging.

Model tests by Tsui (1973) and Tsui and Clough (1974), in which the actual construction process was simulated, confirm a system behavior described in the following sections.

Vertical Distribution of Bending Moments. Figure 9-113 shows the isolated effect of prestressing the first level of anchors, excavation to the second level, and prestressing the second level of anchors, for the vertical anchored wall shown. Bending moments developed in a wall section between two anchor levels is only moderately affected by subsequent excavation and prestressing at lower levels. Indeed, the resultant bending-moment diagram for a vertical section shows a clear tendency to develop strong negative moments at anchor levels and positive moments between these supports. The wall in this case is a flexible model.

Horizontal Distribution of Bending Moments. Likewise, the effect of different events along a horizontal plane between the third and fourth anchor

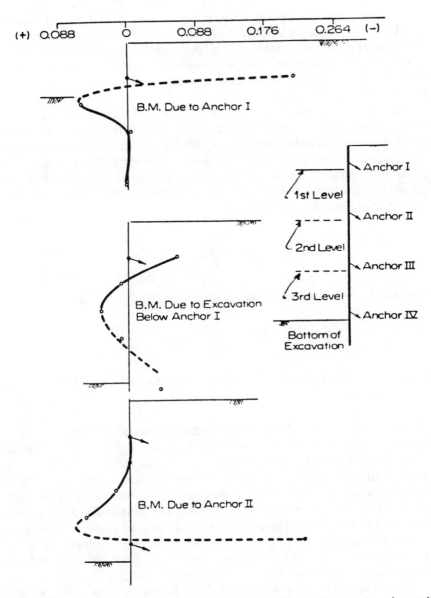

Fig. 9-113 Incremental (isolated) bending moment effect due to prestressing and excavation to level 2. Vertical section through anchors. (From Tsui and Clough, 1974.)

levels is isolated and shown in Fig. 9-114. The individual contributions due to prestressing the left and right anchors as well as subsequent excavation can be seen in parts (a) and (b), respectively. The resultant diagram is shown in (c) for the processes combined. These results clearly indicate a three-dimensional behavior rather than a plane strain loading. Unlike the latter, in which bending moments along the plane of the anchors are assumed constant, the results of Fig. 9-114 show negative moments at the prestressing points changing to positive between supports.

From this pattern we can see that at the level of the anchors the prestress loads dominate the bending response of the wall, and the three-dimensional effect deviates markedly from the assumed plane strain condition. These observations led the investigators to consider the applicability of slabs-on-elastic foundations theory using subgrade modulus back-calculated from observed deformations. A parameter that controls the slab deflection is the characteristic length l_0 defined as follows

$$l_0 = \sqrt[3]{\frac{2D(1 - \mu_s^2)}{E_s}} \tag{9-69}$$

in which E_s and μ_s are the elastic parameters of the soil, and

$$D = \frac{Eh^3}{12(l - \mu)^2}$$

where h denotes the slab (wall) thickness, E the elastic modulus, and μ Poisson's ratio.

System Flexibility and Effect on Three-Dimensional Behavior.

From the theory of slab on elastic foundation, Tsui and Clough (1974) define the flexibility of the wall/soil system in terms of characteristic length. This is considered particularly useful since it accounts for relative effects associated with slab and soil stifness, and can be compared to the spacing of prestress loads. The relevance of l_0 and the horizontal prestress load spacing is demonstrated in Fig. 9-115. The plot is for the theoretical pressure distribution induced by a series of prestress loads spaced at different multiples of l_0. The pressure represents a dimensionless quantity $I_p = l_0^2 p / P$ where p is the contact pressure on the slab and P is the prestress load in the anchor.

From Fig. 9-115 we can see that if the horizontal prestress spacing is greater than $6l_0$, the pressure distribution from one prestress load does not engage in any significant interaction with its nearest neighbors, and thus the plane strain conditions are not met. Indeed, the prestress spacing must be reduced almost to l_0 if the overlapping of pressures from adjoining anchors is to become a model of approximately uniform pressure to justify the assumption of plane strain conditions. Prestress load spacing between the limit

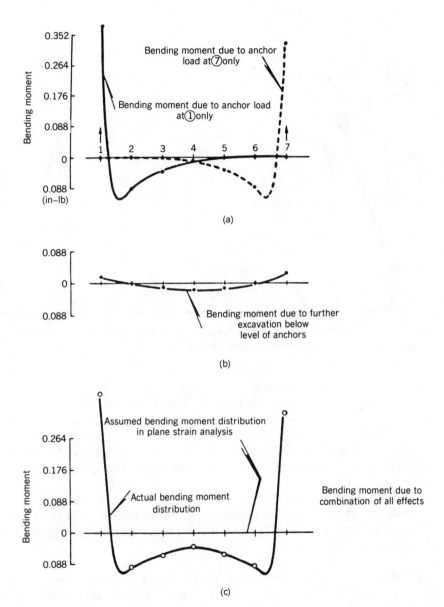

Fig. 9-114 Bending behavior along a horizontal section through third anchor level, wall of Fig. 9-113. (From Tsui and Clough, 1974.)

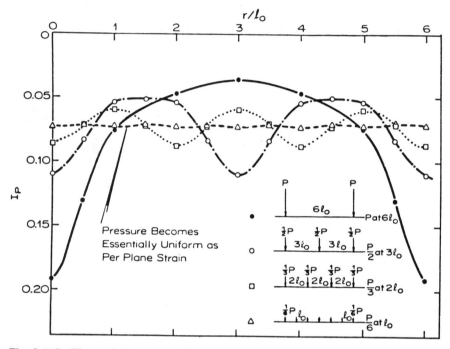

Fig. 9-115 Theoretical pressure distribution for different horizontal prestress load spacing relative to characteristic length. (From Tsui and Clough, 1974.)

values l_0 and $6l_0$ will probably yield conditions at variance with the assumption of plane strain. In this case, the degree of deviation of the actual pressure from the assumed uniform pressure becomes a qualitative index measuring the applicability of the plane strain analysis. Figure 9-116 shows this index as a percent deviation of actual pressure from equivalent uniform pressure, plotted versus the ratio of prestress load spacing and characteristics length s/l_0.

From the plot, the percentage deviation is seen to rise sharply with increasing s/l_0 values. For example, for a prestress load spacing $5l_0$ the deviation is 120 percent, but if the prestress load spacing is reduced to $2l_0$, the deviation is only 15 percent.

Flexibility of Typical Continuous Walls. Walls that can be considered continuous are precast or cast-in-place slurry walls, and certain sheet pile systems. Three typical cases analyzed for plane strain conditions include a 3-ft-thick concrete slurry wall, a 1-ft-thick concrete slurry wall, and an MP-116 sheet pile unit. For each example, l_0 is estimated for soil modulus with values ranging from 50 tons/ft² (soft clay) to 400 tons/ft² (stiff clay) with an assumed Poisson's ratio of 0.3.

For all three examples, variation in characteristic length l_0 with soil modulus is shown in Fig. 9-117. This length is greater than about 6 ft at the lowest

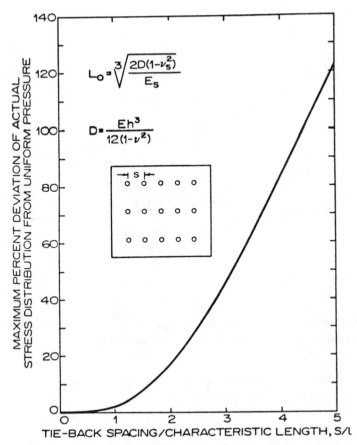

Fig. 9-116 Maximum percent deviation of actual stress distribution from uniform pressure; different ratios of prestress load spacing to characteristic length. (From Tsui and Clough, 1974.)

soil modulus. For a 10-ft horizontal anchor spacing, the ratio is about 0.6, 1.0, and 1.7 for the 3-ft and 1-ft walls and the sheet pile unit, respectively. If the anchor spacing is increased to 15 ft horizontally, these ratios are about 0.9, 1.5, and 2.5. The assumption of plane strain is in this case valid for the concrete slurry walls and acceptable (30 percent deviation) for the sheet pile wall.

For the 1-ft-thick concrete wall, the three-dimensional pressure distribution at the anchor level is compared to the uniform pressure assumed in plane strain analysis, and data are shown in Fig. 9-118. The soil is clay with an assumed modulus 180 tons/ft^2, and the prestress loads are spaced at 10-ft centers. For this condition the ratio s/l_0 is 2, yielding a deviation of 15 percent from plane strain conditions. It appears, therefore, that for the majority of anchored slurry walls plane strain analysis is applicable and valid.

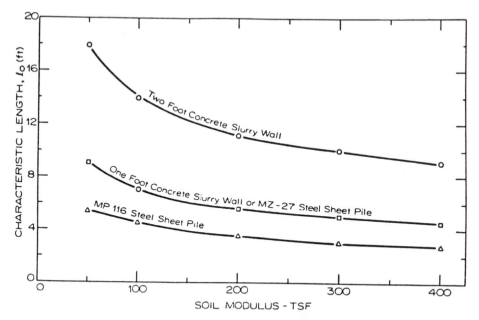

Fig. 9-117 Variation in characteristic length of typical continuous walls and dependence on soil modulus. (From Tsui and Clough, 1974.)

Plane Strain Analysis of Discontinuous Walls. Walls in this group (e.g., soldier piles with lagging) must be represented by equivalent continuous planar units. However, the degree of approximation is not entirely consistent with analytical investigation because of the many practical problems inherent with the construction. A logical approach suggested by Tsui and Clough (1974) is to consider a planar system that has a per foot bending stiffness equivalent to the discontinuous wall. For example, a 12WF120 soldier beam wall with soldier beams spaced on 6-ft centers would be equivalent to a 1-ft-thick concrete wall with a modulus of 3×10^6 psi. However, with this conversion the use of the percent procedure of pressure deviations is valid and applicable only as a lower bound estimate, and the actual deviation should be higher than its planar simulation. For the foregoing example, the soldier pile wall pressure may deviate by as much as 56 percent from the plane strain pressure, compared to a deviation of about 30 percent for the equivalent planar wall. Plane strain analysis is in this case very approximate only.

9-27 NUMERICAL PROCEDURES

Nonlinear analysis of multianchored walls using numerical procedures is suggested by Popescu and Ionescu (1977) and Popescu (1977), based on methods introduced by Haliburton (1968). This procedure is a good alter-

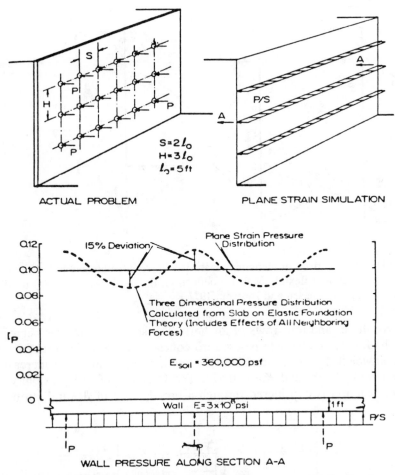

Fig. 9-118 Comparison of plane strain and actual pressure distribution for a 1-ft-thick concrete wall. (From Tsui and Clough, 1974.)

native where the cost of finite-element programs restrict their use. The example presented in this section is for flexible walls, and is limited to anchor force dependence on the main parameters of the design, namely, wall flexibility, anchor stiffness, embedment depth, and initial stress conditions.

The analysis requires repeated trial and adjustment steps until the convergence criterion is met, that is, until two subsequent elastic lines on the wall nearly coincide. A comparison of numerical analysis results with other methods is ilustrated in Fig. 9-119 showing a single-anchored wall analyzed also by finite-element and free earth support methods (Bjerrum, 1972). Computed deformations, earth pressure distribution and bending moments are in fairly good agreement.

A multianchored flexible wall with total depth of 14 m (46 ft) is shown in Fig. 9-120, supporting an excavation 10 m (33 ft) deep. The wall is analyzed

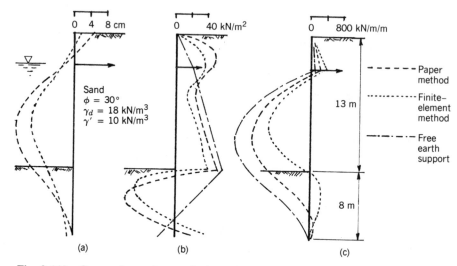

Fig. 9-119 Comparison of numerical analysis with results obtained by other methods; single-anchored flexible wall: (a) deflection, (b) Earth pressures, (c) bending moments. (From Popescu and Ionescu, 1977.)

for two, three, and four rows of anchors. The soil has nonlinear behavior with the following characteristics: zero cohesion, angle of internal friction 30°, unit weight 19 kN/m³, initial tangent modulus in compression increasing linearly with depth, and $K_0 = 0.577$. The influence of wall flexibility on anchor reaction is considered through a broad range of flexural stiffness,

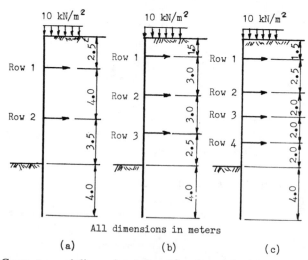

All dimensions in meters

(a)   (b)   (c)

Fig. 9-120 Geometry and dimensions of multianchored flexible wall for two, three, and four rows of anchors: (a) case I; (b) case II; (c) case III. (From Popescu and Ionescu, 1977.)

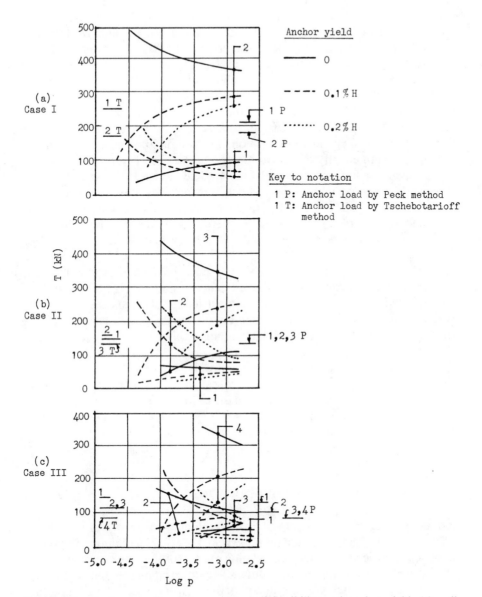

Fig. 9-121 Anchor reaction force versus wall flexibility and anchor yield; (a) wall with two rows of anchors; (b) wall with three rows of anchors; (c) wall with four rows of anchors. Wall of Fig. 9-120. (From Popescu and Ionescu, 1977.)

from 28850 kN/m^2 to 1,920,000 kN/m^2. These values correspond to a 2B-type Larsen sheet pile and a 1.0-m-thick concrete slurry wall.

The results of the analysis are presented in Fig. 9-121(a), (b), and (c), in which the calculated anchor force is plotted versus the flexibility coefficient p and for different anchor yields. The coefficient p is as defined by Rowe ($p = H^4/EI$), where H is the total wall depth.

For zero anchor yield, the lower anchor is the most loaded, and the force value difference between this anchor and the anchors above it is considerable. The more flexible the wall, the less low the anchor force. Higher anchor yield results in a decrease of the lower anchor reaction as well as a decrease of the upper anchor reaction. For the double-anchored wall, the two anchors are equally loaded for a particular value of wall flexibility, but this particular value increases with anchor yield increase. For the walls with three and four rows, there is no particular stiffness resulting in equal anchor force. Anchor loads predicted by apparent pressure diagrams (Tschebotarioff and Peck) are also included in the plots. Evidently, these methods overestimate the upper anchor load, while they underestimate the lower anchor pull.

Anchor stiffness has considerable influence on anchor reaction. Thus, stiff anchors are more unequally loaded than the more flexible tendons. With stiffer units, the lower anchor is the most loaded whereas the upper row carries the least load. This difference may be as high as 80 percent for the flexible walls. With flexible anchor tendons, anchor load approaches a uniform distribution.

Likewise, the initial earth pressure and wall embedment influence anchor load distribution. As embedment increases, the lower anchor reaction decreases with a simultaneous increase in the upper anchor reaction. The effect of K_0 is similar; that is, an increase in K_0 results in a decrease of the lower anchor reaction and an increase of the upper load.

9-28 UNDERPINNING CONSIDERATIONS

Examples of anchored walls used as underpinning are shown in Figs. 8-27, 8-28, and 8-29. The arrangement shown in Fig. 8-27 can eliminate ground movement associated with an excavation except for the displacement caused by the elastic stretching of the tendon. The underpinning scheme shown in Fig. 8-28 can be effective in reducing movement provided the anchors are monitored and restressed to compensate for movement during excavation. The underpinning method shown in Fig. 8-29 is an improved concept of hybridization of anchors and internal bracing (see also Section 9-22) whereby struts are used at the top to prevent initial inward movement and anchors are installed at the lower part of the wall.

In designing an anchored wall for underpinning, two considerations have great importance: (a) the effect of surcharge loads from foundations and existing buildings on the lateral stresses and deformations and (b) effects associated with the mechanics of inclined wall behavior.

Effect of Foundation Weights and Surcharge Loading. Lateral stresses caused by these loads have been investigated by Spangler (1940), Newmark (1942), and Terzaghi (1954b). Solutions are based on elastic theory of stress distribution in a semiinfinite linearly elastic medium, modified for the presence of a rigid wall. Considerable data are available supporting this method, although the reasons for the correlation are not clear.

For surcharge loading there are four basic loading conditions for which solutions of the lateral stresses in elastic medium are readily available. These are (a) point loading, (b) uniform line loading, (c) irregular area loading, and (d) uniform area loading (Goldberg et al., 1976). Solutions of lateral stresses on rigid wall from surcharge of uniform width and infinitely long are also given by Sandhu (1974). A more simplified approach is to apply an earth stress coefficient K to the surcharge loading and consider the surcharge effective within some portion of the cut. This coefficient will range from K_a to K_0.

Loads within a soil mass, such as a foundation or a continuous mat in close proximity to the excavation, are assumed to have a lateral distribution based on the following essential points: (a) the total load of the foundation is reduced by the weight of the overburden, (b) the lateral load is ignored above the level of load application, and (c) the lateral effect diminishes when the edge of the loaded area exceeds a certain distance from the walls. Diagrams for various loading conditions are given by Xanthakos (1979).

For external loads acting either at or below ground surface, the lateral distribution is thus based on both elastic and limiting theory, and very often there is an arbitrary crossover and shifting from one method to the other as the analysis considers more and different types of load. A more realistic procedure is formulated by Haliburton (1968), which takes the analysis in the right direction although it does not eliminate all the problems (Xanthakos, 1979).

Tests on small-scale models simulating strutted excavations supported by flexible walls show that the effect of loads within the soil mass are dependent on the bracing position and excavation sequence (Breth and Wanoschek, 1972). Whereas the magnitude of lateral stresses resulting from foundation loads is governed by the position of the loads and their proximity to the wall, the stress distribution hardly follows an elastic or limiting theory but is decisively governed by the excavation and bracing process.

Inclined Wall Behavior. Two anchored inclined walls are shown in Fig. 9-122. The design assumptions are based on the following pressure diagrams:

1. Rectangular distribution using an earth pressure coefficient

$$K = \frac{K_0 + K_a}{2}$$

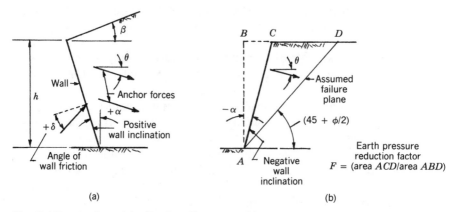

Fig. 9-122 Anchored inclined walls; (a) positive wall inclination (top of wall projection toward excavation); (b) negative wall inclination (top of wall projection toward ground.)

2. Triangular pressure distribution with earth coefficient again

$$K = \frac{K_0 + K_a}{2}$$

3. Triangular pressure distribution as in assumption 2, but multiplied by a reduction factor F.

Assumptions 1 and 2 are for positive wall inclination, and assumption (3) applies to walls negatively inclined. The reduction factor F applied to earth stress coefficient K is (Schnabel, 1971) $F = $ (area ACD)/(area ABD).

Results from model tests carried out by Hanna and Dina (1973) indicate a trapezoidal earth pressure distribution envelope corresponding to the foregoing earth pressure coefficient. For negative inclined walls the Schnabel procedure ensures wall performance closer to the observed behavior. As the excavation is carried down, positively inclined walls exhibit the largest lateral movement near the top. Unlike this pattern, negatively inclined walls register the largest movement near the base. As construction progresses, individual anchor forces undergo changes. For positive wall inclinations more excavation contributes a positive increase to anchor load from the initial value. For negative wall inclinations, on the other hand, the anchors lose some of their load as the excavation becomes deeper.

9-29 LIMIT STATES IN THE DESIGN OF ANCHORED WALLS

When anchored walls or structures therefrom become unfit for use, they are considered to have reached a limit state. We may divide these conditions into three basic groups.

Ultimate Limit States. These may involve structural failure of the system; hence the probability of occurrence should be minimal. Anchored walls may reach an ultimate limit state by one of the following mechanisms:

1. Structural failure of the wall in bending or shear.
2. Failure of the ground mass leading to overall failure and collapse.
3. Rupture of anchors leading to progressive failure by overloading neighboring anchors and eventually leading to generalized failure.
4. Initiation of a plastic response in the ground–wall–anchor system leading to general instability.
5. Base failure beneath the wall because of excessive vertical load.

Serviceability Limit States. These may involve disruption of the functional use of the wall, although collapse or structural failure is not involved. A higher probability of occurrence may be tolerated since the consequences do not include total failure or loss of life. Typical serviceability limit states include (a) excessive outward or inward movement or ground settlement having unacceptable visual effects on the structure or nearby buildings and facilities and (b) excessive crack width at construction joints leading to leakage and visual effects, or gradual deterioration of the wall.

Special Limit States. This group, which may induce damage in conjunction with the presence or occurrence of abnormal conditions or abnormal loadings, includes (a) damage or collapse of the wall or anchors from explosions and vibratory effects, (b) gradual deterioration of the anchorage due to corrosion, and (c) long-term changes in the ground environment leading to a form of instability.

Loads and Load Effects. A distinction must be made between loads and effects therefrom, activating the wall–ground system. Loads constitute external action applied to the wall–ground system. Examples are dead weight or imposed loads, wind forces, or gravitational loads and surcharges generated in a soil mass. Load effects are not applied, but manifested internally. Shear, axial tension, or compression and bending moment in a beam, slab, or column are considered load effects. This distinction excludes soil pressures from the load group; hence these are considered load effects.

Characteristic Values. The term "characteristic value" is related to applicable standards, although in reality it has no direct design significance. As used herein, it is a convenient reference value, for example, a guaranteed minimum structural strength, a maximum code loading, or a statutory value.

The term may refer to (a) characteristic loads such as dead, imposed, and wind loads, defined according to standards and codes; (b) characteristic material properties (strength, stiffness, unit weight, permeability, etc.); and (c) characteristic in situ stresses. The last category includes the initial pore water pressure as the best estimate made from hydrologic data, and characteristic initial vertical and horizontal stresses, especially if expected deformations will not be sufficient for limiting states to develop.

Most Probable Values. For variables that remain constant with time (e.g., dead loads and some material properties), the most probable values are the best estimates of in situ values. For variables that change with time (e.g., live and wind load, or soil strength and stiffness), the most probable values are the best estimates of extreme values anticipated during the life of the structure.

Worst Credible Values. These have an accepted very small probability of occurrence. It is intended that the probability of worst credible loads and worst credible material properties should have a probability index close to 0.1 percent. The worst credible value of any variable is either the maximum or the minimum credible value depending on whether the effect is beneficial or adverse.

For variables such as dead, live, and wind loads and structural material strengths, the worst credible values can be obtained by applying partial safety factors to characteristic values. For soil strength the same approach is sometimes appropriate, but not always.

Partial Safety Factors. These allow for uncertainties inherent in design and construction. The load variation factor γ_{f1} reflects the possibility of unfavorable deviations of loads from characteristic values, whereas the load combination factor γ_{f2} takes account of the reduced probability of loads, stochastically independent, occurring at the same time.

The structural performance factor γ_p considers the following: (a) inaccurate assessment of loading effects and unforeseen stress distribution; (b) variations in dimensional accuracy achieved in construction; (c) the importance of the limit state under consideration; and (d) some systems give warning of approaching a limit state, whereas others will reach it suddenly.

The partial materials factor γ_m applies to the following: (a) the materials in the system may in reality be weaker than assumed and (b) the walls may become weaker by other causes (construction tolerance, defects, etc.).

In considering ultimate limit states, the factors γ_{f1} and γ_{f2} are applied to characteristic values of the loads, and the resulting values are taken as the worst credible loads or load combinations. Likewise, the worst credible material strengths are obtained by dividing characteristic strength values by γ_m.

Rationale of Analysis

Ultimate Limit State. In order to satisfy the requirements of this state, the following must be valid

$$\frac{\text{Worst credible resistance } (R)}{\text{Worst credible load effect } (S)} \geq \gamma_p \qquad (9\text{-}70)$$

where the quantities R and S are calculated using the worst credible values and combinations of loads and material strengths and γ_p is the structural performance factor.

Serviceability Limit State. The requirements of this state are met if it can be shown that movement, distortion, and cracking of the structure, or similar effects on surrounding structures, are acceptable. In general, most probable values of all variables should be considered. However, if the effects under consideration are particularly sensitive to any given variable, it is necessary to assume a more conservative value such as a worst probable value for that variable. If two or more worst probable values are assumed simultaneously, allowance is made for the reduced probability of these occurring simultaneously.

For movement in the soil, two checks are necessary and provide independent results: (a) calculate most probable settlement, heave or other movement directly; and (b) provide evidence that equilibrium can be achieved under soil stresses that are accurately established from experience or are based on a sound theoretical model.

The same rationale applies to a special limit state, but in this case it may be necessary to carry out separate checks for the ultimate limit state. Further caution is necessary if this approach is used for materials and processes that are not fully predictable, such as corrosive tendencies of the ground, resistance to explosions, and long-term soil instability.

REFERENCES

Abbott, M. B., 1960: "One-Dimensional Consolidation of Multilayered Soils," *Geotechnique* (England), **10** (4, Dec.), 151.

Aggson, J. R., 1978: "Coal Mine Floor Heave in the Beckley Coalbed, and Analysis," BuMines RI 8274.

Alberro, J., 1969: "Contribution to Discussion," *Proc. 7th Int. Conf. Soil Mechs. Found. Eng.*, **3**, 349–357.

Aldrich, H. P., 1964: "Precompression for Support of Shallow Foundations," *J. Soil Mech. Found. Div., ASCE.* **91**, (SM2), Proc. Paper 4267, March, pp. 5–20 (see also *Proc. ASCE Conf. on Design of Found. for Control of Settlement*, Northwestern Univ., June 16–19, 1964).

Alyer, A. K., 1969: "An Analytical Study of the Time-Dependent Behavior of Underground Openings," Doctoral Thesis, Univ. Illinois, Urbana.

Armando, E., M. Fornaro, and L. Garrone, 1979: "Examples of Measurements of Dynamic Loads on Anchoring Bolts During Underground Blasting," *Tunnels et Ouvrages Souterrains* (36), 355–358.

Bang, S., 1979: "Analysis and Design of Lateral Earth Support System," Ph.D. Thesis, Univ. Calif, Davis.

Bang, S., and C. K. Shen, 1982: "Soil Reinforcement in Soft Ground Tunneling." U.S. Dept. of Transp., Office of Univ. Research, Washington, D.C.

Bank, S., 1979: "Analysis and Design of Lateral Earth Support System," Ph.D. Thesis, Univ. Calif, Davis.

Bank, S., and C. K. Shen, 1982: "Soil Reinforcement in Soft Ground Tunneling," U.S. Dept. of Transp., Office of Univ. research, Washington, D.C.

Barden, L., 1965: "Consolidation of Compacted and Unsaturated Clays," *Geotechnique* (England), Vol. **15** (3, Sept.), 267.

Barla, G., and C. Mascardi, 1974: "High Anchored Wall in Genoa," *Conf. on Diaphragm Walls and Anchorages*, Inst. Civ. Eng., London, pp. 123–128.

Barron, K., D. F. Coates, and M. Gyenge, 1970: "Artificial Support of Rock Slopes," Dept. Energy and Mines and Resources, Ottawa, Report No. R228, July.

Barton, N., 1973: "Review of a New Shear–Strength Criterion for Rock Joints," *Eng. Geol.*, **7**, 287–332.

Barton, N., 1976: "The Shear Strength of Rock and Rock Joints," *Int. J. Rock Mech. Min. Sci. Geomech.* (Abstr.), **13** (9, Sept.), 255–279.

Barton, N., R. Lien, and J. Lunde, 1974a: "Analysis of Rock Mass Quality and Support Practice in Tunneling, and a Guide for Estimating Support Requirement," NGI Internal Report 54206, 74 pp.

Barton, N., R. Lien, and J. Lunde, 1974b: "Engineering Classification of Rock Masses for the Design of Tunnel Support," *Rock Mech.* (Springer-Verlag, Vienna), **6** (4), 189–236.

Barton, N., R. Lien, and J. Lunde, 1977: "Estimation of Support Requirements for Underground Excavations," *Proc. 16th Symp. Rock Mech., ASCE*, pp. 163–177.

Beck, A., and A. Golta, 1972: "Tunnelsanierungen, der Schweizerischen Bundesbahnen," *Schweiz Bauzeitung*, **90** (36), 857–863.

Benson, R. P., 1970: "Rock Mechanics Aspects in the Design of the Churchill Falls Underground Powerhouse, Labrador," Doctoral Thesis, Univ. Illinois, Urbana.

Bischoff, J. A., and J. D. Smart, 1977: "A Method of Computing a Rock Reinforcement System Which Is Structurally Equivalent to an Internal Support System," *Proc. 16th Symp. Rock Mech., ASCE*, pp. 178–184, Univ. Minnesota.

Bishop, A. W., 1958: "Test Requirements for Measuring the Coefficient of Earth Pressure at Rest," *Proc. Brussels Conf. on Earth Pressure Probl.*, **1**, 2–14.

Bishop, A. W., 1971: "The Influence of the Progressive Failure on the Method of Stability Analysis," *Geotechnique*, **21**, 168–172.

Bjerrum, L., 1967: Seventh Rankine Lecture: "Engineering Geology of Normally-Consolidated Marine Clays as Related to the Settlements of Buildings," *Geotechnique* (England), **17**, (2, June), 83.

Bjerrum, L., 1972: "Earth Pressure on Flexible Structures—A State of the Art Report," *Proc. 5th Eur. Conf. SMFE*, Vol. 1, Madrid, pp. 169–196.

Brady, B. T., and W. I. Duvall, 1973: "Strengthening of Fractured Rock Pillars by the Use of Small Radial Reinforcement Pressures," Report of Investigations No. 7755, U.S. Dept. Interior, Bureau of Mines, Washington, D.C.

Bray, J. W., 1967: "A Study of Jointed and Fractured Rock, II. Theory of Limiting Equilibrium," *Felsmechanik und Ingenieurgeologie* (Rock Mechanics and Engineering Geology), **5**, 197–216.

Brekke, T. L., and T. R. Howard, 1973: "Functional Classification of Gouge Material from Seams and Faults in Relation to Stability Problems in Underground Openings," Final Tech. Report to ARPA, NTIS AD 766 046, 195 pp.

Brekke, T. L., and G. E. Korbin, 1974: "Some Comments on the Use of Spiling in Underground Openings," *Proc. 2nd Int. Cong. Intn. Assoc. Eng. Geology*, Sao Paulo, Brazil, Vol. 2, No. VII-PC-4, Aug.

Brekke, T. L., T. A. Lang, and F. S. Kendorski, 1974: "Some Design and Construction Considerations for Large Permanent Underground Openings at Shallow Depths," *Proc. 3rd Cong. Intern. Soc. of Rock Mech.*, Natl. Acad. Sci., Washington, D.C.

Breth, H., and D. Stroh, 1976: "Ursachen der Verformung im Boden beim Aushub tiefer Baugruben und konstruktiven Moglichkeiten zur Verminderung der Verformung von verankerten Baugruben," *Die Bautechnik*, **51** (3), 81–88.

Breth, H., and H. R. Wanoschek, 1972: "The Influence of Found. Weights upon Earth Pressure Acting on Flexible Strutted Walls," *Proc. 5th Eur. Conf. Soil Mech. Found. Eng.*, Madrid, Vol. 1.

Broms, B. B., and H. Bjerke, 1973: "Extrusion of Soft Clay through a Retaining Wall," *Can. Geotech. J.*, **10** (1), 103–109.

Brooker, E. W., and H. O. Ireland, 1965: "Earth Pressures as Rest Related to Stress History," *Can. Geotech. J.*, **2**, 1–15.

Bureau of Reclamation, 1976: "Design of Gravity Dams."

Casagrande, L., 1964: "Effect of Preconsolidation on Settlement," *J. Soil Mech. Found. Div., ASCE*, **90** (SM5), Proc. Paper 4041, p. 349, Sept.

Castro, G., 1975: "Liquefaction and Cyclic Mobility of Saturated Sands," *J. Geotech. Eng. Div., ASCE*, **101** (GT6), Proc. Paper 11388, June, pp. 551–569.

Cecil, O. S., 1970: "Correlations of Rock Bolt–Shotcrete Support and Rock Quality Parameters in Scandinavian Tunnels," Ph.D. Thesis, Univ. Illinois, Urbana, p. 414.

Chopra, A. K., and P. Chakrabarti, 1970: "A Computer Solution for Earthquake Analysis of Dams," Report No. EERC70-5, Earthquake Eng. Research Center, Univ. Calif., Berkeley.

Christian, J. T., 1980: "Probabilistic Soil Dynamics: State of the Art," *ASCE J. Geotech. Div.* (April), 385–397.

Clough, R. W., 1960: "The Finite Element Method in Plane Stress Analysis," *Proc. ASCE, 2nd Conf. on Electronic Computation*, Pittsburgh, Pa., Sept., pp. 345–378.

Clough, R. W., 1970: "Earthquake Response of Structure," in *Earthquake Engineering*, Prentice-Hall, Englewood Cliffs, N.J.

Clough, G. W., 1973: "Analytical Problems in Modeling Slurry Wall Construction," FCP Res. Rev. Conf., Fed. Hyw. Admin., San Francisco, Sept.

Clough, R. W., and Bureau of Reclamation staff members, 1963: *Earthquake Engineering for Concrete and Steel Structures*, Conf. Proc., Denver, Col.

Clough, G. W., and J. M. Duncan, 1971: "Finite Element Analyses of Retaining Wall Behavior," *J. Soil Mech. Found. Div., ASCE,* **97** (SM12, Dec.).

Clough, G. W., and Y. Tsui, 1974a: "Finite Element Analyses of Cut-and-Cover Tunnel Constructed with Slurry Trench Walls," Duke Univ., Durham, N.C., Soil Mech. Series No. 29.

Clough, G. W., and Y. Tsui, 1974b: "Performance of Tied-Back Walls in Clay," *ASCE J. Geotech. Div.,* **100** (Dec.).

Clough, G. W., P. R. Weber, and J. Lamont, 1974: "Design and Observations of a Tied-Back Wall," *Proc. Spec. Conf. on Performance of Earth and Earth-Supported Structures,* ASCE, Purdue Univ., Lafayette, Ind., Vol. 1 (June), Part 2, pp. 1367–1389.

Cole, K. W., and J. B. Burland, 1972: "Observations of Retaining Wall Movement Associated with Large Excavations," *Proc. 5th Eur. Conf. Soil Mech. Found. Eng.,* Madrid, Vol. 1.

Corbett, B. O., and M. A. Stroud, 1974: "Temporary Retaining Wall Constructed by Berlinoise System at Centre Beaubourg, Paris," *Proc. Conf. on Diaphragm Walls and Anchorages,* Inst. of Civ. Eng., London, pp. 77–83, Sept.

Cording, E. J., and W. H. Hansmire, 1975: "Displacements Around Soft Ground Tunnels," *Proc. 5th Panamerican Congress on SMFE,* Session IV, Nov.

Cording, E. J., A. J. Hendron, and D. U. Deere, 1971: "Rock Engineering for Underground Caverns," *ASCE Symp. on Underground Rock Chambers,* Phoenix, Ariz., pp. 567–600.

Cox, R. M., 1974: "Why Some Bolted Mine Roofs Fail," *Trans. Soc. Mining Eng., AIME,* **256** (2, June), 167–171.

D'Appolonia, E., 1968: "Dynamic Loadings," *Proc. ASCE Conf., Placement and Improvement of Soil to Support Structures,* Cambridge, Mass., Aug.

Davis, E. H., and G. P. Raymond, 1965: "A Nonlinear Theory of Consolidation," *Geotechnique* (England), **15** (2, June), 161.

Deere, D. U., 1963: "Technical Description of Rock Cores for Engineering Purposes," *Felsmechanik und Ingenieur-geologie,* **1** (1), 16–22.

Deere, D. U., A. J. Hendron, F. D. Patton, and E. J. Cording, 1968: "Design of Surface and Near-Surface Construction," in *Rock, Failure and Breakage of Rock,* C. Fairhurst, ed. American Institute of Mineralogy, Metallurgy, and Petroleum Engineering, New York, pp. 237–302.

Deere, D., R. Peck, J. Monsees, and B. Schmidt, 1969: "Design of Tunnel Liners and Support Systems," Univ. Illinois, Urbana, Feb.

deJong, G., 1968: "Consolidation Models Consisting of an Assembly of Viscous Elements or a Cavity Channel Network," *Geotechnique* (England), **18** (2, June), 195.

Department of the Army, Corps of Engineers, 1975: "Rock Reinforcement in Civ. Eng. Works," EM 1110-1-2907.

Dibiagio, E. L., 1966: "Stresses and Deformations Around an Unbraced Rectangular Excavation in an Elastic Medium," Ph.D. Thesis, Univ. of Illinois, Urbana.

Dodd, J. S., 1967: "Morrow Point Underground Powerplant Rock Mechanics Investigations," a Water Resources Tech. Publ., U.S. Dept. Interior, Bureau of Reclamation, Denver, Col., March.

Donovan, N. C., 1971: "A Stochastic Approach to the Seismic Liquefaction Problem," *Proc. 1st Int. Conf. on Applications of Statistics and Probability to Soil and Structural Engineering*, pp. 513–535.

Druss, D. L., 1984: "Development and Applications of the New Austrian Tunneling Method," Thesis, Univ. Pittsburgh.

Duncan, J. M., and C. Y. Chang, 1970: "Nonlinear Analysis of Stress and Strain in Soils," *J. Soil Mech. Found. Div., ASCE*, **56** (SM5), Proc. Paper 7513, Sept., pp. 1625–1653.

Duvall, W., 1976: "General Principles of Underground Opening Design in Competent Rock," *Proc 17th Symp. on Rock Mech.* (Univ. of Utah), Paper 3A1.

Egger, P., 1972: "Influence of Wall Stiffness and Anchor Prestressing on Earth Pressure Distribution," *Proc. 5th Eur. Conf. Soil Mech. Found. Eng.*, Madrid, Vol. 1.

Einstein, H. H., and N. Bischoff, 1977: "Design of Tunnels in Swelling Rock," *Proc. 16th Symp. Rock Mech.*, Univ. Minnesota, ASCE, pp. 185–195.

Einstein, H. H., N. Bischoff, and E. Hofmann, 1972: "Verhalten von Stollensohlen in guellendem Mergel," *Proc. Int. Symp. on Underground Openings*, Lucerne, pp. 296–319.

Einstein, H., and C. Schwartz, 1980: "Improved Design of Tunnel Supports," Vol. I, *Simplified Analysis for Ground Structure Interaction in Tunneling*, Springfield, Va., NTIS PB22154, June.

Fairhurst, C., and B. Singh, 1974: "Roof Bolting in Horizontally Laminated Rock," *Eng. Mining J.* (Feb.)

Fardis, M. N., 1978: "Probabilistic Liquefaction of Sands During Earthquakes," thesis presented to MIT, Cambridge, Mass., in partial fulfillment of the requirements for the degree of Doctor of Philosophy.

Fenner, R., 1928: "Untersuchungen zur Erkenntnis des Gebirgsdruckes," *Glueckauf. Ann. 74*, **32**, Essen, West Germany.

Geoconsult, Pittsburgh, 1982: "Mt. Lebanon Tunnel: Finite Element Analysis," *Outer Lining*, Salzburg, Austria, March.

Glass, C. H., 1973: "Seismic Considerations in Siting Large Underground Openings in Rock," Ph.D. Thesis, Univ. Calif., Berkeley, pp. 1–132.

Goldberg, D. T., W. E. Jaworski, and M. D. Gordon, 1976: "Lateral Support Systems and Underpinning," U.S. Dept. Transp. Fed. Hwy. Adm. Washington, D.C., Vols. I–III.

Golder, H. Q., and A. B. Sanderson, 1961: "Bridge Foundations Preloaded to Eliminate Settlement," *Civ. Eng.* (Oct.), 62.

Golser, J., 1973: "Praktische Beispiele empirischer Dimensionierung von Tunneln," *Rock Mech.*, Suppl. 2, Springer.

Golser, J., 1976: "The New Austrian Tunneling Method," *Proc. Conf. on Shotcrete for Ground Support*, Easton, Md., Am. Soc. Civ. Eng., New York.

Golser, J., 1978: "History and Development of the New Austrian Tunneling Method," *Law/Geoconsult Publ.* (reprinted from a paper presented at the 1978 Eng. Found. Conf., St. Anton, Austria).

Golser, J., 1981a: "The New Austrian Tunneling Method (NATM)," *Proc. Atlanta Research Chamber*, U.S. Dept. Transp., UMTA, March, Washington, D.C.

Golser, J., 1981b: "Tunnel Design and Construction with the NATM in Weak Rock," *Law/Geoconsult Publ.* (reprinted from Int. Symp. on Weak Rock, Tokyo).

Golze, A. R., 1977: *Handbook of Dam Engineering*, Van Nostrand Reinhold, New York.

Goodman, R. E., 1974: "The Mechanical Properties of Joints," *Proc. 3rd Cong. Int. Soc. for Rock Mechanics*, Vol. 1, Part A, pp. 127–140, Denver.

Goodman, R. E., 1976: "Principles of Stereographic Projection and Joint Surveys," in *Methods of Geological Engineering in Discontinuous Rock*, West, St. Paul, Minn.

Goodman, R. E., 1989: *Introduction to Rock Mechanics*, Wiley, New York.

Goodman, R. E., and J. W. Bray, 1977: "Toppling of Rock Slopes," *Proc. Spec. Conf. on Rock Eng. for Found. and Slopes, ASCE*, Boulder, Col., Vol. 2, pp. 201–234.

Goodman, R. E., and G. H. Shi, 1985: *Block Theory and Its Application to Rock Engineering*, Prentice-Hall, Englewood Cliffs, N.J.

Gould, J. P., 1949: "Analysis of Pore Pressure and Settlement Observations at Logan International Airport," *Harvard Soil Mech.* Series No. 34, Harvard Univ., Dec.

Gould, J. P., 1970: "Lateral Stresses on Rigid Permanent Structures," Preprint Proc. ASCE Spec. Conf. on Lateral Stresses in the Ground and Design of Earth Retaining Structures, Cornell Univ., Ithaca, N.Y.

Grant, R., J. T. Christian, and E. H. Vanmarcke, 1972: "Differential Settlement of Buildings," *ASCE J. Geotech. Div.*, **100** (GT9; Sept.), 973–991.

Gray, H., 1945: "Simultaneous Consolidation of Contiguous Layers of Unlike Compressible Soils," *Trans. ASCE*, **110**.

Grob, H., 1972: "Schwelldruck im Belchentunnel," *Proc. Int. Symp. on Underground Openings*, Lucerne, pp. 99–119.

Grob, H., 1974: "Excavation in Rock—Examples from Switzerland," *Rock Mech.*, **6** (1), 3–13.

Haliburton, T. A., 1968: "Numerical Analysis of Flexible Retaining Structures," *ASCE J. Soil Mech. Found. Div.*, **SM6** (Nov.)

Hanna, T., and A. Dina, 1973: "Anchored Inclined Walls—a Study of Behavior," *Ground Eng.* (Nov.)

Harr, M. E., 1966: *Foundations of Theoretical Soil Mechanics*, McGraw-Hill, New York, p. 141.

Hendron, A. J., and A. K. Aiyer, 1971: "Stresses and Strains around a Cylindrical Tunnel in an Elasto-Plastic Material with Dilatancy," Tech. Report No. 10, Omaha District, Corps of Engineers, Omaha, Nebr., Sept.

Heuze, P. E., and R. E. Goodman, 1973: "Numerical and Physical Modelling of Reinforcement Systems for Tunnels in Jointed Rocks," U. S. Corps of Engineers (Army), Tech. Report No. 16.

Hobst, L., and J. Zajic, 1977: *Anchoring in Rock*, Elsevier Scientific Publ. Co., Amsterdam, *Developments in Geotechnical Engineering*, Vol. 13, 390 pp.

Hoek, E., 1974: *Rock Slope Engineering*, Inst. Mining and Metallurgy, London

Hoek, E., and J. W. Bray, 1974, 1977, 1981: "Rock Slope Eng.," Inst. Mining and Metallurgy, London.

Hoek, E., and J. W. Bray, 1977: "Graphical Presentation of Geological Data," in *Rock Slope Eng.*, 2nd ed., Inst. Mining and Metallurgy, London.

Hoek, E., and P. Londe, 1974: "Surface Workings in Rock," *Proc. 3rd Cong. of the Int. Soc. for Rock Mechanics*, Vol. I.

Huder, J., 1965: "The Calculation of Ground Anchors and How They Operate," *Proc Conf. Swiss Soc. of Soil Mechanics*, May.

Huder, J., and G. Amberg, 1970: "Quellung im Mergel, Opalinuston und Anhydrit," *Schweiz Bauzeitung*, **88** (43), 975–980.

Idriss, I. M., and H. B. Seed, 1968: "Seismic Response of Horizontal Soil Layers," *J. Soil Mech. Found. Div., ASCE*, **94** (SM4), Proc Paper 6043, July, p. 1003.

Ishihara, K., and S. Yasuda, 1975: "Sand Liquefaction in Hollow Cylinder Torsion Under Irregular Excitation," *Soils Found.*, **15** (1, March), pp. 45–59

Jaeger, J. C., 1971: "Friction of Rocks and Stability of Rock Slopes," *Geotechnique* **21** (2), 97–134.

James, E. L., and B. J. Jack, 1974: "Design Study of Diaphragm Walls," *Proc. Diaphragm Walls Anchorages*, Inst. Civ. Eng., London.

Janbu, N., 1957: "Earth Pressure and Bearing Capacity Calculations by Generalized Procedure of Slices," *Proc. 4th Int. Conf. Soil Mech. Found. Eng.*, (ISSMFE), **2**, 207–212.

Janbu, N., 1971: "Jordtrykk (Earth Pressure)," Extension Course, Techn. Univ. Norway, Trondheim.

Janbu, N., 1972: "Earth Pressure Computations in Theory and Practice," *Proc. 5th Eur. Conf. Soil Mech. Found. Eng.*, Madrid, Vol. 1.

Jethwa, J. L., and B. Singh, 1984: "Estimation of Ultimate Rock Pressure for Tunnel Linings Under Squeezing Rock Conditions," *Proc. ISRM Symp. on Design and Performance of Underground Excavations* (Cambridge, U.K.), pp. 231–238 (Brit. Geotech. Soc., London).

Johnson, S. J., 1968: "Precompression for Improving Foundation Soils," *Proc. ASCE Conf., Placement and Improv. of Soil to Support Structures*, Cambridge, Mass., Aug.

Johnson, S. J., 1970: "Foundation Precompression with Vertical Sand Drains," *J. Soil Mech. Found. Div., ASCE*, **96** (SM1), Proc. Paper 7019, Jan.

Jonas, E., 1964: "Subsurface Stabilization of Organic Silty Clay by Precompression," *J. Soil Mech. Found. Div., ASCE*, **90,** (SM5), Proc. Paper 4036, Sept., pp. 363–376.

Kalkani, E. C., 1977: "Two Dimensional Finite Element Analysis of Rock Slopes," *Proc. 16th Symp. Rock Mech.*, Univ. Minnesota, ASCE, pp. 15–24.

Kastner, H., 1962: *Statik des Tunnel-und Stollenbaues*, Springer, Berlin/Gottingen.

Kishida, N., 1969: "Characteristics of Liquefied Sands During Mino-Owari, Tohnankai, and Fukui Earthquakes," *Soils Found.* **9** (1, March), 75–92.

Kitajima, S., and T. Uwabe, 1979: "Analysis of Seismic Damage in Anchored Sheet-Piling Bulkheads," Report, Port and Harbor Research Institute, Vol. 18, No. 1, pp. 68–127.

Kjellman, W., 1952: "Consolidation of Clay Soil by Means of Atmospheric Pressure," *Proc. Conf. on Soil Stabilization*, Mass. Inst. Technol. (MIT), June, p. 258.

Korbin, G. E., and T. L. Brekke, 1976: "A Model Study of Spiling Reinforcement in Underground Openings," Technical Report MRD-2-75, Missouri River Div., Corps of Engineers, Omaha, Nebr., Apr.

Korbin, G. E., and T. L. Brekke, 1978: "Model Study of Tunnel Reinforcement," *ASCE J. Geotech Div.* (Sept.), 895–908.

Kovari, K., and P. Fritz, 1977: "Stability Analysis of Rock Slopes for Plane and Wedge Failure," *Proc. 16th Symp. Rock Mech., ASCE*, Univ. Minnesota, pp. 25–32.

Kranz, E., 1953: "Uber die Verankerung von Spundwanden, 2 Auflage," *Mitteilungen aus dem Gebiete des Wasserbaues und der Baugrundforschung*, Vol. 11, Ernst & Sohn, Berlin.

Kruse, G. H., 1970: "Deformability of Rock Structures, Calif. State Water Project, Determination of the In Situ Modulus of Deformation of Rock," ASTM STP 477, American Society for Testing and Materials, pp. 58–88.

Kuesel, T. R., 1981: "On Hard Rock and Hard Progress," *Proc. Atlanta Research Chamber*, U.S. Dept. Transp., UMTA, March, Washington, D.C.

Kuesel, T. R., 1986: "Principles of Tunnel Lining Design," Newsletter, Tunneling Tech. U.S. Natl. Comm., March.

Kureta, S., H. Arai, and T. Yokoi, 1965: "On the Earthquake Resistance of Anchored Sheet Pile Bulkheads," *Proc., 3rd World Conf. on Earthquake Engineering*, New Zealand.

Ladd, C. C., 1959: "Mechanism of Swelling by Compacted Clay," HRB Bulletin No. 245, pp. 10–26.

Lambe, T. W., 1964: "Methods of Estimating Settlement," *J. Soil Mech. Found. Div., ASCE*, **90**, (SM5), Proc. Paper 4060, Sept., pp. 43–67.

Lambe, T. W., 1970: "Braced Excavations," *Proc. ASCE Spec. Conf. Lateral Earth Stresses Earth Retaining Structures*, Cornell Univ., Ithaca, N.Y., June.

Lang, T. A., 1958: "Rock Behavior and Rock Bolt Support in Large Excavations, Snowy Mountains Scheme—Australia, T1 Power Station," *Symp. on Underground Power Stations*, ASCE, New York, Oct. 1957 [Abstr. "Rock Bolting Speeds Snowy Mountain Project," *Civ. Eng.*, **28** (Feb.)].

Lang, T. A., 1959: "Underground Experience in the Snowy Mountains—Australia," *Proc. 2nd Protective Constr. Symp.*, Vol. II, Rand Corp., Santa Monica, Calif., March 24–26, pp. 767–853.

Lang, T. A., 1962: "Theory and Practice of Rock Bolting," *Trans AIME* (Mining), **223**.

Lang, T. A., 1971: "Underground Rock Structures Challenge the Engineer," *Symp. on Underground Rock Chambers*, Am. Soc. Civ. Eng., Jan., pp. 1–20.

Lang, T. A., and J. A. Bischoff, 1981: "Research Study of Coal Mine Rock Reinforcement," U.S. Dept. Interior, Bureau of Mines, Washington, D.C.

Lang, T. A., J. A. Bischoff, and P. L. Wagner, 1979: "Theory and Application of Rock Reinforcement in Coal Mines," U.S. Dept. Interior, Bureau of Mines, Washington, D.C.

Lee, K. L., and H. B. Seed, 1967: "Cyclic Stress Conditions Causing Liquefaction of Sand," *J. Soil Mech. Found. Div., ASCE*, **93** (SM1), Proc. Paper 5058, Jan., p. 47.

Littlejohn, G. S., and I. M. MacFarlane, 1974: "A Case History of Multi-tied Diaphragm Walls," *Proc. Diaphragm Walls Anchorages, Inst. Civ. Eng.*, Lond.

Littlejohn, G. S., P. J. Norton, and M. J. Turner, 1977: A study of rock slope reinforcement at Westfield open pit and the effect of blasting on prestressed anchors, *Proc. Conf. on Rock Eng.*, Univ. of Newcastle Upon Tyne, England, April.

Locher, H. G., 1969: *Anchored Retaining Walls and Cut-Off Walls*, Losinger and Co., Bern, pp. 1–23.

Lombardi, G., 1970: "The Influence of Rock Characteristics on the Stability of Rock Cavities," *Tunnels and Tunneling*, **12** (2, Jan.)

Londe, P., G. Vigier, and R. Vormeringer, 1969: "The Stability of Rock Slopes a Three-Dimensional Study," *Proc., ASCE*, **95** (SM1), 235–262.

Londe, P., G. Vigier, and R. Vormeringer, 1970: "Stability of Slopes, Graphical Methods," *Proc. ASCE*, **96** (SM4), 1411–1434.

Lowe, J. III, P. F. Zaccheo, and H. S. Feldman, 1964: "Consolidation Testing with Back Pressure," *J. Soil Mech. Found. Div., ASCE*, **90** (SM5), Proc. Paper 4058, Sept., pp. 69–86.

Maher, J., and G. Bennett, 1975: "Resin Bolting Developments at the White Pine Mine," *Skillings Mining Rev.* (July), 6–10.

Malijain, P. A., and J. L. Van Beveren, 1974: "Tied-back Excavations in Los Angeles Area," *Proc. ASCE*, **100,** (CO3), 337–356.

Manfredini, G., S. Martinetti, and R. Ribacchi, 1977: "Inadequacy of Limiting Equilibrium Methods for Rock Slope Design," *Proc. 16th Symp. Rock Mech., ASCE*, Univ. Minnesota, pp. 35–43.

Mathews, K. E., and J. L. Meek, 1975: "Modelling of Rock Reinforcement Systems in Cut and Fill Mining," *Proc. 2nd Australia–New Zealand Conf. on Geomechanics*, Brisbane, pp. 42–47.

Monsees, J., 1969: "Design of Support Systems for Tunnels in Rock," Ph.D. Thesis, Univ. Illinois, Urbana (Univ. Microfilms).

Moran, Proctor, Mueser and Rutledge, 1958: "Study of Deep Soil Stabilization by Vertical Sand Drains," Report to Bureau of Yards and Docks, Dept. of Navy, June, pp. 11–113ff (reprinted by Office of Techn. Serv., Dept. Commerce).

Morgenstern, N. R., and Z. Eisenstein, 1970: "Methods of Estimating Lateral Loads and Deformations," *Proc ASCE Spec. Conf. Lateral Stresses*, Cornell Univ., Ithaca, N.Y., (June), pp. 51–102.

Myklestad, N. O., 1944: *Vibration Analysis*, McGraw-Hill, New York.

Nataraja, R., and C. L. Kirk, 1977: "Dynamic Response of a Gravity Platform under Random Wave Forces," *9th Annual Offshore Techn. Conf.*, Paper OTC-2904, Vol. III, pp. 199–208.

Newmark, N. M., 1942: "Influence Charts for Computation of Stresses in Elastic Foundations," Univ. Illinois Bull. 338, Urbana.

Nussbaum, H., 1973: "Recent Development of New Austrian Tunneling Method," *ASCE J. Constr. Div.* (July), **99**, 115–132

Obert, L., and W. I. Duvall, 1967: *Rock Mechanics and the Deign of Structures in Rock*, Wiley, New York, pp. 612–638.

O'Rourke, T. D., E. J. Cording, and M. Boscardin, 1976: "Ground Movements

Related to Braced Excavations and Their Influence on Adjacent Building,'' Report No. DOT-TST-76T-23.

Ortlepp, W. D., 1969: "An Empirical Determination of the Effectiveness of Rock Bolt Support Under Impulse Loading," *Proc. Symp. on Larger Permanent Underground Openings*, Oslo, Norway, pp. 197–205.

Ortlepp, W. D., and J. D. Reed, 1970: "Yieldable Rock Bolts for Shock Loading and Grouted Bolts for Rock Stabilization," *Mining Eng.* (New York), **6** (3), 12–14, 16–17.

Ostermayer, H., 1970: "Erdanker-Tragverhalten und konstruktive Durchbildung," *Vortrage der Baugrundtagung in Dusseldorf*, Essen.

Ostermayer, H., 1974: "Construction Carrying Behavior and Creep Characteristics of Ground Anchors," *Proc. Diaphragm Walls Anchorages*, Inst. Civ. Eng., London.

Otta, L., M. Pantucek, and P. R. Goughnour, 1982: "Permanent Ground Anchors, Stump Design Criteria," *Office of Research and Development, Fed. Hwy. Admin., U.S. Dept. Transp.*, Washington, D.C.

Pacher, F., 1977: "Review and Comment," *NATM Proc. 16th Symp. Rock Mech.*, Univ. Minnesota, ASCE, pp. 223–233.

Panek, L. A., 1955: "Analysis of Roof Bolting Systems," *Mining Eng.*, **7** (10, Oct.), 954–957.

Panek, L. A., 1956: "Theory of Model Testing as Applied to Roof Bolting," BuMines RI 5154.

Panek, L. A., 1962a: "The Effect of Suspension in Bolting Bedded Mine Roof," BuMines RI 6138.

Panek, L. A., 1962b: "The Combined Effects of Friction and Suspension in Bolting Bedded Mine Roof," BuMines RI 6139.

Panek, L. A., 1963: "Design Rationale for Bolting Stratified Mine Roof," 3rd Int. Mining Cong., Salzburg, Austria, 10 pp.

Panek, L. A., 1964: "Design for Bolting Stratified Roof," *Trans. Soc Mining Eng., AIME* (June), 113–119.

Panek, L. A., and J. A. McCormick, 1973: "Roof/Rock Bolting," *SME Mining Eng. Handbook*, Vol. 1, pp. 13-125–13-134.

Parsons, J. D., 1968: "Stabilization and Preparation of Marginal Areas for Building Construction," presented before the Metropolitan Section, ASCE, New York, March.

Peacock, W. H., and H. B. Seed, 1968: "Sand Liquefaction Under Cyclic Loading Simple Shear Conditions," *J. Soil Mech. Found. Div., ASCE*, **94**, (SM3), Proc. Paper 5957, May, p. 689.

Peck, R. B., 1943: "Earth Pressure Measurements in Open Cuts, Chicago Subway," *ASCE Trans*.

Peck, R. B., 1969: "Deep Excavations and Tunneling in Soft Ground," State-of-the-Art Report, 7th ICSMFE, Mexico City, *State-of-the-Art Vol.*, pp. 225–290.

Peck, R. B., 1970: "Observation and Instrumentation, Some Elementary Considerations," Lecture Notes, Met. Section ASCE, Seminar on Field Observations in Found. Design and Construction, New York, April [reprinted in *Hwy. Focus, U.S. Dept. Transp. F.H.W.A.*, **4** (2), 1–5, June, 1972].

Peck, R. B., A. J. Hendron, Jr., and B. Mohraz, 1972: "State of the Art of Soft Ground Tunneling," *Proc. 1st Rapid Excavation and Tunneling Conf.* (AIME) **1,** 259–286.

Penzien, J., 1975: "Seismic Analysis of Platform Structure–Foundation Systems," *7th Annual Offshore Techn. Conf.*, Paper OTC-2352, Vol. III, pp. 153–164.

Pfister, P., G. Evers, M. Guillaud, and R. Davidson, 1982: "Permanent Ground Anchors, Soletanche Design Criteria," *Office of Research and Development, Fed. Hwy. Admin., U.S. Dept. Transp.*, Washington, D.C.

Polshin, D. C., and R. A. Toker, 1957: "Maximum Allowable Nonuniform Settlement of Structures," *Proc. 4th Int. Conf. Soil Mech. Found. Eng.*, Vol. 1.

Popescu, M., 1977: "A Numerical Analysis of Flexible Cast-in-Situ Diaphragm Walls Considering Nonlinear Soil Behavior," *Constr. Rev.* (5), 17–22.

Popescu, M., and C. Ionescu, 1977: "Nonlinear Analysis of Anchor/Ground/Wall System," 9th Int. Conf. Soil Mech. Found Eng., Special Session No. 4, Tokyo.

Priest, S. D., 1985: *Hemispherical Projection Methods in Rock Mechanics*, George Allen & Unwin, London.

Rabcewicz, L. V., 1948: "Patentschrift," Austrian patent No. 165 473.

Rabcewicz, L. V., 1964: "The New Austrian Tunneling Method," *Water Power* (Nov., Dec.)

Rabcewicz, L. V., 1965: "The New Austrian Tunneling Method," *Water Power* (Jan.).

Rabcewicz, L. V., 1969: "Stability of Tunnels under Rock Load," *Water Power* (June, Jul., Aug.).

Rabcewicz, L. V., 1970: "Die halbsteife Schale als Mittel zur empirischwissenschaftlichen Bemessung von Hohlraumbauten," *Rock Mech.*, Suppl. IV.

Rabcewicz, L. V., 1972: "The Importance of Measurements in Cavity Construction, Part I," transl. from *Der Bauingenieur, Law/Geoconsult Publ.* (July, Aug.)

Rabcewicz, L. V., 1973: "Principles of Dimensioning the Supporting System for the New Austrian Tunneling Method," *Water Power* (March).

Rabcewicz, L. V., 1975: "The Importance of Measurements in Cavity Construction, Part III," transl. from *Der Bauingenieur, Law/Geoconsult Publ.*

Rabcewicz, L. V., and J. Golser, 1974: "Application of the NATM to the Underground Works at Tarbela," *Water Power* (Sept., Oct.).

Ranke, A., and H. Ostermayer, 1968: "Beitrag zur Stabilitats-Untersuchung mehrfach verankerter Baugrubenumschliessungen" (A Contribution to the Stability Calculations of Multiple Tied-Back Walls), *Die Bautechnik*, **45** (10), pp. 341–349.

Ranken, R., J. Ghaboussi, and A. Hendron, 1978: "Analysis of Ground Liner Interaction for Tunnels," UMTA-IL-06-0043-078-3, Springfield, Va., NTIS PB294818, Oct.

Sandhu, B. S., 1974: "Earth Pressure on Walls Due to Surcharge," *Civ. Eng., ASCE*, **44** (12, Dec.), 68–70.

Schiffman, R. L., 1958: "Consolidation of Soil Under Time-Dependent Loading and Varying Permeability," *Proc Hwy. Research Board*, Vol. 37, p. 584.

Schnabel, H., 1971: "Sloped Sheeting," *Civil Eng.* (Feb.), 48–50.

Schultze, E., and A. Horn, 1966: "Der Zugwiderstand von Hangebrucken-Widerlagern," *Vortrage der Baugrundtagung in Munchen*, Essen, pp. 125–186.

Seed, H. B., 1968: "The Fourth Terzaghi Lecture: Landslides during Earthquake Due to Soil Liquefaction," *J. Soil Mech. Found. Div., ASCE,* **94** (SM5), Proc. Paper 6110, Sept., P. 1053.

Seed, H. B., 1976: "Evaluation of Soil Liquefaction Effects on Level Ground during Earthquakes," presented at the Sept. 27–Oct. 1 ASCE Annual Convention & Exposition held at Philadelphia, Pa. (Preprint 2752).

Seed, H. B., and I. M. Idriss, 1967: "Analysis of Soil Liquefaction: Niigata Earthquake," *J Soil Mech. Found. Div., ASCE,* **93** (SM3), Proc. Paper 5233, May, p. 83.

Seed, H. B., and I. M. Idriss, 1971: "Simplified Procedure for Evaluating Soil Liquefaction Potential," *J. Soil Mech. Found. Div., ASCE,* **97** (SM9), Proc. Paper 8371, Sept., pp. 1249–1273.

Seed, H. B., and R. V. Whitman, 1970: "Design of Earth Retaining Structures for Dynamic Loads," *Proc. ASCE Spec. Conf. Lateral Stresses,* Cornell Univ., Ithaca, N.Y., June, pp. 103–147.

Selmer-Olsen, R., 1970: "Experience with Using Bolts and Shotcrete in Area with Rock Bursting Phenomena," *Proc. Int. Symp. on Large Permanent Underground Openings,* Oslo, 1969, pp. 275–278.

Sharp, J., 1973: "Use of Artificial Support for Rock Slope Stabilization," Draft Report, Golder, Bawner and Assoc., Canada, May.

Simpson, B., N. J. O'Riordan, and D. D. Croft, 1979: "A Computer Model for the Analysis of Ground Movements in London Clay," *Geotechnique* (England), **29** (2), 149–175.

Sinitsyn, A. P., and S. V. Medvedev, 1972: "Dynamical Stability of Rigid Structures During Earthquakes," *Proc. 5th Eur. Conf. Soil Mech Found. Eng.,* Madrid, Vol. 1.

Skempton, A. W., 1961: "Horizontal Stresses in an Overconsolidated Eocene Clay," *Proc. 5th Int. Conf. Soil Mechs. Found. Eng.,* **1,** 351–357.

Skempton, A. W., and L. Bjerrum, 1957: "A Contribution to the Settlement Analysis of Foundations on Clay," *Geotechnique* (England), **7** (4, Dec.), 168.

Skempton, A. W., and D. H. MacDonald, 1956: "Allowable Settlement of Buildings," *Proc. Inst. Civil Eng.,* Part 3, Vol. 5.

Sokolovski, V. V., 1965: *Statics of Granular Media,* Pergamon Press, New York, 270 pp.

Soos, P. von, 1972: "Anchors for Carrying Heavy Tensile Loads into the Soil," *Proc. 5th Eur. Conf. Soil Mech. Found. Eng.,* Madrid, Vol. 1.

Spangler, M. G., 1940: "Horizontal Pressures On Retaining Walls Due to Concentrated Surface Loads," Iowa Eng. Exp. Stat. Bull. No. 140.

Stamatopoulos, A. C., and P. C. Kotzias, 1985: *Soil Improvement by Preloading,* Wiley, New York.

Stille, H., 1976: "Behavior of Anchored Sheet Pile Walls," Thesis, Royal Institute of Technology, Stockholm.

Stille, H., and B. B. Broms, 1976: "Load Redistribution Caused by Anchor Failures in Sheet Pile Walls," *Proc. 6th Eur. Conf. on Soil Mech. and Found. Eng.,* Vienna, Vols. 1 and 2, 197–200.

Stille, H., and A. Fredricksson, 1979: "Field Measurement of an Anchored Sheet Pile Wall in Clay," *Proc. 7th Eur. Conf. on Soil Mech. and Found. Eng.*, Brighton, Sussex, U.K., Vol. 3, pp. 285–290.

Talobre, J., 1957: *La Mecanique des Roches*, Dunod, Paris.

Taylor, D. W., 1942: "Research on Consolidation of Clays," Publication from Dept. of Civil and Sanitary Engineering, MIT, Serial 82, Aug.

Tcheng, Y., and J. Iseux, 1972: "Full Scale Passive Pressure Tests and Stresses Induced on a Vertical Wall by a Rectangular Surcharge," *Proc. 5th Eur. Conf. Soil Mech. Found. Eng.*, Madrid, Session II, Paper 13.

Terzaghi, K., 1945: "Stability and Stiffness of Cellular Cofferdams," *Trans. ASCE*, **110**, 1083–1202.

Terzaghi, K., 1946a: In *Rock Tunneling with Steel Supports*, R. V. Proctor and T. L. White, eds., Comm. Shearing & Stamping Co., Youngstown, Ohio.

Terzaghi, K., 1946b: "An Introduction to Tunnel Geology," in *Rock Tunneling with Steel Support*, R. V. Proctor and T. L. White, eds., (Ohio: Comm. Shearing & Stamping Co.), Youngstown, Ohio.

Terzaghi, K., 1954a: *Theoretical Soil Mechanics*, Wiley, New York, (7th printing), 510 pp.

Terzaghi, K., 1954b: "Anchored Bulkheads," *Trans ASCE*, **119**, Paper 2720, pp. 1243–1324.

Terzaghi, K., 1955: "Evaluation of Coefficient of Subgrade Reaction," *Geotechnique* (England), **4**, 279.

Tincelin, E., and P. Sinou, 1957: "Deformation Measurements in the Lorraine Iron Mines," *Mine and Quarry* (England) (July), 299–305.

Tschebotarioff, G. P., 1962: *Foundation Engineering*, McGraw-Hill, New York.

Tschebotarioff, G. P., 1973: *Foundations, Retaining and Earth Structures*, McGraw-Hill, New York.

Tsui, Y., 1973: "A Fundamental Study of Tied-back Wall Behavior," Ph.D. Thesis, Duke Univ., Durham, N.C., Dissertation Abstr. Int. Order No. 752435.

Tsui, Y., and G. W. Clough, 1974: Plane strain approximations in finite element analyses of temporary walls, *Proc. ASCE Geotechn. Conf.* Austin, Texas.

Underwood, L. B., and N. A. Dixon, 1976: "Dams on Rock Foundations," *Proc. ASCE Spec. Conf. Rock Eng. Found. and Slopes. Boulder, Col.*, pp. 125–146.

Wang, F. D., and M. C. Sun, 1970: "Slope Stability Analysis by the Finite Element Stress Analysis and Limiting Equilibrium Method," U.S. Bureau of Mines Report of Investigation 7341.

Ward, W. H., 1972: "Remarks on Performance of Braced Excavations in London Clay," *Proc. Conf. Performance Earth Earth-Supported Struct.*, Purdue Univ., Lafayette, Ind.

Ward, W., 1978: "Ground Support for Tunnels in Weak Rock," *Geotechnique* (England), **28** (2, June).

Weidler, J. B., and P. R. Paslay, 1970: "Constitutive Relations for Inelastic Granular Medium," *J. Eng. Mech. Div., ASCE*, **96**, (EM4, Aug.), 395–406.

Whitney, H., and G. Butler, 1983: "The New Austrian Tunneling Method—A Rock

Mechanic Philosophy," *Law/Geoconsult Publ.* (reprinted from *Proc. 24th U.S. Symp. on Rock Mech.*, June).

Wiss, J. F., 1968: "Effects of Blasting Vibrations on Buildings and People," *Civ. Eng.* (July), 46–48.

Wong, I. H., 1971: "Analysis of Braced Excavations," Ph.D. Thesis, MIT, Cambridge, Mass.

Xanthakos, P. P., 1974: "Underground Construction in Fluid Trenches," Colleges of Engineering, Univ. Illinois, Chicago.

Xanthakos, P. P., 1979: *Slurry Walls*, McGraw-Hill, New York.

CHAPTER 10

DESIGN EXAMPLES

10-1 ANCHORAGES AT DEVONPORT NUCLEAR COMPLEX

This project, described in Section 8-16 and shown in Fig. 8-35, covers a site underlain by a series of geosynclical sediments dominated locally by a hard gray–blue-banded slate known as ''shillet.'' The rock dips gently at 3.5°, and has a top layer 0.5–1.2 m thick consisting of weathered and extremely fissile materials. As mentioned in Section 8-16, the initial stage of the project required dredging and dewatering of the site to approximately 15 m (49 ft) below the normal dock water level (Littlejohn and Truman Davies, 1974). This work was protected by a cellular steel sheet pile cofferdam along the south perimeter. Along the other sides, the existing basin walls were strengthened against overturning.

Design Considerations. Essentially, the existing basin walls are gravity structures; hence stability against overturning and sliding is derived from their own weight. Improvement of wall stability under increased overturning forces resulting from water lowering inside the basin was considered under the procedure discussed in Section 9-1. The solution involved the use of 2000-kN-capacity (450 kips) anchors placed at 2.5 m intervals (8 ft) along the back face. This prestressing provides an additional resisting moment approximately 1450 ft-kips per foot of length of wall. At the northwest corner of the basin the rock dips inward, however, manifesting a greater risk of sliding along the wall base or along underlying slate joints. At this location, stability against sliding was provided by casting a thick mass concrete block near the center of the basin, anchoring this block to bedrock using prestressed anchors, and connecting the anchored block to the walls by means of a mass concrete thrust slab (see Fig. 8-35). This scheme was analyzed using the procedure discussed in Section 9-1, downstream stabilization. However, wall stability was essential before dewatering; hence concrete placement together with the installation and stressing of anchors was carried out under 15 m of water.

Overall Rock–Anchor Stability. Uplift capacity of the anchor system was based on the wedge mechanisms discussed in Section 5-3, whereby the system is equated to the weight of a specified rock cone. In this case the inverted cone of rock with an included angle 90° was assumed to fail, and since the load transfer from the tendon to the rock is by bond, the position of the apex of the cone was chosen at the middle of the fixed length, as shown in Fig. 5-4(a). Where the ground is beneath the water table, submerged weight was used. The effect of group action from anchors with overlapping cones was considered as shown in Figs. 5-3(b) and 5-4(b).

A factor of safety of 1.6 was initially considered in analyzing the weight of rock in the assumed pullout zone, but since other rock parameters were ignored, the design allowed anchor working loads to equal the weight of rock using a submerged density of 80 lb/ft^3. Under these criteria the calculated depth through rock to the midpoint of the fixed zone was 12 m. Where anchors had to be spaced at 1-m centers, the depth of alternate units was increased by 2 m (6.6 ft) to spread the zone of load transfer over a greater rock layer and thus improve resistance to laminar failure.

Rock–Grout Bond. Fixed anchor length design was based on the procedures discussed in Section 4-5 for the failure of anchors in rock and for straight shaft. In addition, the design considered the following: (a) core recovery of the shillet at fixed anchor level ranged from 80 to 100 percent; (b) differences in the angle of bedding planes had no significant effect on the shear strength of rock; and (c) for a factor of safety of 3 against failure at the rock–grout interface, the allowable working bond stress should be 0.6 N/mm^2 (85 psi). Under these conditions the assumptions of uniform bond stress distribution along the fixed anchor length was valid. For a factor of safety of 3, a working anchor load of 2000 kN, and a bond stress of 0.6 N/mm^2, the fixed anchor length was estimated at 8 m (26 ft) for a hole with 140 mm diameter.

Tendon–Grout Bond. Based on grout strength 30 N/mm^2 (4350 psi) and the fixed length of 8 m, bond stresses along the grout–tendon interface were well below the safe values recommended in Section 4-3; hence failure at this interface did not control fixed anchor length design.

Tendon Design. Initially, the tendon working stress was specified at 62.5 percent f_{pu}, with a corresponding factor of safety of 1.6 (see also Section

4-2). In order to keep the borehole size to a minimum with a corresponding reduction of drilling costs, low-relaxation Dyform strand was selected, with 15.2 mm diameter and ultimate strength 300 kN, requiring 11 strands per tendon. However, a more conservative design to accommodate forthcoming standards of permanent anchors based on safety factors close to 2 was finally adapted, resulting in 12 strands per tendon and a working stress of 55 percent f_{pu}. The area of steel was checked to ensure that it did not exceed 15 percent the borehole area, and provided grout cover not less than 10 mm.

Test Anchor. A test anchor was used to monitor short-term behavior of the proposed construction, and also confirm design bond values at the rock–grout interface. The hole was drilled under the specified procedure but with a fixed anchor length of 3 m (10 ft). The test tendon consisted of 13 Dyform strands homed and grouted with neat cement grout [water/cement ratio (W/C) = 0.45] during one day. Initial stressing was applied 9 days after installation using a monojack system capable of stressing the complete tendon in one operation.

During the first stage the anchor was stressed to 800 kN and held for 53 h, supplemented by limited monitoring but without recording extensions.

Load–extension graphs for the cyclic loading are shown in Fig. 10-1, and evidently no permanent set is present. For a working bond stress of 0.6 N/mm², anchor behavior was considered satisfactory under cyclic loading.

In the final stage of the test the load was applied in 200-kN increments

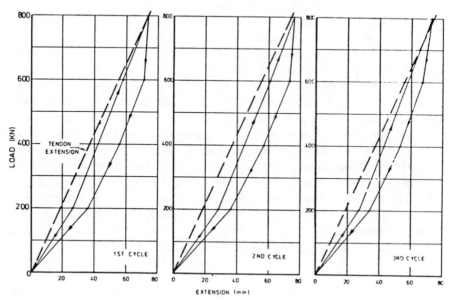

Fig. 10-1 Load extension graphs; test anchor, Devonport Nuclear complex. (From Littlejohn and Davies, 1974.)

up to 2400 kN when the full extent of jack travel was reached. Thereafter the anchor was destressed in 400-kN decrements, also monitoring the extensions at each change of loading and unloading. At the end of this cycle and on completely destressing the anchor, an apparent permanent extension of 8 mm was recorded.

Loading was resumed in 400-kN increments until a load of 2000 kN was reached, and at this point the anchor was locked off. After adjusting the jack, loading was continued in 200-kN increments to the maximum of 3000 kN. At this point anchor extensions reached 285 mm as shown in Fig. 10-2. From these results it follows that with respect to fixed anchor length the factor of safety is 4. Assuming an equivalent uniform bond distribution, the test induced bond stresses of 1.61 and 2.27 N/mm^2 (230 and 325 psi) at the tendon–grout and the rock–grout interfaces, respectively.

10-2 ROCK SLOPE STABILIZATION BY PRESTRESSING

The stability of a potential rock slide mass will be investigated for a highway project, based on the principles discussed in Sections 9-14 and 9-15. Interestingly, the rationale between the plane failure problem and that of wedge failure on two intersecting plane surfaces enables the use of a simple procedure in the solution of the complex three-dimensional wedge problem (Kovari and Fritz, 1977). The same basic procedure can be used to carry out parametric studies and investigate the sensitivity of rock slope construction to variations of individual parameters.

The rock for this example consists of uniform layers of interbedded sedimentary formations with a strike direction approximately parallel to the projected axis of the highway project. The dip α is 40°, and the total weight of the rock mass engaged in the cutting is assumed to be 88,000 Mp. Because of the irregular surface of the rock mass, this volume is divided into 6–8-m strips, and the stability of each strip is analyzed separately. A typical strip with the relevant data is shown in Fig. 10-3.

The design slope angle and also the angle of inclination β of the anchors are fixed by constructional considerations. The shear strength parameters c and ϕ are determined by tests as well as by backward computations from observed slides in the same area. However, because they are subject to uncertainties, an important objective of the investigation is to determine their influence on the calculated factor of safety and the required anchor force. The analysis is carried out with the aid of programmable calculators, and typical results are presented graphically in Figs. 10-4 and 10-5.

The results from Fig. 10-4 indicate the marked effect of the cohesion c and friction ϕ on anchor force T and the factor of safety F_s. For a preselected value of $F_s = 1.5$, the zone of influence of c and ϕ is shown hatched in Fig. 10-5. Minimum values of shear strength parameters determine point A on the T/W axis, whereas maximum shear strength ($c = 0.3$ kp/cm^2 and $\phi =$

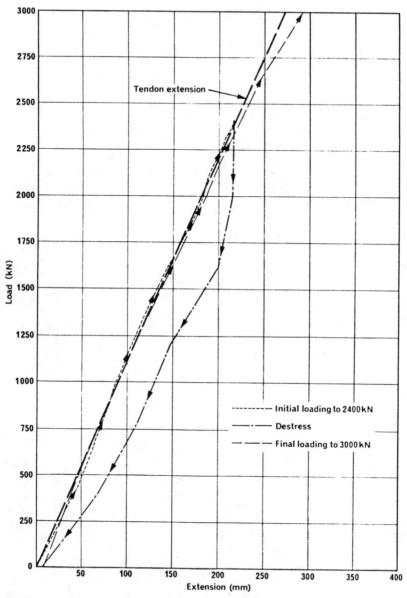

Fig. 10-2 Load extension graph, final stage of test; Devonport Nuclear Complex. (From Littlejohn and Davies, 1974.)

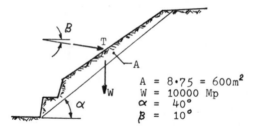

$$A = 8 \cdot 75 = 600\,m^2$$
$$W = 10000\ Mp$$
$$\alpha = 40°$$
$$\beta = 10°$$

Fig. 10-3 Typical strip; side cutting for a highway project, configuration and relevant data. (From Kovari and Fritz, 1977).

36°) yields point D. In terms of cost considerations the existence of cohesion is of practical importance. Selecting the anchor force $T = 0.2\ W$, where W is the weight of sliding mass, the factor of safety is as follows:

$$F_s = 1.65 \quad \text{for} \quad \phi = 36° \quad \text{and} \quad c = 0.3\ \text{kp/cm}^2$$
$$F_s = 1.10 \quad \text{for} \quad \phi = 32° \quad \text{and} \quad c = 0$$

A second example is the analysis of wedge failure according to the method outlined in Section 9-15, shown in Fig. 10-6. In this case considerable water

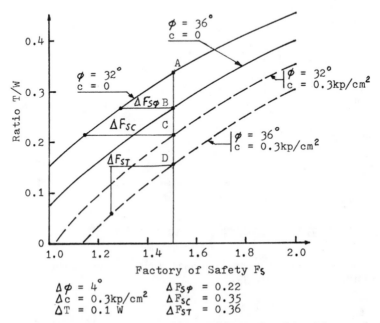

Fig. 10-4 Ratio of anchor force to mass weight as function of factor of safety; variable friction and cohesion. (From Kovari and Fritz, 1977.)

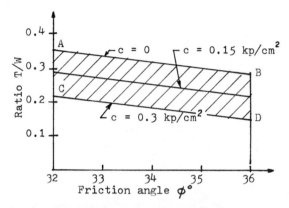

Fig. 10-5 Ratio of anchor force to mass weight as function of shear strength; factor of safety $F_S = 1.5$. (From Kovari and Fritz, 1977.)

pressures acting normal to the sliding planes are assumed in order to demonstrate the effect of complete drainage on anchor force requirements. Using the geometric data of the two sliding planes, the dip α_s of the line of intersection of the two sliding planes and the wedge factor λ are first determined (see also Section 9-15). The anchor force is estimated from the appropriate equation.

Figure 10-7 shows the relationship between anchor force and factor of safety for different cohesion values and for wet and dry slope. It is apparent that with effective drainage it is possible to ensure slope stability without resorting to other means. On the other hand, if water pressures are present, slope stability is achieved only by means of rock anchors.

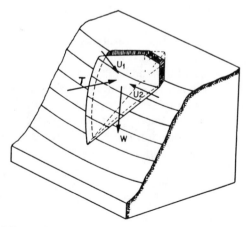

Fig. 10-6 Wedge failure of rock slope with water pressure. (From Kovari and Fritz, 1977.)

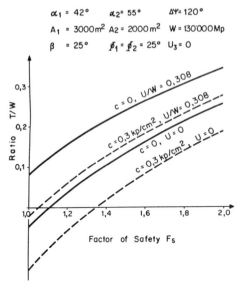

$$\alpha_1 = 42° \qquad \alpha_2 = 55° \qquad \Delta\Psi = 120°$$
$$A_1 = 3000 m^2 \quad A_2 = 2000 m^2 \quad W = 130000 Mp$$
$$\beta = 25° \qquad \phi_1 = \phi_2 = 25° \quad U_3 = 0$$

Fig. 10-7 Relationship between anchor force T and factor of safety for variable cohesion, and for wet and dry slope; U is resultant water pressure equal to $U_1 + U_2$. (From Kovari and Fritz, 1977.)

10-3 DEVELOPMENT OF A DESIGN AND TESTING PROGRAM

This example describes the development of a design to improve flood defences along the north bank of the Thames estuary in Essex, England. The plan consists of a concrete wall constructed to resist the flood and wave effects by the combined action of the soil resistance and permanent ground anchors installed below the river. A full-scale test on a prototype section of the wall supplements the program.

The defence consists of a bank formed of clay materials obtained from the area, surmounted by a short crest wall of sheet piling. Along the protected side heavy industrial development has taken place, with installations extended to the toe of the bank. The existing bank is founded on soft organic clays and silts, and exhibits relatively low factors of safety against seaward slip failure (1.2–1.4). In this case, the primary design requirement of the new defence system is to provide protection against high water levels resulting from high winds and associated surges.

Design Criteria. The first consideration is the restricted space available for new construction, and the low strength of foundation materials that preclude further earthwork as practical possibility. Conventional designs, on the other hand, consisting of sheet pile walls or retaining walls on piles would be impracticable because of the extremely narrow working area and restricted accessibility. A plane retaining wall with integral cutoff along the top of the bank was considered. This wall would add insignificant weight to the existing bank, and therefore would not contribute to the overall low stability against seaward slip. However, the new construction would not provide an adequate factor of safety against a slip toward the landward area under flood conditions.

Selected Scheme. The effect of incorporating anchorages installed in the seaward section of the bank at a low angle was studied since this installation could carry some of the horizontal loading and thus relieve the clay bank of further overstressing conditions. The installation would provide adequate stability, it could be completed with the use of comparatively light equipment, and would have a cost compatible with the economics of the project (Picknett et al., 1974).

Interestingly, the low stability of the banks against seaward slip at low tides precludes anchor prestressing. The anchor load depends, therefore, on the total wall movement and bank deformation under water load. It follows, therefore, that the proportion of the applied lateral load to be resisted by the anchors and the soil could not be calculated with certainty, although preliminary designs were based on the assumption that both passive soil resistance and anchor tension would be mobilized under water loading. Furthermore, the defence program was assigned a 60-year longevity time, implying that the anchorage would have to be protected permanently against corrosion.

Testing Program. It appears from the foregoing remarks that this design methodology was at its conception a new construction process; hence a full-scale test was considered necessary to observe wall, foundation, and anchor behavior under working loads, and document their performance at failure conditions. The layout and details of the test section are shown in Fig. 10-8. We can notice three wall bays: the central test section, and two independent wing bays in order to provide reasonably realistic end conditions. The test program was to subject the wall to a raised water level to reproduce the static effects of a surge, and then apply the design wave loads by means of jacks connected to the wall by bar or strand. Cyclic loading was also induced to represent repeated impacts of the average wave. The design specified anchors of nominal working load of 400 kN, with a factor of safety of 2 against pullout, but with a tendon cross-sectional area larger than necessary to accommodate a higher proportion of load to be resisted by the anchor at small strains as the wall deflected.

The working test was carried out with water levels and loads as shown in Table 10-1. Deflections during the test were well within acceptable limits. For the most severe surge conditions the corresponding displacement was 6 mm, with settlements of 6 mm at the center and 3 mm at the rear of the base. On removal of the load, horizontal recovery was rapid and almost

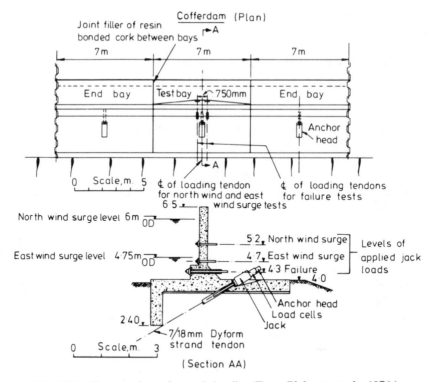

Fig. 10-8 Test section; plan and details. (From Picknett et al., 1974.)

complete, apparently aided by the anchorage. However, vertical recovery was incomplete. With nearly complete horizontal recovery the residual anchor load remained low. Earth pressure measurements were difficult to interpret, but showed that most of the load was transferred to the bank through the base slab rather than by the cutoff. Interestingly, water pressure showed full recovery after the test, and there was no evidence of pore pressure increase due to seepage.

During the dynamic loading anchor response was almost absent, the earth pressure cells registered very small readings, and there was no progressive wall movement.

TABLE 10-1 Working Test: Loads and Water Levels

Test	Water Level, m OD	Jack Load, kN
East wind surge, static	4.750	370
North wind surge, static	6.000	80
East wind surge, dynamic	4.750	0–140 cyclic
North wind surge, dynamic	6.000	0–30 cyclic

From Picknett et al (1974).

For the final phase of the test and with the anchor connected to the wall, the water in the cofferdam was raised to + 6.00 m, with a jack load providing an applied load twice the horizontal working load. Deflection and anchor load increased linearly but remained small. With a further increment of jack load equal to the horizontal component of the anchor load, no significant change in behavior was noticed. Thereafter the load was relaxed, the anchor disconnected, and a loading twice the horizontal working load was repeated. Horizontal wall deflection was further increased by 7 mm.

For the test-to-failure phase the jack load was applied in 100-kN increments with intermediate 10-min pause, until a load of 600 kN was reached and held for 60 min. The loading process was resumed at the same increments up to the maximum 1900 kN. For this test, jack load, anchor load and horizontal deflections are shown in Fig. 10-9. We can observe a relatively small amount of creep while the 600 kN load was held for 60 min. On resuming loading, the stress–strain response remained essentially linear up to 800 kN. Thereafter, the horizontal displacement became progressively nonlinear, and at 1300–1400 kN it increased sharply. When the external load reached 1900 kN, the horizontal deflection reached 90 mm while the anchor load was almost 700 kN. As the external load was held constant for 30 min, the anchor load increased to 800 kN. At this point the displacement was considerable, and the anchor load was twice the working. These conditions terminated the test.

10-4 EXAMPLE OF NATM—THE ARLBERG TUNNEL

This tunnel consists of two single-bore tubes with one lane of traffic in each direction. Each tube has a cross-sectional area of 90–103 m^2 (about 970–1100 ft^2), in a configuration as shown in Fig. 10-10.

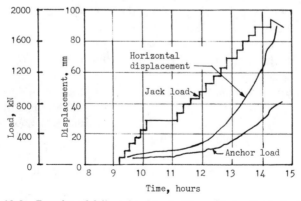

Fig. 10-9 Results of failure load test. (From Picknett et al., 1974.)

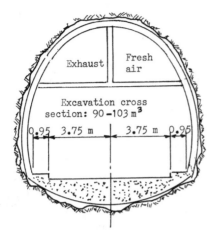

Fig. 10-10 Typical cross section; Arlberg Tunnel. (From John, 1980.)

Geologic Conditions. Initially these were inferred from surface features in conjunction with available geologic maps. Because of the depth of the overburden, prohibiting a complete geologic exploration, test borings were restricted to the portal areas and the upper sections of two access shafts. This information indicated the extent of weathering of the rock, and was supported by geologic data obtained from construction records of an adjacent railroad tunnel (John, 1980).

The tunnel lies close to the contact between two major formations, but the majority of the rock encountered in its path is the feldspathic mica schist. Rock strata are generally dipping at 60° to the horizontal. The mica schists are highly prone to rock falls due to slickensides, mica concentrations, and smooth bedding planes. Several faults were encountered in the zone between the ventilation shafts, and rock pressures generated from excavation caused shear displacements along these faults. Near one end of the tunnel the rock consists of steeply dipping schist layers, large fracture zones, and tectonic faults having their origin in the flow of water. Large overburden stresses in conjunction with weak planes within the schist caused severe squeezing pressures accompanied by large radial deformations.

The design program specified that exact geologic conditions could be established at the time of construction by continuous geologic mapping.

Constructional Considerations. From the brief geologic data we can conclude that tunnel construction would involve excavation in highly deformable hard ground with large overburden. Accordingly, the construction procedure specified the following steps:

1. Excavate short rounds with emphasis on preventing detrimental loosening and disturbance to rock.
2. Install steel ribs for safety purposes.

3. Install bolts or rock anchors close to the face to strengthen the rock mass and produce a self-supported ring.
4. Install a flexible outer lining consisting of slotted shotcrete.
5. Provide continuous geotechnical monitoring through field instrumentation.
6. Use the rock classification system to determine appropriate support system for each ground class encountered.

Because of the complex geologic conditions and variations in rock characteristics, five ground types of rock classes were selected for support design (for a rationale of this design with the Mt. Lebanon Tunnel, see Section 8-11 and Table 8-1). For each ground classification, the contract specifications provided the following information.

1. Description and characteristics of rock.
2. Method of excavation (round length, benches, etc.).
3. Required support elements.
4. Time of installation in relation to the advance of the face.

For each of the five rock classes, the required support system or combination of supports is shown in Fig. 10-11. The rock class most frequently encountered is class IV, characterized by large ground pressures exceeding the compressive strength of the rock. For class IV the maximum allowable run is 1.5 m, and the excavation must be carried out in stages. The support system consists of two layers of shotcrete, wire mesh, and closely spaced

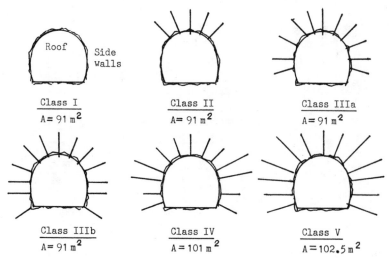

Fig. 10-11 Ground classification for Arlberg tunnel. (From John, 1980).

bolts or rock anchors. Interestingly, the rock classes were determined in the field jointly by the owner's representative and the contractor taking into account the following: (a) results of geologic surveys on site; (b) instrumentation observations; (c) laboratory test results on rock samples; and (d) records of test borings.

Typical convergence measurements are plotted in Fig. 10-12 (John, 1980). We can see that large deformations took place immediately after excavation, probably due to the effect of high overburden pressure and the orientation of the fault zones. The unusually large convergence is the reason for the increased cross-sectional area of the opening required for ground classes IV and V (see also Fig. 10-11). From Fig. 10-12, it becomes evident that initial stabilization of the opening occurred within a period of 40 days. At this point the top heading was completely excavated and the ground support was in place. Convergence, however, resumed with bench excavation and stabilized within 30–40 days after excavation was completed.

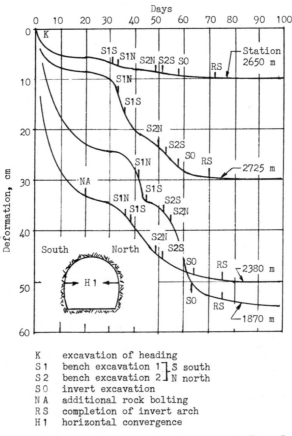

K	excavation of heading
S 1	bench excavation 1 ⎤ S south
S 2	bench excavation 2 ⎦ N north
S 0	invert excavation
N A	additional rock bolting
R S	completion of invert arch
H 1	horizontal convergence

Fig. 10-12 Convergence measurements, Arlberg tunnel. (From John, 1980.)

The rate of convergence or deformation immediately after excavation provided the criterion for determining the necessary support elements. For example, if the rate was less than 25 mm per day no additional bolts or anchors were required. Similar instrumentation results were used in the analysis and design of the inner or final liner, which consists of cast-in-place concrete ring rather than additional shotcrete layers. From deformation readings shown in Fig. 10-12 we conclude that complete stabilization of the opening was attained on completion of the invert (ring closure). Residual deformation thereafter is not indication of potential load increase on the liner, but the effect of structural adjustment of the rock mass outside the tunnel wall.

10-5 STABILIZING EFFECT OF ANCHORING IN A CIRCULAR TUNNEL

According to the theory presented in Section 9-11, anchors placed radially in a circular opening create normal forces within the rock mass and induce radial stress resistance. The associated effect is to prevent a zone of certain width around the cavity (called the "carrying ring") from expanding unnecessarily and partly disintegrating. In this way normal forces are created in the carrying ring either actively by prestressing or passively so that it can supply considerable lining resistance.

From triaxial tests, the values of ϕ, c and σ_{uc} (friction, cohesion, and uniaxial compressive strength, respectively) can be estimated. By using Mohr's circle for a given radial stress σ_r, the corresponding values of σ_θ (tangential stress at carrying ring), τ_r (shear strength of rock) and angle α are determined, and the curve of shear failure can be plotted as shown in Fig. 10-13. The point where the shear plane runs off the border is defined by the angle α, the lower side of which coincides with the direction of movement.

The lining resistance of the carrying ring can be determined from a consideration of the following condition of equilibrium:

$$\frac{\tau_r w \cos \psi/2}{\sin \alpha} = p_i^w \frac{b}{2} \tag{10-1}$$

where p_i^w is the lining resistance of carrying ring and w is the width of the ring. Rewriting Eq. (10-1) we obtain

$$p_i^w = \frac{2\tau_r w \cos \psi/2}{b \sin \alpha} \tag{10-2}$$

For a circular cavity $b/2 = R \cos \alpha$, so that

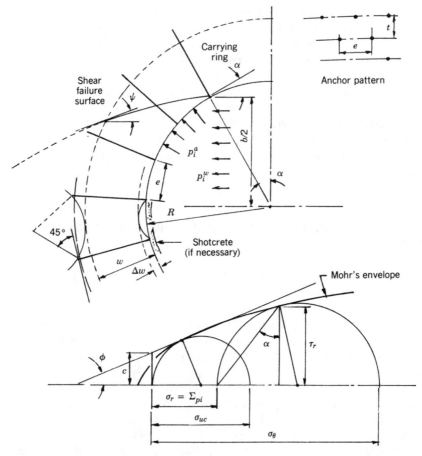

Fig. 10-13 Stabilizing effect on anchoring and relation to lining resistance of the carrying ring. (From Rabcewicz, 1969.)

$$p_i^w = \frac{\tau_r w \cos \psi/2}{R \cos \alpha \sin \alpha} \qquad (10\text{-}3)$$

Example

Given $R = 5$ m, $w = 3.0$ m, $e = t = 1.4$ m, $f_{pu} = 5$ tons/cm², $A_s = 5.3$ cm², $\phi = 30°$, $c = 9.3$ tons/m², $\sigma_{uc} = 32.2$ tons/m².
Compute

$$\sigma_r = p_i^a = \frac{A_s f_{pu}}{et} = \frac{5.3 \times 5.0}{1.4 \times 1.4} = 13 \text{ tons/m}^2 \qquad \text{(from Eq. 9-32)}$$

Also given $\sigma_\theta = 67$ tons/m², $\alpha = 33°$, $\tau_r = 24.7$ tons/m², $\psi = 46°$, $\psi/2 = 23°$. We can now estimate p_i^w with the help of Eq. (10-3).

$$p_i^w = \frac{3.0 \times 24.7 \times 0.92}{5 \times 0.84 \times 0.54} = 30 \text{ tons/m}^2$$

Likewise, if the tunnel is additionally lined with shotcrete, the additional lining resistance can be estimated with the help of Eq. (9-29).

The dependence of lining resistance of the carrying ring on the rock strength (expressed in terms of c and ϕ) and also on the internal pressure resistance provided by the anchor and shotcrete supports is shown graphically in Fig. 10-14. The diagrams are plots of p_i^w versus the friction angle ϕ, and for two support conditions; the first consisting of anchors only ($p_i^a = 13$ tons/m^2) and the second including anchors and shotcrete (total $p_i^a + p_i^s = 46$ tons/m^2). As expected, the lining resistance of the ring increases markedly with increasing rock strength and multiple interior support system.

10-6 ESTIMATION OF ANCHOR CAPACITY BY EQUIVALENT SUPPORT METHOD

From a given structural support we will determine a system of rock reinforcement that is structurally equivalent using the procedure outlined in Section 9-6. The support consists of W12×65 steel sets spaced on 5-ft centers, and fabricated from A36 steel. The rock mass has an effective friction angle $\phi = 36°$.

Relevant parameters are as follows, by reference to Eq. (9-12):

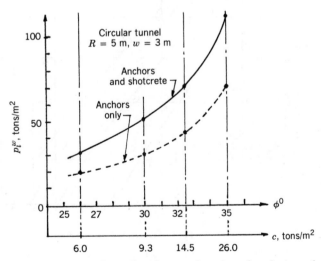

Fig. 10-14 Lining resistance of carrying ring as a function of angle ϕ, rock parameter c, and internal pressure provided by the ground support. (From Rabcewicz, 1969.)

$$A_s = 19.1 \text{ in}^2 \ (123.2 \text{ cm}^2)$$

$$\sigma_s = 36,000 \text{ lb/in}^2 \ (248,000 \text{ kN/m}^2)$$

The thrust capacity of the steel support system is

$$T_s = \frac{36,000 \times 19.1}{5} = 137,500 \text{ lb/foot of tunnel}$$

The rock reinforcement system must increase the thrust capacity of the rock mass by an amount $\Delta T_A = T_s = 137,500$ lb/ft, in order to be equivalent. By reference to Fig. 9-28, the equivalent system of pattern reinforcement consists of 1-in-dia. Grade 60 grouted rebars, spaced at 4-ft centers and 15 ft long. By graphical interpolation, this anchorage will provide approximately the additional thrust capacity of 137,500 lb/ft. Alternatively, the same thrust capacity can be provided by bars 20 ft long, spaced at 4.5-ft centers. The combination to be finally selected will depend on the size of the opening, and the guidelines for minimum anchor length and maximum anchor spacing.

The effect of rock strength on support requirements can be demonstrated if we consider the same problem for rock with $\phi = 55°$. The equivalent rock

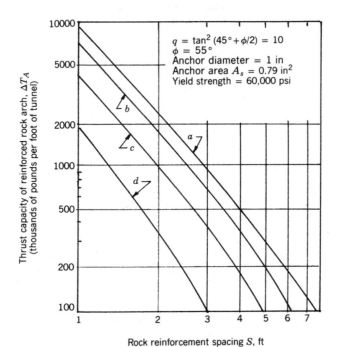

Fig. 10-15 Thrust capacity of reinforced rock arch for various rock reinforcement patterns and $\phi = 55°$: a, anchor length 20 ft; b, anchor length 15 ft; c, anchor length 10 ft; d, anchor length 5 ft. (From Bischoff and Smart, 1977.)

reinforcement is now determined with the help of Fig. 10-15, which, like Fig. 9-28, correlates thrust capacity with anchor length and spacing but a rock friction angle $\phi = 55°$.

For $\Delta T_A = 137,500$ lb/ft, a suitable system of pattern reinforcement consists of 1-in-dia. anchor bars, 15 ft long and spaced at 5.6-ft centers. In quantitative terms, the amount of reinforcing steel for equivalent anchorage support is reduced by more than one-third when the friction angle is increased from 36° to 50°.

10-7 DESIGN OF AN ANCHORED WALL BY LIMITING EQUILIBRIUM METHODS

The anchored wall in this example consists of a soldier beam and lagging system with a reinforced concrete surface poured against the lagging after the anchors are installed, stressed, and locked off. Soil strength is expressed by the Coulomb–Mohr failure criterion, and failure is assumed to occur along a planar sliding surface. The soil is dense sandy silt with the following parameters: $\phi = 30°$, $\gamma = 110$ lb/ft^3, $K_a = \frac{1}{3}$, $K_0 = 1 - \sin \phi = 0.5$, and wall height $H = 33$ ft.

Total horizontal earth thrust is estimated for the K_0 condition as follows:

$$P_e = \tfrac{1}{2}\gamma K_0 H^2 = \tfrac{1}{2}(110 \times 0.5)(33)^2 = 29,950 \text{ lb/ft of wall (say, 30,000 lb/ft)}$$

With these data, the design can now be carried out according to the following steps (see also Section 5-11).

Step 1. Determine anchor vertical spacing. We will assume two rows of anchors, each row located at the $\frac{1}{4}$ point from the top and the bottom, as shown in Fig. 10-16. Each anchor can now be assumed to take a load component from the top or bottom of the wall to the midheight.

Step 2. Anchor horizontal force. We select anchor loads based on uniform horizontal pressure distribution having a resultant equal to 30,000 lb. The horizontal reaction per anchor is, therefore, 15,000 lb/ft.

Step 3. Anchor inclination. We will consider the factors discussed in Section 5-5. In this example the anchors are to be installed in dense sandy silt so that suitable anchoring strata are close to the ground and a steep angle is not necessary. After taking into account possible influence from adjacent utilities, foundations, and wall drawdown induced by the anchor vertical component, we consider an angle of 30° with the horizontal a suitable anchor inclination.

Step 4. Determine anchor load. Given the anchor inclination of 30°, the anchor load is 15/cos 30° = 17.3 kips/ft for both the upper and the lower rows. Next, we select a soldier beam spacing of 8 ft, since this will allow

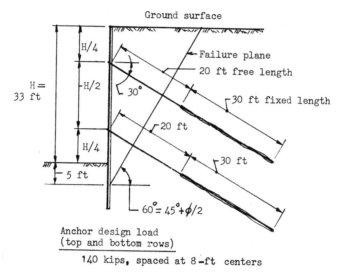

Ground surface

H/4

H = 33 ft

H/2

30°

H/4

5 ft

60°= 45°+φ/2

Failure plane

20 ft free length

30 ft fixed length

20 ft

30 ft

Anchor design load
(top and bottom rows)

140 kips, spaced at 8-ft centers

Fig. 10-16 Wall section and anchor details.

standard lagging lengths and practical beam sizes. The required anchor load (design) is, therefore

$$17.3 \times 8 = 139 \text{ kips} \quad (\text{say, } 140 \text{ kips})$$

At this stage it is necessary to calculate the vertical anchor component to check its effect on wall design. This is $140 \times \sin 30° = 70$ kips. The design of the wall must consider this component and ensure that this load is effectively resisted at the toe of the soldier beam by direct bearing or other suitable means.

Step 5. Determine fixed anchor length. Since the sand is fine-grained but dense, we can assume that cement grout injected under pressure will most likely produce type *B* anchor zone with essentially nonuniform diameter. From this consideration, we base the fixed anchor design on an effective hole diameter of 9 in. In the field, this dimension can be checked from grout intake and in conjunction with ground porosity. For a borehole nominal diameter of 4–6 in, this choice is consistent with the theory presented in Section 4-8. We now estimate $p' = 2 \times 25$, where 25 ft is the assumed height of overburden above the top of fixed anchor. By using Eq. (4-12) and a factor of safety of 2, the fixed anchor length is

$$L = \frac{2 \times 140,000}{50 \times 3.14 \times 9 \times \tan 30°} = 343 \text{ in} = 29 \text{ ft} \quad (\text{use 30 ft})$$

Step 6. Determine failure plane and anchor free length. Using $\phi = 30°$, the failure plane is assumed to begin 5 ft below bottom of final excavation and extending at 45 + $\phi/2$ = 60° as shown in Fig. 10-16. The length from upper anchor entry point to failure plane is 29.75 × sin 30° = 14.9 ft. Allowing 5 ft of penetration, the free length for the upper row is 20 ft. The same free length is used for the lower row to provide the minimum stressing length.

Step 7. Select tendon steel for design load 140 kips. Assuming that a 0.6-in-dia 270-ksi strand will be used, an anchor tendon consisting of 5 strands is selected. For a factor of safety of 2, the required steel area is 2 × 140,000/270,000 = 1.04 in². Each strand has a cross-sectional area 0.215 in², or total steel area provided = 5 × 0.215 = 1.07 in².

At this stage, anchorage design is satisfactory and complete, except for the anchor load capacity to be confirmed and verified by field tests. The design can now proceed with soldier beam and lagging selection.

10-8 DESIGN OF AN ANCHORED SLURRY WALL

The construction of a building will require an excavation 15–20 m (49–66 ft) deep. The architectural design provides that the three sides of the excavation will be braced by the basement floors. The fourth wall, however, is laterally unsupported; hence this wall will be permanently anchored.

Typically, the soil exhibits the following profile:

- Urban fill for approximately the upper 7-m (23-ft) layer.
- Moraine, below the depth of 5 m (16.5 ft).

The moraine is an overconsolidated mix of sandy gravel and cobbles in a silty matrix. From the soil investigation results it appears that the geotechnical parameters of the moraine are essentially constant within a layer, but exhibit a tendency to improve with increasing depth. The water table is not readily detected, but we will assume that some water may infiltrate the soil during protracted rains. A typical wall section is shown in Fig. 10-17.

Wall–Anchor System. The wall system is a reinforced concrete cast-in-place slurry wall 0.60 m (2 ft) thick and keyed 3 m (10 ft) below excavation level. The wall consists of independent 4.5-m-long (14.8-ft) panels excavated in one pass using round-end clamshells. The panels are separated by a 0.5-m gap to provide a gravel filter drain enclosed along the front by cast-in-place concrete cover as the panels are exposed. Four rows of high-capacity prestressed ground anchors are introduced, with two anchors along the same level per panel. Each panel is thus supported by eight anchors.

A service gallery is provided as shown in Fig. 10-17 between the building and the slurry wall, to ensure permanent access to the anchor heads for monitoring and anchor retensioning in the future. In order to prevent the

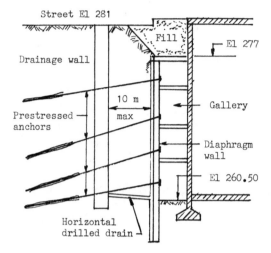

Fig. 10-17 Typical wall section, anchored diaphragm wall supporting a building excavation.

wall from becoming an impervious cutoff along the street (act like a dam), a suitable drainage system will be incorporated and include the following drain components: (a) longitudinal horizontal drain at street level, (b) vertical drains between adjacent wall panels, and (c) drainage walls placed about 10 m from the wall and connected by horizontal drilled drains to the main drainage system.

Soil Data. The following geotechnical parameters will be used:

- Fill: angle of internal friction $\phi = 30°$, cohesion $c = 0$, soil density $\gamma = 125$ lb/ft^3 (2 tons/m^3).
- Moraine: angle of internal friction $\phi = 35°$, cohesion $c = 0$, soil density $\gamma = 125$ lb/ft^3 (2 tons/m^3).

(*Note:* Unite conversion 1 ton/m^3 = 62.5 lb/ft^3 = 9.81 kN/m^3).

Active and passive earth pressure coefficients for the temporary condition (during construction) are as follows:

- Fill: $K_a = 0.33$, $K_p = 5.17$.
- Moraine: $K_a = 0.27$, $K_p = 7.57$.

The wall is also checked for long-term active pressure with $K_a = K_0 - 0.4$.

Wall Analysis. Computations are made for the six representative profiles shown in Fig. 10-18. Each profile corresponds to a particular construction and excavation stage, and for the applicable pressure diagram. For phases

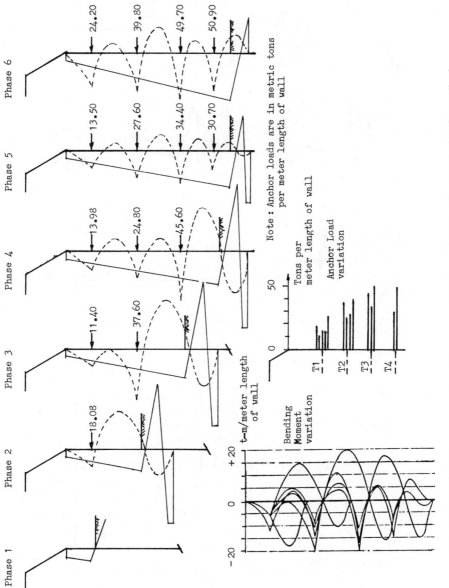

Fig. 10-18 Earth pressures, bending moment diagrams, and anchor loads for the wall of Fig. 10-17; six construction phases.

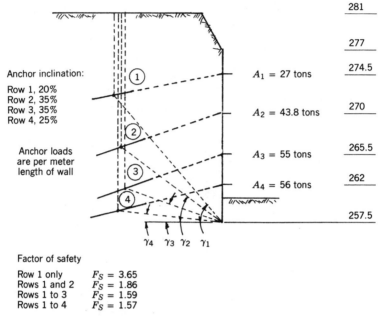

Fig. 10-19 Zone of ground engaged in interaction, and stability of the ground–wall–anchor system.

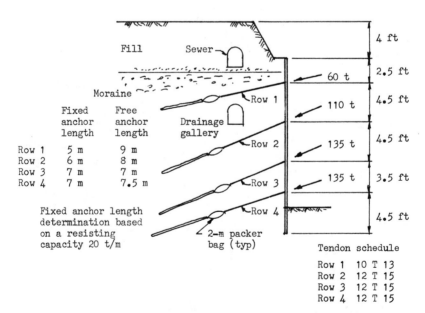

Fig. 10-20 Anchor layout and details; wall of Fig. 10-17.

TABLE 10-2 Technical Data; Anchors of Fig. 10-20

Row	Number of Anchors	Tension T_a (Tons Metric)	Tendon	Total Length of Anchor (m)
1	32	60–90	10T13 and 12T13	15–17
2	34	85–116	12T13 and 12T15	16–17
3	32	100–130	12T15	16.5–17
4	29	110–140	12T15	16.5–18

From Pfister et al (1982).

1 through 5, free earth support conditions are assumed at the base. Phase 6 is analyzed using fixed earth support for more realistic estimation of the key depth. All analyses are carried out using a suitable computer program.

Wall stability is checked according to the Kranz–Ostermayer method (discussed in Section 9-23) for one, two, three, and all four rows of anchors in place. Figure 10-19 shows a summary of the results, including the factor of safety against failure of the soil mass as it interacts with the wall–anchor system.

Anchor Design. Anchor details and layout are shown in Fig. 10-20. Anchor inclination is determined to accommodate the convenience of drilling and homing. Steeper anchor inclination is selected where it is necessary to avoid interference with existing buried obstacles and galleries.

From previous field testing, it is estimated that a resisting force of 20 tons/m (13.5 kips/ft) of fixed length will be mobilized using pressure grouting. The design is based on this criterion, but with 5 m (16.5 ft) specified as minimum. The anchor type selected is the IRP regroutable system discussed in Section 3-4. The tendon consists of 10–12 strands with diameter 13 mm (0.5 in) or 15 mm (0.6 in). Table 10-2 provides technical data for tendons at each anchor level.

Corrosion protection and proof load testing are specified according to Chapters 6 and 7.

10-9 PROCEDURE OF ANCHORED WALL DESIGN BY ELASTIC–PLASTIC METHOD

As mentioned in Section 9-21, elastic–plastic methods of analysis require successful interpretation of results from triaxial compression tests in order to obtain the stress–strain behavior of the soil. Practical limitations are thus imposed that, if properly represented, allow satisfactory results.

Elastic–plastic computer programs usually require additional input, which, from the user's point of view, constitutes a major difference from conventional methods of analysis. Thus, anchor prestress (lockoff) loads, anchor level, inclination, spacing, and stiffness must be included with the input data. The program thereafter calculates deflections, bending moments, shears, and reactions for the wall and the actual anchor loads. A preliminary design is, therefore, necessary using another method before an elastic–plastic analysis can be carried out with reasonable results expected. If the solution appears unsatisfactory (e.g., indicates excessive deflections, anchor load exceeding the allowable, or bending moments exceeding wall capacity), the design must be modified with a second run.

If used successfully, the elastic–plastic method can provide results closer to reality (what is actually measured on site by long-term monitoring programs). A further advantage is that it includes the actual wall and anchor stiffness; hence it provides a sensitivity analysis of the system taking into account all the parameters influencing its performance.

Example of Methodology. A wall–anchor system can be discretized into elements corresponding to different ground layers, projected excavation levels, anchor level location, and methods of lateral supports. Each level is represented by appropriate parameters such as

σ = friction angle

c = cohesion

K_a = coefficient of active pressure

K_p = coefficient of passive pressure

A_a = coefficient of active pressure (effect of cohesion)

A_p = coefficient of passive pressure (effect of cohesion)

K_0 = coefficient of pressure at rest

γ = soil total unit density

γ' = soil submerged unit weight

k_h = coefficient of horizontal subgrade reaction

The analysis for a given wall section begins from an initial phase when the overall system is under a set of specific conditions (e.g., a slurry wall in place with at rest equal pressures on both sides, or modified because of construction operations). The analysis is carried out for each of the following stages:

1. The first excavation stage but without anchors.
2. The first excavation stage with the anchors in place and stressed to a given level.
3. The second excavation level without the second row of anchors (computing deflections, bending moments, shears, rotation, and actual anchor load).

4. The second excavation level with the second row of anchors in place and stressed to the appropriate level.

5. The third excavation stage as in stages 3 and 4.

The anchor stiffness to be entered in the program usually is a tensioning force resulting in a unit length elongation for a tendon of a given elastic modulus, cross-sectional area, and apparent free length.

In this example, the elastic–plastic model is based on the concept that earth pressures acting at any point on the wall are dependent on the strain experienced by the soil as the wall deflects. For each soil element behind the wall, we assume that there exists a characteristic relationship between the ratio K of the horizontal to the vertical stress and strain or relative wall displacement. Such a relationship is shown in Fig. 10-21. The active and passive states constitute the limit cases of the pressure–strain relationship.

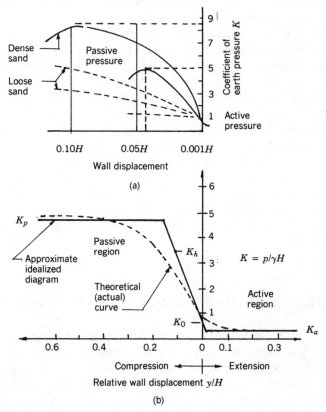

Fig. 10-21 Relationship between pressure and wall displacement: (a) measured data from Terzaghi and Rowe (1955); (b) typical idealized curve used for elastic-plastic analysis for medium dense sand. (From Pfister et al., 1982.)

Between these two conditions, the horizontal earth stress p_h acting on the wall can be approximated by a linear relationship with wall deformation y so that $p_h = k_h y$, where k_h (horizontal subgrade reaction) varies with soil type and strength. In practice, the value of k_h is quite difficult to establish on a rational basis since it depends on the width of the loaded area and on the modulus of subgrade reaction for a horizontal load.

Calculations are made by reiteration using a computer. Usually five iterations will suffice, with 20 reported as maximum. Computations of earth pressures are carried out independently on both sides of the wall. Thereafter, the two stress–strain relationships are combined, and hydraulic pressure is added separately. A system of four equations is developed, relating rotation, bending moments, shear forces, and earth reactions. Starting with the end conditions at the bottom of the wall, we calculate the displacement at the top of the wall, and then determine displacement and earth stresses at each wall level. If at a given point on the wall the displacement falls outside the zone of validity of the stress–strain coefficients, the latter are changed and a new iteration is started.

As the analysis is carried out possible modifications in the course of the program include changed excavation levels (one or both sides), changed water levels, changed external loads on both sides, backfilling, installation of a bracing system, and removal of a bracing system.

Interestingly, an elastic–plastic analysis is likely to yield greater anchor loads than the classical earth pressure method. However, the earth pressure distribution and wall deformation will match observed results better than any other solution, since elastic–plastic analysis considers actual wall stiffness and anchor elasticity. The method, however, is valid in soils that during excavation undergo primarily horizontal movement, such as sands and stiff clays. Elastic–plastic analysis, on the other hand, cannot account for significant vertical soil movement occurring in clay associated with bottom heave, unless the excavation is keyed in stiff bearing layer and base movement is diminished.

10-10 EXAMPLE OF LIMIT STATE DESIGN FOR ANCHORED WALL

Figure 10-22 shows a typical section for a single-anchored wall in loose and dense sand, with water level on either side as indicated. The load variation factor is 1.25, to be applied to the surface surcharge and earth pressure loads (increase for active and decrease for passive), to obtain the worst credible values. Table 10-3 summarizes pertinent material properties and soil parameters, including most probable and worst credible values. Free earth support is assumed for toe penetration.

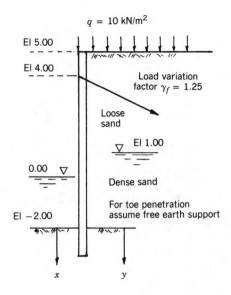

Fig. 10-22 Typical cross section; anchored sheet pile wall.

TABLE 10-3 Soil Data, Most Probable Values, and Worst Credible Values; Wall of Fig. 10-22

Material Property		Most Probable Values	Worst Credible Values
Steel strength		255 MN/m²	$\gamma_m = 1.15$
Steel modulus		207 × 10³ MN/m²	$\gamma_m = 1.00$
Loose sand	γ_L	17.2 kN/m³	18.5 max
			15.7 min
	ϕ	33°	29°
	δ	0.5ϕ	0.5ϕ
	K_a	0.26	0.305
	K_p	5.5	4.3
Dense sand	γ_d	20.6 kN/m³	21.5 max
			19.1 min
	ϕ	35°	31°
	δ	0.5ϕ	0.5ϕ
	K_a	0.23	0.29
	K_p	6.3	4.75
Water	γ_w	9.8 kN/m³	9.8

Step 1. Calculate total pressures:

			Most Probable Values	Worst Credible Values
Active pressure				
At level $+5$	10×0.26	$=$	2.6	3.2
At level $+1$	$(10 + 4 \times 17.2) \times 0.26$	$=$	20.5	25.6
At level $+1$	$(10 + 4 \times 17.2) \times 0.23$	$=$	18.1	22.6
At level $\ \ 0$	$[10 + 4 \times 17.2 +$			
	$(20.6 - 9.8)]$			
	$\times 0.23 + 9.8$	$=$	30.4	37.6
Below -2			$55.0 + 12.3x$	$63.9 + 13.2x$
Passive pressure				
At level -2		$=$	19.6	19.6
Below -2	$x (20.6 - 9.8) 6.3 + 9.8$			
	$(2 + x)$	$=$	$19.6 + 77.8x$	$19.6 + 54.0x$

where x is measured from level -2.

Step 2. Design for ultimate limit state. First we calculate forces using worst credible resistance and worst credible load effects, and then take moments about level $+4$ (anchor level). Note again that x is measured downward from level -2:

Active force $= 189.9 + 63.9x + 6.55x^2$ (kN/m)
Moment $= 713.4 + 63.9x(6 + x/2) + 6.55x^2 (6 + 2/3x)$
Passive force $= 19.6 + 19.6x + 26.9x^2$ (kN/m)
Moment $= 104.5 + 19.6x(6 + x/2) + 26.9x^2 (6 + 2/3x)$

For equilibrium active moment = passive moment, from which we obtain toe embedment $x = 3.2$ m below level -2. For ultimate limit state design, however, Eq. (9-70) applies. If we select a performance factor $\gamma_p = 1.25$ so that $M_P = 1.25M_A$, we obtain $x = 3.9$ m.

Step 3. Anchor load. The horizontal anchor load is estimated as the difference between active and passive force, and for $x = 3.2$ (equilibrium penetration). Thus for $x = 3.2$, anchor load $= F_A - F_P = 103.3$ kN/m. The level of zero shear is estimated from statical analogy at -0.4, and the maximum vertical bending moment is calculated at 256 kN-m. Anchor spacing (horizontal) is estimated by considering the following:

$$\frac{f_{pu}}{\gamma_m} \geq \text{(design load effect)} \times \gamma_p$$

where design load effect = (anchor load/net cross-sectional area) and f_{pu} = 255 MN/m². For anchor rod diameter 50 mm, we obtain anchor spacing s = 2.7-m centers.

REFERENCES

Bischoff, J. A., and J. D. Smart, 1977: "A Method for Computing a Rock Reinforcement System Which is Structurally Equivalent to an Internal Support System," *Proc. 16th Symp. Rock. Mech.*, ASCE, Univ. Minnesota, pp. 179–184.

Druss, D. L., 1984: "Development and Applications of the New Austrian Tunneling Method," Thesis, Univ. Pittsburgh.

John, M., 1980: "Arlberg Expressway Tunnel," *Tunnels and Tunneling* (April, May, June, July).

Kovari, K., and P. Fritz, 1977: "Stability Analysis of Rock Slopes for Plane and Wedge Failure with the Aid of Programmable Calculator," *Proc. 16th Symp. Rock Mech.*, ASCE, Univ. Minnesota, pp. 25–31.

Littlejohn, G. S., and C. Truman Davies, 1974: "Ground Anchors at Devonport Nuclear Complex," *Ground Eng.* (Nov.).

Pfister, P., G. Evers, M. Guillaud, and R. Davidson, 1982: "Permanent Ground Anchors, Soletanche Design Criteria," *Fed. Hwy. Admin., U.S. Dept. Transp.*, Washington, D.C.

Picknett, J. K., D. L. Gudgeon, and E. P. Evans, 1974: "Ground Anchored Sea Walls for Thames Tidal Defenses," Conf., London.

Rabcewicz, L. V., 1969: "Stability of Tunnels Under Rock Load," *Water Power* (July), Part II, 266–273.

Rowe, P. W., 1955: "A Theoretical and Experimental Analysis of Sheet-Pile Walls," *Proc. Inst. of Civ. Eng.* London, Part 1, pp. 32–69.

Terzaghi, K., 1955: "Evaluation of Coefficients of Subgrade Reaction," *Gest*, 5, No. (4), pp. 297–326.

CHAPTER 11

OBSERVED PERFORMANCE OF ANCHORED STRUCTURES

11-1 MONITORING GUIDELINES

Long-term monitoring tests to check service behavior of production anchors and ensure their performance are discussed in Section 7-11 together with provisions for collecting data on loss of prestress, creep displacement with time, and changes in ground characteristics.

For the general structure–anchor system it is mandatory to quantify the accuracy of the design by measuring loads, displacements, and pressures or stresses as a function of time. In certain cases, long-term behavior will require extensive monitoring programs. In this regard previous experiences might be considered as individual case studies, but these should not provide a general basis for interpretation. Whereas the design is indirectly intended to predict the basic behavior of the structure–anchor–ground system, static calculations giving rise to a quantitative solution are based on assumptions considered approximate for several reasons.

For instance, assumed earth coefficients are results of representative soil tests but in many cases they can only be the best estimates or values extracted from tables. Where soil conditions exhibit variations or where soft cohesive soils are present, we can reasonably expect that parameters expressing cohesion or deformation modulus will deviate markedly from in situ values. On the other hand, results of analysis are often approximate since the method of analysis itself has inherent limitations and is approximate only. Deep anchored walls, for example, will register deformations and stress patterns that cannot be predicted on the basis of simplified assumptions and theories, and the same can be said about solid rigid structures or where the natural arrangement of the ground is disrupted or altered by construction operations.

Invariably, anchored structures exhibit behavior depending considerably on anchor performance, both individually and as a group. A sound approach to monitoring must therefore begin with measurements properly sequenced and originating with anchor load fluctuations.

Example of Monitoring Program. Figure 11-1 shows proposed arrangement and extent of monitoring for an excavation supported by a multianchored wall. The monitoring system consists of geodetic measurements for the terrain surface and the top of the wall, deformation measurements of the wall and the soil mass, pore pressure readings, and anchor load measurements.

For this example monitoring is recommended if one of the following conditions exists: (a) the anchored wall is more than 10 m deep and longer than 30 m in granular soil, or more than 7 m deep and longer than 20 m in cohesive soil; (b) the wall supports slopes where rigidity is the determining factor of the analysis; (c) where water table differences on either side exceed 5 m; (d) where soil conditions are uncertain with respect to strength and modes of failure; and (e) where movement or system failure is expected to have effects detrimental to surroundings.

Since time and sequence are of essence, measurements are specified for the following stages: (a) before commencing any field work; (b) at each excavation stage; (c) at least once 3 weeks after final excavation stage, and (d) as many times as will be necessary to establish deformations of structural interpretation, or during activities associated with anchor destressing and

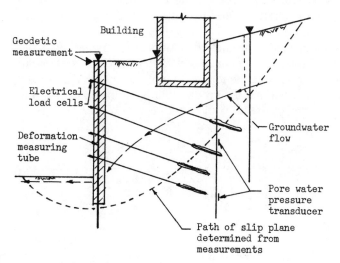

Fig. 11-1 Arrangement of monitoring system for an excavation support by multi-anchored wall.

restressing. Furthermore, if the anchors are permanent, measurements will be required four times during the first year after final excavation, twice yearly during the next 2 years, and once yearly in all subsequent years.

For the same example and in addition to the provisions discussed in Chapter 6, monitoring for corrosion protection will include reference anchors. These will be installed in the same manner and under the same conditions as product anchors, and with the same corrosion protection. Corrosion protection monitoring is enhanced if the reference anchors are of the removable or extractable type.

11-2   EFFECTS OF SINGLE-ANCHOR FAILURE IN A GROUP

If an anchor within a group of anchors fails, the wall will deflect locally and this process will affect the magnitude and distribution of the lateral earth pressures. A redistribution of the lateral earth stresses on the wall will result in the vicinity of the failed anchor, caused by arching in the horizontal and vertical directions and accompanied by an increase in the soil strength. However, only part of the shear strength of the soil is mobilized under working load conditions.

The local displacement of the wall due to failure of a single anchor will cause a local decrease in the lateral earth pressure as shown in Fig. 11-2, and its minimum value will be reached if the shear strength of the soil is fully mobilized ($F_s = 1.0$). Corresponding to this local decrease of the lateral earth pressure there will be an increase in the lateral earth pressure at the adjacent anchors, still in service. From these considerations we conclude that the wall stiffness is relevant to the entire process, and for a stiff rigid wall (e.g., a diaphragm wall) the earth pressure redistribution due to arching will be very small.

Considerable data on anchor failures are provided by Stille (1976) from four sites with anchors inclined at 45°. In a series of tests anchor failure was simulated by releasing the load in one of the anchors in the group, and then measuring the resulting load redistribution among adjacent units. Details of

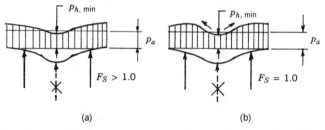

(a)   (b)

Fig. 11-2 Load redistribution on wale beam due to single anchor failure: (a) local decrease in lateral earth stress at failed anchor; (b) corresponding increase in lateral earth stress in adjoining anchors. (From Stille, 1976.)

the test program and the load changes are presented in Fig. 11-3. Interestingly, the total load on a wall panel decreased only where the actual loads were greater than the active earth pressure load. As failure of a single anchor was simulated, the resulting wall movement was very small so that the transfer of load to adjacent anchors was accomplished mainly by structural action rather than through soil arching. Based on the four test cases studied a logical recommendation was to have all top row anchors tested to 1.5 times the load estimated from Rankine's earth pressure theory, and test load the anchors of lower rows to at least 1.3 times the load calculated from the assumed earth pressure distribution. A further recommendation was to increase the horizontal stiffness of the wall. Interestingly, these results supplement the principles of plane strain discussed in Section 9-26.

11-3 LONG-TERM PERFORMANCE OF ANCHORAGE AT DEVONPORT

This project is described in Section 8-16 and reviewed as a design example in Section 10-1. The long term monitoring program focused on two principal aims: (a) investigate the actual anchor loads during basin dewatering and subsequent construction stages and (b) provide a case history of the long-term behavior of the permanent rock anchorage.

Instrumentation Details. Ten anchors were selected for monitoring. Anchor data are given in Table 11-1, and anchor location is as shown in Fig. 8-35. With the exception of anchors 49 and 51, all test anchors were previously stressed by multistrand jacks and therefore were destressed prior to installation of the instrumentation. Each anchor group was also connected to an inclinometer station so that any wall movement could be analyzed and correlated with anchor performance (Littlejohn and Bruce, 1979).

Load measuring devices consisted of vibrating wire load cells, each cell containing three sensing elements arranged at 120° intervals and sampled by means of a portable battery-powered meter. Following removal of the original anchor head plate, where necessary, the surface of the load distribution plate was thoroughly cleaned with a wire brush.

Anchor Stressing. Each anchor was stressed using a monojack. Stressing was carried out in four equal increments per strand in a sequence developed to ensure uniform loading of the cell. In general, the load–extension curves of individual strands were parallel as expected, but at lockoff the total extensions displayed up to 15 percent deviation from the mean, probably reflecting frictional and lockoff losses in the main.

For anchors 220 and 276 the stressing records indicated that one strand from each tendon had experienced grout–steel failure. Both strands, however, were located along part of the tendon circumference, which, owing to

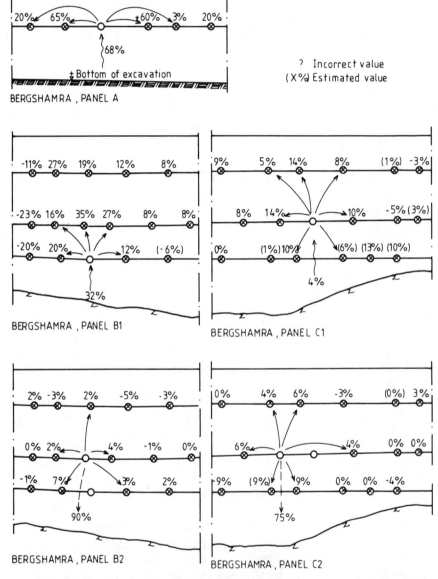

Fig. 11-3 Redistribution of anchor load among adjacent group of anchors, after failure of a single anchor. (From Stille, 1976.)

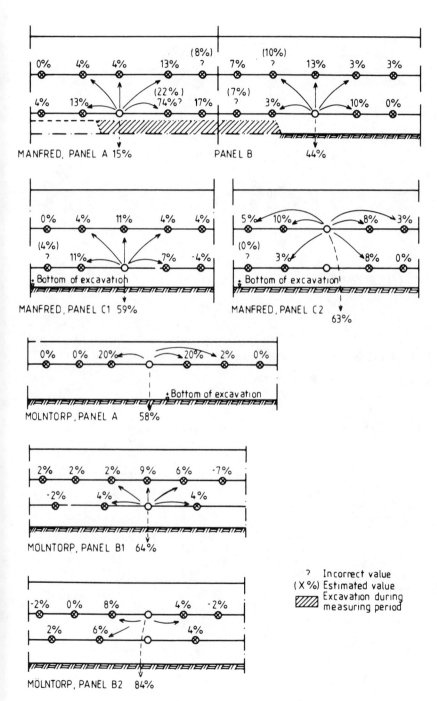

Fig. 11-3 (*Continued*)

TABLE 11-1 Anchor Data; Monitoring Program at Devonport

Anchor No.	Load Cell No.	Wall	Anchor Head Spacing (m)	Anchor Free Length (m)
49	1	West	1.2	26.33
50	2	West	1.2	29.16
51	3	West	1.0	27.16
219	4	North	1.0	27.21
220	5	North	1.0	29.21
221	6	North	1.0	27.21
222	7	North	1.0	29.21
274	8	East	2.0	22.69
275	9	East	2.0	22.69
276	10	East	2.0	22.69

From Littlejohn and Bruce (1979).

TABLE 11-2 Initial Service Loads after Proof Stressing; Anchors of Table 11-1

Anchor No.	Initial Load (kN)	Initial Tendon Stress ($\% f_{pu}$)	Remarks
49	2337	65	Not previously stressed (multijack)
50	2180	61	Previously stressed (multijack) to 2000 kN for 3100 h
51	2172	60	Not previously stressed (multijack)
219	2094	58	
220	2085	63	11 strands effective
221	2110	59	All north wall anchors previously stressed (multijack) to 2280 kN for 1150 h
222	2384	65	
274	2410	73	11 strands effective
275	2250	63	All east wall anchors previously stressed (multijack) to 2280 kN for 1200 h
276	2114	59	

From Littlejohn and Bruce (1979).

borehole inclination, could establish intimate contact with the borehole wall during homing; hence they were readily exposed to contamination resulting in debonding.

A summary of the initial loads for each anchor is shown in Table 11-2. The period of installation, instrumentation and stressing of the 10 anchors was 4 days.

Service Behavior of Monitored Anchors. Up to 24 h, the load on each anchor was recorded at approximately hourly intervals after proof stressing. Rather anomalous patterns were noted for anchors 49, 51, and 274, where the load actually increased by 7.3, 0.6, and 2.9 percent, rspectively.

The period up to 4500 h (approximately 27 weeks after final stressing) covered the critical stage of basin dewatering. Figure 11-4 shows load time records for anchors 50 and 51. The beneficial effect of prestressing (2000 kN for 3100 h, as shown in Table 11-2) on the subsequent performance of anchor 50 is clearly demonstrated (Littlejohn and Bruce, 1979). For the same period wall movements were measured by inclinometers. During dewatering, when 13 m of water head was removed, monitoring for a period of one month indicated that the walls inclined inward from the base to give displacements from 5 to 50 mm at crest level over wall depths of 18–30 m.

For the time period up to 3300 h, the records were essentially influenced by site activities. Following dewatering each cell was read at monthly intervals until 10,000 h after stressing. Thereafter readings were taken once every 4 months up to 18,000 h. A final set followed at 33,000 h (196 weeks). The long-term records for the west wall anchors are shown in Fig. 11-5.

Evaluation of Results. For the first 24 h, 7 of the 10 anchors displayed behavior that is within predictable and widely recorded patterns, specifically, a relatively rapid initial load loss but progressivey reducing in rate. This loss is typically due to relaxation characteristics of the tendon, but also represents movement associated with the anchor-head assembly. Interestingly, the immediate performance of three anchors, displaying an actual load increase, represents an anomaly that remains unexplained.

For the period up to 4500 h and during the dewatering phase, anchor monitoring revealed no major load fluctuations. Any load variation recorded was well within the limits attributed to the influence of ambient-air temperatures on the load sensing devices. It follows therefore that any major change in external loading in the basin on dewatering caused negligible load change in the wall anchors.

From Fig. 11-5, it is evident that the long-term behavior (up to 33000 hours) is characterized by two distinct phases identified in the load–time curve. Initially a rapid loss phase occurs, to be followed by a slower and more uniform reduction of prestress. In six cases the amount of load lost in the initial phase (≤18 weeks) was about 85 percent that measured at the end of the 2-year period. Furthermore, the monitoring confirms that restressing

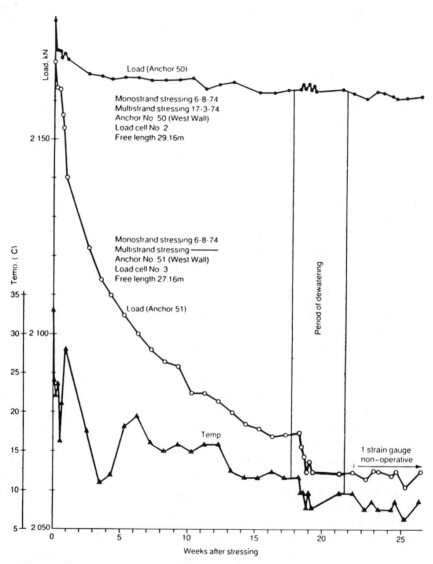

Fig. 11-4 Anchor performance up to 4500 h; monitored anchors 50 and 51, Devonport project. (From Littlejohn and Bruce, 1979.)

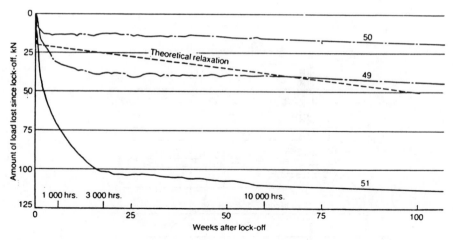

Fig. 11-5   Long-term performance of west wall anchors up to 18,500 h. (From Littlejohn and Bruce, 1979.)

tendons after a certain period reduces the subsequent prestress loss due to relaxation.

11-4   SURVEILLANCE PROGRAM TO CHECK ROCK SLOPE STABILITY

For the rock slope stabilization work of the south portal section of Schallberg tunnel, Switzerland, the monitoring program specified the following:

1. All individual anchors to be used in the surveillance system were installed with the free-length remaining elastic to enable assessment of anchor behavior under the effect of applid load.
2. Two extensometers, 25 and 34 m long, respectively, were installed above and below the road to detect movement of the slope and check rock slope response to construction procedures. These are marked as locations L9 and L12 in Fig. 11-6.
3. For two test anchors with relatively short fixed (bond) lengths, the effective bond resistance at the rock–grout interface was checked by full pullout tests. The boreholes were drilled by rotary coring 116- and 125-mm diameters, respectively.

Results of movement measurements from the extensometers are plotted in Fig. 11-6 as a function of time and for the two measuring positions. For each region the anchor lengths are shown diagrammatically at the left. Movement is typically referred to the deepest measuring point of the extensometers, which is assumed stationary. In the axial direction of the borehole

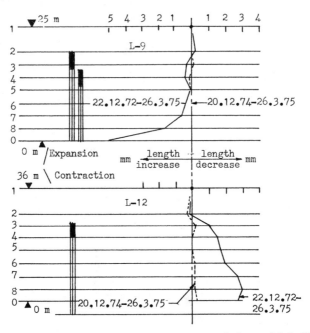

Fig. 11-6 Movement curves for two extensometers; rock slope of Schallberg tunnel. (From VSL–Losinger, 1978.)

movement is plotted normal to the relevant measuring point and these points are then connected together. For the arrangement shown the parts of the curve inclined downwards and to the left indicate expansion.

The solid line represents movement from the start of the measurements as a composite term, whereas the broken line shows the change of movement that occurred between the last two readings.

Figure 11-7 shows movement of the rock surface with time and with reference to the assumed stationary points within the two extensometers. The start of the measurements is December 1972. Expansion movement is indicated by downward plot, and contraction is plotted upward. For the rock surface at the lower measuring point, the curve shows uniform expansion movement until the application of prestressing on day 175. The effect of prestress is considerable contraction as indicated by the steep upward plot, which was almost completely retained until the end of monitoring. A slight expansion detected in the last phase of the program is associated with swelling following heavy snowfall and the resulting saturation.

For the upper section of the slope (monitored by extensometer L9) rock behavior is somewhat different as is shown by the dashed line in Fig. 11-7. The initially stable pattern is probably the result of temporary stabilization of the slip surface in the vicinity of the measuring device. The superimposed expansion developed in conjunction with progressing construction and ac-

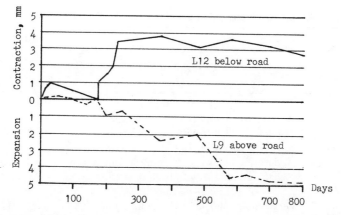

Fig. 11-7 Movement of rock surfaces with time, rock slope of Schallberg tunnel. (From VSL–Losinger, 1978.)

celerated markedly following blasting and loosening of the foundation excavation on the uphill side, as indicated by the steep downward plots. After 575 days movement remained essentially unchanged indicating the stabilizing effect introduced by the anchoring.

11-5 FIELD STUDY OF SPILING REINFORCEMENT FOR ROCK TUNNEL

The behavior of rock tunnel with spiling reinforcement (see also Section 9-9) has been investigated by Korbin and Brekke (1978), and documented in two field studies involving a pilot tunnel for the Burlington Northern Railroad and the Eisenhower Memorial Tunnel on interstate high I-70 west of Denver. More particularly, the field program was developed to focus on topics related primarily to the magnitude, distribution, and time history of deformation-induced tension and bending of spiles resulting from rock excavation.

The Eisenhower Tunnel consists of dual two-lane vehicular openings with maximum overburden 1450 ft (442 m). Opening dimensions are 46 ft (14 m) wide × 40 ft (12 m) high as excavated. Ground conditions along the tunnel alignment show marked variations ranging from massive granite to wide zones of squeezing fault gouge. Rock mass classification is based on joint spacing and degree of alteration (Ranken and Ghaboussi, 1975), from which four rock classes are specified as shown in Table 11-3 (Korbin and Brekke, 1978). Each class is further divided into two categories, indicative of rock quality for each group. An estimate of probable rock loads is also included for comparison with measured loads. These data are formulated for top heading width $B = 46$ ft and height $H_t = 23$ ft.

For top heading and bench excavation, rounds within the top heading were kept from 4 to 8 ft (1.2–2.4 m) long. Spiling reinforcement with the

TABLE 11-3   Rock Class Description and Predicted Rock Loads[a]

Rock Class (1)	Description (6) (2)	Rock Load H_p (6), in feet (meters) (3)
	(a) 0 B–0.25 B	
Ia	Massive to slightly blocky, having a joint	0–12
Ib	spacing > 1.0 ft with no alteration.	(0–3.7)
	(b) 0.13 B–0.35 (B + H_t)	
IIa	Moderately blocky and seamy having a joint	6–25
IIb	spacing > 0.5 ft with little or no alteration.	(1.8–7.6)
	(c) 0.18–1.10 (B + H_t)	
IIIA	Very blocky and seamy, having a joint	12–76
IIIB	spacing < 1.0 ft, moderately to highly	(3.7–23)
	altered with zones of moderate to intense	
	shearing	
	(d) 1.10–2.10 (B + H_t)	
IVa	Squeezing (low to moderate) ground, highly	76–145
	crushed and altered, nonplastic abundant	(23–44)
	clay, joint spacing < 0.5 ft	
	(e) 2.10–4.50 (B + H_t)	
IVb	Squeezing (moderate to high) and swelling,	145–311
	plastic highly altered, mainly clay gouge	(44–95)

Notation: $B = 2 H_t$; height $H_t = 23$ ft (7 m), 1 ft = 0.305 m.
From Korbin and Brekke (1978).

axial pattern shown in Fig. 11-8 was specified for types III and IV ground. Spacing along the circumference of the opening ranges from 2.5 to 5 ft. Radially placed bolts were included for rock type IV and in addition to the diagonal spiles. Occasionally a small stretch of ground type II was reinforced with pattern spiling reinforcement but with twice the axial spacing. Spile and bolt length is 16 and 12 ft, respectively, and both consist of No. 11 reinforcing bars. With the use of percussive drills the reinforcement was placed in predrilled holes filled with bags of polyester resin. The support system within the top heading was completed with the installation of steel sets spaced at 4-ft centers, and a first-stage concrete liner typically placed one month after excavation. Contact grouting was carried out approximately one month after lining.

Test Stations. A total of six test stations consisting of one to five instrumented bars (spiles and bolts) and a strain-gauged steel set were established in ground types Ib to IVa. Following the collection of extensive geologic data it was possible to identify the location of various ground types prior to construction and install the test station in critical areas such as blocky and squeezing rock.

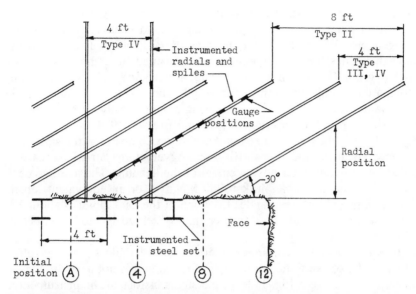

Fig. 11-8 Profile of test station instrumentation arrangement; initial position represents point of instrumented spile installation; circled numbers indicate location of tunnel face relative to initial position measured in feet. (From Korbin and Brekke, 1978.)

Test Results. For each test station three radial distributions of axial stress were obtained at different times and positions of the advancing face. The first distribution was produced 8 ft beyond the initial position (Fig. 11-8). This stage represents the immediate peak response of the reinforcement in terms of deformation-induced tension on excavation (see also Section 9-9).

With continuing advancement of the heading, stress relaxation occurred within the reinforcement, most of which had taken place by the time the face advanced 50 ft from the initial position. A second distribution was obtained at this point. A final distribution was derived from the last set of data recorded at each station, representing the long-term behavior of the reinforced rock mass.

Radial stress distributions are derived from the reinforcement stress–time history by employing a representative function in a least-square routine. Best fit distributions are obtained at different times and positions of the advancing face. Two points are important on the curves, the peak and the inflection point (IP). The peak stress occurs at a given radial position from the opening, and represents the point at which the response of reinforcement to rock deformation is maximum. If we assume a monotonically increasing rock mass strain with decreasing radial distance from the opening, the inflection point marks the breakdown in the compatibility between rock mass and spile strains. For any distance from the opening greater than the inflection point the rock mass–spile system is a continuum, whereas in the zone between the inflection point and the tunnel opening system behavior is essentially inhomogeneous. Within the latter the rock mass displaces farther into the opening than does the reinforcement, accompanied by some loss of rock strength to ground bond. On the other hand, if rock and reinforcement strains are compatible along the entire spile length, the radial stress distribution will follow the rock mass strain, increasing monotonically with decreasing radial position from the opening (Korbin and Brekke, 1978).

Figure 11-9(a) shows radial stress distribution comparison for station 63 + 01, located in ground types IIIb–IVa. A very large stress increase is recorded immediately after advancement of the heading. The average peak stress increase is more than 40 ksi (280 MN/m²), or an equivalent tensile force of 60 kips (270 kN) per spile. Location of the peak point is about 3 ft from the opening, considerably farther than observed in more competent ground.

As face advancement continues, there is considerable stress relaxation but close to the opening the percentage of relaxation is not as great as is at the peak position. From Fig. 11-9(b), the stress history of the instrumented steel set reveals an increase in load followed by stress relaxation as the reinforcement. If we assume that the peak stress drop of about 36 ksi in the

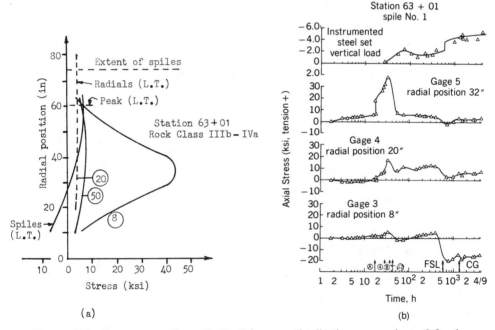

(a)

(b)

Fig. 11-9(a) Test station 63 + 01. Radial stress distribution comparison: 8-ft advance, 50-ft advance, and long term; (b) test station 63 + 01. Stress history of instrumented spile and steel set. Arrows indicate time at which round was advanced. (From Korbin and Brekke, 1978.)

spile steel was released as an equivalent weight of rock requiring support, the instrumented steel set should have shown a corresponding stress increase (about 12 ksi). Since, instead, the set reveals some degree of relaxation, the total stress relaxation in the reinforcement is not directly related to a total load increase within the support.

Instrumented radial bolts installed 50 ft behind the face exhibited a rather small stress increase in the long term, indicating a stable arch. These bolts would be more effective if they were installed closer to the face.

After placing the first-stage concrete liner [FSL in Fig. 11-9(b)], additional relaxation of the ground nearest the tunnel was observed. Strain gauges closest to the opening recorded compression of the reinforcement as the ground loaded the support system. As expected, the magnitude of this relaxation corresponds roughly to the load increase registered within the internal support.

We can conclude from these results that the pattern and extent of radial stress distributions are dependent upon the ground type. In this instance major ground activity occurred within a radius from the opening, and decreased with improved ground type. The force developed within the reinforcement in response to excavation is likewise related to ground type. This relation can be formulated if we present the results in terms of an equivalent pressure, defined as the peak spile force divided over the tributary area reinforced by the bar. As shown in Table 11-4, the lower the rock mass quality the greater the developed pressure. Development of spile force also depends on the time of installation as well as on associated constructional effects.

Interestingly, average bending stresses recorded in all types of ground were small relative to direct tensile stresses. During excavation under spile

TABLE 11-4   Reinforcement System Peak Equivalent Pressure

		Peak Equivalent Pressure, in kips per square foot (kilonewtons per square meter)	
Test Station (1)	Rock Class (2)	8-ft (2.4-m) Advance (3)	Long Term (4)
58 + 86	Ib–IIa	1.9 (91)	1.5 (72)
73 + 42	Ib–IIb	3.1 (148)	2.5 (120)
70 + 14	IIb–IIIa	3.7 (177)	3.0 (144)
72 + 36	IIIb–IVa	6.2 (297)	6.0 (287)
63 + 01	IIIb–IVa	13.2 (632)	1.8 (86)

From Korbin and Brekke (1978).

cover the bending stresses could indicate a concave-deflected shape facing downward. As excavation advanced, the reinforcement tended to revert to the undeflected shape.

Reinforced Arch-Support System Interaction. Stress relaxation may be converted to support load by the equivalent weight of a rock column of height H_p. The rock load H_p is the spile relaxation stress resolved into a vertical force divided by the tributary area reinforced by the spile and the unit weight of rock. Table 11-5 shows a comparison between the calculated rock load H_p (derived from the instrumented steel set and the measured spile stress relaxation) for two critical occasions: before first-stage concrete liner (FSL) and in the long term. Evidently the results show good correlation. In competent ground steel sets received no additional load after liner installation. Ground types IIIb–IVa, however, indicate increased loads irrespective of the stabilizing effect of the support–reinforcement system. This additional loading reflects the nature and quality of the rock mass as well as the stiffness of the support.

Since no direct method is available to measure the capacity of the reinforced arch, we can estimate this capacity assuming that all load measured

TABLE 11-5 Measured, Predicted, and Anticipated Rock Loads on Internal Supports

| | | | Rock Load H_p, in feet (meters) | | | | Spile Stress Distribution Long Term, in feet (meters) | |
| | | | Steel Set | | Spile Relaxation | | | |
Test Station (1)	Rock Class (2)	Anticipated[a] (3)	Before First Stage Concrete Liner[b] (4)	Long Term[b] (5)	Before First Stage Concrete Liner[c] (6)	Long Term[c] (7)	Peak Point (8)	Inflection Point (9)
58 + 86	Ib–IIa	6 (1.8)	0.9 (0.27)	0.9 (0.27)	0.7 (0.21)	1.1 (0.34)	1.3 (0.40)	2.3 (0.70)
63 + 01	IIIb–IVa	60 (18)	6.1 (1.9)	17.4 (5.3)	9.8 (3.0)	17.4 (5.3)	5.0 (1.5)	5.5 (1.7)
70 + 14	IIb–IIIa	12 (3.7)	2.1 (0.64)	—	0.0	—	1.7 (0.52)	2.7 (0.82)
72 + 36	IIIb–IVa	60 (18)	5.1 (1.6)	8.6 (2.6)	4.1 (1.3)	7.9 (2.4)	4.3 (1.3)	5.8 (1.8)
73 + 42	Ib–IIb	9 (2.7)	1.8 (0.55)	1.8 (0.55)	1.9 (0.58)	1.4 (0.43)	1.8 (0.55)	2.6 (0.79)

[a] From Table 11-3.
[b] Measured from instrumented steel sets.
[c] Predicted from measured spile relaxation.

From Korbin and Brekke (1978).

on the internal support is reasonably the load that the reinforcement system cannot sustain. The difference between measured rock loads on the supports and the magnitude of the anticipated loads is an indirect indication of the reinforced arch capacity.

Indication of the overall effectiveness of spiling reinforcement (and, by analogy, system capacity) emerges if we compare measured and anticipated loads in Table 11-5, the latter based on the lower bound values derived from the application of Terzaghi's classification system (Table 11-3). On the average, long-term measured loads are less than 25 percent the anticipated loads, whereas loads on the steel sets along (before FSL) are less than 15 percent. A basic conclusion is therefore that the rock mass–reinforcement system is the primary factor of the permanent stabilization of the tunnel opening. However, in view of the conservative nature of the assumptions the actual effectiveness of the arch may be overemphasized, although it does not alter the basic conclusion.

Design Concepts. From the foregoing comprehensive field study, Korbin and Brekke (1978) set forth basic design concepts. It appears that the formation of a stable reinforced rock arch is a self-equilibrium process engaging a specific zone within the immediate vicinity of the opening. The thickness requirements of the arch depend on the ground type and the construction process, and relate to a lesser degree to the opening size, shape, and depth.

Arch capacity, as developed by the forces induced in the reinforcement, is influenced by opening size, shape, and initial state of stress. Increased capacity is possible through an increase in the size or number of the reinforcement bars. Interestingly, the rationale of this process with the rock yielding theories discussed in Section 9-11 is obvious. Thus, in the interaction between ground and tunnel support time is a prime factor.

Deformation-induced tension within the spiling reinforcement is the prime mechanism by which this element contributes to the initial as well as the permanent stabilization of the opening. On the other hand, bending stresses resulting from excavation are negligible compared to tensile stresses. Hence, a major difference is obvious betwen spiling reinforcement and forepoling. Distributions of stress or strain increase are similar in form to bell-shaped probability distributions. However, positions of the peak and inflection points are strongly dependent on ground type: the more competent the ground, the closer these points are to the opening.

11-6   FIELD STUDIES ON FREEZING AND THAWING EFFECTS

An example of wall anchored in ground which may become partly frozen is given by McRostie and Schriever (1967), and involves the excavation for the National Arts Center in Ottawa. This excavation was exposed for two winters, and observations were made on wall and anchor behavior under frost action. Average excavation depth was 15 m (49 ft), with the upper 7 m in soft ground and the remainder in rock. Because of local reported failures due to frost pressure, the wall was protected by enclosing the heat for the entire length except for a short test section.

For the three monitored anchors, horizontal movement of the upper ends caused by frost action reached 22 mm at the time when the maximum frost penetration was 1.2 m (4 ft). Interestingly, the anchor loads were only 15 to 25 percent the values that should be expected if all the structural units in the anchor and the soil behaved elastically. A summary of the measured and the calculated stress levels in the tendons is shown graphically in Fig. 11-10. Although the investigators have suggested that these differences are due primarily to creep, it is interesting to note that average lateral earth pressures on the wall were in some cases only 50 percent the soil pressures predicted in the design. Noting that the frost pressures are higher if anchor yield is very small, it was concluded that both frost pressures and earth pressures can be increased or decreased by preventing or permitting lateral displacement of the support, respectively.

Interesting measurements of anchor loads on 300 units in various anchorage installations in Sweden are reported by Sahlstrom (1969). Results for two anchored sheet pile walls are shown in Fig. 11-11. The load–time diagrams reveal that anchor loads reach maximum values during the winter, but revert to minimum values in the summer. In general, the maximum loads are not caused only by combined earth and water pressures, but abnormal forces can be induced also by alternating freezing and thawing.

The anchored retaining wall shown in Fig. 11-12(a) supports ground consisting of clay and moraine. During the fall season the groundwater flow is quite heavy. During the winter season 1968–1969 anchor loads increased more than 200 percent from the calculated values. Some anchors failed, and

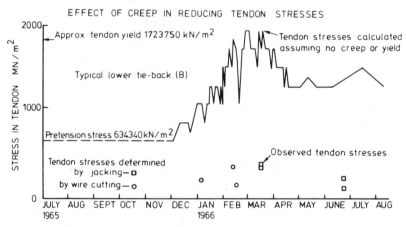

Fig. 11-10 Comparison of estimated and measured anchor stress. (From McRostie and Schriever, 1967.)

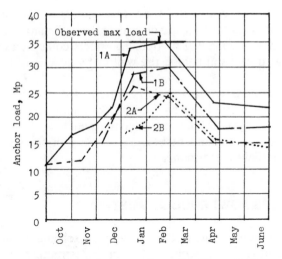

Fig. 11-11 Anchor loads in sheet pile walls. (From Sahlstrom, 1969.)

the wall came close to collapse. However, the broken anchors were immediately replaced and failure was prevented. The problem of overloading
was probably caused by the freezing effects shown in Fig. 11-12(b). A remedial measure was the introduction of heat along the sheet pile wall aimed
at reducing the freezing effect. After 24 h of heating, anchor loads were
reduced by about 50 percent from the value before heating.

Examination of boreholes behind the sheet pile wall revealed that the
frozen zone was 5 ft thick, whereas the frozen zone under the surface was
only 2 ft under the beneficial effect of snow cover.

Further Swedish experience of the frost problem is provided by Sandgrist

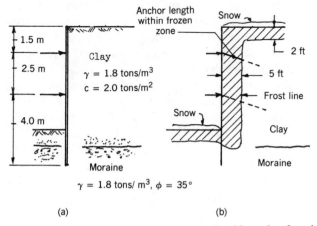

Fig. 11-12 (a) Section through sheet pile anchored wall subjected to freezing effects;
(b) extent of frozen zone. (From Sahlstrom, 1969.)

(1972) and Sandegren et al. (1972). Several failures of anchored sheet pile walls caused partly by freezing are also reported by Broms and Stille (1976). Presently, anchor overloading by freezing effects could be avoided by sufficient drainage of the soil behind the wall or by heating of the wall during the winter.

In northern Canada grouted anchors in permafrost are becoming routine experience. Work carried out by Ladanyi and Johnston (1973, 1974) and Johnston and Ladanyi (1974) has provided the guidelines for creep settlement predictions and time-dependent bearing capacity of frozen ground to be checked against field performance.

11-7   THE KOPS POWERHOUSE, AUSTRIA

The Kops powerhouse (part of the hydropower scheme of the Voralberger Illwerke A.G. in Austria) is one of the earliest structures of this size for which the excavation was carried out with prestressed anchors, bolts and shotcrete as the only means of structural support (Rabcewicz, 1969). To confirm the adequacy of its design, an extensive monitoring program was developed and implemented in the field.

The dimensions of the main powerhouse are $28 \times 30 \times 70$ m ($92 \times 95 \times 230$ ft). Although the construction method is suitable for the in situ geologic conditions, unavoidable openings in the cavern walls for access and busbar tunnels and other connecting galleries to the distribution manifold were designed so as not to impair the stability of the system as one unit.

The powerhouse is located in sound, well-bedded amphibolite striking 50° to the axis with a dip 35°. The thickness of the bedding varies within narrow limits, and the average overburden is about 200 m (655 ft). Two joint systems are identified, the more pronounced striking 126° to the axis and dipping 75°. With few exceptions, the joints are well closed. However, in the lower sections of the powerhouse the rock exhibits a strong tendency to separate along bedding planes, although the presence of two small fault zones filled with clay and crushed rock crossing the structure at 90° and dipping steeply did not cause instability problems.

The excavation sequence, rock strengthening and lining installation were carried out according to the scheme shown in Fig. 11-13. In particular, the time sequence of the various working stages for the top excavation was given special consideration. Thus, the side galleries (phase 1) were driven almost simultaneously, and each new exposed rock area of the roof was immediately protected by expansion bolts (stage 1a) and covered by 10 cm of shotcrete (stage 1b) to be followed by prestressed Perfo bolts (stage 1c). The excavation of the center pillar (phase 2) followed 30–50 days later. Bolting (stages 2a, 2c) and shotcreting (stage 2b) were carried out on the same day with the side galleries.

The combined effect of prestressed bolts and shotcrete apparently re-

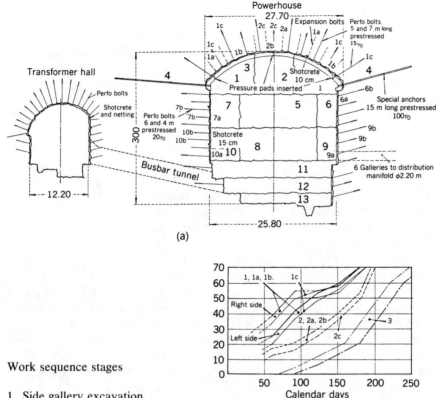

(a)

(b)

Work sequence stages

1. Side gallery excavation
1a. Side gallery strengthening by expansion bolts
1b. Side gallery strengthening by shotcrete (10 cm) and netting
1c. Side gallery strengthening by Perfo system bolts
2. Center pillar excavation
2a, 2b, 2c. Center pillar strengthening as 1a, 1b, 1c
3. Concrete roof arch (abutments, shuttering, and concreting in sections of 5 m)
4. Special anchors, 15 m long, prestressed to 100 tons
5. Bench upper part, center portion, excavation
6, 7. Bench upper part, side portions, excavation
6a, 7a. Bench upper part, side portions, strengthening by shotcrete (15 cm) and netting
6b, 7b. Bench upper part, side portions, strengthening by Perfo bolts
8. Bench, middle part, center portion, excavation
9, 10. Bench, middle part, side portions, excavation
9a, 9b, 10a, 10b. Bench, middle part, side portions, strengthening by shotcrete, netting and Perfo bolts
11. Bench, lower part, excavation and strengthening of walls by shotcrete, net, and Perfo bolts
12. Botton upper part, excavation and strengthening of walls by shotcrete, net, and Perfo bolts
13. Bottom lower part, excavation and strengthening of walls by shotcrete, net, and Perfo bolts

Fig. 11-13 Kops hydropower scheme; (a) cross section through the powerhouse and the transformer hall, also showing the connecting tunnel; (b) time schedule for the construction of the top section. (From Rabcewicz, 1969.)

651

sulted in a state of permanent equilibrium. However, since the degree of safety could not be established accurately it was decided to insert a relatively thin concrete roof arch (phase 3). This work was carried out by forming the abutments on both sides of the roof, followed by placing the shuttering and concreting the arch in sections 5 m long. The construction of the upper part of the cavern was completed by installing special anchors close to the abutments, 15 m long and prestressed to 100 tons (phase 4).

The core and the bottom section of the cavern were excavated in the sequence defined by phases 5 through 13, Fig. 11-13. The walls were likewise stabilized with prestressed Perfo anchors and shotcrete, including the end walls of the excavation. The time sequence required stabilization of one section before excavation of the lower section could be carried out. For the excavation of the surface of the walls the presplitting method was used to keep overbreak to a minimum.

The long-term monitoring program was based essentially on anchor load measurements. Anchors subjected to testing were 10 m long, 26 mm in diameter, and provided with electrical measuring strips (Rabcewicz, 1969). The program involved 93 anchors, 16 installed to support the roof, and the remaining anchors placed in the side and end walls. Pressure pads were also placed at the top and the sides of the roof arch. Interestingly, the measured stresses were very small. This, combined with the extremely small vertical displacement observed during monitoring, indicates that the portion of the rock load resisted by the concrete arch was very small.

Results of measurements show that the total settlement of the crown of the roof amounted to a maximum of 14 mm at one point, and averaged close to 3.2 mm elsewhere. Wall deformations showed a tendency for the sides of the cavern to move inward with an average movement of 5 mm and a maximum of 17 mm. The cavern approached full stability with all deformations gradually diminishing shortly after the excavation was completed.

11-8   BEHAVIOR OF ANCHORED WALLS IN LONDON CLAY

Neasden Underpass

The Neasden underpass in north London is largely in cut up to 10 m (33 ft) deep, with diaphragm walls supporting the excavation in stiff, silty, and highly fissured clay. At a typical section excavation depth is 8 m (26 ft) with the wall embedded 5 m (16.5 ft) into the stiff clay. At the site London clay exists to a depth of about 30 m (98 ft) and is underlain by the much stronger Woolwich and Reading beds which have an undrained shear strength 300 kN/m^2 (about 6260 lb/ft^2). Average undrained shear strength for the clay is 100 kN/m^2 close to the ground surface, and increases to nearly 200 kN/m^2 (4200 lb/ft^2) at a depth of 25 m. At the average fixed anchor length the undrained shear strength is 150–175 kN/m^2. This range is essentially the same as measured by Hooper and Butler (1966) for London clay.

All anchors have underreams (see also Sections 3-3 and 4-9). In order to establish the working load 10 test anchors were used. These were constructed vertically and with the number of underreams varying from seven (for 500-kN working load) to three (for 200-kN working load). All multibell anchors have shaft diameter 175 mm, underream diameter 535 mm, and underream spacing 1150 mm. Typical test results indicated an initially linear load–displacement curve with yielding beginning at a load approximately proportional to the number of underreams (Sills et al., 1977). Interestingly, soil characteristics and anchor details are essentially similar to the data analyzed in Section 4-9 and shown in Table 4-11 and Fig. 4-28(b) (Bastable, 1974). For the same soil strength and underream dimensions, we can estimate the factor of safety associated with the seven- and the two-underream anchor. From a backanalysis of the data shown in Fig. 4-28(b) and Table 4-11, ultimate loads are 650 and 1600 kN for the two- and the seven-underream anchor, respectively, giving factors of safety of 1600/500 = 3.20 and 650/200 = 3.28.

A typical wall section with anchor location and instrumentation details is shown in Fig. 11-14. The diaphragm wall is 0.6 m (2 ft) thick and is braced with four rows of anchors. Prior to tensioning to working load, each anchor was loaded to 77 percent f_{pu} for a 5-min period. After reloading the load at each strand was checked and the anchor destressed. Thereafter, anchors were restressed to 115 percent the design working load, and checked 24 h later for relaxation losses. If these exceeded 5 percent, the load was restored and rechecked 24 h later before lockoff.

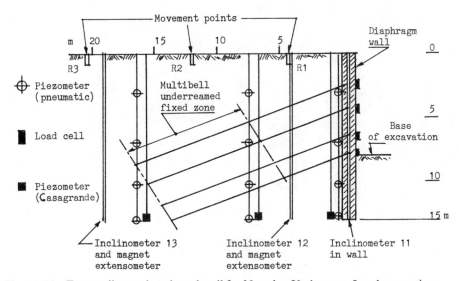

Fig. 11-14 Excavatiion and anchored wall for Neasden Underpass, London; section showing anchors and instrumentations. (From Sills et al., 1977.)

Excavation Sequence. The slurry trench for the diaphragm wall of the test section was excavated in January 1972, and the concrete wall was in place in the same month. The interior excavation for the underpass was carried out in stages as reproduced in Fig. 11-15, showing approximate excavation depths, the time sequence and the phasing.

Surface Movement. Figure 11-16 shows the pattern and magnitude of surface movement normal to the wall at the inclinometer and movement point locations as functions of time. Initially the movement is displacement toward the excavation. Vertical settlement begins to occur as the excavation is completed. At all points the horizontal movement is greater than the settlement. Within 14 months of the end of excavation all movement appears to have stabilized and practically ceased. At this stage maximum horizontal and vertical displacement is 50 and 30 mm (2 and 1 ¼ in), respectively. Approximately ⅓–½ of that value occurred by the time the excavation was completed. The vertical displacement of the wall indicated by inclinometer 11 must have been caused largely as a result of the vertical component of the anchor loads and in spite of the relatively flat profile.

Time-dependent movement has clearly taken place following the excavation period and in spite of the relatively shallow depth. Considerable movement extends back beyond the anchorage zone.

Wall and Ground Movement. Composite lateral wall–ground movement for various excavation stages is shown in Fig. 11-17. The displacement of the vertical profiles along the wall and at 4 m behind the wall exhibit the same pattern. The movement is essentially lateral translation during excavation followed by some rotation after excavation is completed. At a vertical profile 19 m behind the wall the movement is entirely translational.

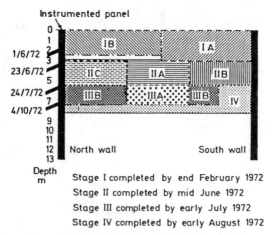

Fig. 11-15 Excavation progress and sequence; Neasden Underpass. (From Sills et al., 1977.)

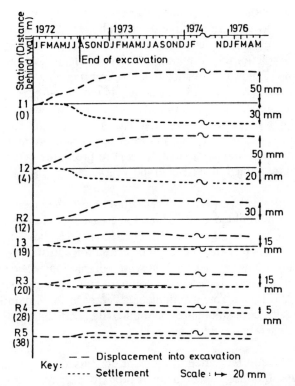

Fig. 11-16 Surface movement behind anchored wall; Neasden Underpass. (From Sills et al., 1977.)

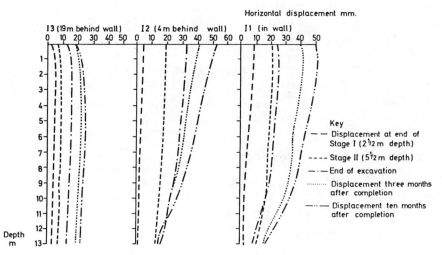

Fig. 11-17 Lateral wall and ground movement, Neasden Underpass. (From Sills et al., 1977.)

Pore Water Pressure Measurements. Pore pressures were recorded for the period of excavation and for the period following the end of excavation. Pore pressures began to drop at the beginning of excavation and continued to fall after completion of excavation, but more slowly. A minimum value was reached 8–10 months after excavation. Subsequently there has been a gradual increase in pore pressure, although 44 months after excavation pore pressure values were still considerably lower than the original values. Similar effects have been observed by Burland and Hancock (1976) around a deep excavation in London clay at Westminister. Vaughan and Walbancke (1973) note that in some clay excavations pore pressures may not revert to their steady-state values for some 40–60 years later.

Anchor Loads. The two top rows of anchors have shown similar behavior under excavation and load. Anchor load decreased as the row beneath was stressed followed by recovery, and thereafter nearly constant load values (about 430 kN and 500 kN for the top and second rows, respectively). Anchor loads in the third row decreased following completion of excavation, and then reached constant values 480 kN compared with the initial load 500 kN. The loads in the fourth row continued to decrease for 8 months after excavation, reaching constant values of about 370 kN compared with the initial 440 kN.

Evaluation of Results. When movement decreased and eventually stabilized, the ratio of its horizontal to vertical component at the top of the wall was about 2, whereas the same ratio at the bottom of the wall was slightly in excess of 1.

Appreciable horizontal extension developed in the 4-m zone behind the wall between vertical profiles 11 and 12. Furthermore, the settlement of profile 11 was greater than that of profile 12. If the equivalent extensions along the anchors associated with the composite ground movement are calculated, it can be seen that these are of the order of only 2–3 mm. In this context, anchor loads in the three upper rows have changed very little; that is, ground movement in this case is consistent with satisfactory anchor performance (Sills et al., 1977). These observations are correlated with results from finite-element analysis discussed in Section 11-10.

Guildhall Precincts Redevelopment

Figure 11-18(a) shows a typical section through the diaphragm wall supporting the excavation for this project. Also shown are soil data, anchor details, and lateral earth pressure reactions per unit length of wall used to calculate prestress loads. The excavation depth is 10.4 m (34 ft), and the wall is braced at two levels.

Lateral wall movement at various excavation stages is shown in parts (b) to (i) (Littlejohn and MacFarlane, 1984). The displacement profiles show a

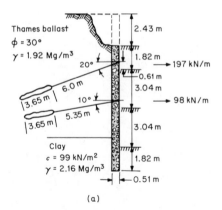

(a)

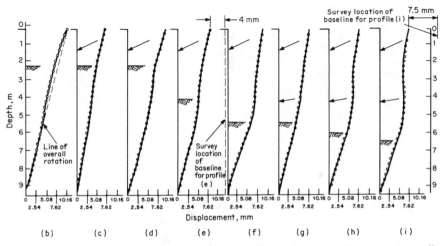

Fig. 11-18 Guildhall Redevelopment, London; (a) section through diaphragm wall; (b) through (i) observed wall movement during excavation. (From Littlejohn and MacFarlane, 1974.)

logical progression of soil–structure interaction. They are plotted relative to the toe of the wall, however, so that they do not provide overall wall movement except for profiles (e) and (i) where movement was measured during a general wall survey. Pertinent data on bending moments were obtained by graphical differentiation of the displacement profile gradients.

Profile (b) shows overall wall rotation toward the excavation that is superimposed on the cantilever effect above the first excavation level (4.25 m or 14 ft). The maximum differential deflection (lateral) between the top and the toe occurred at this stage and measured 10 mm, corresponding to a rotation or slope of 1/970. The effect of prestressing the top anchor is revealed in profile (c), showing the wall drawn back toward its initial shape. The difference between profiles (c) and (d) reflects time effects of 4 days, and indicates shear movement in the fixed zone as anchor resistance is mo-

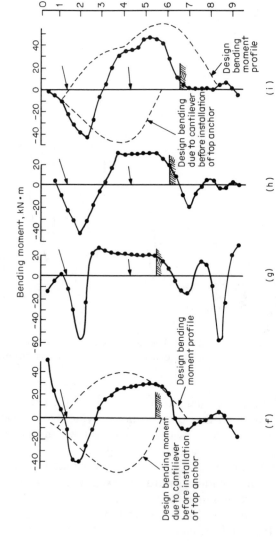

Fig. 11-19 Bending moments for the wall of Fig. 11-18, 0.3-m strip; excavation stages (f) through (i). (From Littlejohn and MacFarlane, 1974.)

bilized. During the same period an overall movement occurred toward the excavation, accompanied by a bulging deflection below excavation level, accentuated in profile (e). This bulging continued as excavation was taken down to the second level shown in profile (f). However, the bulging is seen to have disappeared when the second row of anchors was installed and prestressed as shown in profile (g). Some bulging reappeared when the excavation reached the final level as can be seen in profiles (h) and (i).

The measured vertical settlement indicates negligible scattered changes adjacent to the wall and the ground surface. During the period for which the upper anchors were monitored (103 days) the prestress was essentially stable, indicating the absence of any significant movement of the wall relative to the fixed anchor zone.

Four bending moment profiles, marked (f) through (i), are shown in Fig. 11-19. The flexural deformation fits consistently the movement profiles shown in Fig. 11-18. As should be expected from the wall deformation profiles, the bending moment at the second (lower) anchor level is positive although this level is statically the interior support of a continuous beam.

A basic point of interest in this analysis is that the soil stress distribution is triangular rather than rectangular and that the wall yields progressively as the excavation proceeds so that the statical continuity of a beam in flexure is disrupted when the beam yields at the supports. We can understand this concept using a step-by-step consolidation of the multianchored system into a repetitive single-anchor wall design of a model allowing the wall deformation to be estimated in a manner that satisfies rotational as well as horizontal equilibrium.

Profiles (h) and (i) in Fig. 11-19 are compatible with the single-anchored wall response, and the support of the two anchors appears to be consolidated at a point between them so that the bending moment is negative there and positive elsewhere. Further consolidation of the moments by differentiating the curve in conjunction with the deformation diagrams results in a lateral pressure pattern that is more triangular than trapezoidal.

Keybridge House, Vauxhall

Figure 11-20(a) shows a typical wall section supporting an excavation that is 14.5 m (48 ft) deep. Also given are anchor details and soil data. Wall embedment below excavation level is 2.2 m (7.2 ft). The upper ground layer consists of gravel underlain by London clay stiffening progressively from an undrained shear strength $45 kN/m^2$ near the ground level to $135 kN/m^2$ at the depth of the cut.

Lateral wall movement at various excavation stages is shown in Fig. 11-20(b)–(i). As indicated by profile (b), the wall rotated about its toe, responding to soil that becomes progressively stiffer. The installation and stressing of the upper anchor to 450 kN drew the wall back as shown in (c); 29 days later the wall had reversed to the shape of profile (b) as shown in (d), although no major change in anchor load occurred.

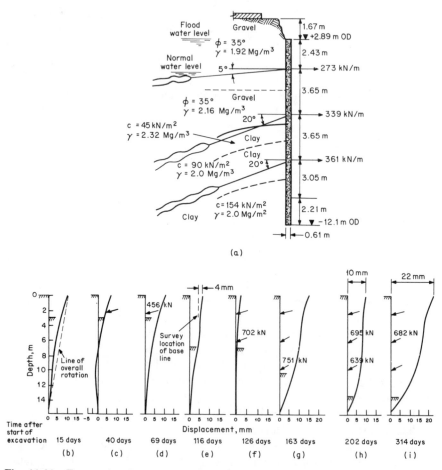

Fig. 11-20 Excavation for the basement of the Keybridge House, Vauxhall, London; (a) section through diaphragm wall; (b) through (i) observed wall movement in various construction stages. (From Littlejohn and MacFarlane, 1974.)

When the excavation reached the second level (6.8 m) the wall deformation below that level shown in profile (e) was essentially the same as in profile (d), but an inward toe displacement 2.5 mm occurred toward the excavation caused by the consolidation of clay on the cut side. Beam action below excavation level was combined with the considerable wall stiffness and resulted in very small bending. With two anchor levels stressed as shown in (f), the difference in displacement between the top and the toe became even smaller. However, excavation to the third level (10.4 m) caused the movement shown in (g) in which the displacement at the top exceeded the displacement at the toe by 13 mm (0.5 in), and this in spite of the stressing of all three anchor levels. At the final excavation level the toe was displaced 1.5 mm toward the cut. At this stage the wall underwent some rotation about

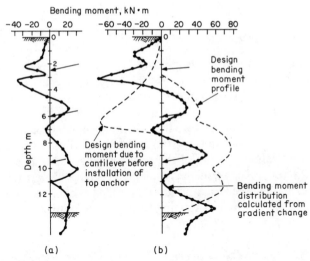

Fig. 11-21 Bending moments for the wall of Fig. 11-20, based on a 0.3-m strip; final excavation stage. (From Littlejohn and MacFarlane, 1974.)

a point between the two upper anchors, which reduced the differential deflection between the top and the toe to 10 mm (0.4 in). The vertical settlement at the final stage measured at the top of the wall was 12 mm (0.5 in), caused by the vertical component of the anchor load.

Time effects on wall movement become evident from profile (h) to (i), and reflect a period of 3 months. During the same period the differential displacement between top and toe more than doubled, and this despite a small loss of prestress seen in the middle anchor. This indicates the consolidation of the stressed soil in the fixed anchor, and also a broad movement of the entire ground mass behind the wall as we have noted with the Neasden Underpass.

Bending moment diagrams are shown in Fig. 11-21 for the final excavation stage and for a 0.3-m wall strip. The profile in (b) also shows the design and measured moments for the cantilever portion of the wall. The maximum magnitudes of the bending moments (design and measured) are in fair agreement, although there is an evident variation in their distribution. Interestingly, measured moments represent normal seasonal groundwater level, whereas design bending moments are based on flood level (Littlejohn and MacFarlane, 1974).

Bending-moment discontinuities at anchor locations are slightly lower than predicted, probably because of anchor inclination and overdig before anchor installation. The shape of the moment curve above the top anchor approaches a sinusoidal curve and shows the effect of the stiff guide wall at the rear of the crest.

11-9   FIELD STUDIES OF SPECIAL ANCHORAGES

Inflated Cylinder in Soft Ground

Essentially, this is a modified version of type B pressure-grouted anchor. After the hole is drilled as shown in Fig. 11-22(a), a pressure tube and grouting pipe are inserted as shown in (b). As the annular space betwen the tube and the hole is filled with grout, the tube is laterally inflated by hydraulic pressure. This process consolidates the cement grout which is tightly pressed against the walls as shown in (c). After the tube is withdrawn the tendon is inserted and the center portion of the borehole is grouted as shown in (d). The resulting fixed anchor zone has a diameter slightly larger than the original borehole, and is essentially nonuniform. This anchor has been installed in various geologic formations ranging from rock to alluvial deposits, but is more suitable in soft ground (Mori and Adachi, 1969).

Anchorage for Iwaki Transmission Towers, Tokyo. The tower foundations for this project required 336 anchors installed vertically to resist uplift caused by overturning effects and eccentric loads. Pullout tests were carried out on three anchors summarized in Table 11-6(a). Anchors marked 1 and 3 had inflated grout columns produced by hydraulic pressure 5 kg/cm² (about 70 psi or 0.5 N/mm²). Anchor 2 was installed using conventional methods.

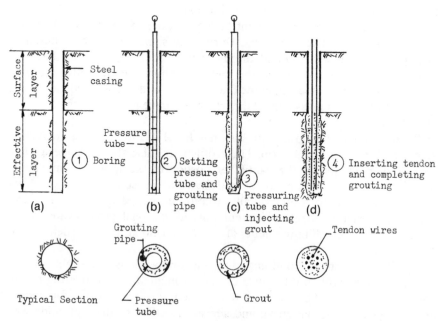

Fig. 11-22   Inflated cylinder in soft ground; schematic presentation of anchor installation. (From Mori and Adachi, 1969.)

TABLE 11-6   (a) Test Anchor Data; Iwaki Transmission Towers

PS Anchor No.	Length (m)	Diameter (cm)	Direction	PC Strand Wire	Descriptions
1	6.0	20	Vertical	16φ 12.8 mm	Prestressed anchor inflated by effective pressure 5.0 kg/cm²
2	6.0	20	Vertical	16φ 12.8 mm	Not inflated mortar pile compared with No. 1
3	8.0	20	Vertical	16φ 12.8 mm	Prestressed anchor inflated by effective pressure of 5.0 kg/cm²

(b) Standard Penetration Test

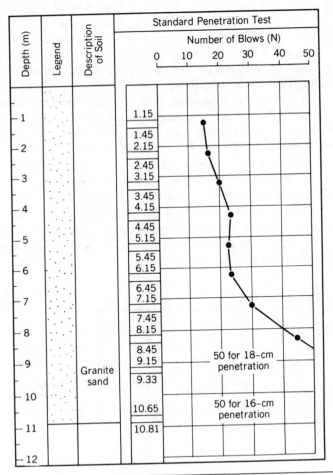

From Mori and Adachi (1969).

The ground at the site consists mainly of sandy soil decomposed from granite. The sand is relatively well graded with moisture content 15–25 percent. Based on the known texture comprising 75 percent sand and 25 percent gravel, this soil can be described as gravelly sand with relatively high permeability. Results and N values (blow count) from standard penetration tests are shown in Table 11-6(b).

Cement mortar with water/cement (W/C) ratio 0.42 and sand/cement ratio 1.0 was used for grouting. The 14-day strength of this mortar is 300 kg/cm² (about 4200 psi). The grout was allowed to cure for more than 2 weeks before pullout tests commenced. Anchor load was applied in increments with 8-min pause, and was repeated five times in a cyclic pattern. Maximum load was reached at the end of the fifth cycle.

Figure 11-23 shows anchor head displacement plotted versus anchor load for anchors 1 (inflated) and 2 (conventional). For anchor 1 maximum pulling resistance is 70 tons (metric) or twice the maximum load 36 tons sustained by anchor 2. Maximum load was reached at a displacement 2.0 cm for anchor 2, and 2.7 cm for anchor 1. Shear resistance per unit surface area is shown in Fig. 11-24 for the inflated anchor 1, and is plotted versus displacement and at various fixed anchor depths.

The results of Fig. 11-24 indicate that average maximum shear values close to 2kg/cm² (28 psi) were induced at relative displacement 0.5 mm. Thereafter, shear resistance remained constant or decreased slightly.

The initial pressure 5 kg/cm² (70 psi or about 0.5 N/mm²) is within the pressure range intended to produce increased anchor load resistance, but the manner in which this pressure was applied clearly gives no indication of what portion remained locked in. From the principles discussed in Section 4-8, we can conclude that a portion of the initial hydraulic pressure used to inflate the grout column dissipated, but locking-in and wedging effects in the inflated anchor were caused by dilatancy as the pull was applied and helped increase pullout resistance.

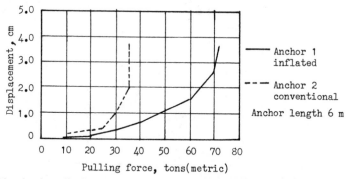

Fig. 11-23 Anchor displacement versus anchor load; anchors 1 and 2. (From Mori and Adachi, 1969.)

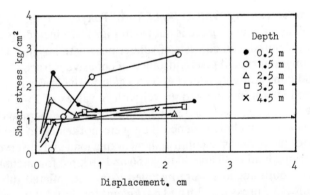

Fig. 11-24 Shear stresses versus displacement as function of depth; anchor 1 (inflated). (From Mori and Adachi, 1969.)

We can estimate the magnitude of p_g (residual grout pressure normal to the walls) from a backanalysis with the help of Eq. (4-12a). Given the measured pullout load $T_f = 70$ tons $= 70.000$ kg, $D = 20$ cm, $L = 6$ m $= 600$ cm, and $\tan \phi = 0.62$ (assumed $\phi = 32°$), we calculate

$$p_g = \frac{70,000}{3.14 \times 20 \times 600 \times 0.62} = 3 \text{ kg/cm}^2$$

which is nearly two-thirds the initial pressure. Alternatively p_g can be estimated as normal effective stress acting at the interface, in which case it can be related to the measured shear stress $\tau = 2$ kg/cm² and the friction angle δ between the soil and the grout. Setting $\delta \approx \phi$, then $p_g = \tau/\tan \delta = 2.0/0.62 = 3.2$ kg/cm², which is reasonably close to the value obtained from Eq. (4-12a).

For anchor 2 (not inflated) the maximum shear stress can be calculated from the maximum measured load and the fixed anchor dimensions, assuming uniform distribution. This stress is $36000/3.14 \times 20 \times 600 = 1.03$ kg/cm², or about one-half the maximum friction developed for anchor 1 as would be expected. For an anchor diameter 20 cm results from this test cannot be correlated with the data shown in Fig. 4-19. For anchor 2, however, good correlation is obtained between maximum load or shear resistance (skin friction) and N values plotted in Figs. 4-20 and 4-21. Using $N = 20$ obtained from Table 11-6(b), $L = 6$ m, and soil type gravelly sand, we can enter the graphs of Fig. 4-20 to obtain ultimate load-carrying capacity of approximately 310 kN $= 70$ kips $= 32$ tons, which is fairly close to the measured value 36 tons. Likewise, for $N = 20$ and with the help of Fig. 4-21 we obtain maximum skin friction about 1.2 kg/cm², which compares satisfactorily with the measured shear resistance.

Anchorage for Retaining Wall at Ikebukuro, Tokyo. A typical cross section of the wall and the anchorage is shown in Fig. 11-25 including soil data. The retaining anchored wall consists of soldier piles with lagging. The upper row anchors (marked type *A*) were installed in cohesive terrace clay of high sensitivity and cohesion values of 0.4–0.65 kg/cm². The lower-row anchors have their fixed length in sand and gravel with *N* values of 25–50.

The two anchors selected for testing were to become permanent units after the tests were concluded; hence they were not tested to failure. The main purpose of the test was to confirm the working load and observe anchor behavior. Working load is 30 and 40 tons (66 and 88 kips) for anchors *A* and *B*, respectively. both anchors were installed using the inflated tube procedure, under effective pressures 2.0 and 6.0 kg/cm² for *A* and *B*, respectively. Anchor load was applied incrementally until it exceeded the working load by 10 tons. The maximum load was reached after three and four cycles. Interestingly, the measured factor of safety is 1.33 and 1.25 for units *A* and *B*, respectively.

Results from the test show that for anchor *A* displacement occurred at increasing rate when the load exceeded 30 tons, and creep displacement increased in the same context. For anchor *B* displacement likewise occurred

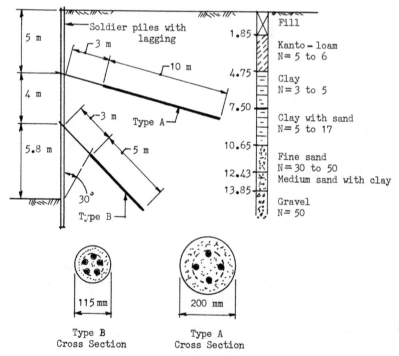

Fig. 11-25 Typical cross section; anchored soldier pile with lagging wall. (From Mori and Adachi, 1969.)

at increasing rate when the load reached 30 tons, but creep increased after the load exceeded 40 tons.

Shear stresses versus displacement at various depths are plotted for anchor A in Fig. 11-26. Shear stresses increase linearly with displacement in the initial stage of loading, reach maximum values at displacement 3–5 mm, and as displacement continues they stabilize at 60–80 percent the maximum values. Average stabilized shear strength values extraced from Fig. 11-26 are approximately 0.5 kg/cm², and compare very closely to the average shear strength (about 0.52 kg/cm²) of the clay obtained from laboratory tests. In this case, the adhesion factor a in Eq. (4-16) is almost unity. It appears, therefore, that the inflated tube procedure combined with the small pressure 2 kg/cm² compensated for any softening effects resulting from the drilling operations. This influence should be more beneficial in clays of relatively medium strength (say, 1400–1500 lb/ft²), whereas in stiff to hard clay a portion of the undrained shear strength should be expected to be permanently dissipated.

Anchoring in Soil Employing Special Drilling Methods

Where the ground exhibits variations in stratigraphy, alternating layers and high water table, drilling without casing is impracticable. If the advantage of high grouting pressures is to be utilized, it is also necessary to use casing.

Different methods have been developed for drilling with interior extension

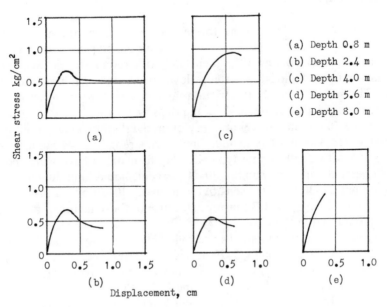

(a) Depth 0.8 m
(b) Depth 2.4 m
(c) Depth 4.0 m
(d) Depth 5.6 m
(e) Depth 8.0 m

Fig. 11-26 Shear stresses versus displacement as function of depth; anchor A (inflated). (From Mori and Adachi, 1969.)

drill steel rods and exterior casing working simultaneously. These methods can be used in ground containing stones and boulders of hard type rock, encountered frequently in Sweden (Moller and Widing, 1969). With one method, the casing has a cutting shoe at its lower end for drilling the hole. The casing and the conventional drill rods with the bit are rotated and hammered simultaneously.

With another method, an eccentric bit is attached to the lower end of the drill. After drilling is completed the rods and the drill bit can be withdrawn through the casing. With a third method, the casing has also a cutting shoe at its lower end. This ring bit is locked to the drill rod by drilling. In this manner the drill rods and the cutting shoe are advanced as one unit. The casing is nonrotating.

In Sweden anchorages are typically installed at sites where soft clay on top of firm friction soil necessitates anchor lengths of 60–70 m (195–230 ft) in order to locate the anchor zone in competent ground. Typical maximum anchor loads attained in sand and gravel are 175 tons metric (400 kips). Examples of ground types where anchors have been used includes the following.

Moraine. This is firmly pressed soil of the glacier age, consisting of clay, silt, sand, and gravel with a large content of boulders (10–15 percent). Moraine has high strength, and is normally impervious and often cohesive. Usual density is 1.7–2.2 tons/m^3. Invariably all drilling methods have low to medium capacaity.

Sand and Gravel. These were formed in rivers during the glacier age or in the sea during a later age. In the former case mixed sand and gravel exist with a silt content below 10 percent and with round stones. Sea sand is normally pure soil with rather steep sieve curve: fine, medium, and coarse sand. The soil is typically the friction noncohesive type with permeability 0.01–2 cm/s, friction angle 30–35°, and density 1.7 tons/m^3.

Moraine, sand, and gravel are very often overlaid by clay settled in stationary water. The clay has an undrained shear strength of 1–2 Mp/m^2, slowly increasing with depth. Quite typically clay formations have water content 50–80 percent, fineness number 30–60 percent, liquid limit 30–60 percent; plastic limit 25–30 percent, sensitivity normally 10–20 but occasionally up to 200, and density 1.5–1.7 tons/m^3. Because of these features, anchorages are seldom fixed in the clay.

A typically soil profile shows the following stratigraphy: at the top is a loose layer of clay, normally 5–20 m thick but often underlain by a layer of sand usually 5 m thick. Below that exists a stony gravel stratum 5–10 m thick. Nearest to the rock exists moraine 1–3 m thick and with large content of boulders. In general, the groundwater level is only 1–3 m below ground surface.

Typical Anchor Installation Procedure. In these soil conditions an anchorage installation is typically carried out in the following steps:

1. Drilling, simultaneously with extension rods and casing, to the intended depth. Water flushing with water pressure 10 kp/cm^2. In impervious soil (clay, silt, fine sand, and moraine) the flushwater returns in the casing together with the drill cuttings. In previous soil (sand and gravel) the flushwater disappears.
2. In the fixed anchor zone the casing is moved up and down until it can be easily moved in the soil.
3. The extension rods with pilot bits are extracted, but the casing remains in the ground.
4. A flushing hose is lowered to about 1 ft above the bottom end of the casing, and water flushing begins and continues until the returning water is clean.
5. The entire casing is filled with grout with W/C ratio 0.38–0.40.
6. The tendon is inserted in the grout in the casing (strands, wires, or steel rod).
7. The casing is withdrawn by 1 m, and thereafter the withdrawal continues slowly while grout is pumped simultaneously and under a pressure of 3–10 kp/cm^2. The fixed anchor zone usually is from 5 to 7 m.
8. When the low end of the casing has reached the upper end of the fixed zone a final grouting is repeated with a pressure of 2–10 kp/cm^2 depending on the perviousness of the soil.
9. The grouting pressure is relieved, and the casing is withdrawn 1–2 m.
10. Water flushing is introduced in the casing to remove away any grout from the free anchor length.
11. The remaining casing is withdrawn.
10. After 7 days a tension test is carried out and the anchor is prestressed.

In the past anchors installed according to the foregoing procedure have had tested loads 45 Mp in coarse silt, 60 Mp in fine sand, 100 Mp in coarse sand, and 165 Mp in gravel (Moller and Widing, 1969).

11-10 ANALYSIS OF GROUND MOVEMENT IN LONDON CLAY BY FINITE-ELEMENT METHODS

The behavior of deep excavation for the Neasden Underpass discussed in Section 11-8 has been analyzed by Simpson et al. (1979) using a nonlinear finite-element program in which very small strains are simulated. This program has been used successfully in predicting braced wall behavior in stiff London clay. The program simulated excavation for the following steps:

1. Estimate initial effective stress in the ground adjacent to excavation.
2. Proceed with the first excavation stage to a depth of 3 m.
3. Install the first row of anchors, test and prestress.
4. Proceed with the second excavation stage to a depth of 5.5 m.
5. Install, test, and prestress the second row of anchors.
6. Excavate to 8.5 m depth, and install the third row of anchors (tested and prestressed).
7. Install, test, and prestress the fourth row of anchors.

Figure 11-27 shows predicted versus wall displacement observed for steps 4 and 7. The results are in fair agreement. Predicted settlements however are considerably greater than observed vertical displacement especially at some distance from the wall, probably because the soil is in undrained condition.

Pore water pressures around the excavation were also determined and are shown in Fig. 11-28. Initially, negative pore water pressures were predicted in the center part of the excavation with positive values relatively close to the wall behind the excavation and on the passive side. In the anchor zone behind and below the wall an area of relatively high pore pressures can be distinguished caused by the simulated loading of the anchors.

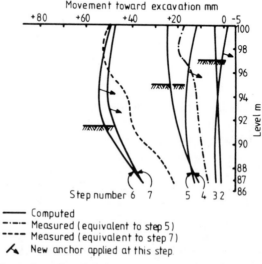

Fig. 11-27 Comparison of predicted and measured wall performance; Neasden Underpass, London. (From Simpson et al., 1979.)

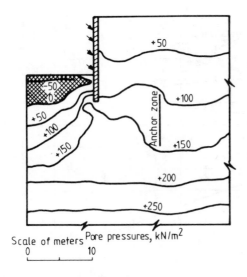

Scale of meters
0   10

Pore pressures, kN/m²

Fig. 11-28 Predicted pore pressures around the excavation; Neasden Underpass, London. (From Simpson et al., 1979.)

11-11 OBSERVED PERFORMANCE OF RIGID WALL AT WORLD TRADE CENTER

Figure 11-29(a) and (b) shows bending moments and lateral stresses for panel W35 of the World Trade Center in New York (Saxena, 1974). The moment diagrams have been correlated with the measured curvature of the wall and the actual flexural deformations. These values correspond closely to moments computed from dynamometers installed on the reinforcement bars (Gould, 1970) so that the two sets of observations actually represent wall behavior. The general shape and magnitude of the diagrams appear reasonable, although slope observations are not sufficiently sensitive to register abrupt moment changes at anchor locations. The horizontal pressures at the end of excavation (August 26, 1968) shown in (b) are estimated from tie-rod load cells and from moment configurations.

Although there is a variance between this pressure diagram and the one shown by Gould (1970), the moment diagrams for this panel developed by the two investigators are in good agreement. The moment curve between depths of 20 and 40 ft suggests the consolidation of several anchors into a single support. In the early stages of excavation the moments are largely positive, and have a pattern typical with anchored bulkheads. As excavation continues negative moments are developed at the region of the four center ties as these begin to act as a single support, and thus a large positive moment remains in the lower part. The insertion of anchor T_6 causes a discontinuity in the moment curve but does not reduce its magnitude appreciably. The final large bending moments near the base of the wall indicate a considerable horizontal subgrade reaction along the contact zone between the rigid wall and the hard rock. These moments are in conformity with the steep curvature

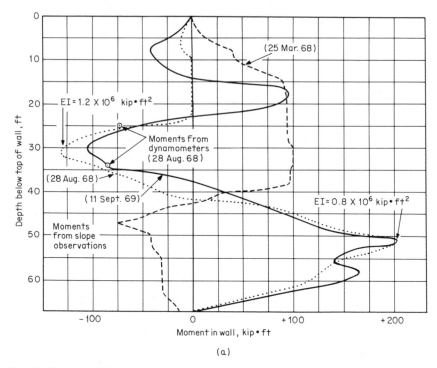

Fig. 11-29 Panel W35 of the World Trade Center, New York: (a) bending moments; (b) lateral stresses and tieback loads. (From Saxena, 1974.)

of the wall just above the rock level since it could not rotate about its tip because of the considerable fixity.

Design and measured lateral pressures are correlated approximately only. Although the diagrams in Fig. 11-29(b) show the original at rest pressure redistributed to take into account anchor prestressing, pressure diagrams obtained from field measurements show less pressure in the upper wall section and much higher pressure in the lower part of the wall.

11-12 INCLINED WALL FOR THE MUNICH SUBWAY

The inclined (positive inclination) bored pile wall shown in Fig. 8-29 was constructed to support the excavation and simultaneously underpin buildings located within the limits of the proposed structure. Interestingly, this solution was competitive with the cost of conventional underpinning since the inclined wall did not cost substantially more than a vertical system.

Because of the severe restraints in the available construction space, the inclined wall was inserted with minimum clearance to the footing of the adjacent store building. The wall consists of secant piles installed with the

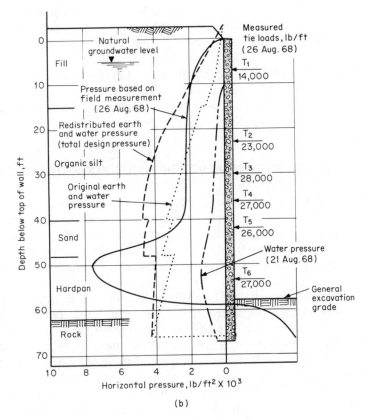

Fig. 11-29 (*Continued*)

Benoto boring ring. This machine (Xanthakos, 1979) operates from a piling ring. During excavation a steel casing is forced into the ground by means of hydraulic rams, while a secondary transverse ram imparts a twisting motion to the casing. A grab removes earth material from inside the casing, which is always kept ahead of the grab. Because this wall is fairly watertight the groundwater level was not lowered; hence detrimental consolidation settlement was avoided.

The installation exemplifies the principles of hybrid construction. The wall is braced at the top by internal struts, at the bottom by sufficient embedment in dense sand, and at the center by two rows of tiebacks. Because the wall was rigid and braced at the top and the bottom by essentially nonyielding supports, lateral movement was limited to inconsequential values. Interestingly, the secant-pile wall supported the temporary street decking and all moving loads in addition to the vertical component from the anchor load. Since lateral deformations were not allowed or predicted, the wall was designed for earth pressures at rest, full hydrostatic pressure, and lateral stresses imposed by the continuous footing load in the earth mass. In the embedded portion the wall was considered elastically supported, whereas at the anchorage points elastically yielding supports were simulated (Linden, 1969).

The excavation was carried down in steps. Initially it reached the first excavation level to allow installation of the upper row of anchors. When the second excavation level was reached, the lower row of anchors was installed. Thereafter both rows of anchors were prestressed.

11-13   OBSERVED PERFORMANCE OF HYBRID CONSTRUCTION

The concepts suggested in Section 9-22 to improve control of ground movement by hybridizing the bracing system are exemplified in the design and construction of a diaphragm wall to support a basement excavation in the central district of London (Hodgson, 1974). The bracing systems in this example included preloaded structural steel struts, ground anchors, and a reinforced concrete waling slab.

The development consists of two blocks, B and C as shown in Fig. 11-30. The two blocks are connected at first basement level. The plan shows the location type, and extent of each bracing system. A typical cross section through block C is shown in Fig. 11-31. Three methods or combinations therefrom were used for temporary support of the wall. Anchorages fixed in the gravel and clay strata were used generally, except for one section of block B where Macalloy bars were employed to facilitate destressing and removal. In the area adjacent to Westminster Cathedral structural steel struts were considered more practical. The reinforced concrete waling slab was incorporated in the third basement of block C.

For the top section of the wall in both blocks B and C the bracing consisted of 288 anchors with a calculated maximum working load 56 tons (metric) along the section adjacent to Victoria Street, and 45 tones elsewhere. For both cases the free length was 5 m with a fixed anchor zone of 4.3 and 3.7 m, repsectively. Most anchors were installed normal to the wall, with an inclination from the horizontal of 25–30°. In the third basement on block C, 28 anchors were used in the lower row with 55-ton working loads. In the gravel section five test anchors were placed and stressed, whereas two anchors in the clay were tested to failure. The failure loads of the test anchors in clay are shown in Table 11-7.

The third basement excavation extends 7 m (23 ft) below the level of the second basement slab, giving a total depth from street level of 14 m (46 ft). Because of the proximity to existing buildings and concern for movement the diaphragm wall at this location was laterally supported with a concrete waling slab at second basement level. Concrete subpiers were augered from the second basement level, having a cutoff level at the underside of the third basement raft slab. A steel casing was left around each subpier and used as

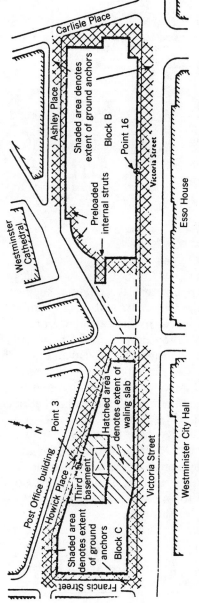

Fig. 11-30 Plan showing location of two construction blocks; type and extent of bracing. (From Hodgson, 1974.)

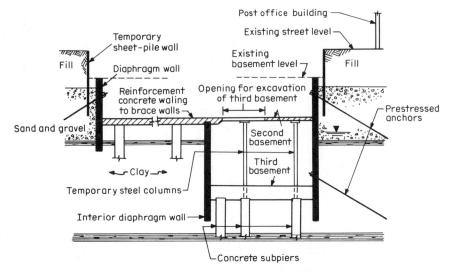

Fig. 11-31   Section through block C, construction of Fig. 11-30. (From Hodgson, 1974.)

access to position and install temporary steel columns supporting the second basement slab as shown in Fig. 11-31. Excavation of the third basement was then carried out as a mining operation through an opening in the slab of the second basement level. Following excavation of 4 m of ground the bottom anchors were installed and fixed in stiff clay. Excavation was then continued to final level, and the third basement raft slab was placed for the full width of the basement so that the horizontal thrusts from the diaphragm walls on each side counterbalanced each other. With the third basement raft slab in place, the clay anchors were destressed.

Observed Movement. Because of the considerable live load surcharge imposed by heavy street traffic alongside the wall, anchor loads were much greater than the component due to lateral earth stresses only. This extra load therefore compensated for a great portion of movement expected to occur during excavation.

TABLE 11-7   Failure Loads of Test Anchor in Clay; Victoria Street Project

Length of Drilled Hole (m)	Free Length (m)	Failure Load (tons)
24.5	20	145
18	14	93

From Hodgson (1974).

The clay anchors installed in the third basement of block *C* were monitored for a period of 3 months. During this time relaxation due to creep was negligible. Wall movement, on the other hand, was checked and measured during all construction stages. Maximum recorded wall deflection at the top was 3 mm, and clearly inconsequential. Interestingly, during piling operations a movement of 25 mm was observed at the adjoining wall section associated with vibratory effects. After investigation it was concluded that the frequency of the vibration used in pile work had coincided with the natural frequency of the gravel so that the wall and the retained material had moved as one body. This movement stopped as soon as this vibrator was replaced.

REFERENCES

Bastable, A. P., 1974: *Multibell Ground Anchors in London Clay*, 7th FIP Cong., New York, pp. 33–37.

Broms, B. B., and H. Stille, 1976: "Failure of Anchored Sheet Pile Walls," *Proc. ASCE*, **102**(GT3), 235–251.

Burland, J., and R. Hancock, 1976: *Inst. Struct. Eng.*, Special Presentation.

Gould, J. P., 1970: "Lateral Pressures on Rigid Permanent Structures," *Proc. ASCE Spec. Conf., Lateral Earth Stresses Earth Retaining Structures*, Cornell Univ., Ithaca, N.Y., June.

Hodgson, T., 1974: "Design and Construction of Diaphragm Wall on Victoria Street, London." *Proc. Diaphragm Walls Anchorages*. Inst. Civ. Eng., London.

Hooper, J., and F. Butler, 1966: "Some Numerical Results Concerning the Shear Strength of London Clay," *Geotechnique* (England **XVI**(4).

Johnston, G. H., and B. Ladanyi, 1974: "Field Test on Deep Power-Installed Screw Anchors in Permafrost," *Can. Geotech. J.*, **11**(3), 248–358.

Korbin, G. E., and T. L. Brekke, 1978: "Field Study of Tunnel Reinforcement," *J Geotech Div. ASCE* (Aug.), 1091–1108.

Ladanyi, B., and G. H. Johnston, 1974: "Behavior of Circular Footings and Plate Anchors Embedded in Permafrost," *Can. Geotech. J.*, **11**(4), 531–553.

Linden, J. V., 1969: "Inclined Walls for the Munich Subway," *Proc. 7th Int. Conf. Soil Mech. Found. Eng.*, Special Sessions 14, 15, Mexico City, pp. 102–103.

Littlejohn, G. S., and D. A. Bruce, 1979: "Long Term Performance of High Capacity Rock Anchors at Devonport," *Ground Eng.* (Oct.), London.

Littlejohn, G. S., and I. M. MacFarlane, 1974: "A Case History Study of Multi-tied Diaphragm Walls," *Proc. Diaphragm Walls Anchorages*, Inst. Civ. Eng., London.

McRostie, G. C., and R. W. Schriever, 1967: "Frost Pressure in the Tieback System at the National Arts Center Excavation," *Eng. J., Eng. Inst. Can.* (March), 17–21.

Moller, P., and S. Widing, 1969: "Anchoring in Soil, the Swedish Method," *Proc. Int. Conf. Soil Mech. Found. Eng.*, Special Sessions 14, 15, Mexico City, pp. 184–190.

Mori, H., K. Adachi, 1969: "Anchorage in an Inflated Cylinder in Soft Ground," *Proc. Int. Conf. Soil Mech. Found. Eng.*, Special Sessions 14, 15, Mexico City, pp. 175–183.

Rabcewicz, L. V., 1969: "Stability of Tunnels Under Rock Load, Part III," *Water Power* (Aug.) 298–302.

Ranken, R. E., and J. Ghaboussi, 1975: "Tunnel Design Considerations: Analysis of Stresses and Deformations Around Advancing Tunnels," report FRA-OR&D75-84, Dept. Transp., Fed. Railroad Admin. Washington, D.C.

Sahlstrom, P.D., 1969: "Forces in Sheet Pile Tiebacks," *Proc. Int. Conf.,* Int. Soc. Soil Mech. Found. Eng., Special Sessions 14, 15, Mexico City, pp. 211–213.

Sandegren, E., P. O. Sahlstrom, and H. Stille, 1972: "Behavior of Anchored Sheet-Pile Wall Exposed to Frost Action," *Proc. 5th Eur. Conf. Soil Mech. Found. Eng.*, Madrid, **102**(GT3: 1976), pp. 235–251.

Sandgrist, G., 1972: "Anchored Sheet Pile Walls In Cohesionless Soils. Deformation and Tie Forces in the Walls During Pile Driving Inside the Excavation and During Frost Action," Byggforskningen, Report No. R38, 31 pp.

Saxena, S. K., 1974: "Measured Performance of Rigid Concrete Walls at the World Trade Center," *Proc. Diaphragm Walls Anchorages*, Inst. Civ. Engl, London.

Sills, G. C., J. B. Burland, and M. K. Czechowski, 1977: "Behavior of an Anchored Diaphragm Wall in Stiff Clay," *Proc. 9th Int. Conf. Soil Mech. Found. Eng.*, Tokyo, Vol. 2, pp. 147–155.

Simpson, B., N. J. O'Riordan, and D. D. Croft, 1979: "A Computer Model for the Analysis of Ground Movement in London Clay," *Geotechnique* (England) **29**(2), 149–175.

Stille, H., 1976: "Behavior of Anchored Sheet Pile Walls," Thesis, Royal Institute of Technology, Stockholm.

Vaughan, P., and J. Walbancke, 1973: "Pore Pressure Changes and the Delayed Failure of Cutting Slopes in Over Consolidated Clay," *Geotechnique* (England) **23**(4).

VSL–Losinger, 1978: "Soil and Rock Anchors, Examples from Practice," April, Bern, Switzerland.

Xanthakos, P. P., 1979: *Slurry Walls*, McGraw-Hill, New York.